Geological Applications of Wireline Logs II

Geological Society Special Publications
Series Editor J. BROOKS

GEOLOGICAL SOCIETY SPECIAL PUBLICATION NO 65

Geological Applications of Wireline Logs II

EDITED BY

A. HURST
Unocal
UK

C. M. GRIFFITHS
University of Trondheim
Norway

P. F. WORTHINGTON
BP Research Centre
UK

1992

Published by

The Geological Society

London

THE GEOLOGICAL SOCIETY

The Society was founded in 1807 as the Geological Society of London and is the oldest geological society in the world. It received its Royal Charter in 1825 for the purpose of 'investigating the mineral structure of the Earth'. The Society is Britain's national learned society for geology with a Fellowship of 7000.

Fellowship is open to those holding a recognized honours degree in geology or cognate subject and who have at least two years relevant postgraduate experience, or have not less than six years relevant experience in geology or a cognate subject. A Fellow who has not less than five years relevant postgraduate experience in the practice of geology may apply for validation and subject to approval will be able to use the designatory letters C. Geol. (Chartered Geologist). Further information about the Society is available from the Membership Manager, Geological Society, Burlington House, London, United Kingdom W1V 0JU.

Published by the Geological Society from:
The Geological Society Publishing House
Unit 7
Brassmill Enterprise Centre
Brassmill Lane
Bath
Avon BA1 3JN
UK
(*Orders*: Tel. 0225 445046)

First published 1992

Distributors

USA
  AAPG Bookstore
  PO Box 979
  Tulsa
  Oklahoma 74101-0979
  USA
  (*Orders*: Tel: (918)584-2555)

Australia
  Australian Mineral Foundation
  63 Conyngham St
  Glenside
  South Australia 5065
  Australia
  (*Orders*: Tel: (08)379-0444)

British Library Cataloguing in Publication Data
A catalogue record for this book is available from the British Library
ISBN 0–903317–80–X

Printed in Great Britain by
Galliard (Printers) Ltd, Great Yarmouth

Contents

Physical properties

Mineralogy and geochemistry

Introduction

A. HURST[1], C. M. GRIFFITHS[2] & P. F. WORTHINGTON[3]

[1] *Unocal UK Ltd, 32 Cadbury Road, Sunbury-on-Thames TW16 7LU, UK (Present address: Department of Geology and Petroleum Geology, Aberdeen University, Aberdeen AB9 2UE, UK)*

[2] *Institute of Petroleum Engineering and Applied Geophysics, University of Trondheim, N-7034 Trondheim, Norway (Present address: BP/Statoil Alliance, Ranheimsveien 10, 7400 Trondheim, Norway)*

[3] *BP Research Centre, Chertsey Road, Sunbury-on-Thames TW16 7LN, UK (Present address: Consultant, 23 Woodlands Ride, Ascot, Berkshire SL5 9HP, UK)*

The contents of this volume provide extensive documentation of the growing diversity of geological problems which can be successfully addressed using wireline log data. From the evidence of this volume, the traditional applications of wireline logs in stratigraphy and sedimentology have continued to grow during the three years since *Geological Applications of Wireline Logs* (GAWL I) in 1988. In contrast, the level of original work using geochemical logs seems to have declined sharply in industrial applications, although remaining significant in the Ocean Drilling Program (ODP). Since GAWL I, high-density/high-resolution log measurements now appear to be an integral part of the characterization of petroleum reservoirs, a trend which, due to the development of slim-hole technology, has been successfully followed by the ODP. Most striking, however, is the increased level of activity focusing on the physical properties of rocks, and the interpretation of measurements made on them. This activity clearly reflects the importance of understanding the interaction between the rock mass and the measuring device before uncritical interpretations are made. It also points to the significance of integrating measurements made at several scales of investigation so as to reveal more of the character of the sub-surface. Finally, there is abundant evidence in this volume for the importance of integrating data of different origin and the potential of multi-disciplinary effort to advance our knowledge. This is of course not unique to geological applications of wireline logs, but an important trend common to all fields of scientific research where rapid progress and discovery are being made.

We hope that this volume will act as a small milestone which records the status of geological applications of wireline logs in the early 1990s.

Demand and scientific advances permitting, a future volume in the mid-1990s will provide a further opportunity to evaluate the progress in this exciting field of the geological sciences.

Sedimentology and stratigraphic correlation. **Bourke** documents a procedure for successful sedimentological interpretation of high-density resistivity measurements. **Cameron** presents a methodology for deriving sedimentary orientations from dipmeter data. **Ruhovets et al.** document the use of interpretation models for logs with different vertical resolution for their integration with logs with high vertical resolution. Three ODP papers (two by **Salimullah & Stow**, and **Pezard et al.**) have the interpretation of high-resolution resistivity measurements as key information in their unravelling of the sedimentary history of volcaniclastic facies and regional sedimentological trends respectively. **Slatt et al.** illustrate the value of outcrop studies when attempting subsurface correlation and **Hatton et al.** illustrate a simple method for correlation in clastic sequences, which include volcaniclastic sediments which may throw new light on the prospectivity of some areas. The remaining paper in this section (**Bølviken et al.**) presents quantitative methods for automated identification of sedimentary facies and lithologies from wireline logs.

Fractures and stress. Several scales of investigation are described by papers in this section. Basinal, or regional, trends in present day in situ stress orientation derived from borehole breakouts are documented in three papers: **Hillis & Williams** describe stress regimes in the Timor Sea from breakout data; **Yassir & Dusseault** describe similar data from the Ontario cratonic area; **Cowgill et al.** provide some preliminary results from the tectonically complex area of the

Witch Ground Graben. Identification of fractures and their character in boreholes is investigated by four papers. Fracture identification, formation and permeability in basement are investigated by **Jackson** *et al.*, **Goldberg** *et al.* and **Bremer** *et al.* In the fourth paper, distinction of open fractures and non-sealing faults from breakout data is attempted by **Bell** *et al.* **MacLeod** *et al.* demonstrate the potential for identifying tectonic rotations by integrating high-resolution borehole images and core data, so permitting core orientation without resort to palaeomagnetic data. **Hornby & Luthi** demonstrate the potential for enhancing interpretation of open fractures by integrating measurements made at different scales of investigation using borehole scans and Stoneley waves. Finally, **Adams** *et al.* give an example from the Niger Delta of how the structural interpretation of seismic data may be enhanced by using dipmeter data.

Physical properties. Excluding the papers on in situ stress, seven papers address diverse aspects of other physical properties of rocks. Two papers investigate seismic properties. **McCann & Sothcott** report the laboratory measurement of velocity and attenuation using a single-frequency pulsed sine wave. **Raaen** demonstrates derivation of shear velocity from pseudo-Rayleigh waves. Aspects of thermal properties are the basis of two papers, first the presentation of temperature decay logs (**Alm**), and second an evaluation of the prediction of thermal conductivity from wireline log suites (**Griffiths** *et al.*). In a geotechnical evaluation, **McCann & Entwisle** discuss the in situ determination of Young's modulus of the rock mass from wireline logs. **Ølgaard** presents a new rationale for the interpretation of nuclear log measurements. Data from a regional North Sea well log database, calibrated with known pore fluid compositions, provide the basis for a study of salinity variations in various oil fields and allow inferences to be made regarding basin fluid dynamics (**Gran** *et al.*).

Mineralogy and geochemistry. Despite the relatively recent advent of wireline logs that allow downhole measurement of geochemical parameters (see GAWL 1), only a single ODP paper (**Harvey & Lovell**) investigates the interpretation of these data, drawing attention to the problem of compositional co-linearity in mineral inversion. The other papers investigate mineralogical and geochemical problems without resort to downhole geochemical parameters. **Selley** identifies the potential for detailed sandstone diagenetic studies using logs, **Myers & Jenkyns** describe an efficient and accurate, field-tested method for determining the total organic carbon content of hydrocarbon source rocks which uses only conventional log data, which is apparently more accurate than alternative methods using geochemical log data, and **Cheshire & Sellwood** presents a forward modelling approach to the determination of the diagenetic characteristics of sandstones.

We would like to thank the independent referees and authors who undertook refereeing as part of their commitment to the publication of this volume. Without the help of the following we could not have attained the high scientific quality which was our goal: J. T. Adams, A. Adamson, J. S. Bell, L. T. Bourke, R. Brereton, E. Bølviken, G. I. F. Cameron, T. J. Chapman, M. J. Cheshire, M. B. Dusseault, C. J. Evans, E. Fjær, D. Goldberg, P. K. Harvey, C. Hermanrud, M. M. Herron, R. Hillis, R. Holt, B. E. Hornby, J. J. Howard, P. D. Jackson, J. L. Jensen, J. D. Kantorowicz, D. Kassenaar, M. King, J. Kulenkampf, S. R. Lawrence, M. A. Lovell, S. Luthi, P. Lysne, K. Mandziuch, R. Marsden, M. Matthews, C. McCann, D. M. McCann, D. Moos, B. P. Moss, K. J. Myers, O. Olsson, J. Owens, L. M. Parson, V. Patel, P. A. Pezard, A. M. Raaen, S. Raikes, M. H. Rider, M. Sams, R. C. Selley, R. M. Slatt, H. Smith, B. Steingrimsson, A. E. Stocks, D. A. V. Stow, J. Tittman, B. Twombley, J. R. Underhill, J. J. Walsh, J. Ward, H. Williams, M. Worthington, N. Yassir and G. Yielding.

Sedimentology and stratigraphic correlation

Outcrop gamma-ray logging to improve understanding of subsurface well log correlations

ROGER M. SLATT,[1] DOUGLAS W. JORDAN,[1] ANTHONY E. D'AGOSTINO[2]
& ROBERT H. GILLESPIE[1]

[1] *ARCO International Oil and Gas Company, 2300 West Plano Parkway, Plano, Texas 75075, USA*

[2] *ARCO Oil and Gas Company, 2300 West Plano Parkway, Plano, Texas 75075, USA*

Abstract. Reliable, reproducible gamma-ray logs of outcrops have been generated by two techniques which aim to improve the visualization of interwell-scale lateral continuity (and discontinuity) of strata and to demonstrate the reliability and potential pitfalls of subsurface wireline log correlations. One innovative technique was developed which uses a standard gamma-ray sonde run from a logging truck in order to log vertical cliff or quarry faces. A second technique employs a hand-held gamma-ray scintillometer to log more easily accessible outcrops.

Examples are presented from the Jackfork Group (Pennsylvanian), Arkansas, USA, of outcrop gamma-ray logging of both laterally continuous and discontinuous turbidites in structurally simple and complex settings. Because the strata from which the logs were measured can be visualized and discussed at the outcrop, these examples can clearly illustrate the following aspects of wireline log correlation: (1) reliability can be greatly improved by understanding expected subsurface depositional geometries and lateral facies changes; (2) wireline log correlations in stratigraphically and structurally complex settings, such as many oil or gas fields, may not be reliable without sufficient coring and special logging in addition to well testing; (3) erroneous correlations can result from the common practice of inferring three-dimensional rock geometries from two-dimensional well log data. Additional examples demonstrate that correlations and interpretations of subsurface wireline logs in both oil/gas fields and in exploration areas can be improved by comparing the subsurface logs with outcrop gamma-ray logs of nearby analogous strata.

In this paper we present some results of research we have been conducting on gamma-ray logging of rock outcrops in order to visualize better interwell-scale lateral continuity (and discontinuity) of strata and to demonstrate the reliability and potential pitfalls in subsurface well log correlations to engineers and geoscientists on training exercises and field trips. We have utilized two different techniques for outcrop gamma-ray logging under different conditions. The first technique employs a standard scintillometer tool (sonde) run from a logging truck. This innovative technique is used to log cliff or quarry faces that are inaccessible by foot. To our knowledge, this method of outcrop logging has not been previously published. The second technique uses a commercial, hand-held gamma-ray scintillometer to log readily-accessible outcrops. This technique has been used on occasion by geologists to evaluate lithology and organic content of beds, as well as to correlate strata between outcrop sections and subsurface wells (Provo *et al.* 1977; Chamberlain 1984; Myers & Wignall 1987; Howe 1989; Myers & Bristow 1989).

The outcrop examples presented in this paper are mainly from the Pennsylvanian Jackfork Group in Arkansas. Superb exposures in quarries and dam sites contain sheet-like, laterally continuous (on the scale of the outcrop) to lenticular, laterally discontinuous turbidite strata. A review of the geological framework of the Jackfork Group is given by Jordan *et al.* (1991).

Outcrop logging techniques

Truck-mounted gamma-ray sonde

The gamma-ray sonde that we use is about 2.1 m long and weighs approximately 40 kg. This sonde is sufficiently sturdy so that the source crystal can withstand the impacts it receives as the tool is lowered and raised along the rock face.

The logging truck is driven to the top of the quarry or cliff to a pre-selected logging site, which is, wherever possible, a smooth, vertical cliff with few obstructions. A small tripod with a

From HURST, A., GRIFFITHS, C. M. & WORTHINGTON, P. F. (eds), 1992,
Geological Applications of Wireline Logs II. Geological Society Special Publication No. 65, pp. 3–19.

Fig. 1. Gamma-ray sonde (large arrow) being raised from the base of the west wall of Hollywood Quarry by a cable (small arrow) attached to the logging truck (not in view). The person at the base of the quarry wall is using a rope attached to the sonde to guide it past rock protuberances as it is drawn upward. This logging run produced the gamma-ray log that is superimposed upon the quarry wall in the photograph. Note the scale marked in 0.3 m increments.

sheave-wheel on top is positioned at the cliff edge to guide the logging cable over the edge. The sonde is lowered down the vertical face by cable to the base of the cliff (Fig. 1). The sonde is then raised at a constant rate (3 m/min) and gamma-ray measurements are continuously recorded. The logging rate is slower than conventional (borehole) rates since, on a cliff face, the gamma-ray tool is exposed to less than half of the rock mass normally encountered in a borehole.

A rope is attached to the base of the sonde and is used to guide the sonde past protuberances as it is drawn upward (Fig. 1). This procedure is also necessary to ensure that the tool is kept in close proximity to the cliff face.

Conventional downhole gamma-ray measurements can be affected by minor borehole washouts (especially when used wth a heavy mud); also, extraordinarily large cave-ins or solution caves can significantly decrease the count rate. Tests at the outcrop were performed to determine the maximum distance away from a cliff face that the gamma-ray tool could be positioned before the count rate is adversely affected. Repeated measurements of a single sample site

Fig. 2. Gamma-ray logs obtained with the hand-held scintillometer at 0.6 m (smoother curves) and 0.15 m sample spacings along the base of the west wall of DeGray Lake Dam Spillway, and photographically superimposed upon the wall perpendicular to dip. The logged interval extends from 101 m (top of the upper sandy interval) to 143 m (base of the lower sandy interval). Stratigraphic top is toward the left. Numbers refer to measured gamma-ray log intervals (in metres). As in all subsequent figures, cps refers to 'counts per second' measured with the hand-held scintillometer.

show that the sonde may be up to 0.6 m away from the face and have little effect on the count rate. However, the variance in count rate increases significantly beyond this distance.

Other steps used to ensure quality outcrop logs include: (1) having photographs of the cliff available during logging, (2) hanging a well marked scale over the cliff during photographing and logging (Fig. 1), (3) providing walkie-talkie communication between the logging truck and the ground observer, and (4) marking on the sonde the exact gamma-ray recording point (midway down the length of the sonde) so that the observer on the ground can mark on the photographs, and/or relay to the logging truck via walkie-talkie, the exact location on the cliff of any key measurement site. Upon completion of logging, the data can be displayed in a standard log format (Fig. 1).

Hand-held gamma-ray scintillometer

For this type of logging, a lightweight, portable Scintrex™ GIS-5 integrating gamma-ray spectrometer is used. This instrument has the capability of measuring either total radiation or individual radioelement concentrations (Myers & Wignall 1987). Typically, total radiation measurements are made at 0.6 m stratigraphic intervals. Selected stratigraphic sections have been logged at 0.15 m intervals to provide more detail. An example of both logs is shown in Fig. 2. The recording diameter of the instrument equals 0.3 m (Scintrex users manual), so that gamma-ray measurements made on thinner beds will be influenced to some degree by gamma-radiation from adjacent strata.

Field tests conducted to determine the optimal sampling rate and number indicated the most efficient method of obtaining data is to take five total count readings for three seconds each, then to discard the highest and lowest values and average the other three readings. The data can then be displayed as a standard well log curve as a function of stratigraphic depth.

Log correlation of laterally discontinuous, lenticular strata

Big Rock Quarry in North Little Rock, Arkansas contains a 1000 m wide by 60 m high quarry face composed predominantly of laterally discontinuous sandstone and shales that were deposited mainly by turbidity currents and debris flows in depressions on a continental slope or base-of-slope environment (Jordan *et al.* 1991). Three gamma-ray logs obtained with a logging truck, and spaced about 150 m and 185 m apart (Fig. 3), were measured along the northeast quarry wall. A close-up view of the stratification and correlative gamma-ray log at station No 1 (Fig. 3) is shown in Fig. 4. Comparison of the log with the stratification illustrates the consistently good log response to gross lithologic variability that was encountered.

Description of a detailed lithologic section was not possible because of the steepness of the quarry wall. However, Fig. 5 compares Log No 1 with a subsurface core of the upper part of the same interval taken approximately 15 m behind the quarry wall several years ago by Shell Oil Company (Link & Stone 1986; Stone pers. comm. 1990). The correlation between the log response and gross lithology is good. The exception to this occurs in the thinly-bedded interval between 8 and 14 m. The log response depicts this interval as a thicker sandstone sequence than is actually the case.

The three logs (Fig. 3) taken at this site are used to illustrate potential problems in well log correlation of laterally discontinuous strata. On several field trips, participants have correlated these logs prior to examining the outcrop. Typically four or five distinct stratigraphic intervals are correlated on the logs (Fig. 6A). However, when the quarry is examined, most beds are not laterally continuous at the 150 and 185 m spacing of the logs. A better correlation, made after studying the quarry face, is shown in Fig. 6B. In comparing this correlation to that shown in Fig. 6A, it is particularly noteworthy that the

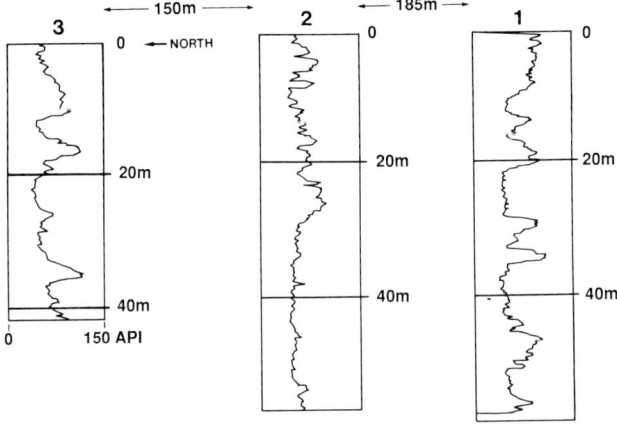

Fig. 3. Gamma-ray logs obtained with a logging truck along the northeast wall of Big Rock Quarry. Datum is the top of the quarry.

Gamma-Ray Log #1

0　　　　150 API

-20m

-40m

Fig. 4. Comparison of gamma-ray log No 1 at Big Rock Quarry (Fig. 3) with the quarry wall at the logging site. Arrows point to individual beds depicted on the log.

thick sandstone from 27–52 m on Log No 2 is a discontinuous sandstone slump upon which younger sandstone beds onlap, and (2) the cleaning-upward sandstone at 0–8 m on Log No 3 is a laterally discontinuous sandstone interval near the top of the quarry face. In the subsurface, these beds, though thick, would not be continuous between typically spaced development wells, so knowledge of the expected depositional geometries would help clarify the degree of uncertainty associated with well log correlations.

Log correlation of laterally continuous, sheetlike strata

The DeGray Lake Dam Spillway, near Arkadelphia, Arkansas contains a 300 m thick vertical succession of steeply-dipping turbidite strata. These strata record repeated cycles of turbidite sedimentation in a deep marine basin (Jordan et al. 1991). The succession is superbly exposed and easily accessible along two spillway walls (east and west) spaced approximately 90 m apart. The entire vertical section has been measured in detail by Morris (1977) (Fig. 7) and Breckon

(1988); individual measured segments have been published by Moiola & Shanmugam (1984). The overall vertical sequence grades upwards from thin-bedded sandstones and shales at the base, through thicker-bedded sandstones, into massive pebbly sandstones/conglomerates at the top (Fig. 7). Major cycles of sandstone deposition are separated by 3 to 9 m thick shales of hemipelagic and debris flow origin.

Lateral continuity and correlation of stratigraphic intervals

The pebbly sandstone/conglomerate beds at the top of the sequence form resistant ridges which can be traced continuously along strike over a distance of several kilometres in the field and on air photographs (Breckon 1988), indicating the sequence is laterally continuous for long distances. Figure 7 shows gamma-ray logs obtained with the hand-held scintillometer at 0.6 m sample spacings for the east and west walls of the spillway. Thick stratigraphic intervals are easily correlated on gamma-ray logs over the 90 m separating the east and west walls.

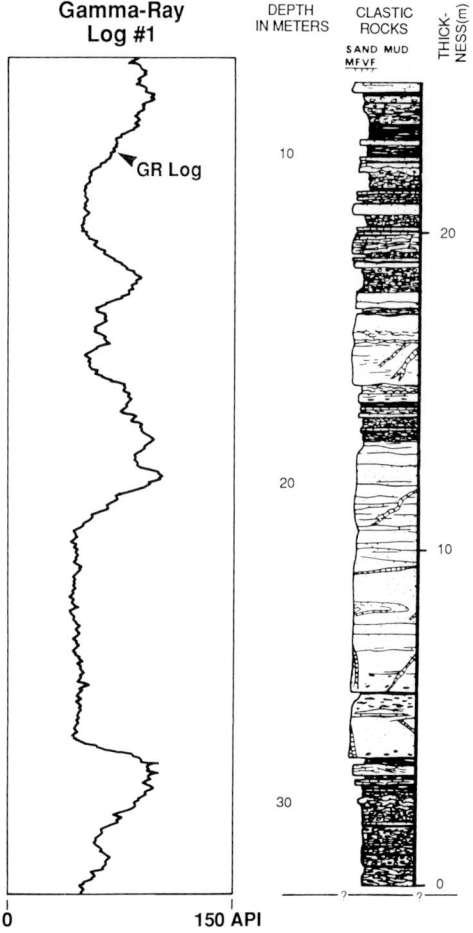

Fig. 5. Comparison of gamma-ray log No 1 at Big Rock Quarry with a Shell Oil Company core taken 15 m behind the quarry face at the logging site (Stone, pers. comm. 1990). Core description is after Link & Stone (1986).

Lateral continuity and correlation of individual beds

In order to address the issue of lateral continuity and well log correlation of individual beds, we measured 110 sandstone and shale beds that could be correlated on both sides of the spillway on the basis of distinctive sedimentary features and their position within the entire vertical sequence. The beds that were measured occur within two stratigraphic intervals along both sides of the spillway: (1) on the eastern wall (DeGray East) and western wall (DeGray West) of the spillway at 98–107 and 101–110 m (log depth), respectively (Figs 7 & 8), and (2) on the

eastern wall and western wall of the spillway at 111–140 and 114–143 m, respectively (Figs 7 & 9). These beds have been estimated to be laterally continuous for at least hundreds of metres (Jordan *et al.* 1991), so they would probably be continuous in the subsurface between typically spaced development wells and, consequently, potentially correlative on gamma-ray logs.

Gamma-ray logs were generated from these intervals with the hand-held scintillometer at both 0.6 m and 0.15 m sampling intervals. The log obtained for the upper sequence at a 0.6 m sample spacing (Fig. 8) only resolves two sandstone intervals (labelled A and B) which can be identified and correlated, indicating bed resolution at this sample spacing is on the order of 2–3 m. However, at 0.15 m spacing, the beds can be further subdivided into individual correlative packages (A1, A2, B1, B2, and B3) and in some instances into individual sandstone beds (a, b, c, and d) (Figs 2 & 8), indicating bed resolution at 0.15 m sample spacing is on the order of 0.5 m. A similar bed resolution of about 0.5 m occurs for the lower sequence of rocks, which also correlates across both sides of the spillway on gamma-ray logs sampled at 0.15 m spacing (Figs 2 & 9). These results are used to demonstrate that sequences of sheetlike, laterally continuous beds can be readily correlated on gamma-ray logs although individual beds probably cannot be distinguished nor correlated.

Depositional lobes, channels and compensation cycles

Turbidite depositional sequences that appear on well logs (and in core and outcrop) to become thicker-bedded, coarser-grained, and cleaner (lower API counts) stratigraphically upward are often interpreted as 'lobes' formed by continual progradation of sediments during a depositional cycle (Selley 1979; Shanmugam & Moiola 1988, 1991). By contrast, those that appear to become thinner-bedded, finer-grained and dirtier stratigraphically upward are often interpreted as 'channel-fill' deposits formed by successive upbuilding and backfilling of progressively finer-grained channel-fill as a channel goes through a cycle of formation, sedimentation and channel-filling/abandonment (Selley 1979; Shanmugam & Moiola 1988, 1991).

Rider (1990) lists a number of factors other than grain-size and clay content that affect gamma-ray log response, and suggests that interpreting depositional facies from gamma-ray log shapes is tenable only under certain circumstances. Based upon our study of the DeGray Lake Dam Spillway section, we suggest a factor

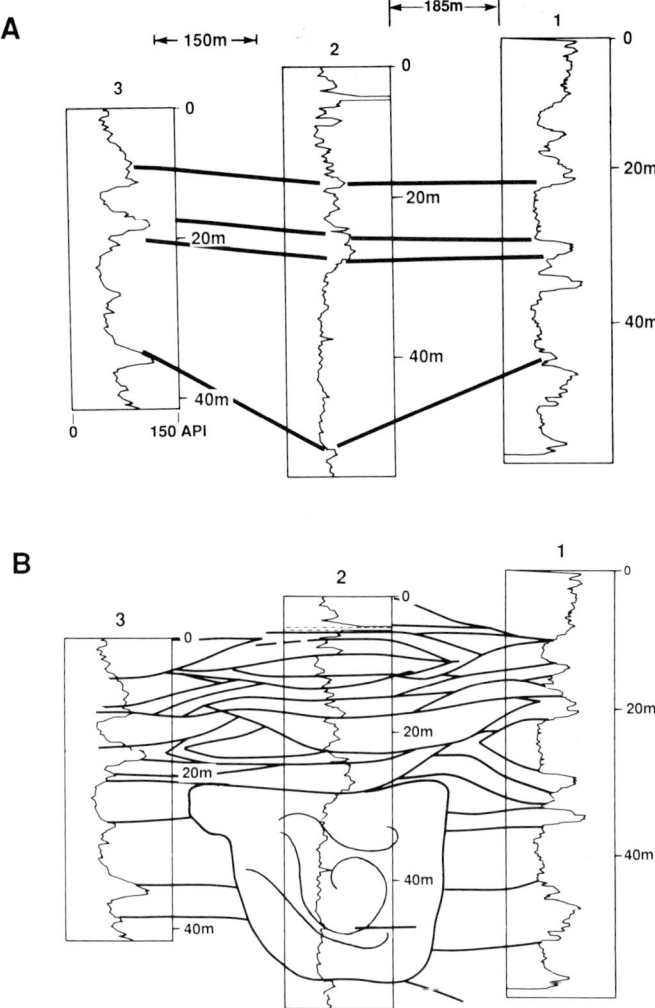

Fig. 6. Gamma-ray log correlations from Big Rock Quarry (A) before and (B) after examining the quarry wall.

that can be added to Rider's (1990) list is the inconclusive, but common, practice of interpreting or inferring three-dimensional facies architectures ('lobes' and 'channel-fills') from two-dimensional log data. For example, the interval at 111–140 m/114–143 m on the east and west walls of the spillway appear on the 0.6 m log (Fig. 7) to become thicker-bedded and cleaner (lower gamma-ray counts) stratigraphically upward, as would be typical of a 'lobe' deposit. However, the greater detail of the 0.15 m logs (as well as the stratigraphy) shows that the sequence is really a composite of several thinner cyclical sequences (Fig. 9) that could be interpreted as either thickening/cleaning-upward or thinning/dirtying-upward depending upon where one wishes to place boundaries. Figure 9 shows that for individual sandstone beds correlated on both sides of the spillway, the side of each bed which is thickest alternates vertically between the east and west walls of the spillway in a fairly consistent manner. This alternation suggests that cyclicity of sedimentation occurs in a lateral as well as vertical fashion owing to back-and-forth switching of depositional axes with time. This switching would occur when one turbidite bed is deposited on the sea floor and the next younger bed is laid down in the adjacent topographic

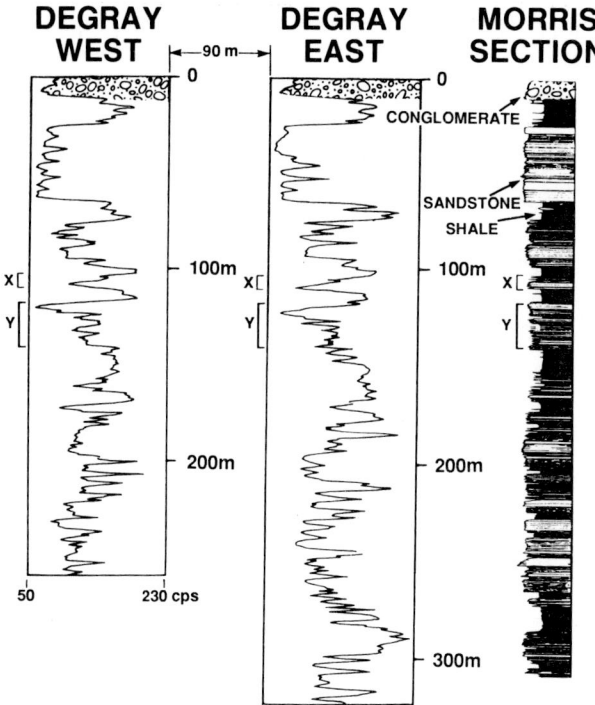

Fig. 7. Gamma-ray logs obtained with a hand-held scintillometer (0.6 m sample spacing) along the east and west walls of DeGray Lake Dam Spillway. The two walls are spaced about 90 m apart. Morris's (1977) measured section is shown for comparison. The intervals labelled X and Y refer to the stratigraphic intervals in Figs 8 & 9, respectively.

depression created by the first bed. Switching back-and-forth of individual bed axes in this manner produces 'compensation cycles' (Fig. 10) (Mutti & Normark 1987) which appear on gamma-ray logs (and outcrops) as a complex succession of both thickening/cleaning-upward and thinning/dirtying-upward sequences (Fig. 9).

Correlation of compensation cycles

Figure 10 shows a hypothetical sequence of turbidite strata (after Mutti 1985) and gamma-ray logs through different parts of the sequence. This hypothetical example demonstrates the difficulty that would be encountered in attempting to correlate depositionally related beds that vary in thickness laterally unless a well defined stratigraphic log marker were present which could be used as a datum. In this instance, as in many subsurface examples, without such a datum, slipping the logs to match thick sandstones and/or thin sandstones (or shales) (for example logs C and A), or interpreting facies based upon log

character (for example a 'channel' sandstone at B which cuts through 'lobes' between A and D), would lead to erroneous correlations.

Figure 11 demonstrates the same principle for groups of strata, in this case the 111–140 m/ 114–143 m spillway section (Fig. 9) and another measured section, called the DeGray Lake Dam Intake, located 1500 m west of the spillway. Our correlations show that some intervals become sandier and thicker toward the east while other intervals become sandier and thicker toward the west in a compensatory manner.

We use these two examples to illustrate on the outcrop that without some knowledge of depositional geometries and an understanding of lateral facies changes (i.e. thick-bedded sandstone grading laterally into thin-bedded sandstones and shales) proper correlation of stratigraphic intervals or depositional cycles at the scale of individual beds or groups of beds would be difficult. In the vertical dimension, if the hypothetical sequence (Fig. 10) or the spillway-intake sequence (Fig. 11) were to be misinterpreted as laterally continuous 'lobes' and later-

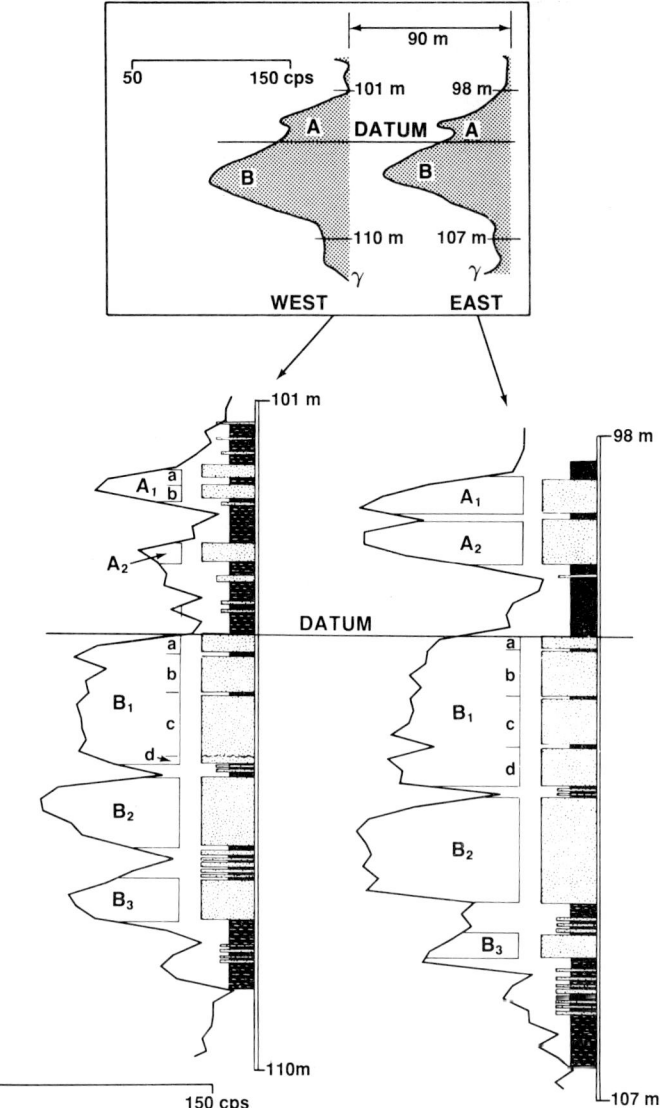

Fig. 8. Detailed stratigraphy and gamma-ray logs (0.15 m sample spacing) for the interval 98–101 m and 101–110 m (log depth) on the eastern and western walls, respectively, of DeGray Lake Dam Spillway. Figure 2 shows the outcrop section and Fig. 7 shows the position of the interval (X) within the entire spillway section (compare measured depths in metres). The inset shows the log interval obtained at 0.6 m sample spacings. Detailed stratigraphy of the two major intervals, subdivided into smaller packages (A1, A2, B1, B2, and B3) and sandstone beds (a, b, c, and d), was interpreted by D. R. Lowe (in Jordan *et al.* 1991).

ally discontinuous 'channel-fills' rather than as 'compensation cycles', a completely different correlation scheme would result.

Correlation of logs in a complex geological setting

At Hollywood Quarry, located about 11 km west

of the DeGray Lake Dam Spillway, 15 closely-spaced gamma-ray logs were obtained with the logging truck around the periphery of the quarry (Fig. 12). This quarry contains both lenticular and sheet-like turbidite strata; in addition, some faults cut through the quarry. This logging exercise was designed to develop a scaled-down representation of well logs and spacing patterns as

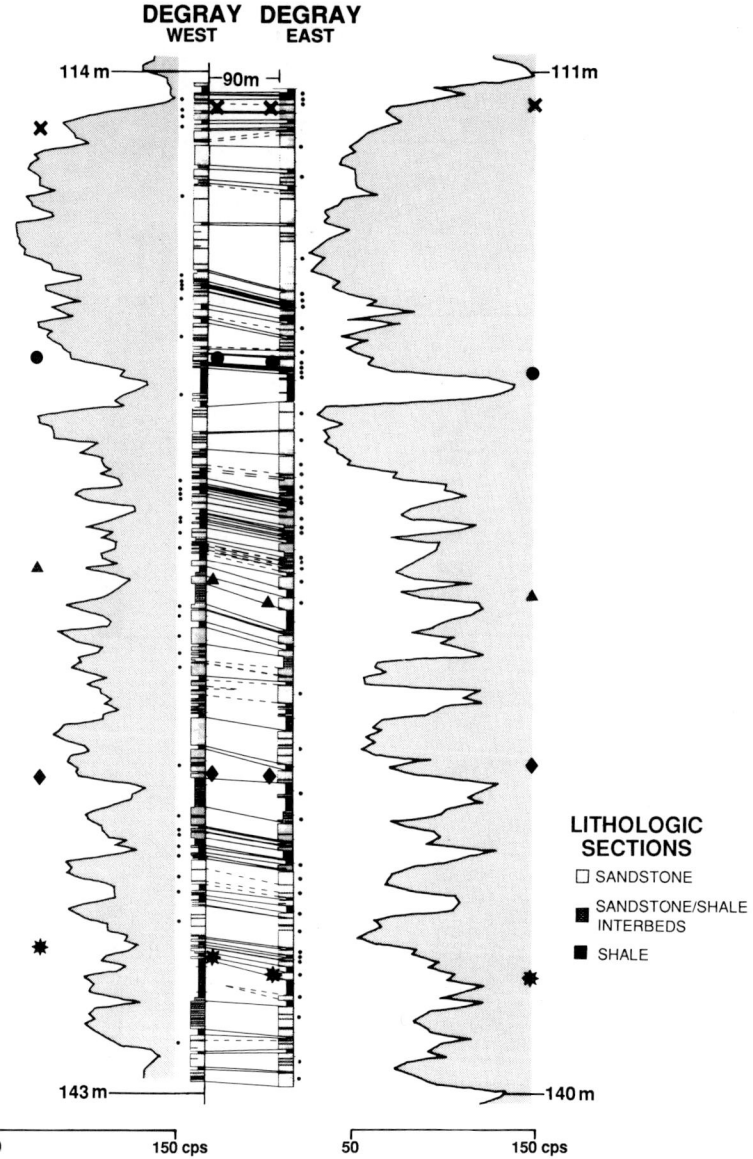

Fig. 9. Detailed stratigraphy and 0.15 m gamma-ray logs for the interval 111–140 m and 114–143 m (log depth) along the eastern and western walls, respectively, of DeGray Lake Dam Spillway. Figure 2 shows the outcrop section and Fig. 7 shows the position of the interval (Y) within the entire spillway section (compare measured depths in metres). The large symbols refer to some individual beds correlated between the outcrops and logs. Small dots are located adjacent to the side of the bed (on the east or west wall of the spillway) that is thickest.

might occur in a subsurface oil or gas field. With such a representation, it is possible to demonstrate to others at the outcrop the potential correlation pitfalls which might be encountered in fields with these types of geological complexities.

Figure 13 shows the log correlations at Hollywood Quarry. The beds strike about N35°E and dip 7° to the northwest so that the beds on the east wall of the quarry dip beneath those on the west wall. The upper part of the southern

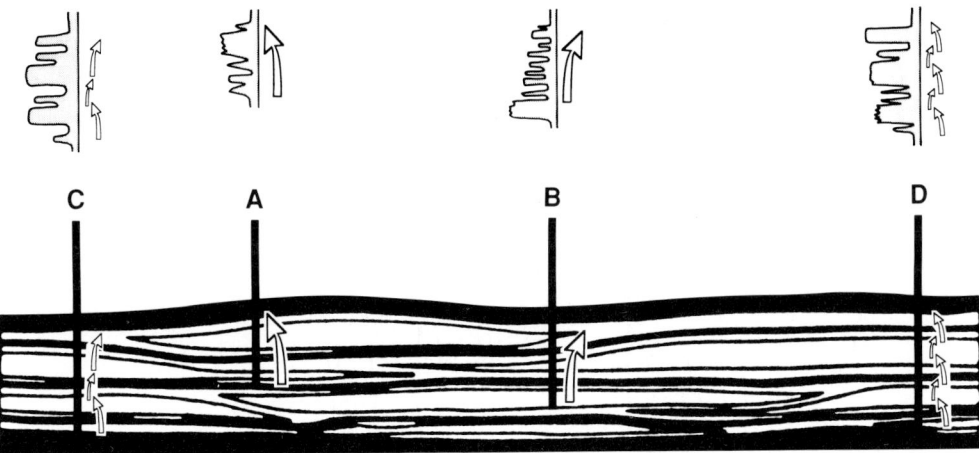

Fig. 10. Schematic two-dimensional view of turbidite geometries which illustrates the development of compensation cycles (Mutti & Normark 1987) as an individual bed is deposited in the adjacent topographic depression created by deposition of the immediately preceding bed. Some hypothetical gamma-ray logs are shown to illustrate how thinning/dirtying- and thickening/cleaning-upward vertical sequences can be formed by these compensation cycles. Lower diagram is modified from Mutti (1985). No scale is implied.

half of the west wall comprises laterally discontinuous strata. The northern half of the west wall contains more laterally continuous, but highly faulted beds. The east wall contains thick, lenticular sandstones and pebbly sandstones. In this example, if only logs were available, the correlations shown in Fig. 13 most likely would not have been made. Considering that these logs are relatively closely spaced, this exercise demonstrates some of the uncertainty in well log correlations that might be anticipated in oil and gas fields of complex facies architecture and structure. It also helps to explain, particularly to engineers, the need to obtain core in development wells for detailed sedimentologic and stratigraphic analysis, in addition to FMS or dipmeter logs to distinguish beds.

Comparing subsurface gamma-ray logs of strata with logs of outcrop analogs

Perhaps the most important application of outcrop gamma-ray logging is the ability to objectively improve subsurface well log interpretations and correlations by comparing the logs with those obtained from nearby analogous outcrops. Outcrop gamma-ray logs are particularly useful in areas where there are few wells, but abundant outcrops (Chamberlain 1984). Two examples are presented which demonstrate this application.

The first example is from the Jackfork Group. The Shell Rex Timber No 1–9 wildcat well was drilled in the mid-1980s 9.5 km southwest of the DeGray Lake Dam Spillway. Sedimentological examination of the outcrop revealed the presence of a distinctive conglomerate bed at the top of the DeGray sequence (Fig. 14). Based upon examination of the mudlog for this well, we interpreted the conglomerate bed to occur at 817–824 m in the well. The base of the conglomerate bed was used as a datum to tie the outcrop gamma-ray log with the well gamma-ray log (Fig. 14). Once this was accomplished, it was possible to correlate other intervals between the well and the outcrop, and from that, to predict other lithologies in the well on the basis of our knowledge of the outcrop lithologies.

The second example is from the Miocene Rincon and Topanga formations exposed along the southern California coast at Point Mugu. Here, a 350+ m thick, flat-lying clastic sequence is composed of highstand shelf sandstones and shales overlain by lowstand submarine canyon turbidite- and slump-fill (Jordan & Marquard 1990). The hand-held scintillometer was used to generate a gamma-ray log (Fig. 15A) which was calibrated to facies and systems tracts interpreted from the outcrop. This calibrated log was then utilized to interpret facies and systems tracts from subsurface gamma-ray logs in a well 13 km away, in the adjacent offshore Santa Barbara Channel (Fig. 15B). Once the potential hydrocarbon-bearing zone (i.e. lowstand turbidite facies) was identified on other well logs in the offshore (and tied to seismic sections), the

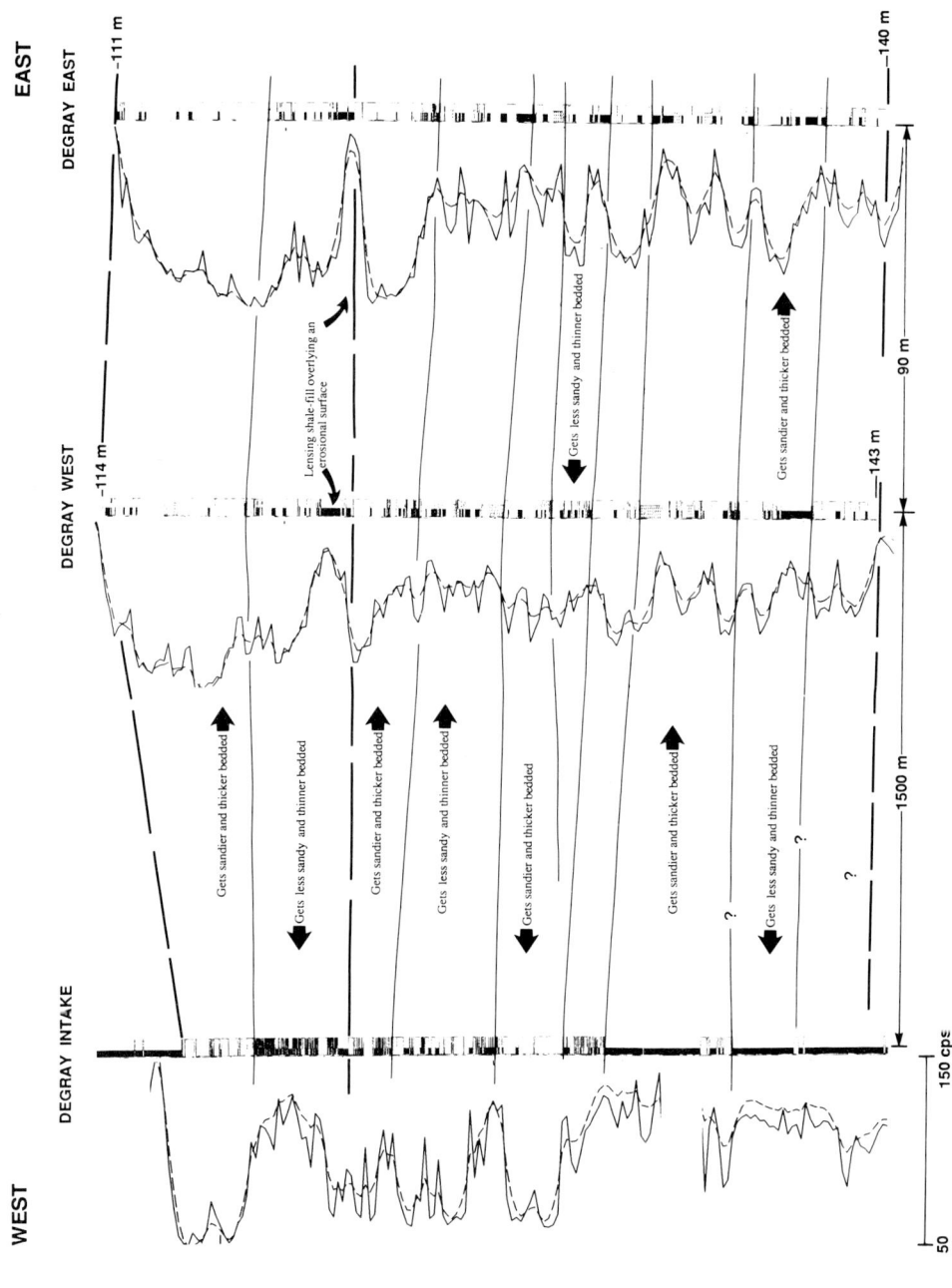

Fig. 11. Correlations of the interval 111–140 m/114–143 m at DeGray Lake Dam Spillway with the stratigraphically equivalent interval at DeGray Lake Dam Intake, 1500 m away, showing how individual sequences vary laterally to form compensation cycles. The solid log curve represents actual data measured at 0.6 m increments, and the dashed curve has been smoothed.

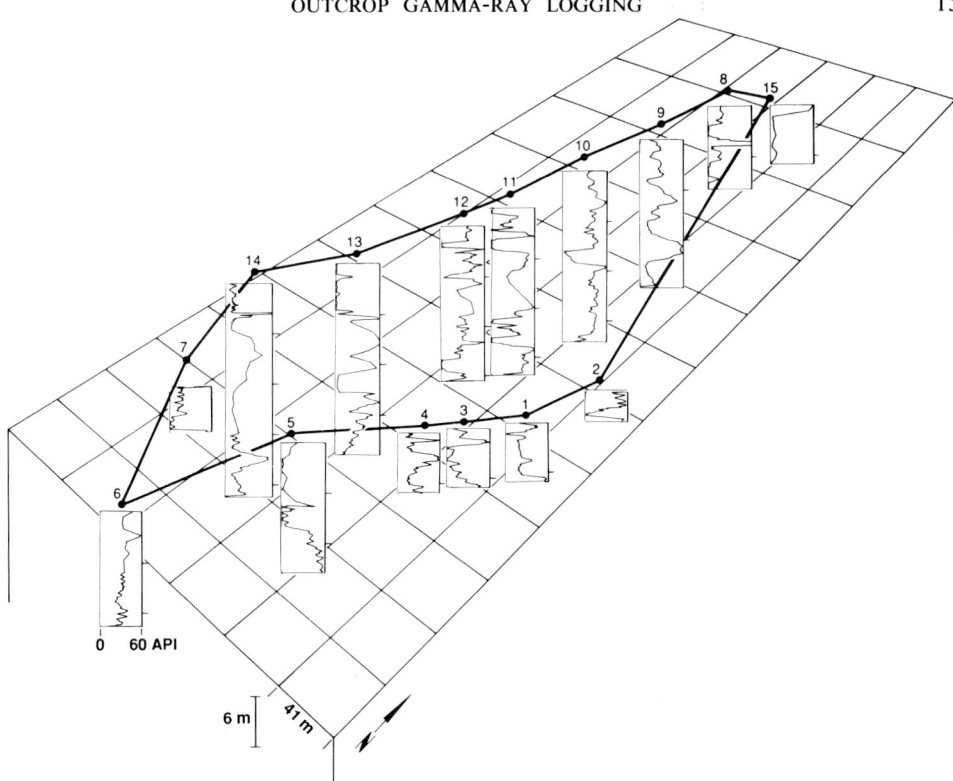

Fig. 12. Location of the fifteen gamma-ray logs obtained with the logging truck at Hollywood Quarry. Heavy line connects the top of the quarry wall.

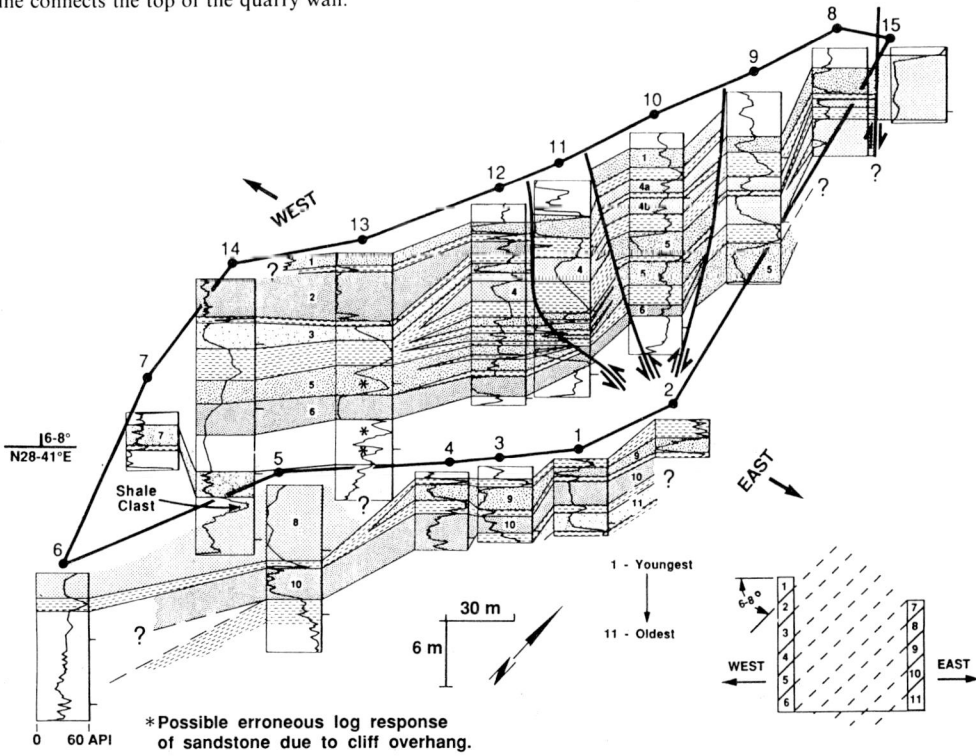

Fig. 13. Gamma-ray log correlations at Hollywood Quarry. The inset shows the attitude of the beds; beds 7–11 on the east wall dip beneath beds 1–6 on the west wall of the quarry.

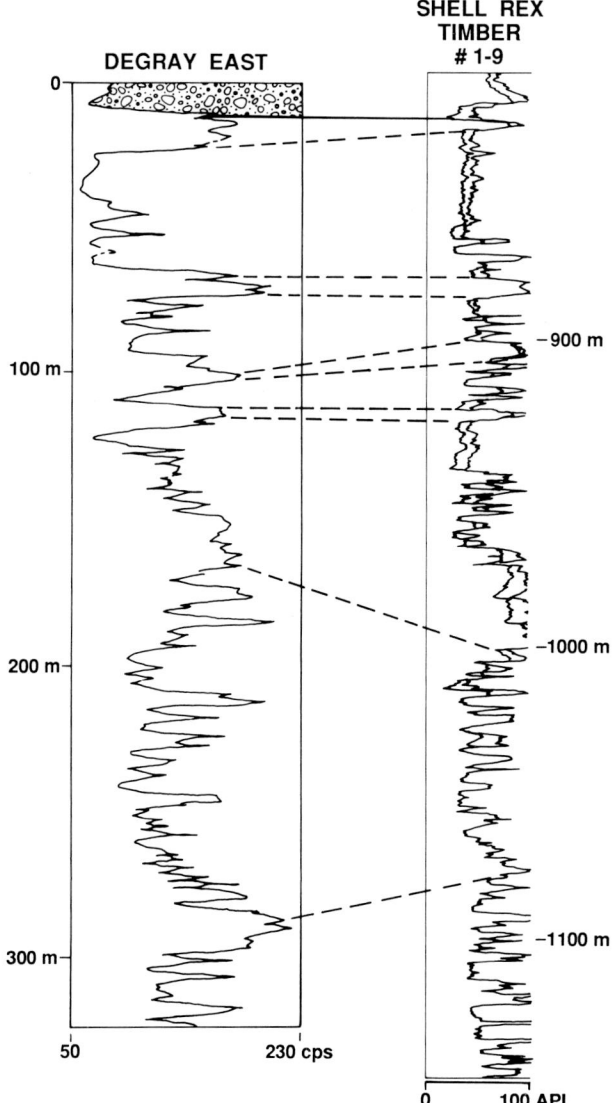

Fig. 14. Correlation of the Jackfork Group rocks in the Shell Rex Timber No 1–9 (Clark County, Arkansas) with the outcrop gamma-ray log from the east side of DeGray Lake Dam Spillway. Correlations are based upon comparisons of outcrop and mudlog lithologies and upon comparisons of outcrop and wireline log curve shapes.

interval could be mapped in order to delineate potential exploration targets.

Conclusions

1. Two techniques have been utilized which produce reliable gamma-ray logs of outcrops. One innovative technique was developed which uses a standard gamma-ray sonde run from a logging truck to log vertical cliff or quarry faces. The second technique employs a hand-held gamma ray scintillometer to log easily accessible outcrops. These logging techniques are useful for demonstrating to engineers and geoscientists, at outcrop, the reliability, potential pitfalls and degree of uncertainty associated with correlating

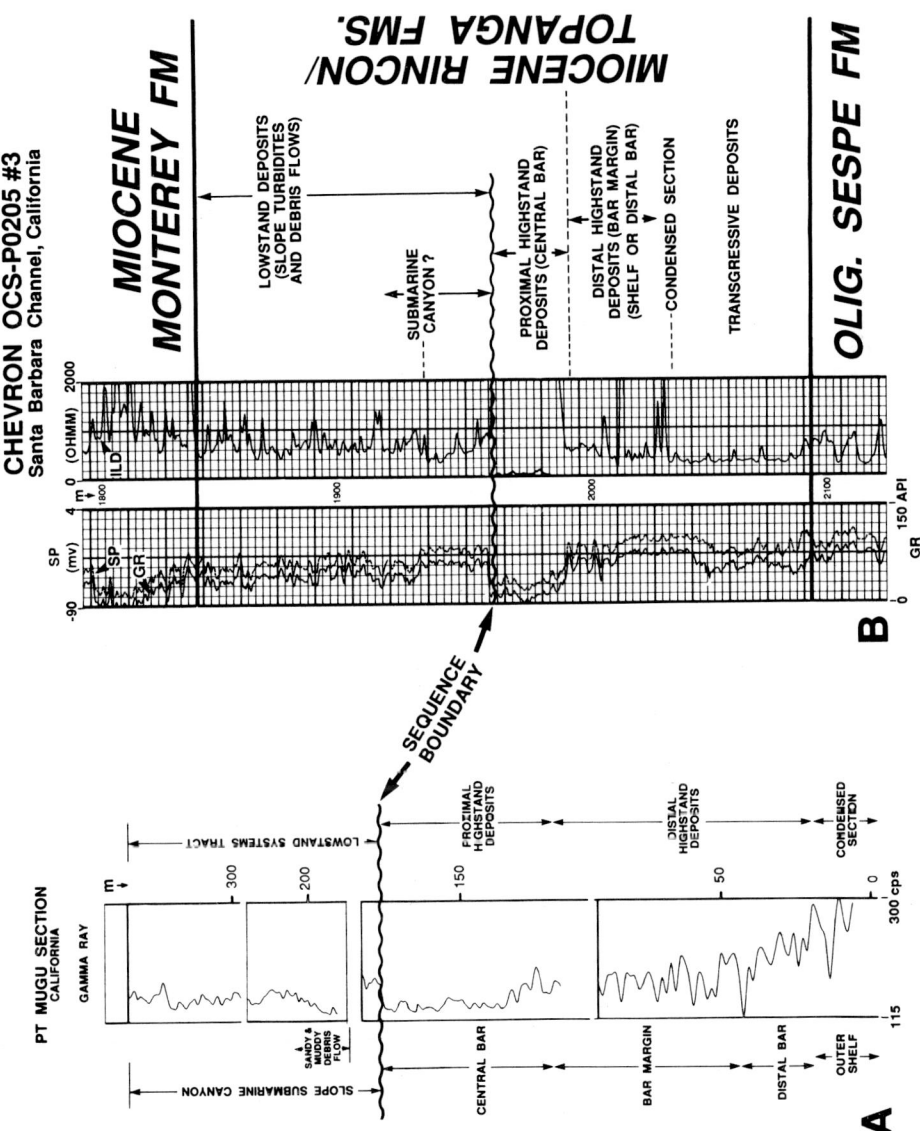

Fig. 15. (A) Measured gamma-ray log from Pt. Mugu, California section. Log depths (m) refer to measured outcrop section. Gaps denote covered or inaccessible portions of the section. Interpretations of facies and systems tracts are based upon outcrop examination. (B) Wireline logs from the same stratigraphic interval in an offshore well in the Santa Barbara Channel, California 13 km from the outcrop. Stratigraphic units were inferred from the log pattern, but only after the log patterns from the coastal exposures were understood and interpreted Log depths are in metres. Note two scales for the Pt. Mugu section.

subsurface gamma-ray logs, as well as the nature of stratification and vertical bed cyclicity depicted on logs.

2. Outcrop examples are used to demonstrate to others on the outcrop that a knowledge of the expected depositional geometries and lateral facies changes of subsurface strata will improve subsurface well log correlations, or at least reduce the degree of uncertainty associated with correlations. This is particularly true where beds are laterally discontinuous at the scale of typical development well spacings, where individual beds are thin, and where there are significant lateral facies changes. In stratigraphically and structurally complex settings, such as many oil and gas fields, proper stratigraphic correlations may not be feasible using only well logs. Substantial coring, special logging, and well testing are required.

3. Our examples are also used to show that the common practice of inferring three-dimensional rock geometries from two-dimensional well log data can result in erroneous interpretations of depositional geometries and, consequently, of well log correlations and lateral bed continuity.

4. Outcrop gamma-ray logging is particularly applicable to improving interpretations and

correlations of subsurface well logs in oil and gas fields as well as in exploration areas by comparing the subsurface logs with outcrop gamma-ray logs of analogous strata.

5. Well log correlation problems of the type presented in this paper are more clearly explained and understood when the strata from which the logs were measured can be visualized and discussed at the outcrop. Because of this, we have found outcrop gamma-ray logging to be an excellent training tool for well log correlation and interpretation.

We wish to express our appreciation to Andrew M. Slatt, W. J. Ebanks Jr and Mark H. Scheihing for field assistance in obtaining some of the logs, Charles G. Stone for field guidance and for freely discussing his knowledge of the Jackfork Group, Frederick B. Keller for supplying a mudlog of the Shell Rex Timber #1–9 well, Donald R. Lowe for sharing his sedimentological insights and providing measured sections in Fig. 8 and Malcolm H. Rider and an anonymous reviewer for providing constructive criticism which improved the quality of this paper. ARCO Oil and Gas Company and ARCO International Oil and Gas Company kindly allowed us to publish this paper and the drafting departments performed their usual high-quality work.

References

BRECKON, C. E. 1988. *Sedimentology and facies of the Pennsylvanian Jackfork Group in the Caddo Valley and DeGray Quadrangles, Clark County, Arkansas.* PhD dissertation, University of Tulsa.

CHAMBERLAIN, A. K. 1984. Surface gamma-ray logging: a correlation tool for frontier areas. *AAPG Bulletin,* **68**, 1040–1043.

COWAN, D. R. & MYERS, K. J. 1988. Discussion-Surface gamma-ray logs: a correlation tool for frontier areas. *AAPG Bulletin,* **72**, 634–663.

HOWE, D. 1989. Surface gamma-ray profiling technique applied to Cretaceous Ferron Sandstone, east-central Utah (abs.). *AAPG Bulletin,* **73**, 365.

JORDAN, D. W., LOWE, D. R., SLATT, R. M., STONE, C. G., D'AGOSTINO, A. E., GILLESPIE, R. H. & SCHEIHING M. H. 1991. Scales of geological heterogeneity of Pennsylvanian Jackfork Group, Ouachita Mountains, Arkansas: Applications to field development and exploration for deep-water sandstones. *Dallas Geological Society Field Guidebook, 1991 American Assocation of Petroleum Geologists Annual Convention, Dallas.*

——— & MARQUARD, R. S. 1990. Integration of sequence stratigraphy and process sedimentology: Miocene Rincon and Topanga Formations, Santa Barbara Basin near Point Mugu, California (abs.) *AAPG Bulletin,* **74**, 689.

LINK, M. H. & STONE, C. G. 1986. Jackfork Sandstone

at the abandoned Big Rock Quarry, North Little Rock, Arkansas, *In:* STONE, C. G. & HALEY, B. R. (eds) Sedimentary and igneous rocks of the Ouachita Mountains of Arkansas. *Guidebook for the Geological Society of America Annual Meeting, San Antonio,* 1–8.

MOIOLA, R. J. & SHANMUGAM, G., 1984. Submarine fan sedimentation, Ouachita Mountains, Arkansas and Oklahoma. *Transactions of the Gulf Coast Association of Geological Societies,* **34**, 175–182.

MORRIS, R. G. 1977. Flysch facies of the Ouachita Trough—with examples from the spillway at DeGray Dam, Arkansas, *In:* STONE, C. G. (ed.) Symposium on the Geology of the Ouachita Mountains, *Arkansas Geological Commission,* **1**, 158–168.

MUTTI, E. 1985. Turbidite systems and their relations to depositional sequences, *In:* ZUFFA, G. G. (ed.) *Provenance of Arenites,* Reidel, Dordrecht, 65–93.

——— & NORMARK, W. R. 1987. Comparing examples of modern and ancient turbidite systems: Problems and concepts, *In:* LEGGETT, J. K. & ZUFFA, G. G. (eds) *Marine Clastic Sedimentology: Concepts and Case Studies,* 1–38.

MYERS, K. J., & BRISTOW, C. S., 1989. Detailed sedimentology and gamma-ray log characteristics of a Namurian deltaic succession. Part II. Gamma-ray logging, *In:* WHATELY, K. & PICKERING, K. T.

(eds) *Deltas: Sites and Traps for Fossil Fuels*. Geological Society, London, Special Publication, **41**, 81–88.

MYERS, K. J. & WIGNALL, P. B., 1987. Understanding Jurassic organic-rich mudrocks—New concepts using gamma-ray spectrometry and palaeoecology: Examples from the Kimmeridge Clay of Dorset and the Jet Rock of Yorkshire. *In:* LEGGETT, J. K. & ZUFFA, G. G. (eds) *Marine Clastic Sedimentology: Concepts and Case Studies*, 172–189.

PROVO, L. J., KEPFERLE, R. C. & POTTER, P. E., 1977. Three Lick Bed: useful stratigraphic marker in the Upper Devonian Shale in Eastern Kentucky: *Energy Research and Development Administration, Morgantown Energy Research Center*, CR-77-2, 56.

RIDER, M. H. 1990. Gamma-ray log shape used as a facies indicator: critical analysis of an oversimplified methodology. *In:* HURST, A., LOVELL, M. A., & MORTON, A. C. (eds) *Geological Applications of Wireline Logs*. Geological Society, London, Special Publication, **48**, 27–37.

SELLEY, R. C. 1979. Dipmeter and log motifs in North Sea submarine-fan sands. *AAPG Bulletin*, **63**, 905–917.

SHANMUGAM, G. & MOIOLA, R. J. 1988. Submarine fans: characteristics, models, classification, and reservoir potential. *Earth-Science Review*, **24**, 383–428.

—— & MOIOLA, R. J. 1991. Types of submarine fan lobes: models and implications. *AAPG Bulletin*, **75**, 156–179.

Techniques and applications of petrophysical correlation in submarine fan environments, early Tertiary sequence, North Sea

I. R. HATTON,[1] M. REEDER,[1] M. St J. NEWMAN[1] & D. ROBERTS[2]

[1] Kerr-McGee Oil (UK) plc, 75 Davies Street, London W1Y 1FA, UK

[2] ResTech Europe Ltd, 5 Red Lion Court, Off Alexander Road, Hounslow, Middlesex TW3 1JS, UK (Present address: Roberts Associates, Charnwood, Mayfield Road, Weybridge, Surrey KT13 8XB, UK)

Abstract. Accurate stratigraphic delineations and correlations are key factors in exploration and development of early Tertiary reservoirs of the North Sea. In the submarine fan environment, application of biostratigraphic data can be complicated by stacking of sand sequences. A sand indicator is derived from a wireline log computation. The sand intervals are then edited out to generate a log profile of the background sedimentary sequence. Additional refinement is obtained by use of cross plot analysis to define petrophysical attributes of selected stratigraphic units. An example is described where such analysis has been applied to the Balder Formation where a 'tuff profile' has been established. Further examples illustrate how the methods serve to extrapolate detailed biostratigraphic data of wells into adjacent wells with limited or no data.

The mature status of the North Sea Basin requires the resolution of complex correlation problems on the path to accurate delineation and evaluation of discoveries, and in the development of new exploration concepts.

The techniques of petrophysical correlation described in this paper were developed in response to practical problems encountered in evaluation of early Tertiary reservoirs. These reservoirs were formed by the stacking of mass gravity flow sands. High-resolution correlations within the subsurface utilizing the methodology described herein, supported by detailed quantitative biostratigraphy; and examination of considered analogue sequences at outcrop in the Late Eocene of the Southern Maritime Alps, suggest little or no erosive downcutting has taken place. Accordingly, some conclusions derived from the techniques described would not be valid for other depositional environments or locations; however, the general method may prove applicable.

Deposition of Tertiary reservoir sequences

The reservoir sequence as seen in cores and wireline log suites consists of interbedded sands and shales that build to thicknesses of several hundred feet. The location and geometry of an individual sand body is controlled by pre-existing basin-floor architecture, determined by tectonic processes and the form of preceding sand deposits. The sand bodies are volumetrically significant but represent a near-instantaneous event in comparison to the background hemipelagic shale deposition.

The problems resulting from this situation are two-fold. Firstly, the structural elevation of major constituent sand bodies may vary considerably within a defined stratigraphic unit. Frequently, seismic data can only resolve the top and bottom of said unit, defining only the *structural* element of the overall trap geometry. Well sections, by use of petrophysical correlation techniques in association with biostratigraphic control, reveal the temporal and spatial relationships of sands, and in so doing the *stratigraphic* element is accounted for. This is crucial in situations where the hydrocarbon column extends only part way through the section; inside the gross volume of the trap, younger sands tend generally to be structurally elevated, and occur above the hydrocarbon water contact. Older sands can be deeper and may be water saturated (Fig. 1). Secondly, the hemipelagic shales form potential intra-reservoir barriers, thus knowledge of their specific distribution may be exploited in reservoir management.

The following methodology was developed to resolve such problems within the Late Palaeocene–Early Eocene Balder Formation which contains significant volumes of bedded tuffaceous material. The petrophysical properties of the tuff have been exploited to aid correlation.

From HURST, A., GRIFFITHS, C. M. & WORTHINGTON, P. F. (eds), 1992,
Geological Applications of Wireline Logs II. Geological Society Special Publication No. 65, pp. 21–30.

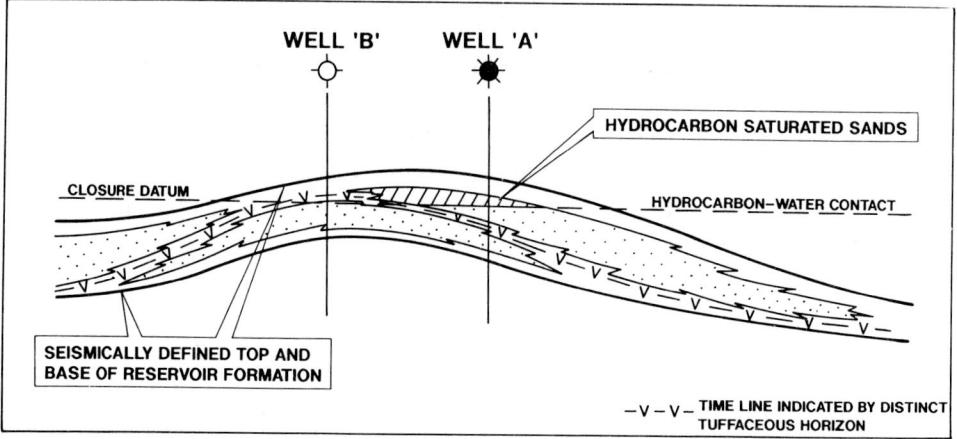

Fig. 1. In an appraisal situation, older sands within a seismically defined sequence may occur below the hydrocarbon–water contact and/or outside of trap closure. Alternatively, an exploration well drilled at location B could condemn a trap as 'water-wet'.

Development of regional depositional models and analogues combined with high-resolution seismic data may result in a re-appraisal of exploration potential.

Petrophysical correlation technique

Objectives and general procedures

The Balder Formation exhibits characteristic log responses that can be correlated between wells. Additionally, there is a characteristic change in these log responses from the top to the bottom of the Balder Formation that lends itself to the correlation of individual members. To enhance the correlation, each well is 'de-sanded': sands are first identified and then removed from the logged interval, and the remaining shales and tuffs are joined together. This effectively produces a re-constituted Balder Formation interval, together with the surrounding shales and siltstones. The procedure used to identify the sands is described in the section 'Discrimination of Sand'.

For each well, all available log data are presented, both before and after de-sanding. The data displays before de-sanding highlight the identified sand bodies that are to be removed. The data displays after de-sanding include annotation of original depths and the locations and sands removed. Correlations made on the de-sanded data can, accordingly, be related to individual sand bodies.

The before and after de-sanding log data displays are identical in format and may include dipmeter, conductivity and computed dip results, in addition to all the regular log data. A key element in these displays is the tuff indicator.

The tuff indicator is built from a combination of density neutron and gamma ray logs and is described in more detail in the section 'Delineation of tuff'. The tuff indicator is designed to enhance the characteristic difference in log response of tuff, compared to shales and siltstones. It is presented on multi-well cross sections as the prime characteristic for well to well correlation.

Log database, data quality and correlations

Although only certain log curves are finally utilized in the tuff and sand delineation, all available log data, including dipmeter, are displayed in the before and after de-sanding data displays. This is done in order that any data that may be of potential benefit in correlation are readily available in the databases.

In some wells, the interval analysed spans two logging runs. In these cases, log data are checked for consistency between the two runs and, where appropriate, normalization corrections are applied, e.g. gamma ray normalization. The log data are then merged at the optimum depth.

For all wells, log data are depth matched to within ± 0.5 feet, using the resistivity log as the depth reference.

For wells with spontaneous potential (SP) measurement, corrections for baseline drift are applied and, where necessary, the SP is filtered to remove mechanical noise.

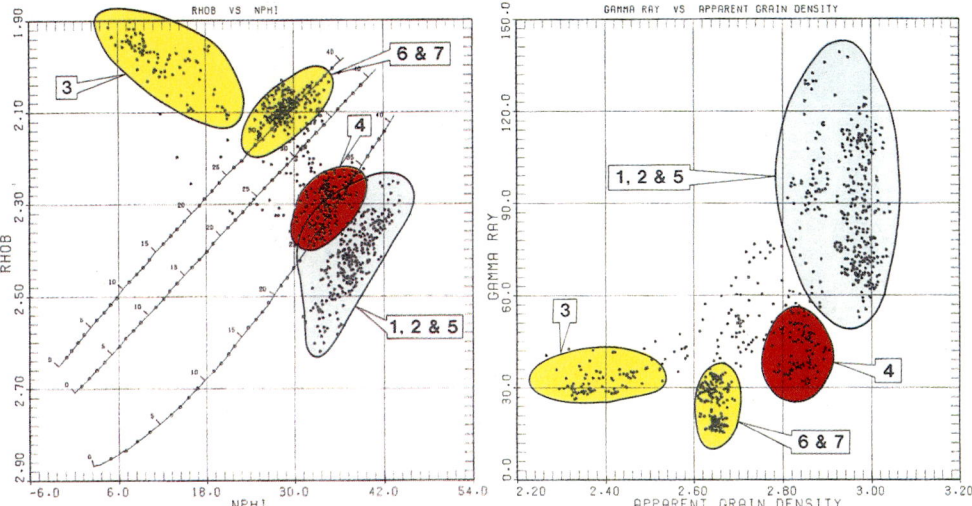

Fig. 2. Cross plot analysis of well 16/7–2 (Fig. 3). (a) RHOB vs NPHI. In computing apparent grain density (RGHA) pososity effects are removed to highlight lithological variances. (b) Gamma Ray vs Apparent Grain Density. To correctly identify sands to be removed from the log, a discriminator based on apparent grain density is used. One based upon gamma ray only would remove tuffaceous intervals as well. Key to numbered fields: (1) claystone/shale; (2) claystone/shale, minor tuff; (3) sand, gas bearing; (4) massive tuff; (5) claystone/shale; (6) sand, oil-bearing; (7) sand, water-bearing.

In some wells, both open hole and through casing density/neutron/gamma ray data are available. Appropriate shifts are applied to the cased hole logs to maximize the data available for analysis.

Environmental corrections for hole size, mud weight, etc., are applied to the gamma ray, density and neutron logs.

Characteristics of tuff

Part of our experience includes detailed histogram and crossplot analysis of log responses in extensive, continuously cored tuffaceous and non-tuffaceous intervals (Fig. 2).

Tuff is consolidated pyroclastic rock deposited as an ash fall erupted from volcanoes. It is composed of igneous rock fragments, volcanic glass and fragments of quartz, biotite, plagioclase and hornblende. There is a complete gradation from tuff to detrital sandstone. Fining-up sequences within tuffaceous bands are indicative of probable reworking by turbidity currents.

Due to its potential variable composition and compaction, exact logging tool response characteristics for tuff have not been defined in the literature.

However, previous studies have revealed some log relationships that are particular to igneous rocks. A thorium/uranium ratio close to 4 has

been observed in most intrusive volcanic rocks. A similar ratio for extrusive volcanic material such as tuff has not been reported in the literature. This is possibly because extrusive rocks have much lower concentrations of both thorium and uranium, (cf. intrusive rock), resulting in poor definition of the ratio. Definition of tuff through the analysis of thorium, uranium and potassium log responses in the Eocene has so far proved inconclusive.

However, other combinations of log responses have been observed that are particular to tuff and different from those encountered in non-tuffaceous intervals. In general terms, tuff has high apparent neutron and sonic porosities together with a medium density and a low to medium gamma ray response. This combination of responses could be described as a *shale-like* porosity tool response together with a definite *non-shale* response on the gamma ray. These general log characteristics of tuff were used to distinguish tuff from shale and from sand. (Described in more detail in sections 'Delineation of tuff' and 'Discrimination of sand'.)

Delineation of tuff

The characteristics of tuff described in the previous section were used to build a log model for delineation of tuff from sand and shale. The log model is based on the comparison of V_{shale} computed from two different log indicators. These two log indicators were selected such that one was sensitive to tuff and the other not.

(i) V_{shale} *indicator, sensitive to tuff.* The distinctive difference in log response between tuff and shale is the gamma ray. The tuff gamma ray response is low to medium and variable, compared with the typically high and generally consistent response in shale. Accordingly, gamma ray was selected for the V_{shale} indicator that is sensitive to tuff.

(ii) V_{shale} *indicator, insensitive to tuff.* Apparent grain density (RHGA) was found to be an excellent V_{shale} indicator that is insensitive to tuff. RHGA is computed from a combination of density and neutron log readings, using fixed assumed values for the density and neutron fluid, matrix and shale parameters. The calculation of RHGA is detailed in the Appendix. Because tuff has very similar density and neutron log responses to the surrounding shales, RHGA also has a very similar value in both tuff and shale.

For each well, V_{shale} was computed from both the gamma ray and the apparent grain density RHGA:

From GR : VSHALE_GR =
$$\frac{(GR - GR\_CL)}{(GR\_SH - GR\_CL)}$$

From RHGA : VSHALE_RHGA =
$$\frac{(RHGA - RHGA\_CL)}{(RHGA\_SH - RHGA\_CL)}$$

where:

VSHALE_GR	=	V_{shale} computed from GR
GR	=	GR log value
GR_SH	=	GR in shale
GR_CL	=	GR in clean formation
VSHALE_RHGA	=	V_{shale} computed from RHGA
RHGA	=	RHGA log value
RHGA_SH	=	RHGA in shale
RHGA_CL	=	RHGA in clean formation

The V_{shale} 'clean' parameters for both gamma ray and RHGA were selected to achieve good agreement between both indicators in clean water bearing sand intervals. The V_{shale} 'shale' parameters were similarly selected to achieve agreement in the shales. Thus, comparison of the two V_{shale} curves gave general agreement in both sand and shale, but produced a significant de-

parture in the tuff intervals. An overlay display of these two Vshales, with cross-hatched shading whenever V_{shale}_GR < V_{shale}_RHGA is used for the tuff indicator on the well data displays. Additionally, the actual difference between the two V_{shale} indicators is presented as a single curve Tuff flag.

An added benefit of using a tuff delineation model based on V_{shale} type computations is that log normalization is inherent in the technique—V_{shale}_GR is a normalized GR, V_{shale}_RHGA is a normalized RHGA. Accordingly, the tuff delineation results will, in principle, be directly comparable on a multi-well basis.

The tuff delineation model was developed further to address wells where density and neutron logs were unavailable or of poor quality: Spontaneous Potential (SP) was found to be a good V_{shale} indicator that is insensitive to tuff. This allows V_{shale}_SP to be used as a direct replacement for V_{shale}_RHGA in the model. For some wells, a combination (maximum value) of V_{shale}_SP and V_{shale}_RHGA was used for tuff delineation in bad hole conditions. (A maximum value was used because bad hole results in a reduction of RHGA.)

A negative aspect of using the SP for tuff delineation is that it has poor vertical resolution, which can lead to erroneous indications in thin beds. Additionally, the SP does not always have good sensitivity between sand and shale (is dependent on the salinity of the drilling mud), and is also prone to external noise and drift. However, careful treatment to correct for drift and noise in most cases resulted in an adequate delineation of the main Balder Tuff event.

Discrimination of sand

In order to remove the sands, a discriminator was needed that would distinguish the sands from the shales and tuffs. Due to the nature of the log responses in the Balder Tuff, the gamma ray is not an appropriate log for delineating sands (a gamma ray discriminator will remove most of the tuffs together with the sands). The V_{shale} indicator that is insensitive to tuff was the logical choice for a sand discriminator:

$$V_{shale}\_RHGA \text{ and/or } V_{shale}\_SP$$

The value of RHGA in both tuff and shale is around 2.95 to 3.00 g/cm^3 compared with a value in water-filled sandstone of 2.65. Oil has a very similar response to water: RHGA in oil-bearing sandstone is just marginally lower than the 2.65 water-filled value. When gas is present, the RHGA value can go below 2.00 g/cm^3.

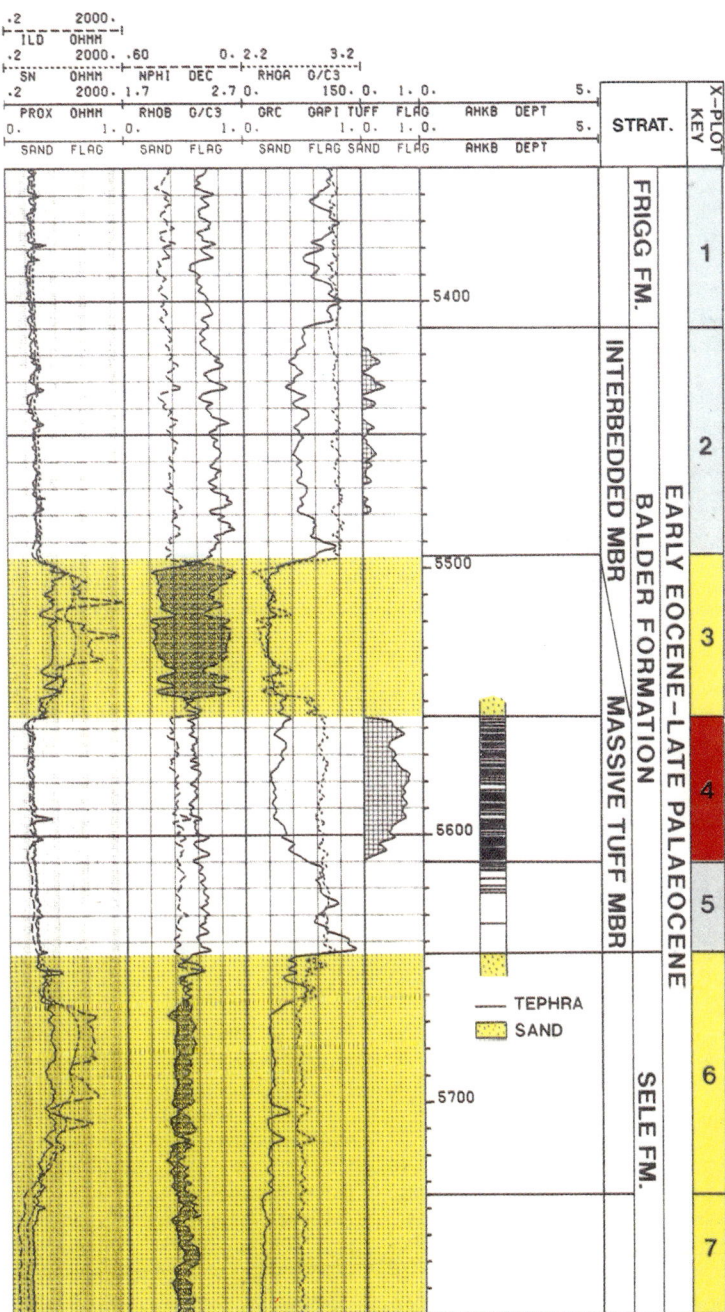

Fig. 3. Comparison of cored interval, well 16/7a–2, and log derived tuff indicator. Core description from Morton *et al.* 1990. The interval 5408′–5492′ has poor biostrat. control for a definitive age date. However, justification for inclusion within the Balder Formation is provided by correlation to a nearby well with suitable control (Fig. 5). Logs indicate an interbedded tuff sequence which would correspond with zone 2C of Knox *et al.* 1988, and is similar to known cored intervals not currently in the public domain. The interval 5557′–5610′, the massive tuff member, is zone 2B of Knox *et al.* 1988. The numbered intervals correspond to the datafields plotted on Figs 2a and 2b.

Accordingly, RHGA exhibits a very good sensitivity to sand versus shale/tuff, and hydrocarbon effects enhance the sand discrimination. An initial cut-off level of V_{shale} RHGA = 70% was selected for sand discrimination. However, final selection of sand intervals was made manually, using the discriminator as a guide. This was found to be necessary in order to ensure that shale/sand boundary effects were removed from the de-sanded data.

Displays of all the data before de-sanding, highlight the sand intervals with stippled shading. Sand intervals were then removed from each well, and the data re-displayed.

The criteria used for sand removal were as follows:

> sands thicker than 4 feet were removed;
> shales thinner than 4 feet occurring within a thick sand body were removed with the sand.

The 4 feet thickness criterion was based on the vertical resolution of the density, neutron and gamma ray logs: it was observed that these tools generally required a bed thickness of at least 3 feet for full development of the log response.

Presentation of results

The following displays are used to present the results of the tuff delineation and de-sanding analyses:

(1) Composite data displays before de-sanding;
(2) Composite data displays after de-sanding;
(3) Multi-well cross sections of tuff flag and gamma ray before de-sanding;
(4) Multi-well cross sections of tuff flag and gamma ray after de-sanding;

Comments on results

The tuff indicator has been successfully used to both delineate and characterize the occurrence of tuff in this area of the South Viking Graben. However, valid interpretational application of the tuff indicator requires an awareness of potential anomalies inherent in the computation: spikes on the tuff indicator (tuff indications of a short depth duration) are frequently an artifact of the computation and do not relate to the occurrence of tuff. When the tuff indicator is computed from a combination of three radioactive measurements (density, neutron and gamma ray logs), spurious spikes may occur due to a combination of the statistical variations of each measurement. Additionally, at bed boundaries and over thin beds, the different vertical resolution of each tool can produce (spurious) spikes on the tuff indicator. This is particularly true when the SP and gamma ray combination is used for tuff delineation. Accordingly, any spiked tuff indications of less than 1–2 feet should be ignored.

Despite these limitations, the tuff delineation technique has proven to be extremely successful in identifying the main internal characteristics of the Balder Formation over a wide area of the South Viking Graben. It has been observed that the tuff profile is remarkably consistent in thickness and character over quite large geographical areas. This has enabled detailed well to well correlation of several distinct features within the Balder Tuff formation. Such detailed correlation has been of particular benefit in wells where the Balder Tuff has been segmented within multiple thick sand bodies: after removal of the sands, the characteristic profile of the tuff and the background sedimentation becomes remarkably clear.

Figure 3 shows the technique applied to the Balder sequence of well 16/7a–2, the cored interval of which was described by Morton *et al.* (1990).

Examples of application

Sand stratigraphy, Balder Formation

Figure 4 demonstrates a correlation exercise for three Balder Formation sections in block 9/18a. Figure 4(a) comprises the de-sanded profiles for gamma ray and the tuff indicator flag. Five subsequences are shown based upon the log character. Only the gross sequence is controlled by biostratigraphic data.

Figure 4(b) shows the sands restored to the previous diagram. It is immediately evident that there are at *least* ten episodes of sand deposition represented in these wells. In a field appraisal or development scenario such detailed correlations can be employed in the prediction of sand fairways and thence reservoir internal geometry.

Fig. 4. Correlation of Balder Formation, Block 9/18a, Viking Graben, UKCS. (a) The 'de-sanded' section broken down to five divisions on the basis of the tuff flag character. Star flags denote position of removed sands. (b) The 'sand restored' section (see text for further comments).

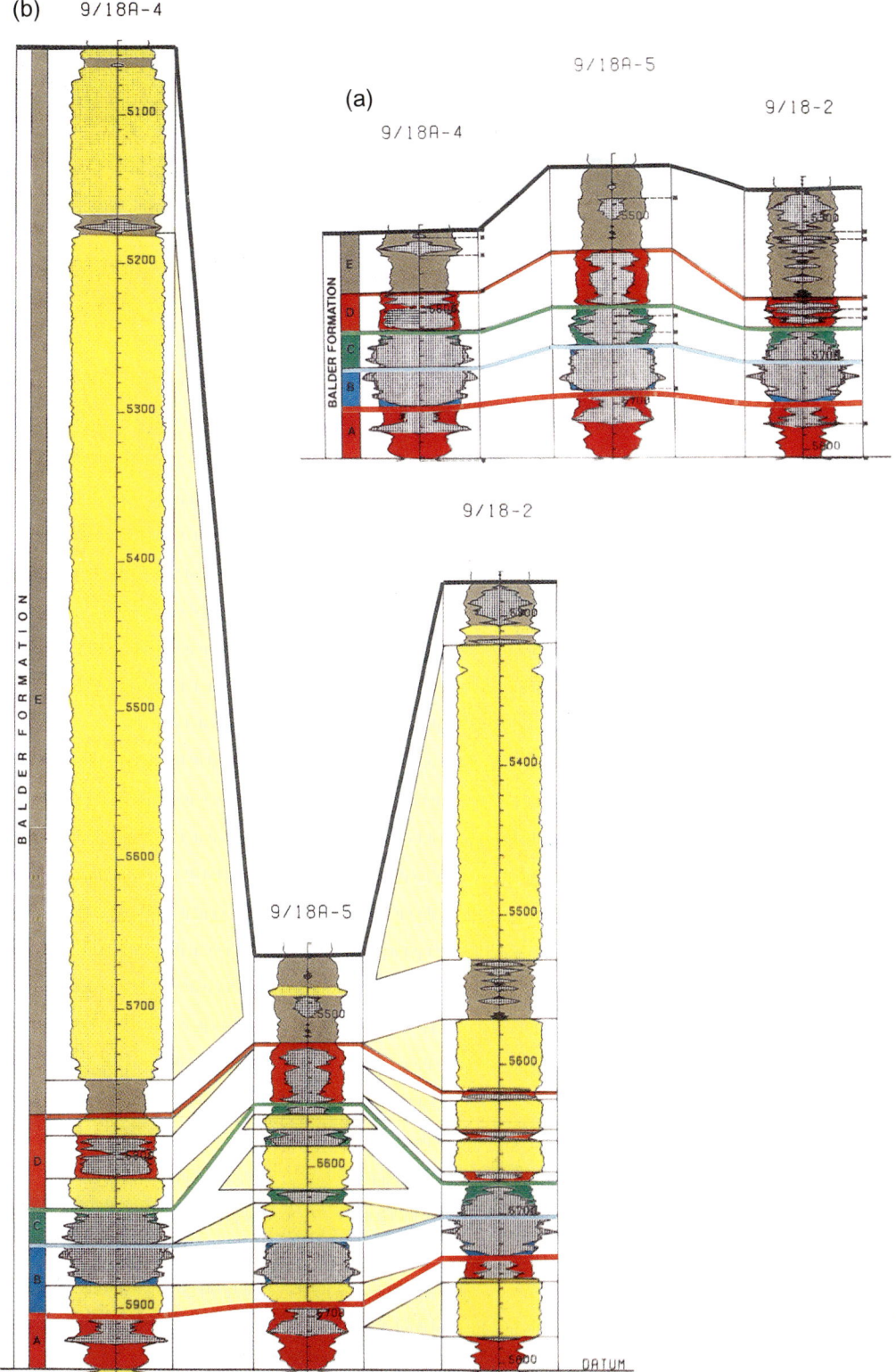

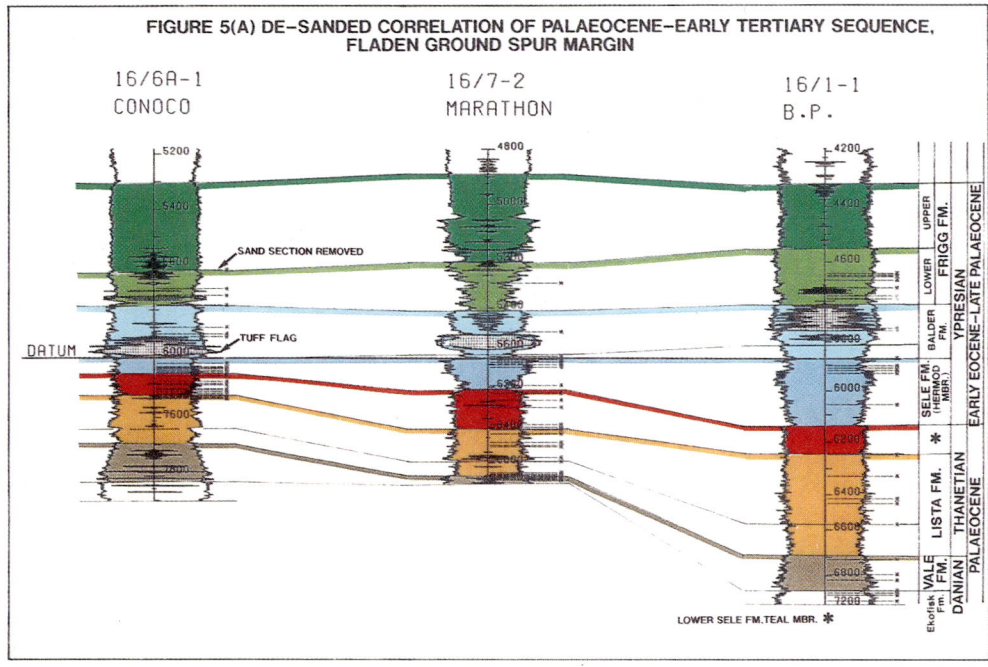

Fig. 5. Correlation of Palaeocene to Early Eocene sequence, Quadrant 16, Fladen Ground Spur, UKCS. (a) 'De-sanded' correlation of gamma ray log. Star flags denote position of removed sands. (b) (opposite) 'Sand restored' correlation. This figure demonstrates an exploration application of the 'de-sanding' process. In addition to the major sequence boundary correlations, other correlative features are revealed which can be checked iteratively with the biostratigraphic data for significance. Well 16/1–1 was initially correlated without biostratigraphic data and subsequently found to be correct when data became available. This demonstrates a further application of the technique when such biostratigraphic data is unavailable.

Exploration application, Palaeocene–Early Eocene sequence

The above-mentioned techniques of petrophysical correlation and de-sanding can be applied in exploration for subtle stratigraphically constrained prospects in the submarine fan or other environments.

Figure 5(a) shows three Palaeocene–Early Eocene well sections from the Fladen Ground Spur Margin. The logs shown are gamma ray logs in 'Mae West' configuration with sands removed. Biostratigraphic data have been applied to this set of curves enabling significant shifts in parameter measurements to be judged in terms of floral and faunal events.

Figure 5(b) illustrates the restored well sections showing the correlations made on the de-sanded diagram. From this diagram a more precise definition of the temporal distribution of sands can be made.

The spatial distribution of sands can be determined by comparing the thickness of background shale deposition, which is frequently consistent within the locale of defined structural blocks. Variations in the mapped thickness of a given stratigraphic unit can be ascribed to sand deposition. The success of this approach is dependent on having correctly isolated that unit in the seismic interpretation process, and being aware of potential additional shale section missing from drilled locations due to faulting or condensation.

Extrapolation of biostratigraphic events

A further application of the de-sanding process has been utilized where the detailed biostratigraphic analysis required for seismic–stratigraphic work has been lacking. The de-sanded logs have facilitated direct comparison between wells with detailed biostratigraphic analysis

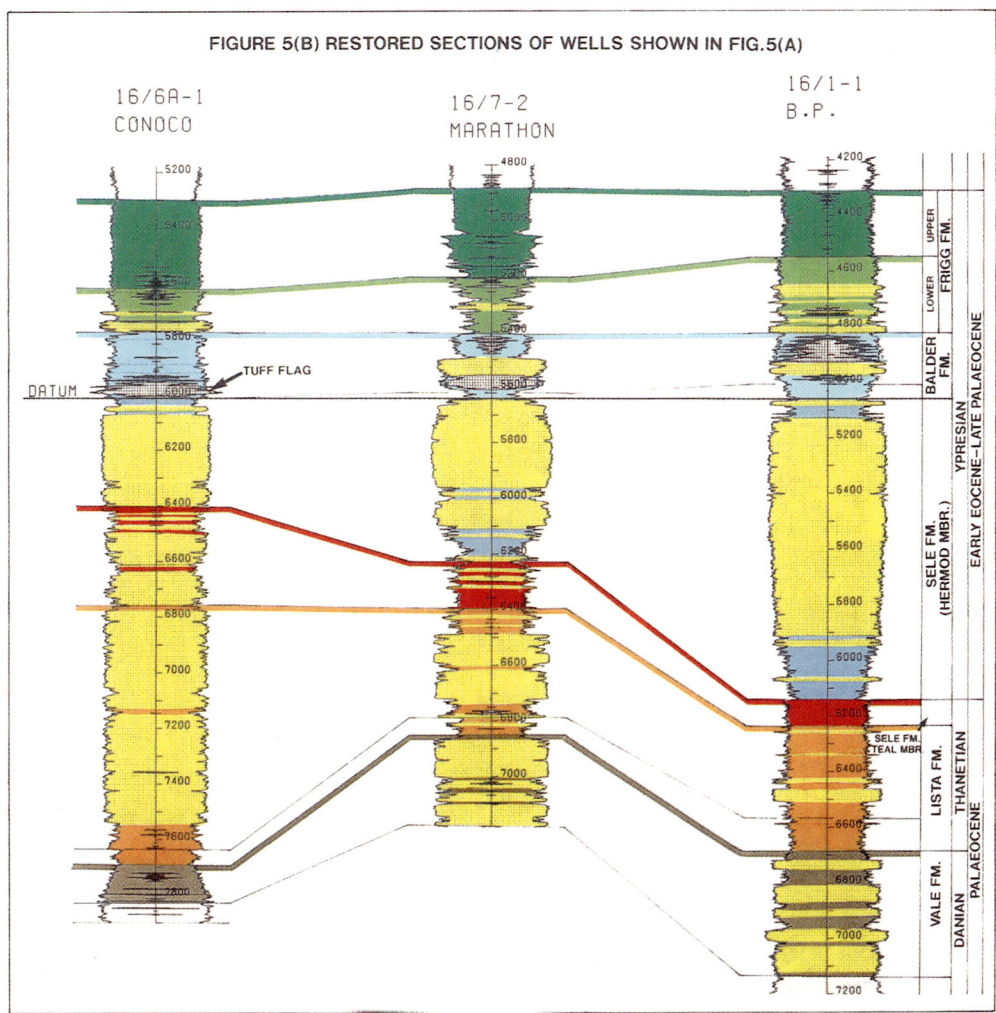

FIGURE 5(B) RESTORED SECTIONS OF WELLS SHOWN IN FIG.5(A)

and those without. This method was applied to well 16/1–1 (Fig. 5(a), (b)). Subsequent biostratigraphic analysis confirmed the derived correlation.

It is perceived that this technique can be used in the future to revive the exploration value of wells whose sample sets are depleted.

Conclusions

High-resolution correlation problems can be solved by integrated use of biostratigraphic data and manipulation of wireline data.

In the submarine fan environment, the temporal and spatial distribution of sands can be determined with considerable accuracy. The practical benefits arise in appraisal and development drilling decisions, reservoir management, and extrapolation from known fields to exploration scenarios, particularly in the review of failed exploratory wells.

Correlation techniques as described in this paper will prove useful as well sample sets become depleted. Detailed sequence stratigraphy can, using these methods, be projected to well sections where sample sets are depleted, and/or detailed biostratigraphy is not available.

The techniques described in this paper were developed jointly by the staff members of Kerr-McGee Oil (UK) plc and ResTech Europe Ltd. The authors thank their respective managements for the authority to publish this work.

Appendix. Apparent grain density algorithm

Input data:

RHOB   = Log measurement of bulk density, (environment corrected);
NPHI   = Log measurement of neutron porosity, (environment corrected);
RHOB_F   = Fluid density = 0.96 (assumes a mixture of water and OBM).

Apparent grain density:

$$RHGA = \frac{RHOB - (PHIX * RHOB\_F)}{(1 - PHIX)}$$

where:

$$PHIX = \frac{(PHID\_A * NPHI) - (PHID * PHIN\_A)}{(PHID\_A - PHIN\_A)}$$

$$PHID = \frac{(2.71 - RHOB)}{(2.71 - RHOB\_F)}.$$

PHID_A, and PHIN_A are calculated according to the relative magnitude of NPHI and PHID:

for NPHI < PHID: PHID_A = 1.0
                       $PHIN\_A = -(1.17 + 2.06 * NPHI) + 10^{-(0.4 + 16 * NPHI)}$

for NPHI > PHID: $PHID\_A = \frac{2.71 - 4.0}{2.71 - RHOB\_F}$

                       $PHIN\_A = 0.7 - 10^{-(5 * NPHI + 0.16)}$

References

KNOX, R. W. O'B. & MORTON, A. C. 1988. The record of early Tertiary N. Atlantic volcanism in sediments on the North Sea Basin. *In*: MORTON, A. C. & PARSON, L. M. (eds) *Early Tertiary Volcanism and the Opening of the N.E. Atlantic*. Geological Society, London, Special Publication, **39**, 407–419.

MORTON, A. C. & KNOX, R. W. O'B. 1990. Geochemistry of late Palaeocene and early Eocene tephras from the North Sea Basin. *Journal of the Geological Society, London*, **147**, 425–437.

Sedimentological borehole image analysis in clastic rocks: a systematic approach to interpretation

LAWRENCE T. BOURKE

Reservoir Studies Group, Schlumberger Data Services, Woodlands Drive, Dyce, Aberdeen AB2 0ES, UK

Abstract. Detailed study over the four years since the introduction of the Formation MicroScanner Tool* (FMS) has resulted in the evolution of a systematic interpretation approach to the extraction of a formal sedimentological description of extensive reservoir intervals using borehole images in clastic sediments. This approach takes account of thorough log quality control and the recognition of non-geological artifacts and involves the integration of the available open hole logs to establish lithology, along with the incorporation of core or sidewall cores as 'hard' lithological and textural data to ensure interpretation confidence.

The approach is practical and straightforward. However, a workstation provides significant advantages to the skilled interpreter compared with a paper-based analysis. After quality control and recognition of artifact images, five interpretation steps are suggested. (i) Fine-calibrate open hole logs (conventional lithology indicators) from sidewall and conventional core samples. Enhance lithology recognition using dipmeter resistivity curves. (ii) Annotate the above lithological information onto FMS images. (iii) Constrain feature recognition from images using local geological knowledge. (iv) Pick and classify feature dips, then check dip category validity by generating dip and azimuth histograms. Reclassify as necessary. (v) Extract relevant sedimentary and lithological data from images to construct a conventional sedimentological facies description.

One of the key aspects of borehole image analysis is the development of an interpreter's skill in recognizing geological features from microresistivity images. Finally, aspects of core reconciliation such as depth matching are considered.

The Formation MicroScanner* (FMS) Tool (Lloyd *et al.* 1986) has been in use in the U.K.C.S for five years. An interpretation approach was outlined in Harker *et al.* (1990), based on initial North Sea interpretation studies which were largely confined to comparisons within cored intervals. While the basic interpretation steps outlined in that paper were sound, new developments and applied experience have significantly enhanced the quality and detail of interpretation results obtained.

Detailed sedimentological studies involving the integrated interpretation of FMS images, conventional dipmeter, open hole logs and cores have been carried out in more than 100 wells in mainly clastic sediments from the UKCS, Norwegian Sector, onshore Europe and the Far East since late 1986. These studies were diverse and varied from mature fields to rank wildcat wells, and included a comprehensive range of depositional and stratigraphic settings. This comprehensive experience base has allowed the evolution of an integrated interpretation approach to

the extraction of sedimentological information from borehole images. This approach is sufficiently flexible to cater for the typically incomplete nature of hard control data (core), and for the systematic recognition of non-geological image artifacts, which is essential to the extraction of sedimentologically and structurally useful borehole image data.

Interpretation approach

The interpretation flow diagram presented in Fig. 1 combines the original interpretation approach presented in Harker *et al.* (1990) together with additional improvements presented in this paper. Refinements to the original interpretation approach can be attributed to four factors: the large number of wells studied, artifact image recognition, workstation benefits and increased borehole coverage.

The additional experience gained from analysis of a large number of wells, many of which offered no core coverage, has reinforced bore-

---

* Mark of Schlumberger

From HURST, A., GRIFFITHS, C. M. & WORTHINGTON, P. F. (eds), 1992,
Geological Applications of Wireline Logs II. Geological Society Special Publication No. 65, pp. 31–42.

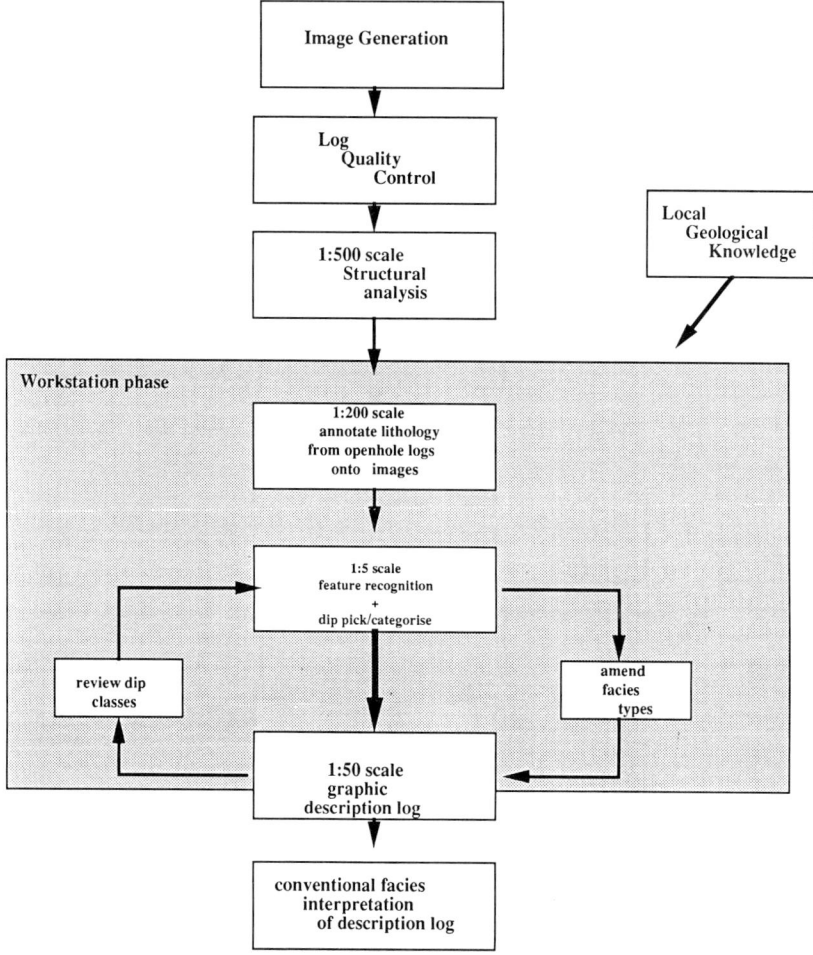

Fig. 1. Formation MicroScanner Tool interpretation flow diagram.

hole image interpretation methodology. The lack of conventional cores in many of these studies provided an impetus to push the interpretation of borehole images to their logical limit. The compilation of a systematic review of artifact images from microresistivity imaging tools and their recognition as presented in Bourke (1989) was an important addition to the three-level image-grading scheme proposed in Harker *et al.* (1990). The introduction of the four-pad Formation MicroScanner Tool at the start of 1989 has further improved interpretation confidence. By ensuring 'four-way' borehole coverage, this tool has further improved the visualization of many medium-to-large-scale features which were ambiguous with only two-pad

orthogonal coverage (Fig. 2). This trend in increasing microresistivity-image borehole coverage has continued with the introduction in 1991 of an eight-pad tool, the fullbore Formation MicroImager* (FMI). In addition to improvements in borehole coverage afforded by new tools, the introduction of borehole image-interpretation workstations has arguably been the most significant advance in sedimentological FMS image interpretation. While a workstation approach to borehole image interpretation is not indispensable (all background work for the Harker *et al.* (1990) paper was carried out on paper image-plots), the benefits of workstation input to image analysis greatly aids interpretation in terms of data manipulation and accur-

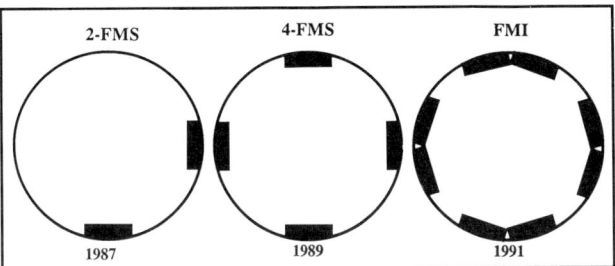

Fig. 2. Comparison of borehole pad distribution for the two-pad and four-pad FMS tools and the Formation MicroImager.

acy. It must be stressed, however, that the use of a workstation is not a substitute for the skills and knowledge of the interpreter, but an additional aid to the interpretation process.

Additions to the original interpretation methodology are as follows:

workstation data manipulation and enhancement;

integration of open hole and image data;

accurate workstation derived, feature-classified dip data;

re-evaluation of dip classes;

graphical display of image interpretation.

Workstation data manipulation and enhancement

A number of workstations suitable for borehole image analysis are now available. The workstation used in this work was the Formation MicroScanner Image Examiner* (Boyeldieu & Jeffreys 1988). During the data review, analysis and data presentation stages of a borehole image interpretation, a workstation provides significant advantages over manual interpretation of paper plots. Specifically, the ability to change presentation scale at will is a significant benefit to the visualization of sedimentary features over a wide scale range (Fig. 3). Additionally, the ability to interactively enhance features of interest is of crucial importance, particularly where interlamination of beds with strongly contrasting resistivity or cementation causes a loss of image contrast (Fig. 4). This loss of contrast is characteristic of dynamically normalized images which are the standard FMS display option for large-scale paper image logs.

Integration of open hole and image data.

Microresistivity imaging tools have extremely good vertical resolution, with typical vertical sampling rates of 0.1 inch (0.25 cm). By virtue of the nature of dipmeter resistivity measurement combined with measurement range control characteristics, they are very good at highlighting subtle resistivity variations arising from sedimentary detail but have limited gross lithological sensitivity. For example similar resistivity shading can be observed in images from adjacent sands and shales depending on the composition and fluid content of these sediments. It is therefore essential to derive lithology from other wireline logs (e.g. gamma ray, density neutron etc.) as a supplement to image interpretation. These logs, however, have vertical resolutions which are 1–2 orders of magnitude less than that of FMS images, and must be correlated with care.

High sample-rate, open hole logging (e.g. 1.2 inch compared with the standard 6 inch vertical sample-rate) and application of enhanced resolution processing techniques are today increasingly improving the effective vertical resolution of lithology tools. However, the borehole image interpreter is still generally limited to conventional open hole logs. The suggested approach to annotating lithological determinations from (low-resolution) open hole logs onto (high-resolution) microresistivity images is hierarchical. Firstly boundaries between major lithological or depositional units are identified and annotated on images. Secondly, smaller beds, which approach or are smaller than the vertical resolution of open hole logs, can be seen as incomplete log inflections. Utilizing local depositional knowledge and the nature of the sedimentary detail seen on images, the smaller beds can usually be accurately determined. For example, the transitional shallow marine sandstone in Fig. 5 exhibits a gradual upward decrease in gamma ray, the neutron porosity and density logs show an inflection towards clay. It is not possible to determine from open hole logs whether this is a thinly bedded or bioturbation mottled assemblage. The

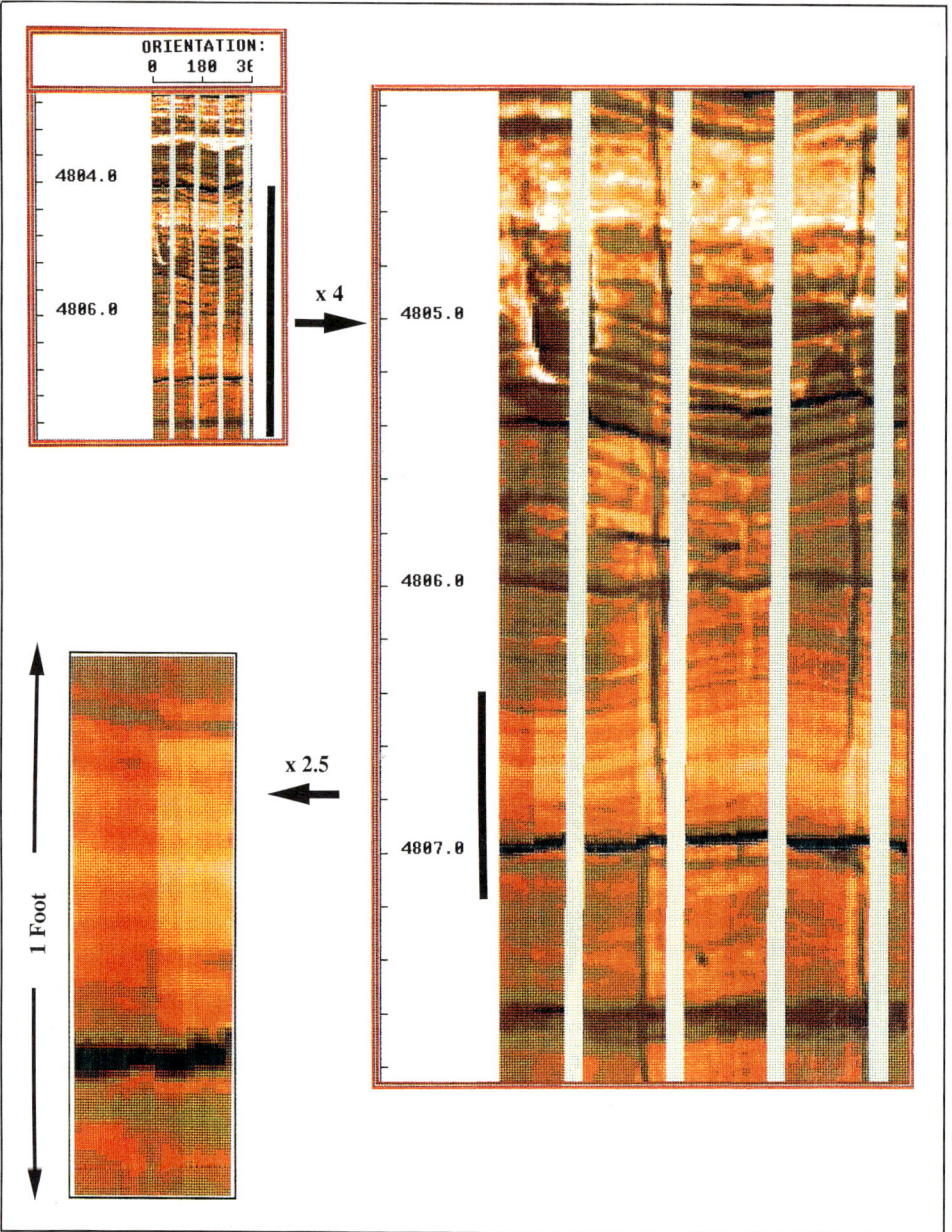

Fig. 3. Illustration of image magnification from a workstation. The magnified interval is indicated with a black vertical bar. Depths are in feet.

corresponding FMS images reveal both thin beds and mottling consistent with bioturbation. In this example, the precise clay content for the bed is not interpreted and the top and bottom contacts are transitional. Cores obtained over this interval would provide a similar description and the core log would record a transitional facies from sand to bioturbated shaly sands.

Dipmeter resistivity traces (acquired simultaneously with the FMS images), form an effective 'scale bridge' between open hole logs and images. Such data can reveal subtle bed bound-

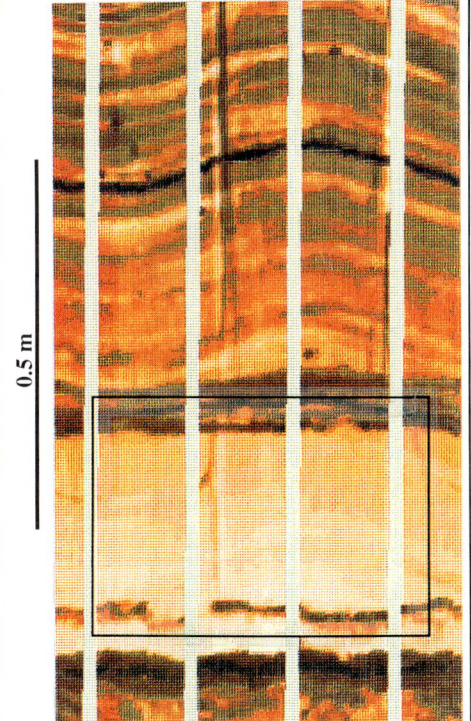

0.5 m

The resistive (light shaded) bed above shows little primary detail.

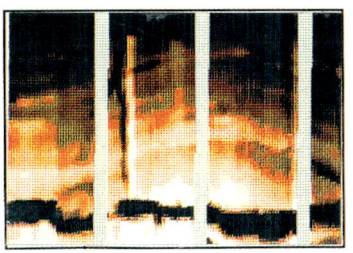

After workstation enhancement more bedding is visible.

Fig. 4. Example of image enhancement in a resistive (light shading), calcite cemented, cross-bedded interval.

aries and vertical trends which are not recorded by the open hole logs and such small scale vertical resistivity trends (Fig. 6) may be seen on dipmeter curves but may not be obvious on images.

Accurate workstation derived, feature classified dip data

A significant advantage of a workstation approach to borehole image interpretation is the high accuracy of interactively derived dip data. The manual approach to determining dip from inclined planar sufaces which are seen on the borehole wall is to use a best fit curve match with a set of 'overlay' sine curves representing a range of dip magnitudes from 0–90°. The accuracy of the overlay technique is poor, particularly in azimuth, and serves only as a quick-look approach. By contrast, dip accuracy derived on a workstation can be considered equivalent to the orientation specifications of the imaging tool, which are very precise in azimuth. Interpreter repeatability can be identified as the main source of error and is discussed under 'measurement repeatability'.

In addition to accurate dip determination, a workstation also allows the interpreter to categorize dip data into separate bedding categories. Accuracy of dip measurement together with ease of dip classification are at the heart of recent interpretation improvements. Dips picked interactively and saved on a workstation in specific bedding categories provide an important reinforcement to feature interpretation and also assist in subsequent facies classification. After interactive dip analysis is completed, individual dip categories can be selected by depth zone or lithological unit and viewed in sedimentological context. Generally most useful for single bedding sets are azimuth frequency histograms and dip magnitude frequency plots. These give the interpreter an immediate impression of azimuth modality and dispersion and the range of dip magnitude. If the data spread and character are consistent for the bedding type classified then such data may have directional significance. When multiple bedding categories are displayed, colour coded Schmidt plots or Wulff plots are extremely valuable in illustrating, for example, bedding dip relative to fractures.

A trough cross-bedding example (Fig. 7) provides an illustration of this approach. An azimuth histogram of picked trough cross beds from a single sedimentary facies would be expected to have a high degree of azimuth dispersion but an obvious unimodal dominant trend. Clearly, when such an azimuth pattern is seen, the original feature classification gains even greater interpretation confidence and a reliable transport direction indication. The palaeocurrent significance of dip azimuth patterns for specific bedding types will vary with depositional environment (Fig. 8).

Re-evaluation of dip classes

A great strength of the workstation dip picking and classification approach is the re-evaluation

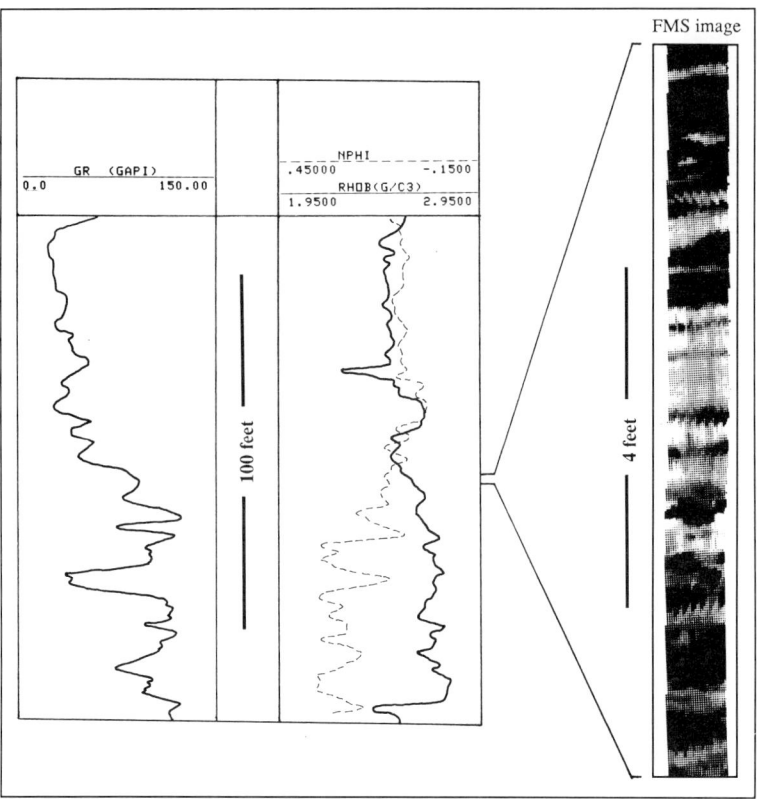

Fig. 5. Images reveal thin bedded sands, sandy lenses and bioturbation mottling. These thinly bedded lithologies are beyond the resolution of the open hole logs.

of dip. Where bedding type is ambiguous, dips can be saved in an 'uncertain dip' category and then subsequently reclassified on the basis of similarity to other sets (in the same lithological setting), or reclassified on the basis of azimuth and dip magnitude character.

In conventional dipmeter analysis, even with optimized processing selection and a sound approach to feature selection, there is still a high degree of abstraction in the analytical approach. The interpreter can never have complete confidence in feature recognition, and there are usually additional dip data arising from geological 'noise', i.e. unknown geological features and spurious dips. With borehole images, a relatively small manually derived data set can have considerable significance since only confident dips are retained in the dataset. Dip data picked from borehole images are frequently not detected by conventional dip processing at least in the following cases.

(a) The bedding feature is not prominent on the borehole wall, for example, in poorly consolidated, high permeability sands. Thick mudcake and formation smearing give rise to mottles on borehole images which are the dominant borehole feature, and which produce scattered and spurious computed dip. On borehole images the interpreter can recognize the fainter geological detail beyond the borehole wall artifacts to produce reliable dip data. Examples of these features are presented in Adams *et al.* (1990).

(b) Highly non-planar features may not be detectable by conventional dipmeter processing, but using images, an approximate fit can be obtained to derive a dip and azimuth. In the case of non-planar beds such as slump or scour features, a 10 cm diameter core offers little bedding surface from which to determine the dip of such medium scale features. When correctly identified on the borehole wall, these can be extremely useful palaeocurrent indicators.

(c) Features which vary in resistivity character across the borehole will probably not be detected in conventional computation; a meaningful dip

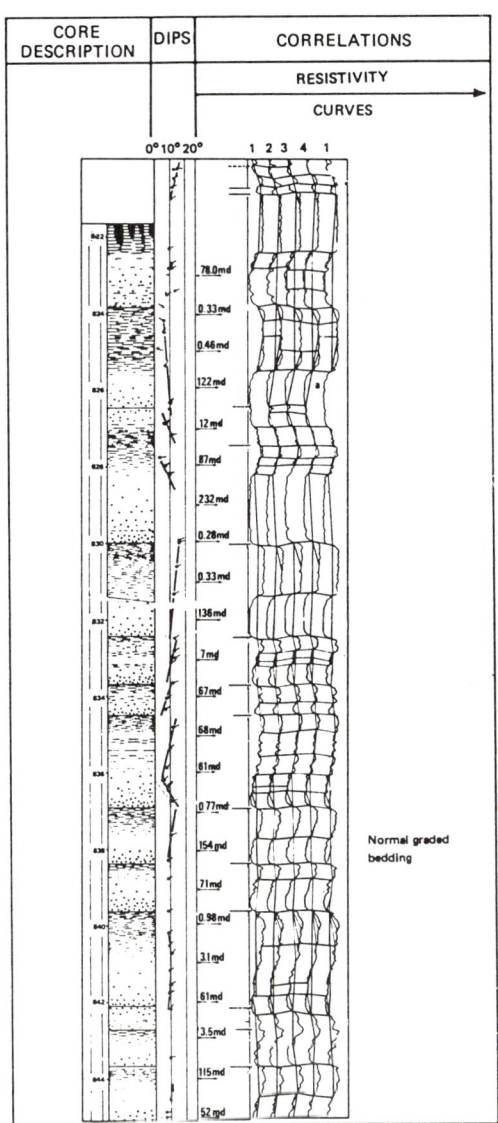

Fig. 6. Comparison between grain size detail from core (left) and a corresponding section of dipmeter resistivity data after Serra (1985).

can be derived from images of a thin bed which pinches out across the borehole but would not be determined by dipmeter. Similarly, highly irregular or wavy beds can most easily be recognized and picked from borehole images.

Graphical display of image interpretation

This is a synthesis of all previous interpretative steps and is the borehole image equivalent of a core description log. The important steps are:

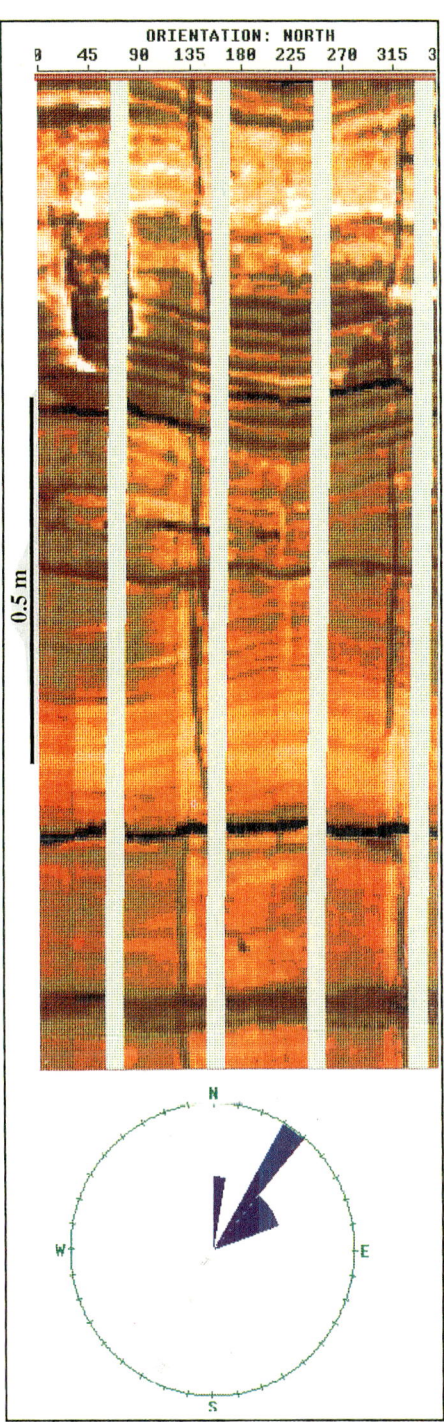

Fig. 7. FMS images of cross bedding. The workstation derived dip data (below) shows unimodal, moderately scattered azimuths consistent with trough cross bedding.

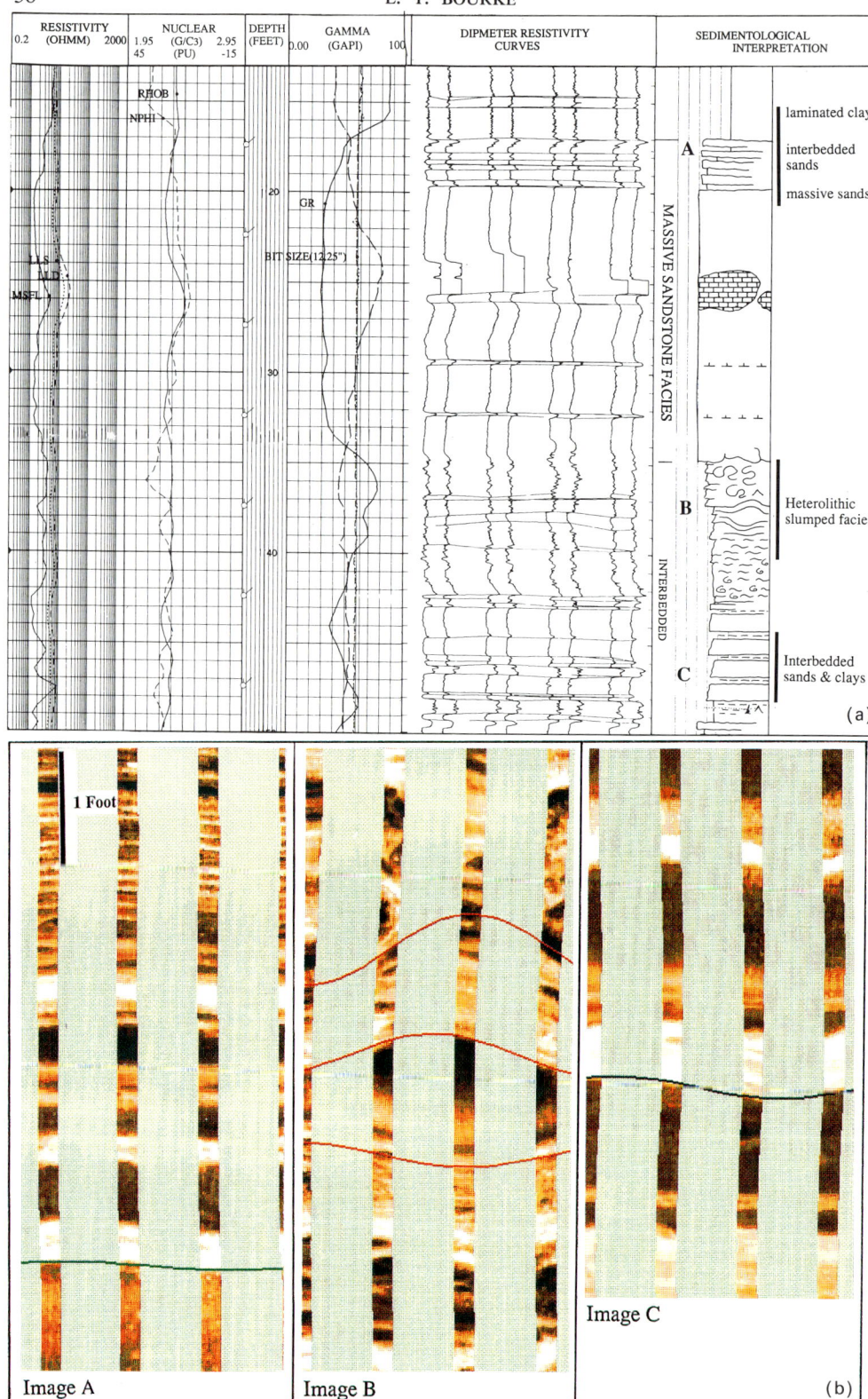

(a)

Image A

Image B

Image C

(b)

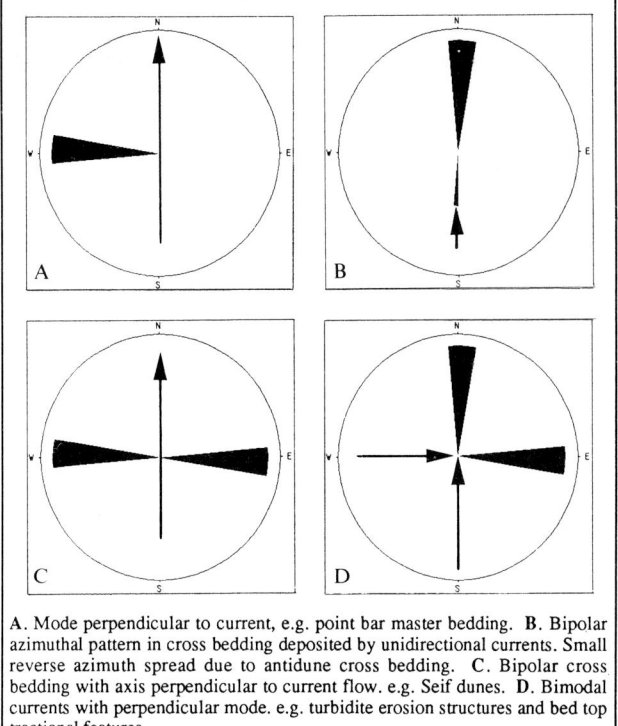

A. Mode perpendicular to current, e.g. point bar master bedding. **B.** Bipolar azimuthal pattern in cross bedding deposited by unidirectional currents. Small reverse azimuth spread due to antidune cross bedding. **C.** Bipolar cross bedding with axis perpendicular to current flow. e.g. Seif dunes. **D.** Bimodal currents with perpendicular mode. e.g. turbidite erosion structures and bed top tractional features.

Fig. 8. Typical azimuth histogram patterns associated with major depositional settings and their relationship to current direction (after Selley 1976).

bed-by-bed lithology definition; feature recognition; dip determination and classification; summary of observations on a graphical sedimentary description log.

The lower limit on vertical resolution of resistive features is approximately 1 cm, which implies that laminated features such as current ripples with set heights of less than 3 cm are difficult to visualize due to their very fine internal lamination. By contrast, discrete burrows, minor slumps and injection features a few centimetres in height are generally recognizable from their outline form. The recognition of individual sedimentary features is aided by delineation of lithology, recognition of artifact features and the limitation of interpretation possibilities by use of local depositional knowledge.

A 1:50 scale final sedimentary interpretative log is probably optimum given the resolution limitations of images and the practicality of describing long imaged intervals. A composite display should have the lowest resolution logs on the left with increasing vertical resolution logs towards the right (Fig. 9).

Grain size information will generally be lacking from such an interpretative log. It is always possible to distinguish sands, silts and clays derived from open hole logs particularly where images have been used to reinforce the textural character of the sequence. Percussion sidewall cores can offer very valuable calibration points in complex lithologies where the presence of clays, micas or potassium feldspars may cause uncertainties in lithological determination.

Some inferences regarding grain size trends can be added to the lithology curve in the sedimentological interpretation track. In cored intervals in the study well, or from an adjacent well, it

Fig. 9. Typical sedimentological description possible from a submarine fan sequence using borehole image data. 1:50 scale is a suitable reproduction scale for such composite presentations, 'grain size' curve is notional. Figure 9b illustrates extracts from the FMS images associated with this interpreted sequence.

may be possible to establish a 'grain-size rule base' where, for instance, vertical trends in porosity at some facies boundaries can be confidently tied to a vertical grain-size trend. Similarly, in some depositional environments, resistivity ramps on dipmeter curves can relate to vertical grading. This interpreted grain-size information should be used cautiously since over-interpretation can greatly reduce the confidence of subsequent workers in the original interpretation. When there is significant doubt, grain size inferences should be strictly avoided.

The final graphical log will generally provide a faithful reproduction of medium- to large-scale sedimentary features and some small sedimentary details. All lithological units should have been recognized and bad data zones highlighted. Large scale features are generally more readily recognized from FMS images than cores due to the 'aspect' of borehole images in relation to cores (Harker *et al.* 1990). Accurate, feature specific, dip data from images can greatly reinforce feature recognition. In combination with the graphic description log, this information ensures a thoroughly integrated (conventional logs, images, cores) sedimentary log with useful bedding- and facies-specific directional data.

Measurement repeatability

Interactively picked dip data from FMS images by workstation provide dip information which has excellent azimuth accuracy. The main source of error arises from the ability of the interpreter to pick dip planes consistently.

To quantify the accuracy of dip data computed manually from FMS images using the workstation, a series of tests were run (J. Adams pers. comm.). These involved three experienced FMS interpreters each independently computing dips on the same bed boundaries within a cross-bedded sequence, and repeating the operation. Both a 'best-case' and a 'worst-case' example were tried over the same sequence: the best case using 70% borehole coverage from four runs of a two-pad tool; the worst case using only one run of the two-pad tool, with 20% coverage (Fig. 2).

The best case gave rise to a reproducibility error of $\pm 2.6°$ on dip magnitude, $\pm 5.7°$ on dip azimuth (mean values of standard deviations of results from one bedding plane, computed six times; Table 1), the worst case has an error of $\pm 2.5°$ on dip magnitude, $\pm 7.8°$ on azimuth (Table 2). This range of values is similar to reproducibility error values quoted for core goniometry (Nelson *et al.* 1987). However, as the FMS tool inclinometry and positioning accuracy

is very high (maximum variation $\pm 0.2°$ dip, $\pm 2°$ azimuth), the overall error in dip computation can be considered to be within tolerable limits for detailed palaeocurrent analysis.

Core reconciliation

Microrestivity borehole imaging affords the opportunity to reduce coring and thereby save costs. While this does occur in some circumstances, this trend has not been observed in North West Europe. In clastic reservoirs where FMS is being run, there is no obvious reduction in coring. Indeed the FMS log is being used in conjunction with cores to extend reservoir characterization beyond cored intervals. The opportunity therefore exists to usefully combine both datasets.

The combination of sedimentological information from conventional core with that from images is a highly complementary process. Core provides an effective means of "calibrating" FMS images to increase interpretation confidence beyond cored intervals. In the other direction, the FMS provides absolute accuracy of core orientation. The reconciliation of core is usually quite straightforward. However, in massive sands or in zones of highly fragmented core recovery, reconciliation is simultaneously more challenging and more useful. A brief discussion of some of the problems of FMS image and core reconciliation is presented in Adams *et al.* (1990).

Where conventional core is completely absent, or where data infill outwith cored intervals is necessary, the use of percussion sidewall cores can provide not only sound lithological control but also textural and sedimentary structure indications. Sidewall cores are of course routinely described for lithology by biostratigraphic companies and often also by the wellsite geologist. These descriptions do not usually record textural or sedimentary structure details and subsequently, most of the sample is taken for palynological or micropalaeontological analyses. The remaining sample is usually small and often the least representative part of the original plug. It is often possible for a sedimentologist to extract considerable non-destructive information from complete sidewall core samples (even when recovery is fragmented).

Conclusions

One of the major conclusions of this work is that a suitable geological workstation is an essential part of the complete borehole image evaluation study where interpretation confidence is maxi-

Table 1. *Manually derived cross-bedding dip/azimuth data used in the dip reproducibility test: 'best case' example*

1	2	3	4	5	6	SD dip	SD azimuth
27/193	32/128	30/178	31/177	30/179	28/203	1.169	9.87
35/216	33/205	36/221	36/219	27/194	24/202	4.67	9.84
25/209	26/208	25/210	27/217	26/213	25/211	0.74	2.98
26/200	25/198	26/205	28/198	27/203	26/206	0.94	3.19
24/198	28/199	29/201	30/197	28/197	26/198	1.98	1.37
42/199	35/211	38/211	38/198	35/208	30/198	3.68	5.93
20/197	23/204	22/210	22/202	19/201	21/205	1.34	3.97
32/227	35/218	30/222	30/224	30/212	28/216	2.19	5.04
31/200	26/206	25/216	34/204	27/208	30/213	3.13	5.37
25/201	25/178	25/210	28/205	25/198	22/182	1.73	11.73
17/208	30/203	21/216	25/209	32/206	20/205	5.39	4.14
22/200	26/210	26/202	27/209	24/202	24/203	1.67	3.77
37/191	27/201	39/211	37/206	37/200	26/194	5.24	6.75
Mean values of standard deviations						2.64	5.68

Table 2. *Manually derived cross-bedding dip/azimuth data used in the dip reproducibility test: 'worst case' example*

1	2	3	4	5	6	SD dip	SD azimuth
33/189	38/185	33/181	34/184	29/187	30/187	2.91	2.56
39/178	35/180	29/181	31/171	32/171	33/169	3.18	4.79
21/205	22/206	22/197	21/201	21/203	19/215	1.00	5.53
26/191	27/204	24/209	30/186	25/207	28/197	1.97	8.43
28/185	31/200	31/203	29/204	28/177	27/188	1.53	10.12
21/191	18/200	30/199	18/222	19/215	21/194	1.26	11.10
24/189	24/185	22/178	24/177	23/189	23/175	0.75	5.73
33/206	29/188	25/162	26/162	20/174	29/211	4.04	19.60
26/206	28/217	24/208	21/198	33/216	32/218	4.23	7.20
37/206	27/202	31/209	28/206	27/202	29/206	4.07	2.47
Mean values of standard deviations						2.49	7.76

mized. Such a system should include image enhancement capability to optimize image quality as well as depth and scale manipulation of image data in real time. The workstation must also incorporate interactive manual dip computation with the capability to store picked dip data in user defined categories, and the ability to display this data within the system. Finally, the ability to faithfully reproduce picked images and a continuous log of interpreted images from such a system allows proper documentation of the analysis.

As with sedimentological analysis of core data, the systematic description of borehole images can provide a detailed graphical sedimentary interpretive log. Unlike cores there can be areas of increased uncertainty in recognizing small features and determining grain size, however, in compensation for this shortcoming, given good hole conditions, such images generally provide complete reservoir coverage and high-feature orientation accuracy. The important underlying message here is that the resulting sedimentological interpretation, as with core interpretation, is in isolation no more than a description. Exactly the same sedimentological skills are required to generate a subsequent facies breakdown from the resulting data and for production or refinement of a conceptual depositional model.

The time required to extract useful sedimentological information from borehole images is comparable to that needed for effective description of cores, even where a workstation is employed. It is conceptually important to note that there are circumstances where, just as with cores, a quick answer can be provided by a glance at the data. Detailed sedimentological interpretation requires an equally systematic and rigorous interpretation approach.

Particular acknowledgement should go to the clients of Reservoir Studies Group who have provided the opportunity to carry out detailed work in a great variety of depositional and structural settings. I extend personal thanks to my colleagues, John Adams and Stuart Buck who have added their own perspectives and experience in borehole image analysis over the last two years.

References

ADAMS, J. T., BOURKE, L. T. & BUCK, S. G. 1990. Integrating Formation MicroScanner images and cores. *Oilfield Review*, January.

BOURKE, L. T. 1989. Recognizing artifact images of the Formation MicroScanner Tool. *Transactions of SPWLA 30th Annual Logging Symposium*. Paper WW.

BOYELDIEU, C. & JEFFREYS, P. 1988. Formation Micro-Scanner: New Developments. *Transactions of SPWLA 11th European Evaluation Symposium*. Paper G.

HARKER, S. D., McGANN, G. J. BOURKE, L. T. & ADAMS, J. T. 1990. Methodology of Formation Microscanner Tool image interpretation in Claymore and Scapa Fields (North Sea). *In*: HURST, A., LOVELL, M. A. & MORTON, A. C. (eds) *Geolo-gical Applications of Wireline Logs*. Geological Society, London, Special Publication, **48**, 11–25.

LLOYD, P. M., DAHAN, C. & HUTIN, R. 1986. Formation imaging with electrical scanning arrays. A new generation of stratigraphic high resolution dipmeter tool. *Transactions of SPWLA 10th European Formation Evaluation Symposium*. Paper L.

NELSON, R. A., LENNOX, L. C. & WARD, B. J. Jr., 1987, Oriented core: its use, error and uncertainty. *AAPG Bulletin*, **71**, 357–367.

SELLEY, R. C. 1976. *An Introduction to Sedimentology*. Academic, London.

SERRA, O. 1985. *Sedimentary Environments from Wireline Logs*. Schlumberger Publication M-081030/SMP-7008.

Evolution of the Izu–Bonin intraoceanic forearc basin, western Pacific, from cores and FMS images

PHILIPPE A. PEZARD,[1] RICHARD N. HISCOTT,[2] MICHAEL A. LOVELL,[3] ALBINA COLLELA[4] & ALBERTO MALINVERNO[5]

[1] *Département de Génie Océanique, Institut Méditerranéen de Technologie, 13451 Marseille, France*

[2] *Earth Sciences Department and Centre for Earth Resources Research, Memorial University of Newfoundland, St John's, Newfoundland A1B 3X5, Canada*

[3] *Geology Department, Leicester University, University Road, Leicester LE1 7RH, UK*

[4] *Dipartimento di Scienze della Terra, Università della Calabria, 87030 Arcavacata, Cosenza, Italy*

[5] *Marine Geology and Geophysics, Lamont-Doherty Geological Observatory, Palisades NY 10964, USA*

Abstract. One of the objectives of Ocean Drilling Program (ODP) Leg 126 was to investigate the origin and evolution of the Izu–Bonin arc and forearc, both products of the subduction of Pacific lithosphere under the Philippine Sea Plate. Within the forearc basin, a full set of downhole measurements was recorded in two deep holes (792E and 793B). In addition, borehole electrical images were obtained (for the first time in the context of ODP) with the Formation MicroScanner (FMS*).

The main result of the drilling is that the forearc basin formed between 31.0 and 24.0 Ma by separation of a formerly contiguous frontal and outer arc high. The cored material shows a characteristic pattern of volcanogenic input, from turbidites and debris flows produced by volcanism and erosion of surrounding highs. The short rifting period is characterized by high sedimentation rate (300 m/Ma). In this context, the high resolution of FMS images was used to analyze the sedimentary processes associated with the deposition of deep-water volcaniclastics. The images reveal fine details of turbidite sequences that dip at low angles due to recent tectonics.

An FMS-based sedimentary log was calibrated from cores and prepared for each of the two holes, providing continuous bed-by-bed sections and permitting the investigation of trends in bed thicknesses. Palaeocurrent data were obtained from the analysis of ripple marks. During early basin history (30.2 to 29.5 Ma), the main sediment source was located to the east, in the vicinity of the modern outer arc high, with a secondary transport-mode oriented northward, along the basin axis. In the shallower section, emplaced at a lower rate from 28.9 to 27.3 Ma, axial transport from the north dominates a small component of flow from the western margin of the basin.

Initiated by rifting of the arc during the Oligocene time, basin development was followed by periods characterized also by extensional tectonics. Postdepositional extensional deformations such as normal microfaults, conjugate high-angle fractures, and dewatering veinlets were identified in the core and on FMS images. The orientation of the stress field within the arc and forearc was obtained from the analysis of borehole ellipticity. The results confirm models of stress distribution in forearc–arc–back arc regions. In particular, a rotation of the maximum horizontal stress trajectory in the overlying plate was observed, in a direction orthogonal to the plate boundary. In spite of a 90° clockwise rotation of the Philippine Sea plate since Oligocene time, the orientation of the stress field seems to have remained stable with respect to the trench axis over this period.

Sites 792 and 793 were drilled in the forearc basin of the Izu–Bonin Arc (Fig. 1). The thickest stratigraphic unit at all three sites is an upper Oligocene turbidite succession derived from the contemporary Izu–Bonin intra-oceanic arc (Fujioka & Saito in press; Hiscott & Gill in press). The Oligocene basin formed by extension (Leg 126 Shipboard Party 1989) in a direction parallel to the volcanic arc, with a width of about 60 km. The eastern margin of the basin consisted of Eocene volcanics, including boninites (Matsuda 1985; Fryer, Pearce *et al.* 1990).

---

* Mark of Schlumberger

From HURST, A., GRIFFITHS, C. M. & WORTHINGTON, P. F. (eds), 1992,
Geological Applications of Wireline Logs II. Geological Society Special Publication No. 65, pp. 43–69.

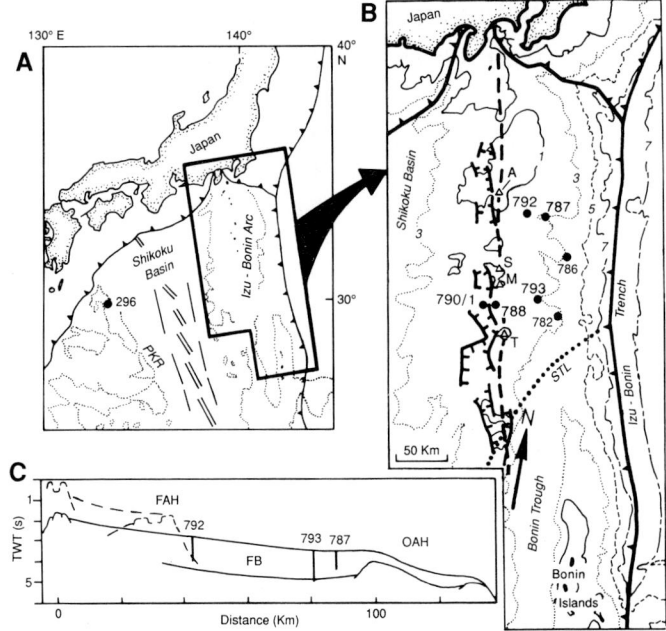

Fig. 1. Location of DSDP-ODP drillsites. (A) Position of the Izu–Bonin Arc at the eastern edge of the Philippine Sea plate; DSDP Site 296 on the Palau–Kyushu Ridge and the location of part B are indicated. (B) bathymetry of the northern part of the Izu–Bonin Arc with the location of Leg 126 Sites 787, 792, 793, and Leg 125 Sites 782 and 786. The Sofugan Tectonic Line (STL) is also indicated. Trenches are outlined by barbed lines, and backarc basins by hashured lines. The position of the modern volcanic front is marked by a heavy dashed line. (C) Projection of Leg 126 forearc basin sites into a single synthetic cross section: FAH, frontal arc high; FB, forearc basin; OAH, outer arc high.

The western margin has been split into two since the Oligocene. One of them is now located under the modern volcanic front and the other, the Palau–Kyushu Ridge (Fig. 1), was separated from the modern arc by Miocene opening of the Shikoku basin (Kobayashi & Nakada 1979). Lower Oligocene rocks as old as 33.6 ± 1.2 Ma have been recovered from a seamount near the modern volcanic front (Yuasa *et al.* 1988), but Eocene rocks from the same area have never been sampled.

Palaeomagnetic results from Leg 126 (Koyama *et al.* in press) confirm a clockwise 90° rotation of the Philippine Sea plate since the Oligocene. In this paper, however, modern geographic coordinates are used for features located within the Izu–Bonin volcanic arc, because the rotation has not affected the relative position of the arc, forearc basin, and trench.

The Oligocene basin fill shows a general thickening from north to south above irregular basement (Taylor, Fujioka *et al.* 1990). Although the ODP sites were not drilled in the thickest areas, seismic character suggests that the oldest,

unsampled forearc-basin deposits are similar to those cored by Leg 126 (Taylor, Fujioka *et al.* 1990). Water depths of about 4–5 km (Kaiho in press) prevailed on this elongate basin plain, which was devoid of large-scale channels or submarine valleys (Taylor, Fujioka *et al.* 1990; Klaus & Taylor 1992).

Worldwide, sandy turbidite facies from intra-oceanic forearc basins have not been widely studied. The Leg 126 results, therefore, provide an important addition to global understanding of forearc sedimentology. As well as obtaining a relatively continuous core record, the Leg 126 forearc sites were the first ODP boreholes to be imaged by the Schlumberger Formation Micro-Scanner tool (FMS). The combination of high-quality core and high-resolution electrical images has allowed us to construct two bed-by-bed sedimentary columns through vertical sections of the forearc basin fill. Such a detailed and continuous vertical coverage is normally only available from outcrop studies.

The aim of this paper is thus to provide a detailed description of the deep-water, Oligo-

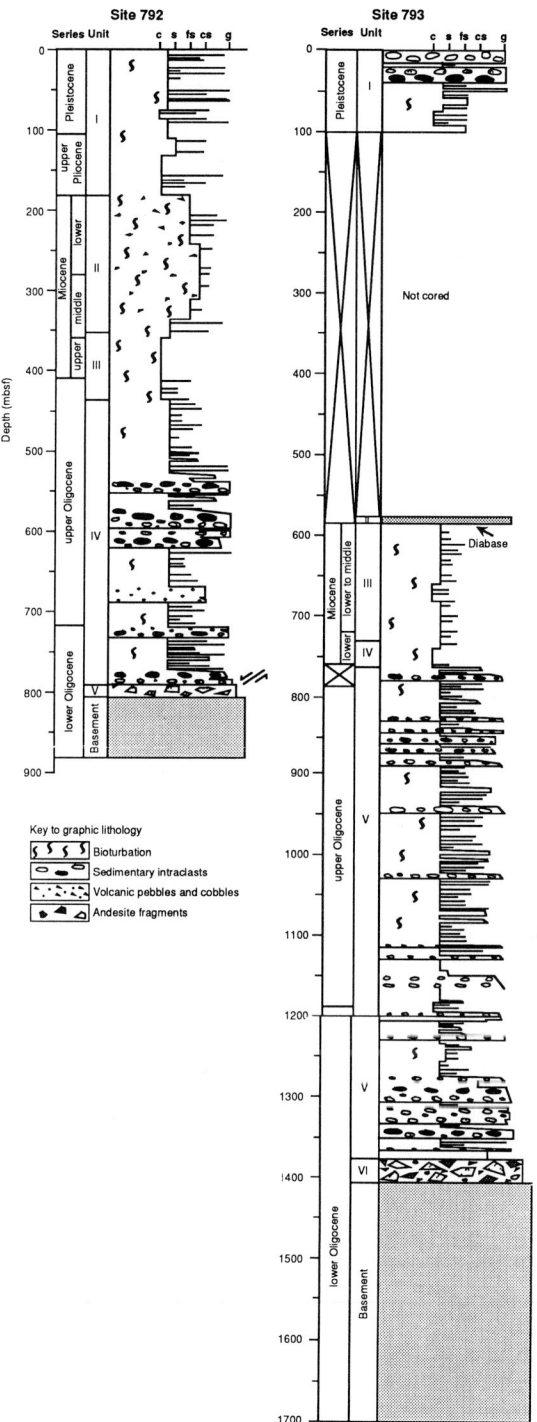

Fig. 2. Graphic sedimentary sections for the two logged forearc sites (792 and 793), showing depth (mbsf) and age. (c, clay/claystone, s = silt/siltstone, fs = fine-grained sand/sandstone, cs = coarse-grained sand/sandstone, and g = gravel/conglomerate.

cene, volcanic turbidite succession in the forearc basin from core data and the analysis of FMS images. Attention will be paid to both sedimentary and tectonic features outlined by the images, and inferences will be drawn for the geodynamic characteristics of this intra-oceanic plate boundary.

Lithostratigraphy of the forearc basin

The basement of the forearc basin sedimentary sequence was cored at both sites, and consists of andesite and basaltic andesite (Taylor *et al.* in press). Most of the overlying sediment (Fig. 2) is a thick Oligocene turbidite succession deposited at un-decompacted rates of about 100–300 m/ Ma (Taylor, Fujioka *et al.* 1990; all rates are here uncorrected for post-depositional compaction). The oldest recovered sedimentary rocks (about 31.0 Ma) are at Site 793, near the basin axis. Site 792 is located near a basement high that did not become buried by turbidites until about 29.5 Ma.

The two sites have thus relatively complete records of Late Oligocene sedimentation (about 29.0 to 24.0 Ma). The sedimentary facies are mainly turbidites and submarine debris-flow deposits. Coarse-grained, pebbly sandstones and conglomerates are older than 27.5 Ma, while younger deposits are generally finer-grained, burrowed, with calculated sedimentation rates of less than 20 m/Ma. The lower to middle Miocene is represented by fine-grained, deep-water sediments, with sedimentation rates ranging from 5 to 15 m/Ma. This age was sampled at Site 792 only, with sandy mudstone and muddy sandstone intervals. The only significant terrigenous component in the Miocene sediment is clay. The Pleistocene and Pliocene sections were deposited in deep-sea valleys cut into the forearc basin deposits and consist, in whole or in part, of mixtures of pumiceous and scoriaceous sand and gravel.

FMS images and downhole measurements

FMS sensor and data processing

The FMS creates a picture of the borehole wall by mapping its electrical conductivity using an array of small, pad-mounted electrodes (Ekstrom *et al.* 1986; Luthi & Banavar 1988). The configuration developed for ODP (Pezard, Lovell *et al.* 1990) uses four arrays of 16 electrodes each to generate the images. A single pass of the sensor maps about 30% of a 25.4-cm-diameter borehole. During logging, a current flows from each electrode to a single return electrode located at the top of the sonde. Because of electrode geometry, the sensor has a very shallow depth of investigation (a few centimetres beyond the borehole wall).

A constant difference of electrical potential is held between each of the electrodes and the remote return. As the conductivity of the rock formation varies in front of each electrode during recording, a variable current is injected for each electrode into the rock to satisfy the constant potential condition. This current is measured and transformed into an electrical conductivity by Ohm's law. As the 'constant' difference of potential is regulated in the tool to vary, in order to match the electrical conductivity of the rock and generate the best possible image, the images do not represent a direct map of electrical conductivity of the borehole wall. Such a map can be obtained only after calibration of the FMS images with calibrated electrical resistivity measurements with a similar depth of penetration from the borehole into the rock (such as that of the LLs or SFL). The button current is thus recorded as a series of curves that represent relative changes of microconductivity in the rock caused by either (a) electrolytic conduction in the pore space or (b) cation exchanges on the surfaces of clay minerals.

The current intensity is converted to variable-intensity grey images, in which black is the highest (relative) conductivity and white the lowest. The sampling rate of the FMS is about 2.5 mm. However, the FMS might detect thinner features that have high conductivity contrast to their surrounding. Each electrode is oriented in space with three-axis accelerometers and fluxgate magnetometers, making it possible to derive the strike and dip of geological structures. In this paper, the analysis is restricted to the Oligocene sections that were successfully logged with the FMS during Leg 126. The present FMS analysis of sedimentary structures is restricted to two intervals that span (Fig. 3) from 4023 to 4330 metres below rig floor (mbrf) (Hole 793B), and from 2232 to 2441 mbrf (Hole 792E). Analysis of the stress field utilizes FMS images from the basement of 793B.

Downhole measurements of physical properties

In each of the two holes, a complete set of physical properties measurements was recorded. The two intervals where FMS images were analysed are presented in Figs 4 & 5, for Holes 793B

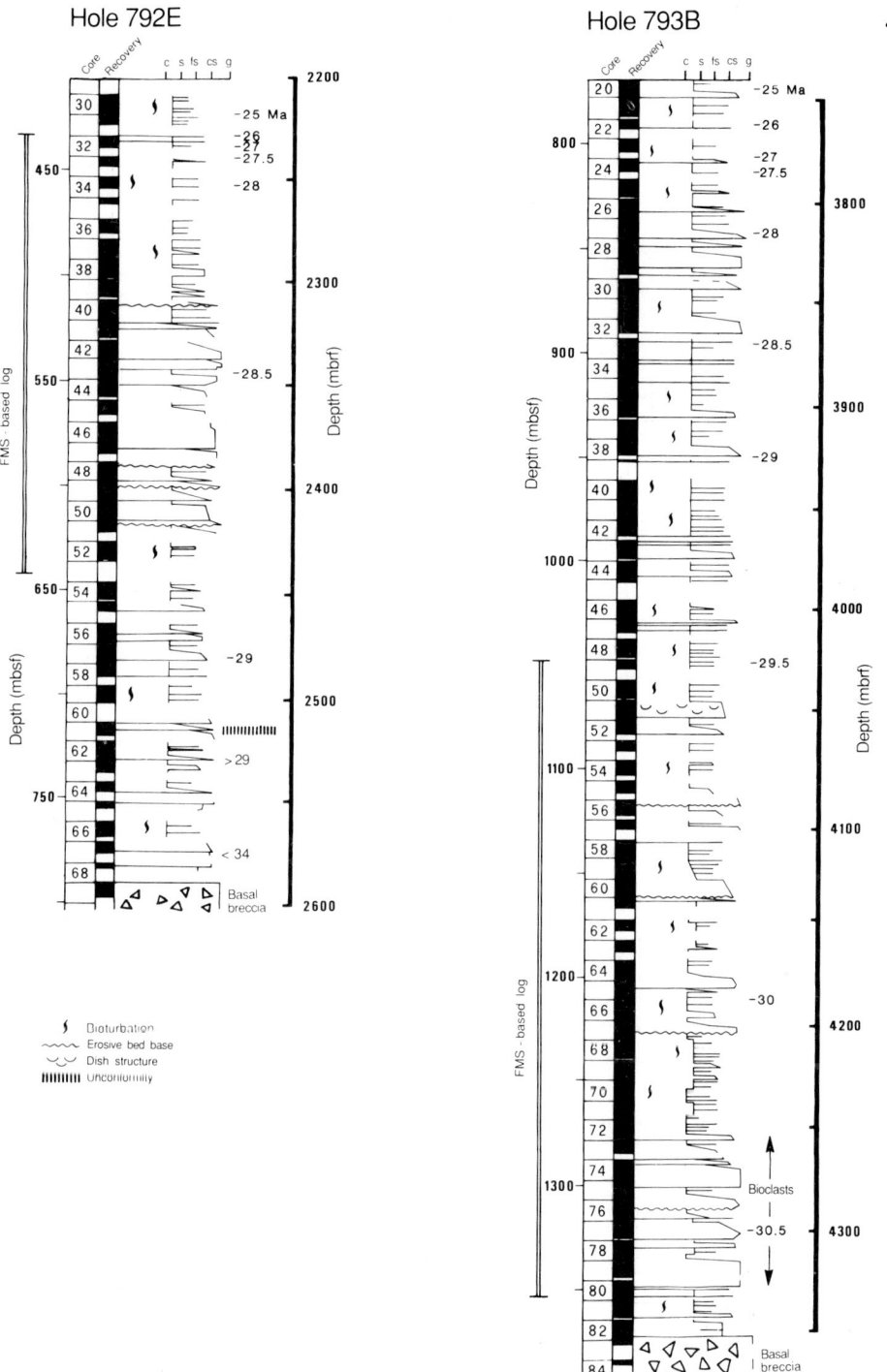

Fig. 3. Simplified sedimentary columns, based on core descriptions, for Holes 126-792E and 793B. The columns indicate grain size, bedding characteristics, occurrence of large bioclasts, core numbers, recovery, depths (in metres) below rig-floor (mbrf) and sea-floor (mbsf), selected age-picks (e.g. *27.0 Ma), and the location of the FMS-based logs of Figs 20 and 21. Sediment grain size description as in Fig. 3. Information is only shown where a core was recovered.

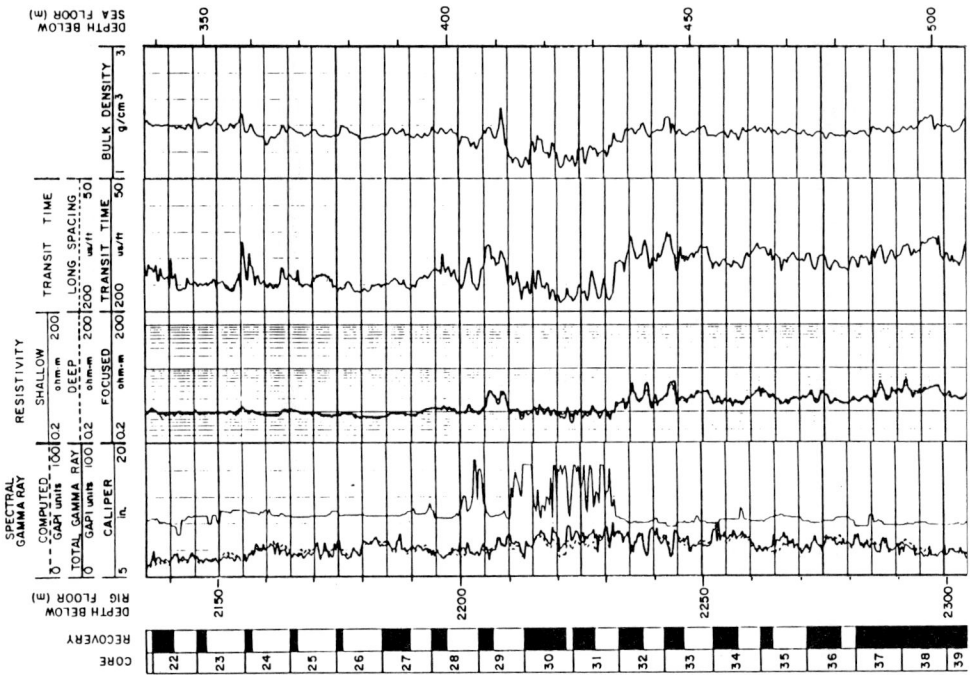

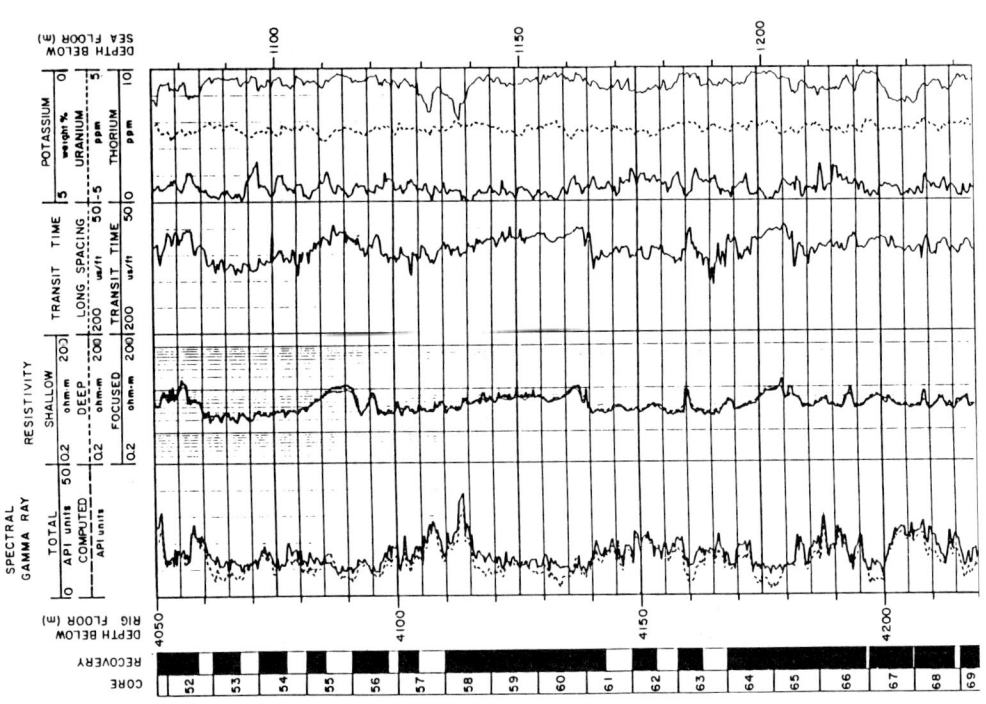

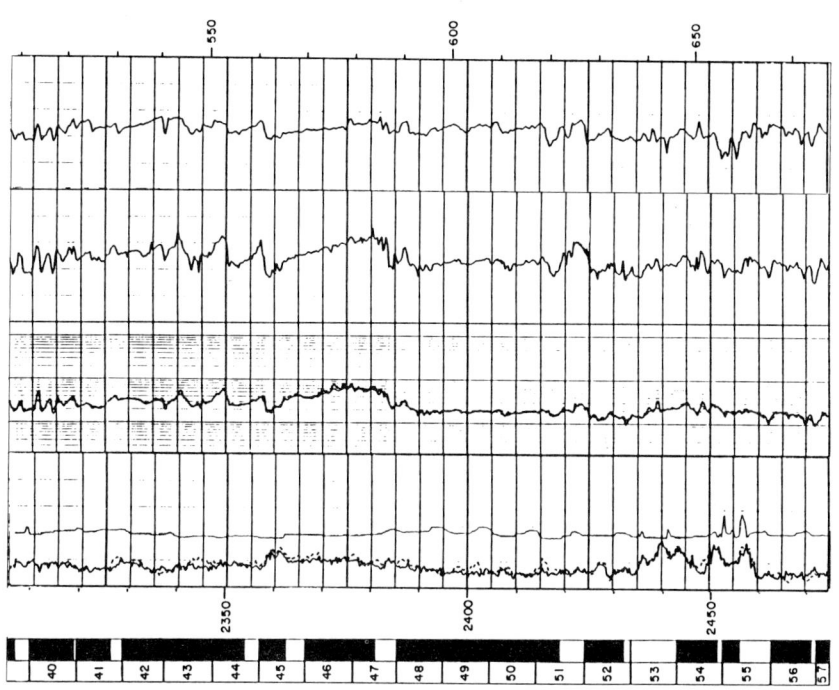

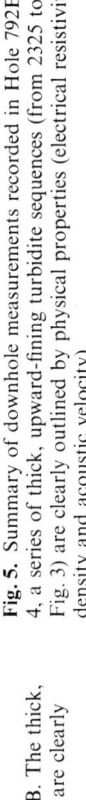

Fig. 5. Summary of downhole measurements recorded in Hole 792E. As in Fig. 4, a series of thick, upward-fining turbidite sequences (from 2325 to 2385 mbsf; Fig. 3) are clearly outlined by physical properties (electrical resistivity, bulk density and acoustic velocity).

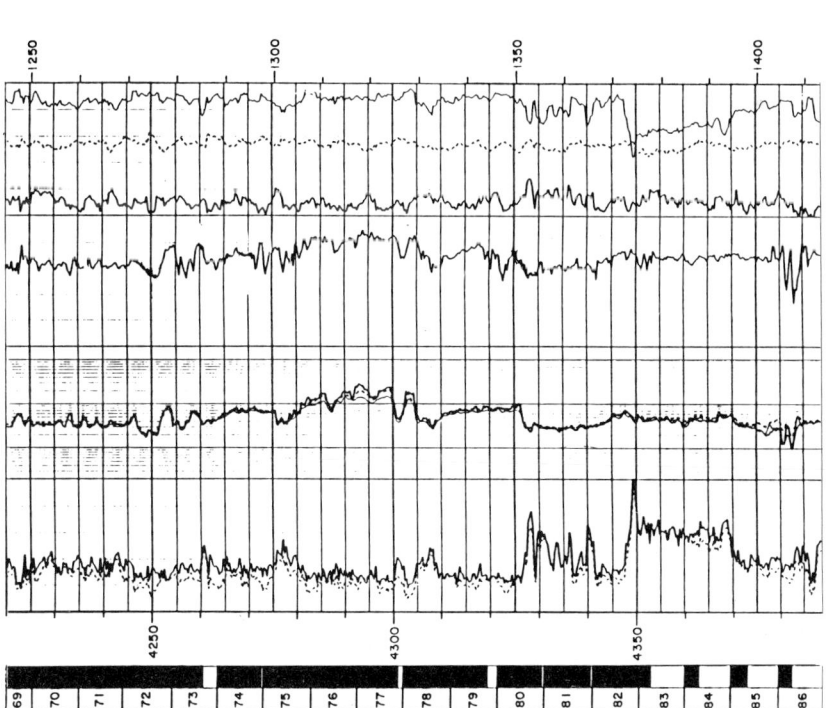

Fig. 4. Summary of downhole measurements recorded in Hole 793B. The thick, upward-fining turbidite sequences (from 4275 to 4325 mbsf; Fig. 3) are clearly outlined by the electrical resistivity and the acoustic data.

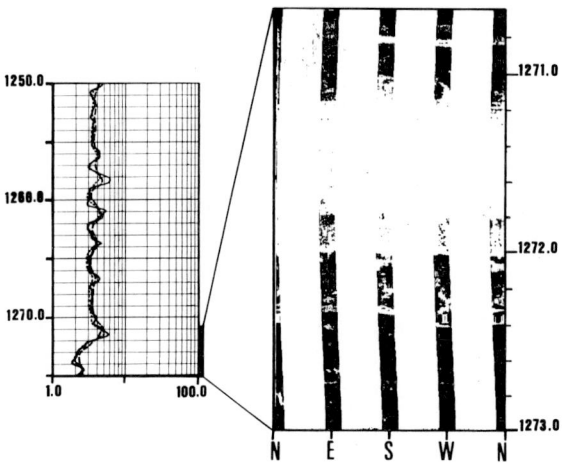

Fig. 6. Resolution of the large-scale resistivity downhole data compared with that of FMS images from Hole 793B. The white part of the image represents the more resistive interval, and the dark one the more conductive. Fine details from within a turbidite sequence (bedding dip at 1270.5 m, erosional base at 1272.4 m) are averaged by the Phasor Induction and revealed by the electrical images. A near-vertical, mineralized, fracture is observed at about N270. It developed as the result of palaeo-extensional stresses in the conglomeratic base of the turbidte sequence (from 1271.0 to 1272.0 m), aligning with bedding when reaching the more conductive, plastic, and clay-rich beds at 1271.1 m. Depths are in mbsf and electrical resistivity in Ω m.

and 792E respectively. The electrical resistivity data were recorded with an induction device, while acoustic measurements were made with a long-spaced sonde. The results pertaining to basin development are outlined here.

Sedimentary structures are detectable at scales larger than the resolution of the various logging tools (of the order of a metre). For example, the thick, fining upward turbidite sequences from 4275 to 4325 mbsf in Hole 793B (Fig. 3) clearly appear in electrical resistivity and acoustic data. Similarly, in Hole 792E, a series of thick, fining upward turbidite sequences (from 2325 to 2385 mbsf; Fig. 3) are visible on electrical resistivity, bulk density and acoustic velocity logs. The conglomeratic base of each turbidite series is resistive, with a high bulk density and acoustic velocity.

The FMS images confirm the observation in Fig. 6, where the resolution of large-scale resistivity downhole data are compared with that of FMS images from Hole 793B. The white part of the image represents the more resistive interval, at the base of the turbidite sequence. Fine details (bedding dip at 1270.5 m, erosional base at 1272.4 m) are averaged by the induction sensor and revealed by the electrical images. A near-vertical, mineralized, fracture is observed with N270° orientation.

Generation of FMS-based sedimentary sections

For rocks of similar mineralogy, cementation and fluid type, the pixel tone on the FMS images is thus observed to be a function of grain size, with the darker tones representing finer sedimentary units. At a superficial level, therefore, the FMS images provide a strip record of texture, and sharpness of contacts between beds of different grain size. The approximate correspondence between beds in FMS images and equivalent beds in cores was obtained by converting log to core depths in each hole. The thicker, and therefore more unusual, sandstone beds in cores could in most cases be matched with the same beds in the FMS images, where the sandstones appear light in tone.

Based on these first matches between cores and images, the lightness of the tones on the images was calibrated against grain size reported in shipboard visual description. Thinner beds were then matched using distinctive sequences of bed thicknesses. This was done for all recognizable beds, down to a thickness of about 1 cm. For parts of the succession where core recovery was incomplete, the FMS images alone, combined with the knowledge in near-by rock of the correspondence between grain size and image tone, were used to construct the bed-by-bed sections of Figs 7 & 8. The depths of all bed contacts were entered into a computer data file. The total number of sharp-based sandstone beds imaged by the FMS over the analysed intervals are 807 and 447 in Holes 793B and 792E respectively.

On Figs 7 & 8, positions of core tops from the drilling record, and the result of the match between visual description and FMS log depths are indicated. Sedimentary structures and large clasts, most of which can be recognized on the images and confirmed by examination of core photographs are included for thicker sandstone beds. Ages indicated come from the age-depth plots of Taylor, Fujioka *et al.* (1990). There is a remarkable similarity between the FMS-based sedimentary logs (Figs 7 & 8) and downhole measurements of resistivity (Figs 4 & 5). The most spectacular example of this relationship is the thick debris-flow deposit with its base at 4090 mbrf in Hole 793B. This bed has a sharp basal contact in the FMS data, and a gradational transition to more conductive deposits with finer grain sizes toward the top. Knowing that the resistivity data so accurately reflect the combined FMS and core data is important since it encourages the study of depositional processes in sections not imaged with the FMS.

Images of turbidites and related facies

The generalized sedimentary logs for the Oligocene sections cored at sites 792 and 793 are presented in Fig. 3. Note that the top of the FMS-based section in Hole 793B is slightly older than the base of the FMS-based section in Hole 792E. Furthermore, all sediments in the FMS-based section at Hole 793B are older than the basal sediments at Hole 792E. The parameters used to distinguish the sedimentary facies described below are grain size, bed thickness and sedimentary structure. Bed thicknesses for the thickest beds cannot be obtained from cores because single depositional units commonly span core intervals, which boundaries could contain bed boundaries; continuous FMS images do not have this limitation.

Sedimentary facies

Mudrocks and graded siltstone beds. Mudrocks generally occur as interbeds between coarse-grained units and form 10.0% and 5.0% of the FMS-based sections in Holes 793B and 792E respectively. Mudrocks are in some cases finely laminated, but are generally burrow mottled. Siltstones, mainly as very thin to thin graded

beds, form 8.3% and 11.0% of the FMS-based sections in Holes 793B and 792E, respectively. The beds have sharp bases, and are parallel and/or cross laminated, as described by Bouma (1962).

Very thin and thin graded sandstone beds. Very thin and thin graded beds, with bases of generally very fine-grained to fine-grained sandstone, form 7.0% and 4.0% of the FMS-based sections in Holes 793B and 792E respectively. Bouma sequences prevail (Fig. 9). Except for coarser grain size, these beds are similar in appearance to many of the graded siltstone beds.

Medium to thick graded sandstone beds. Medium to thick graded sandstone beds, with bases of generally fine- to coarse-grained sandstone, form 30.0% and 36.0% of the FMS-based sections in Holes 793B and 792E respectively. Most of the thicker beds have a basal division that is structureless or parallel laminated (Fig. 10), and Bouma sequences prevail here also (Fig. 11).

Very thick sandstone, granular sandstone and pebbly sandstones. Most beds of this facies consist of medium- to coarse-grained, sand-sized detritus (Figs 7 & 8). Some of the beds have conglomeratic basal divisions; we arbitrarily restrict this facies to beds with less than 1 m of conglomerate. Basal erosional relief of a few centimetres can be seen in some cores and FMS images. Beds of this facies form 38% and 34% of the FMS-based sections in Holes 793B and 792E, respectively. At both holes, a few exceptional beds greater than 10 m thick account for a significant part of the section.

Essentially all beds thinner than 3 m, and some beds as thick as 10 m, are normally graded throughout, with tractional structures like parallel lamination and cross lamination in their upper parts. Other beds, generally thicker than 3 m, are characterized by very poor sorting, poorly developed clast fabric, and poorly developed grading that, if present, is restricted to the top of the bed. These beds may contain upper divisions of parallel or ripple lamination, and concentrations of pebble-to-boulder-sized intraclasts.

Conglomerates. Conglomerates form about 10% of the Oligocene section at Sites 793 and 792. Some very thick sandstone and pebbly sandstone beds have thin, basal, conglomerate divisions here omitted from the conglomerate facies. The conglomeratic parts of the beds have mean sizes in the fine pebble or granule range. They are generally structureless, except for local

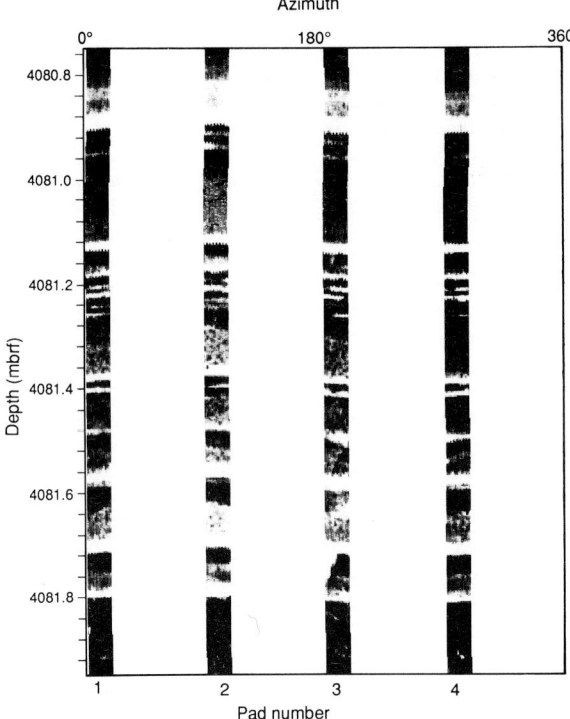

Fig. 9. FMS image of very-thin and thin, sharp-based, graded beds from Hole 793B. Lighter tones indicate coarser-grained sediments. Darker vertical features at N200 are fractures.

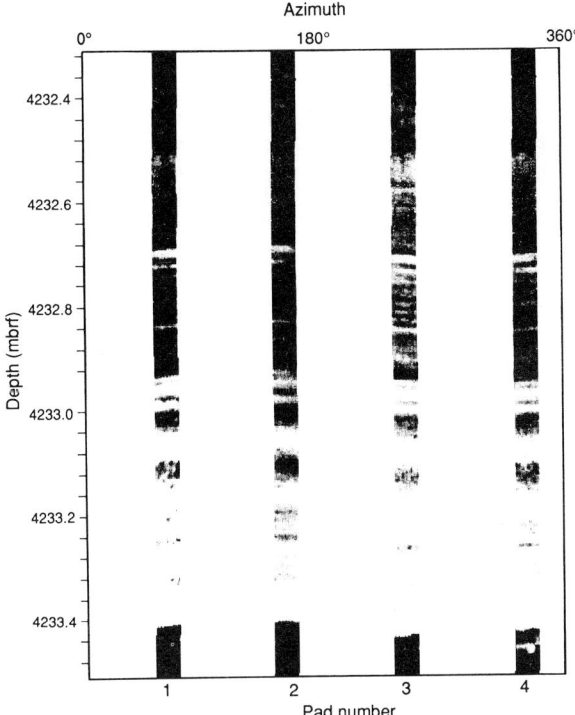

Fig. 10. FMS image of a single, thick, parallel laminated, graded sandstone bed from Hole 793B (base at 4233.4 mbrf and top at 4232.4 mbrf). Lighter tones indicate coarser-grained sediment.

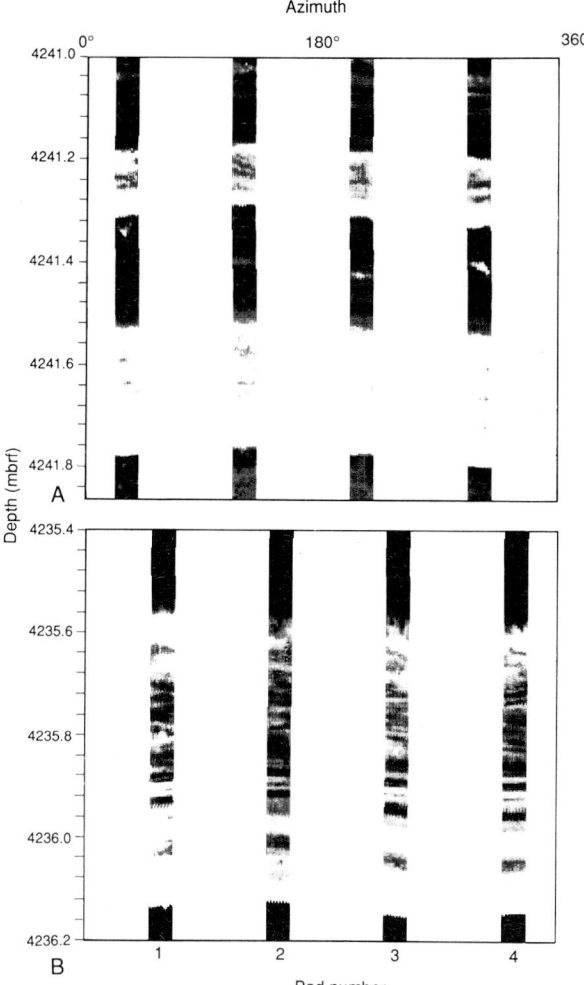

Fig. 11. FMS images of sedimentary structures and grading in medium to thick sandstone beds of Hole 793B. Lighter tones indicate coarser-grained sediments. (A) Two graded beds are both cross-laminated. The cross-lamination is most visible in the upper bed at 4241.25 mbrf. (B) Parallel lamination and wavy lamination in a sharp-based, graded sandstone bed.

concentrations of fine-grained, mainly mudstone cobble-and-boulder-sized intraclasts similar to those found in the very thick bedded sandstones.

Depositional processes

All siltstones, sandstones and conglomerates were deposited by sediment gravity flows, ranging in character from turbidity currents of both high and low concentration, to debris flows. This variability diluted the hemipelagic input of nannofossil-rich mud. There is insignificant hemipelagic sediment where sedimentation rates exceed about 250 m/Ma.

The graded sandstones with Bouma (1962) sequences, mostly pebbly sandstones, and graded to graded-stratified conglomerates were deposited by turbidity currents (Walker 1975). The very thick, mainly structureless beds of sandstone characterized by poorly developed fabric and grading, and large mudstone intraclasts well above the base of the bed, are interpreted as debris-flow deposits.

There is no seismic evidence that the very thick beds of sandstone, pebbly sandstone and conglomerate were deposited in major channels (Taylor, Fujioka *et al.* 1990; Klaus & Taylor 1992). Very thick beds younger than about

30.0 Ma are interbedded with thin to medium bedded turbidites, and seem to represent unique, high-volume events, rather than channel-fill deposits. Some of these events were debris flows that may have been generated by catastrophic failure of parts of the coarse-grained apron surrounding the arc volcanoes (Moore *et al.* 1989).

The general setting, apparently devoid of major channels, is considered to have been an oversupplied basin plain (Pickering 1982), similar to the example described by Ricci Lucchi & Valmori (1980). This interpretation is supported by dispersed palaeocurrent directions (see below). Unlike some ancient basin plains (Ricci Lucchi & Valmori 1980; Hiscott *et al.* 1986), the thickest beds do not have impressive mudstone caps that would indicate deposition within a confined basin. Other controls on sediment failure were probably seismic activity, and bottom slopes of perhaps a few degrees across the margins of the forearc basin.

Bed-thickness statistics

A cursory examination of the FMS-based logs (Figs 7 & 8) shows that the number of sandstone beds is inversely proportional to bed thickness. This means that many thin beds and few thick ones were imaged. If this observation corresponds to a power law distribution, the number $N(T)$ of beds with thicknesses greater than T is proportional to T^{-b} (b is a positive constant). Many geological and geophysical quantities (earthquake magnitudes, volumes of volcanic eruptions) follow such a power law distribution (Turcotte 1989). Turbidite bed thicknesses have also been proposed to be distributed in this way (Hsü 1983).

Figure 12a shows logarithmic plots of the number $N(T)$ of sandstone-shale couplets whose thickness is greater than T, versus T; plots using sandstone bed thicknesses only are essentially the same. The data come directly from the files

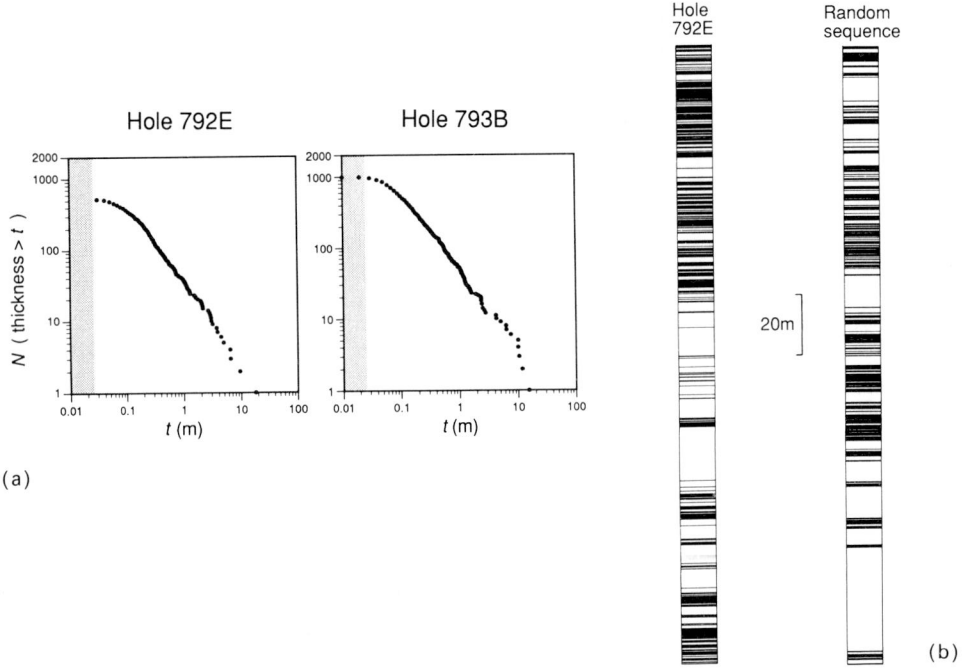

(a)

(b)

Fig. 12. (A) Logarithmic plot of the number of sandstone-shale couplets in the FMS-based section of thickness greater than T versus T (in metres). As predicted for a power-law distribution, the plot follows a straight line whose slope is the characteristic exponent (slightly larger than 1 here). The stippled area defines the lower limit for this analysis, as constrained by the resolution of the FMS sensor (i.e. 2.5 cm). (B) Plot of the bases of sandstone beds in the FMS-based plot for Hole 792E compared to a plot of a random sequence. The random sequence contains the same number of bed boundaries, and the bed thicknesses are distributed according to a power law with a characteristic exponent of 1.0.

that were used to generate the FMS-based sedimentary log. In log–log space, a power law distribution plots as a straight line with slope *b*. Here, in both cases, the slope is slightly larger than 1.0 for bed thicknesses ranging from 10 cm to more than 10 m (three orders of magnitude). For beds thinner than 10 cm, the number of beds is smaller than what would be expected from extrapolation of the slope defined by thicker beds, which can be interpreted in two ways. Either this fact is genuine, or beds just thicker than the resolution limit of the FMS (2.5 cm) were not counted correctly. As the resolution limit is bound to have some influence on the measurements, we prefer the latter interpretation.

Bed-thickness trends

The FMS-based sedimentary logs (Figs 7 & 8) are continuous bed-by-bed sections that are ideal for investigation of trends in bed thickness. It is readily apparent that some part of each log consists mainly of very thick beds, but there are no particularly consistent trends within, or across these boundaries. Globally, the very thick beds in the Oligocene succession seem to be scattered more-or-less at random in a background of thinner beds. The impression of random occurrence of the very thick beds is strengthened by the resemblance of the succession of beds in Hole 792E to a random sequence containing the same number of beds (Fig. 12b). The bed thicknesses in the random sequence have a power law distribution with a slope of 1.0. The phenomenon of clustering of thicker beds observed in Hole 792E is also present in the

deeper part of the random sequence. The occurrence of such a cluster is, in fact, inherent to the random nature of the distribution and referred to as 'random Cantor dusts' (Mandelbrot 1983).

In a geological context, however, such clusters must reflect local or regional changes in the volumes or pathways of sediment gravity flows. It is these changes that may occur randomly. Because the clusters of thicker beds at Sites 793 and 792 cannot be correlated, we cannot isolate local from regional controls on cluster occurrence.

Bed-thickness frequencies (recurrence intervals for sediment gravity flows)

The data files used to plot the FMS-based sedimentary logs were sorted to determine the number of sandstone beds in each bed thickness class of Ingram (1954). The FMS-derived sedimentary log for Hole 793B (Fig. 7) represents a time interval of about 1.15 Ma, with relatively constant sedimentation rate. The section from Hole 792E includes strata deposited at a decreasing rate, from 28.0 Ma. For the purpose of general assessment of average bed-thickness frequencies, the sequence younger than 27.5 Ma has consequently been ignored. A time-span of 1.55 Ma is assumed for the deposition of the beds described in the FMS-based sedimentary section from Hole 792E, which equates to depths from 2240 to 2440 mbrf (Fig. 8).

Knowledge of the number of beds in each thickness class, and the total time represented by each log permits calculation of the number of small to very large sediment gravity flows to reach a particular location per unit time

Table 1. *Bed-thickness frequency data from FMS-based sedimentary logs*

Hole	Thickness (cm)	Number of beds	Number (per Ma)	Approximate recurrence interval (years)
792E	1–3	7	4.5	222,200
	3–10	133	86	11,600
	10–30	197	127	7,900
	30–100	83	54	18,500
	100–300	19	12	83,300
	300–1000	7	4.5	222,200
	>1000	1	.65	1,538,500
793B	1–3	82	71	14,100
	3–10	325	283	3,500
	10–30	268	233	4,300
	30–100	93	81	12,400
	100–300	28	24	41,700
	300–1000	7	5	200,000
	>1000	4	3.5	285,700

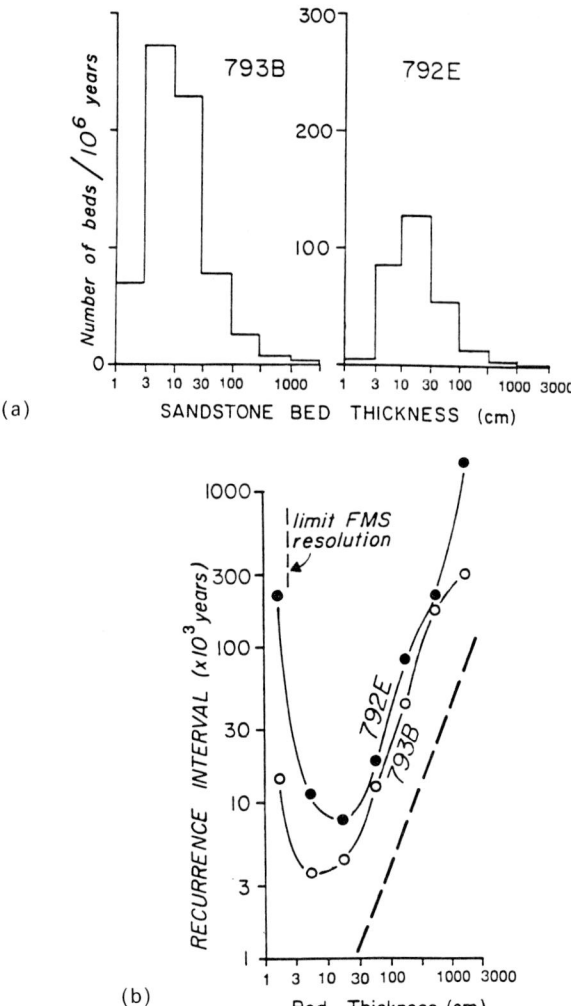

Fig. 13. (A) Histogram of sandstone and conglomerate bed thickness from FMS images, normalized to a time interval of 1.0 Ma. The values for beds 1–3 cm thick are probably too low, because these beds are at or near the resolution of the FMS sensor. (B) Recurrence interval for beds of different thickness, derived from (A). The heavy dashed line has a slope that would indicate a recurrence interval proportional to bed thickness.

(Fig. 13a; Table 1). The vertical sampling rate of the FMS (2.5 mm) means that an unknown number of very thin beds (<2.5 mm) may have been missed during analysis of the images. At the other end of the spectrum, beds thicker than 10 m are so rare that recurrence intervals for the exceptional events that deposited them must have considerable uncertainty.

A derivative parameter is the recurrence interval for beds of different thicknesses (Fig. 13b; Table 1). As before, probable failure to recognize all very thin beds leads to artificially high recurrence intervals for these beds. For other

thickness classes, recurrence intervals range from about 5.0 Ka to about 1.5 Ma. The power law distribution with an exponent of 1.0 of sandstone bed thicknesses implies that the recurrence interval is simply proportional to bed thickness. This prediction is illustrated by the dashed line of unit slope on Fig. 13b.

Triggering processes

The detritus in the turbidite and debris-flow deposits is almost exclusively volcaniclastic, probably derived from failure of unconsolidated or semi-consolidated ejecta and epiclastic material around the active arc volcanoes. Possible triggers of failures in this setting are oversteepening of slopes due to increases in volcanic output (Moore *et al.* 1989), seismic activity associated with major eruptions, subduction-related earthquakes, and longer-term increased mobilization of material due to fall in relative sea level (Mutti 1985). Important proposed excursions in global sea level in the Oligocene are a drop of about 150 m at 30.0 Ma, and a rise of about 100 m at 25.0 Ma (Haq *et al.* 1987). The rise at 25.0 Ma was probably accentuated by subsidence resulting from the initiation of rifting of the volcanic arc to form the Shikoku Basin (Hussong & Uyeda 1981).

From a statistical point of view, it is probably unjustified to look for different triggering mechanisms for thicker versus thinner beds. In such a case, the frequency of different processes would have to scale exactly to generate a population of bed thicknesses distributed according to a power law of about 1.0 (Fig. 12). This seems a very unlikely coincidence.

In the case of triggering from large earthquakes associated with subduction processes, magnitudes greater than 5.0 are needed, to allow for the sediment liquefaction necessary to trigger the downslope movement of large volumes of sediment (Keefer 1984). The recurrence interval of such event is, however, of the order of 25 years near Japan (Mogi 1990), which is far too little for any bed thickness according to Fig. 13b. This mechanism appears thus also as a very unlikely generating process, although very large earthquakes might be necessary to create the very thick sandstone beds.

Clearly, factors other than the availability of triggering events must be important in determining whether or not exceptionally large volumes of material can be mobilized. Also, the emplacement of a large flow depends on the presence of large, unstable accumulations of loose materials. In the Izu–Bonin Arc, the presence of the very thickest beds might be due to the coincidence in space and time of an unusually large and unstable mass of volcaniclastic material, at times of an unusually large magnitude earthquake in the immediate vicinity.

Palaeocurrents

Methods and limitations

Two types of palaeocurrent information were obtained, with critical orientation of the FMS. The first is grain fabric in parallel laminated sandstone beds, and the second is ripple migration directions. Both types of data have relatively large sampling errors, but there are no other palaeocurrent data for these Izu–Bonin forearc deposits.

The orientation technique for samples extracted from the core is described in Hiscott *et al.* (in press). Ripple-scale lamination is visible on FMS images of many turbidite beds. This cross lamination is more subtle than parallel lamination, and could only be appreciated and measured with confidence on the workstation. An example of the presence of such cross lamination, with interpretation, is given in Fig. 14a. In the four FMS images, each mapping about 7% of the surface of the borehole wall, cross lamination will be intersected at an unknown angle from the direction of maximum foreset dip, giving an apparent dip only. A complicating factor is that ripple-scale cross lamination generally consists of troughs, so that the direction of maximum foreset dip at any point may not be parallel to the palaeoflow direction. Such geometric irregularities could not be compensated for in the analysis of FMS images. Parallel foresets were hence assumed, unless proven to the contrary.

From the analysis of FMS images, a total of 21 possible examples of cross laminations were identified in Hole 792E, and 34 in Hole 793B. Due to variable image quality and low resistivity contrasts, the cross lamination was not generally visible in all four image strips. Where visible, a sinusoid was fitted to the dipping lamination and forced through the same stratigraphic level on the opposite side of the borehole, thus defining a plane with a dip parallel to the apparent dip of the cross lamination on the borehole wall. When possible, this procedure was repeated for the same ripple train using apparent dips from cross lamination visible in other image strips, preferably located at 90° from the first. The planes defined by these apparent dips were plotted on a Schmidt stereonet, corrected for local bedding dip, and the true dip and dip azimuth of the foreset plane were determined.

The angle of repose of fine-grained sand, under water, is about 25°. It is therefore expected that maximum foreset dips will be in this range, although smaller dips will characterize the lower parts of trough fills, and higher dips might occur as the result of minor wet-sediment deformation. The accuracy of individual foreset dip directions is fairly poor, mainly because of our lack of ability to account for complex foreset shapes. For foresets defined with more than one apparent dip, the result may be within ±25° of the true migration direction of the ripples. For steep, single determinations, the accuracy falls to within ±40° of the reported value (Fig. 14b).

AZIMUTH

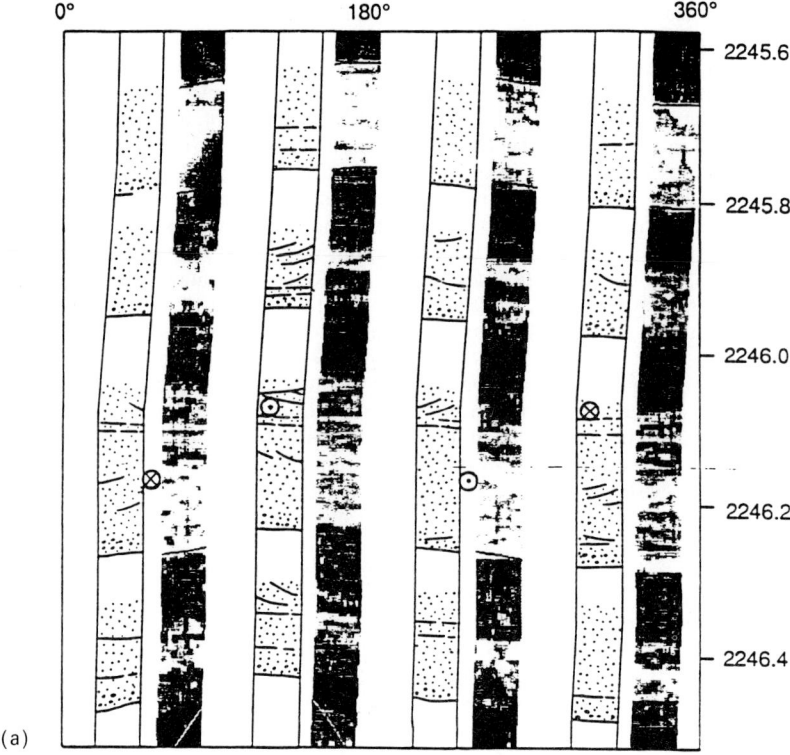

(a)

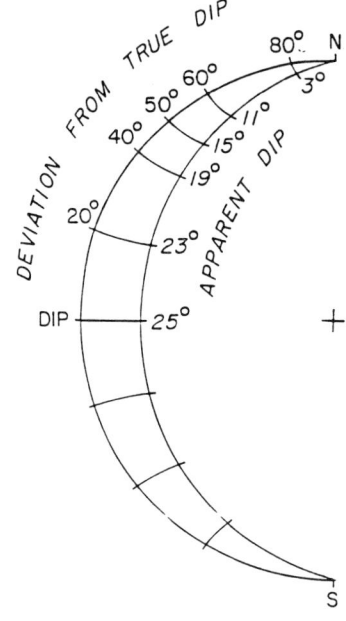

(b)

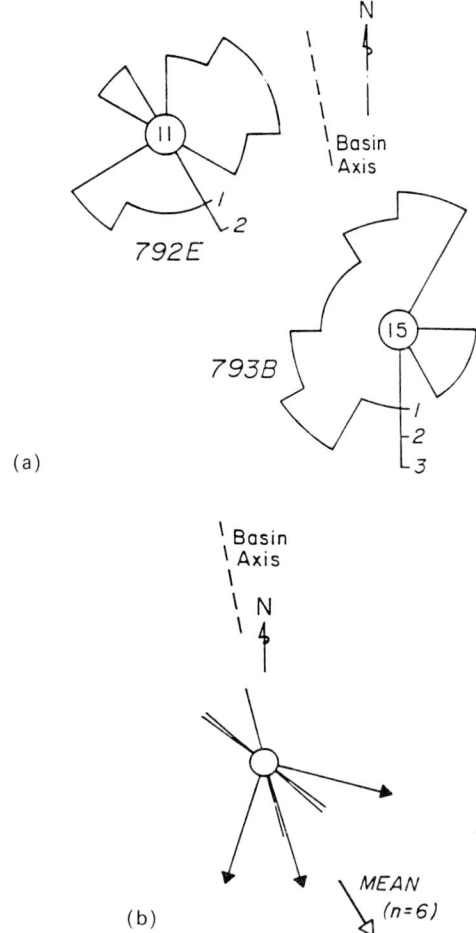

(a)

(b)

Fig. 15. (a) Equal area rose diagrams summarizing migration directions deduced from FMS images analysis in Holes 792E and 793B. The data are plotted relative to true north, and the trend of the axis of the forearc basin (Fig. 1). (b) Grain fabric results in Hole 792E (Table 1). Imbrication measurements were successful in determining flow direction for the three results plotted with solid arrow heads. Other long-axis measurements, plotted on both sides of the origin, only give line-of-motion information. The mean of all six fabric determinations, assuming the southeast quadrant to include the downcurrent direction, is N148 (arrow with open head).

Fig. 14. (a) Analysis of a FMS image through beds with apparent cross lamination, Hole 792E. The sketches are correctly positioned relative to compass directions, whereas the FMS images are not. The elevation of the bases of the beds differs from pad to pad because the beds are dipping to the northwest. Apparent cross-laminations at 2245.8–2245.95 mbrf is, in fact, much steeper than the angle of repose. Acceptable results were obtained from two levels in the underlying, probably composite bed; circled dots and circled crosses represent, respectively, the heads and tails of arrows aligned with the inferred ripple migration directions. For reasons given in the text, the quality of these images is much poorer than that available at the Schlumberger workstation. (b) Margin of a Schmidt stereonet, and position of a great circle for a plane dipping at 25°, which is the appropriate angle of repose for fine sand under water. In a direction located 45° away from the true dip direction, the apparent dip is still large (about 17°).

Even if measurement errors could be accounted for, palaeocurrents obtained from the upper parts of turbidites are known to show more scatter than data from the sole markings and basal grain fabric, mainly due to meandering of the decelerating tails of the currents (Parkash & Middleton 1970). This factor contributes additional uncertainty to the usefulness of the ripple migration directions in specifying dispersal patterns in the basin.

Grain fabric data

The results of grain fabric data from the six samples recovered in Hole 792E (where bedding dips were sufficient to orient the core by the mean of the FMS) are summarized with Fig. 15. A mean of 148° points toward a palaeoflow to the southeast, slightly to the east of the axis of the forearc basin. Data within the depth range of the FMS-based sedimentary log (Fig. 8) are plotted adjacent to the sample position.

Ripple orientation

Inferred ripple migration directions of highest quality are plotted on the FMS-based sedimentary log (Figs 7 & 8) and summarized as rose diagrams on Fig. 15. The two sites give somewhat different results, with considerable scatter in both cases. For the older (30.2–29.5 Ma), more basinward section (Site 793), there is a strong component of flow toward the southwest, away from the outer arc high and transverse to the basin axis (Figs 1b and 15a). There is also a mode toward the north, approximately along the basin axis. A significant gap in the data spans two classes from 030–090°; this may point in the upcurrent direction.

For the younger section at Site 792, close to the modern volcanic arc (Figs 1c and 15a), transport directions are, on the average, transverse to the basin axis. There are modes of approximately equal size indicating flow to the southwest and the northeast. Data are scarce in the region 240–360°. If there was a single foreset dip in the class 120–150°, then there would be a basis for suggesting that the data define a very broad distribution with flow away from, and on either side of about 310°. Such a dispersed palaeoflow distribution, with mean toward the southeast, would be consistent with the more reliable grain-fabric data from the same part of the section in Hole 792E (Fig. 15b).

Oligocene basin evolution

The oldest deposits (about 30.5 Ma) drilled by Leg 126 were found in Hole 793B. The basal sediments in this part of the forearc show a concentration of very thick beds of coarse-grained sandstone, pebbly sandstone, and fine pebble conglomerate. Pebbles here were partly derived by erosion of (a) a volcanic terrane including basalt and rhyolites, and (b) a shallow marine shelf colonized by large foraminifers and calcareous algae. Some of the fossil material is older than the forearc succession (middle to upper Eocene), due to re-sedimentation in the basin.

The limited palaeocurrent data from ripples in Hole 793B suggest a component of transport away from the eastern margin of the basin from 30.2 to 29.5 Ma, and a secondary mode oriented northward, along the basin axis. During this time, at Site 792 which corresponds to the western margin of the basin, no volcanic sediments were being deposited onto the 34.0–35.0 Ma basement (Taylor *et al.* in press). These observations suggest that the main sediment source may have been located in the vicinity of the modern outer arc high, where Eocene volcanics are known to occur (Fryer, Pearce *et al.* 1990).

The concentration of coarse pebble beds at about 30.5 Ma, and the inclusion of bioclastic debris in these beds and overlying sandstones as young as 30.0 Ma (Fig. 3), might be a reflection of the proposed major drop in sea-level at 30.0 Ma (Haq *et al.* 1987). This rapid lowering of base level around the volcanoes would have rapidly made available large amounts of unconsolidated debris, including erosional products from older rocks, and bioclasts from shelf deposits. This drop may also have promoted volcanic activity due to unloading of the flanks of the arc volcano (Wallmann *et al.* 1988). Tectonic influence is also possible, with formation of the extensional forearc basin. Uplift of the flanks, deepening of the basin, increased seismic activity and, eventually, increased volcanic activity would then have contributed to the coarse-grained and very thickly bedded interval located at the base of the unit.

Depositional rates at Site 793 remained relatively constant until about 29.0 Ma (Fig. 3), and sedimentation was characterized by mixed thin and thick bedded, basin-plain turbidites, punctuated periodically by deposits of large-volume, high-concentration turbidity currents and debris flows. Ripple data suggest palaeoflow away from the eastern margin, at least until 29.5 Ma. Subsequently, the first turbidites began to accumulate at Site 792. The rate of accumulation at Site 792 exceeded that of Site 793 until about 28.0 Ma, and palaeocurrent data (grain fabric and ripples) suggest a change to predominantly southward axial transport, with a small component of flow away from the western margin of the basin.

Other local and regional data support the conclusion that the main, latest Oligocene source area was to the west. At Leg 126 Site 787 (Fig. 1), depositional rates are lower and $CaCO_3$ contents in mudstones higher than at Site 792 located to the west. Further west, at DSDP Site 296 on the Palau–Kyushu Ridge which formed part of the Oligocene western margin of the forearc basin, equivalent upper Oligocene rocks are tuffs, lapilli tuffs and volcanic sandstones and siltstones of more proximal character than the Leg 126 successions (Karig, Ingle *et al.* 1975). In a west–east transect from the Palau–Kyushu Ridge to Site 787 (Fig. 1), therefore, there is a consistent decrease in proximal character and coarseness of the uppermost Oligocene

deposits, indicating that the main sediment source at this time was to the west.

Sediment input from the eastern margin of the basin may have continued after 29.5 Ma, until the latest Oligocene or early Miocene, but at a diminishing rate. After this, the eastern margin became completely buried by marine sediments and continued to subside to its current depth (Honza & Tamazaki 1985; Taylor, Fujioka *et al.* 1990; Klaus & Taylor 1992). This hypothesis is based on the ambiguous nature of the mainly axial palaeocurrent data from Hole 792E (28.8–27.3 Ma). The palaeocurrent data suggest that the slope of the sea-floor changed from northward (31.0–29.5 Ma) to southward after 29.0 Ma. General thickening of the Oligocene section from north to south in the basin (Taylor, Fujioka *et al.* 1990) suggest that the southward

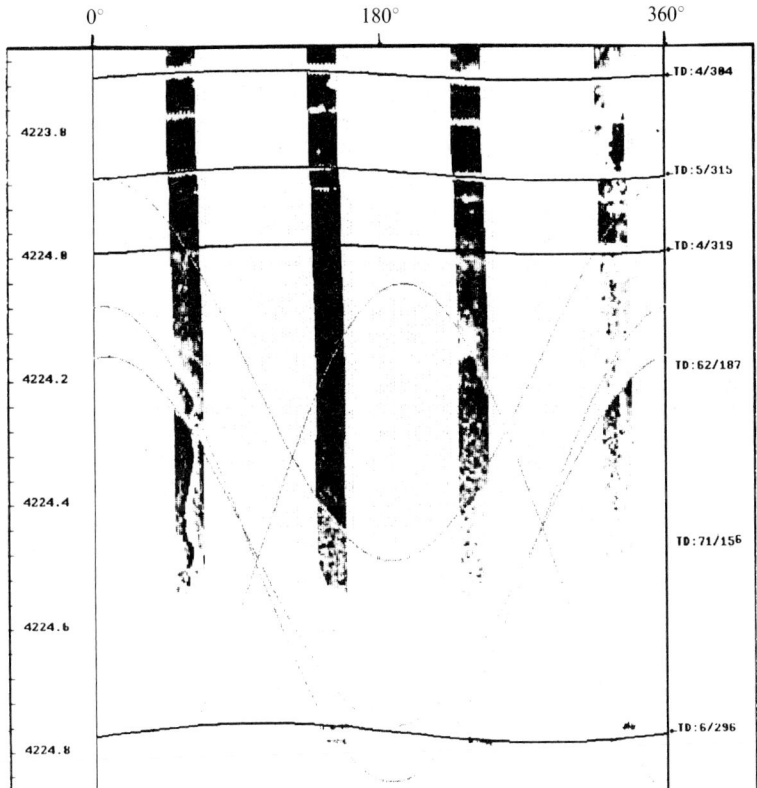

Fig. 16. Example of computer-fitted planes to near-horizontal bedding planes and steep, resistive, conjugated fractures. The presence of dipping events focuses the measuring current in certain wedges of formation, creating zones of apparent conductivity larger (with darker images) than reality (from 4224.1 to 4224.5 mbrf for each of the four images). The steep fractures appear generally slightly more resistive than the surrounding matrix, hence proving mineralized. Similarly, the near-vertical fracture imaged at about N060 has a conductivity similar to that of the host rock. This fracture is due to palaeo-extensional stresses, elsewhere traced in the core.

tilt has remained from 29.0 Ma to the present. None of the very thick beds has large-volume mudstone caps, which may indicate that the basin was never closed by a cross-basin high or sill. Alternatively, the volcanoes may have supplied predominantly sand-sized detritus.

There is a concentration of very thick beds of coarse-grained sandstone, pebbly sandstone, and minor conglomerate with ages of about 28.8 to 27.8 Ma (Figs 2 & 3). The proposed fluctuation in global sea level is about 50 m at this time (Haq *et al.* 1987), which may explain the fluctuations in grain size and bed thicknesses, according to Mutti (1985). Alternatively, a tectonic origin with variation of the volcanic source elevation or intensity may also be the source of these fluctuations (Cloetingh 1986).

The last important stage in the evolution of the forearc basin is the sharp decline in sedimentation rate from 27.0 to 23.5 Ma, with an increasing abundance of mudstone. This fine-grain blanket effectively completed the filling of the extensional forearc basin. Either a sharp rise in global sea level (Haq *et al.* 1987) drowning the subaerial part of the volcanic chain, hence reducing the supply of volcanic detritus, or a fast decline in volcanic activity of the arc due to the opening of the Shikoku Basin, splitting the arc in two (Kobayashi & Nakada 1979) might explain the abrupt decrease in sedimentation rate.

Structure and stress orientation in the arc

Structural interpretation of Sites 792 and 793 is mostly based on analysis of the FMS borehole electrical images on a graphic workstation, as well as on core observation and downhole measurements. When unwrapped in a depth-azimuth representation (Fig. 16), plane features intersecting the borehole appear as sinusoids. The lowest point of the sinusoid gives the dip direction and the amplitude is related to the dip angle. Figure 16 includes near-horizontal bedding planes dipping to the northwest (4223.8 mbrf), steep conjugate fractures dipping to the south (4224.2 mbrf), and a near-vertical fracture located at about N060 (4244.4 mbrf). Throughout the two holes, the fractures were observed to be mostly steep, with little conductivity contrast with the surrounding matrix. Firstly, the steep character confirms the extensional setting of the environment. Secondly, the low-conductivity contrast points toward systems of mineralized fractures. A consequence was the relative difficulty to detecting such fractures on the FMS images. Fortunately, their steep character tends to distort the current lines used to generate the images, concentrating them in

some wedges of formation (4224.3 mbrf at N330 or 4224.4 mbrf N150, for example, on Fig. 16) and leaving them absent from others. This purely geometrical effect creates zones with apparent conductivity larger than the surrounding media (darker), and other with lower apparent conductivity (lighter). As these two domains are separated by the fracture trace, this effect was often used to detect and map steep mineralized features.

Structure at ODP Site 793

Site 793 is located in the centre of the Izu–Bonin forearc sedimentary basin, 125 km to the west of the Izu–Bonin Trench, in an interchannel area of the southern side of the broad Sumisu Jima canyon (Taylor, Fujioka *et al.* 1990). Hole 793B penetrated 1404.0 m of sediments (with 74% recovery), and 278.0 m into basement (with 33% recovery). The downhole measurements are in general agreement with the core measurements, outline the lithostratigraphic boundaries derived from core analysis (Figs 2 & 4), although the measurements had to be made in stages due to the presence of several bridges requiring clearing with the drill pipe, and leading to only a partial coverage of the cored interval.

Over 1500 planar features were mapped from the FMS images in Hole 793B. Figure 17 reports a series of rose diagrams of azimuth directions for these computer-fitted planes. Bed boundaries from within the sedimentary columns (from 4150 m to 4025 m) shows a slight (about 5°) structural tilt, extremely well confined to the west (about N285; Fig. 17a). Figure 17b represents bed boundaries from within a longer interval in the sedimentary column (from 4340 to 4025 m). In the deeper section (4340 to 4150 m), the dip azimuth tends to rotate toward the NW (about N305; see also Fig. 14). Conjugated sets of steep, commonly mineralized fractures are mapped throughout the sedimentary column and in the basement (from 4025 to 4525 m). The dominant mode, oriented N025–N205 (see also Fig. 14), and the two weaker modes (oriented N115–N295 and N165–N345) are present throughout.

Fine details of microstructures related to the extensional environment under which the fore-arc basin developed are also revealed by the images (Fig. 18). FMS image of a steep (75°), SE-dipping (N120), reverse normal fault intersected in Hole 793B at 4075.4 mbrf. The (reverse) vertical offset is of the order of 5 mm, as constrained by consecutive conductive and resistive layers located on each side of the fault, as visible on one of the four images at about N210.

plot mode: ↻ Azimuth
534 dips processed

plot mode: ↻ Azimuth
902 dips processed

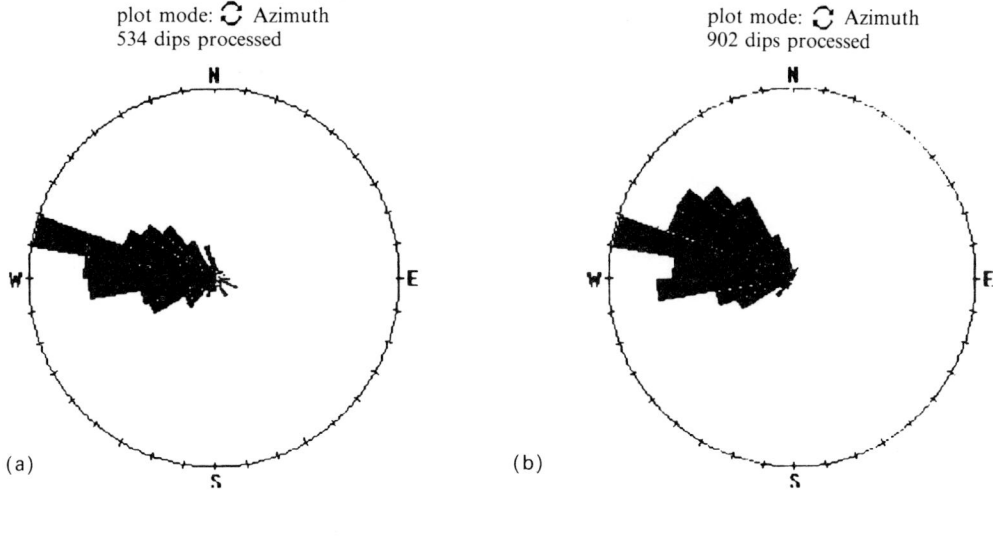

(a)

(b)

plot mode: ↻ Azimuth
549 dips processed

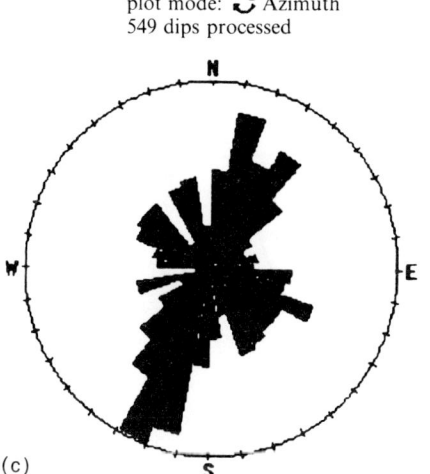

(c)

Fig. 17. Rose diagrams of azimuth directions for computer-fitted planes derived from FMS images in Hole 793B. (a) Bed boundaries from within the sedimentary columns (from 4150 m to 4025 m) show a slight (about 5°) structural tilt to the west (about N285). (b) Bed boundaries from within the sedimentary columns (from 4340 to 4025 m). In the deeper section (4340 to 4150 m), the dip azimuth tends to rotate toward the NW (about N305; see also Fig. 14). (c) Conjugated sets of steep and mostly mineralized fractures mapped both in the sedimentary columns and in basement (from 4025 to 4525 m) The dominant mode, oriented N025–N205 (see also Fig. 14), and the two weaker modes (oriented N115–N295 and N165–N345) are present throughout.

At the same depth, the image located at N300 also shows two vertical, mineralized (resistive), normal faults with a similar offset. A more complete analysis of such microstructures in cores and FMS images would be indicative of post-depositional extensional deformations to which the arc was subjected.

Structure at ODP Site 792

Site 792 is located on the western half of the Izu–Bonin forearc basin (Fig. 1), 170 km to the west of the Izu–Bonin Trench, where the forearc sediments lap onto the edge of a basement high. Hole 792E penetrated 804.0 m of sediments

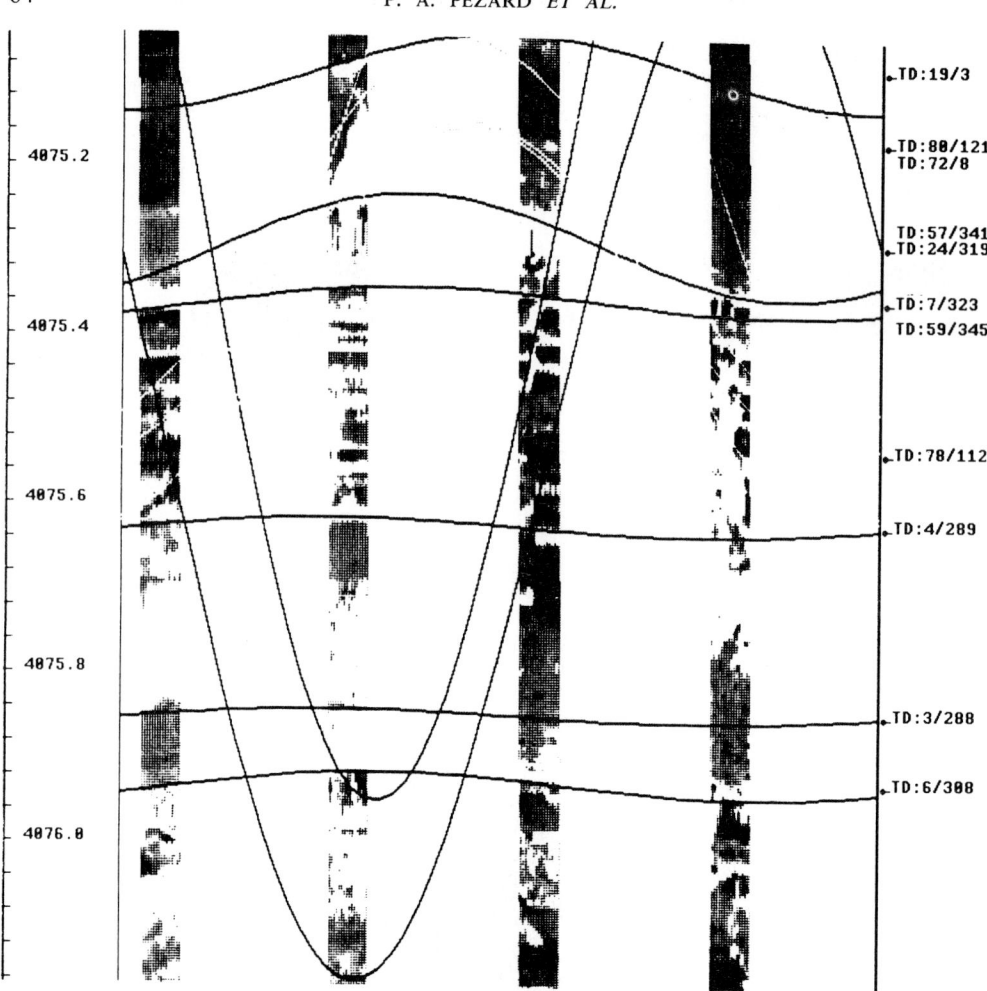

Fig. 18. FMS image of a steep (75°), SE-dipping (N120), reverse normal fault intersected in Hole 793B at 4075.4 mbrf. The (reverse) vertical offset is of the order of 5 mm, as constrained by consecutive conductive and resistive layers located on each side of the fault, as visible on one of the four images at about N210. At the same depth, the image located at N300 also shows two vertical, mineralized (resistive), normal faults with a similar offset.

(with 54% recovery), and 81.9 m into basement (with 16% recovery). The downhole measurements are in general agreement with the core measurements, outline each of the lithostratigraphic boundaries derived from core analysis (Figs 2 & 5), and confirm the presence of two fault zones which are the sites of fluid circulation traced in the temperature record (Taylor, Fujioka *et al.* 1990). One of them constitutes the sediment–basement interface at present, and the other one is located at 2200 mbrf (400 mbsf), at the boundary between Late Oligocene and Early Miocene (Fig. 2).

Over a thousand planar features were mapped from the FMS images in Hole 792E. The rose diagrams of azimuth directions for these computer-fitted planes are presented on Fig. 19. The bed boundaries from the upper part of the hole (from 2205 m to 2090 m) outline the presence of an unconformity located at 2154 mbrf (Fig. 19a). Above, the bedding dips to the east (about 15°) whilst, below, a northerly trend and smaller dip angles (about 5°) are picked up. This unconformity is also emphasized by a change in physical properties (Fig. 5), and appears to dip about 20° to the south (N200). Conjugate sets of steep and mostly mineralized fractures map throughout the sedimentary column, with two orthogonal

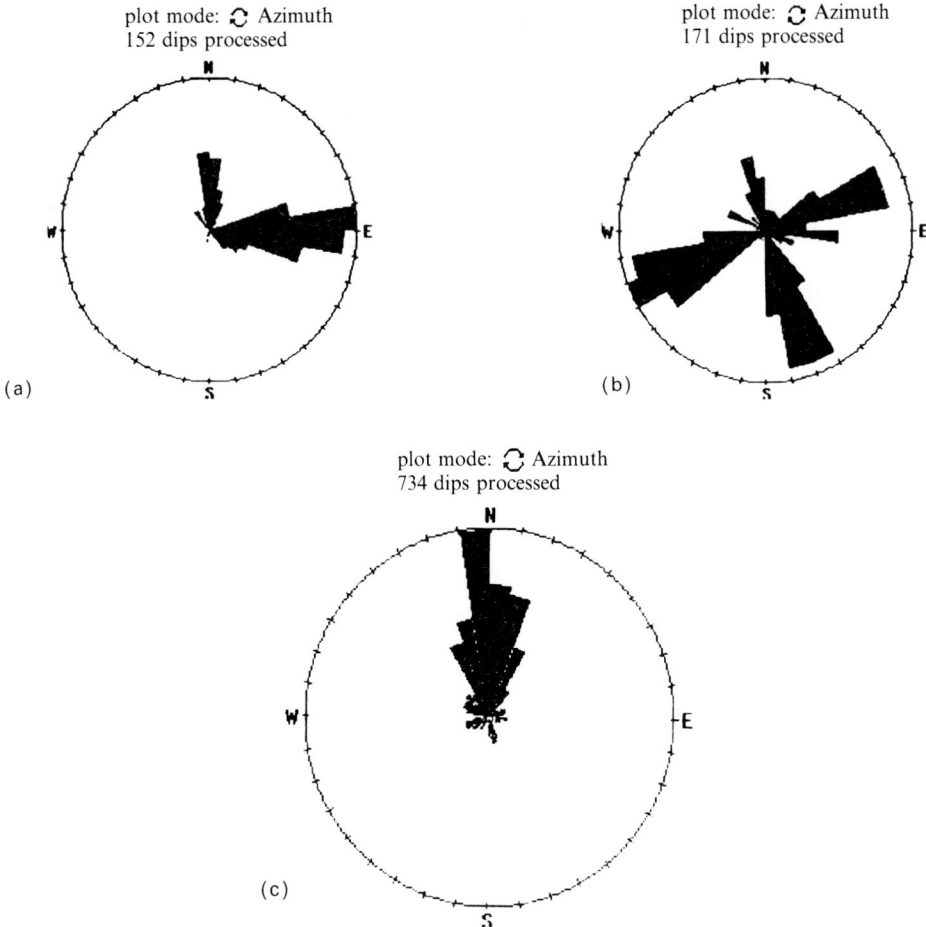

plot mode: ↻ Azimuth
152 dips processed

plot mode: ↻ Azimuth
171 dips processed

(a)

(b)

plot mode: ↻ Azimuth
734 dips processed

(c)

Fig. 19. Rose diagrams of azimuth directions for computer-fitted planes derived from FMS images in the sediments of Hole 792E. (a) The bed boundaries from the upper part of the hole (from 2205 m to 2090 m) outline the presence of an unconformity located at 2154 mbrf. Above, the bedding dips to the east (about 15°) whilst, below, a north trend and smaller dip angles (about 5°) are picked up. This unconformity is also emphasized by a change in physical properties (Fig. 5) and appears to dip about 20° to the south (N200). (b) Conjugate sets of steep and mostly mineralized fractures mapped throughout the sedimentary column. Two orthogonal modes oriented N065–N245 and N165–N345 are present. (c) The bed boundaries covering the imaged interval located below the unconformity show a regional, structural tilt of about 10° to the north.

modes oriented N065–N245 and N165–N345 (Fig. 19b). The bed boundaries covering the imaged interval located below the unconformity show a regional, structural tilt of about 10° to the north (Fig. 19c).

Present stress field orientation in the Izu–Bonin Arc

The present stress field orientation in the Izu–Bonin Arc is revealed at each of the drillsites by either the development of borehole breakouts in the direction of minimum horizontal stress, or the presence of drilling-induced vertical fractures on each side of the borehole, in the direction of maximum horizontal stress. The FMS images recorded in Holes 792E and 793B were consequently analysed for the presence of such near-vertical fractures, and comparison with borehole elongation provided by the simultaneous measurement of two orthogonal calipers. In doing this, care must be exercised as near-vertical, conductive fractures might well be oriented in the long axis of the borehole (Fig. 20). In this

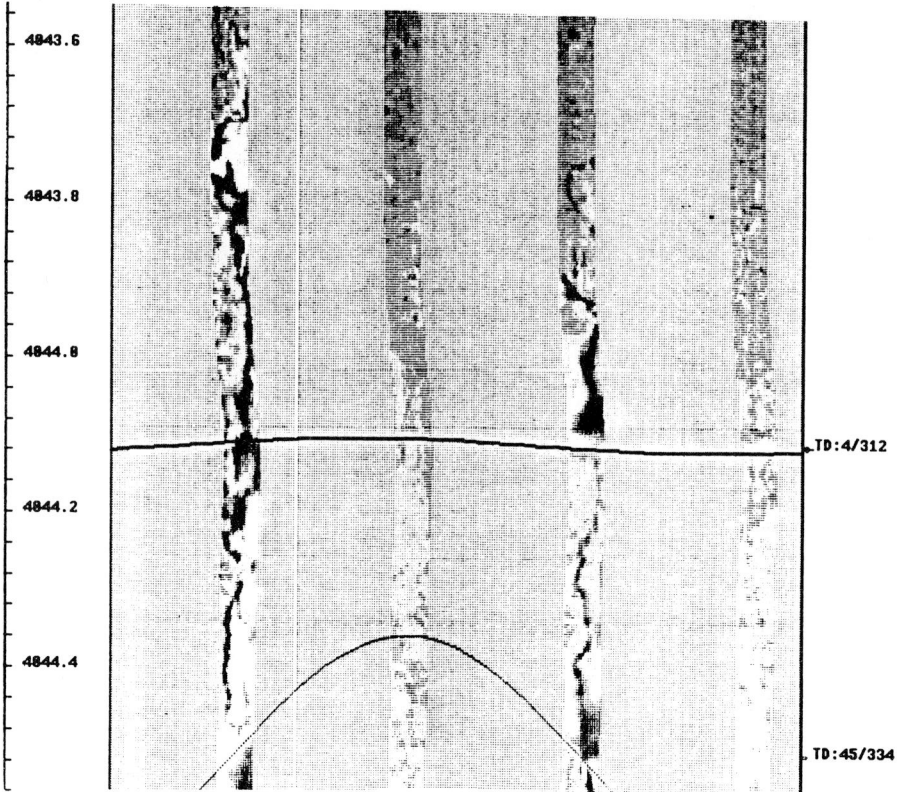

4843.6

4843.8

4844.0

TD:4/312

4844.2

4844.4

TD:45/334

Fig. 20. FMS images of near-vertical, conductive fractures oriented in the long-axis of the borehole, here slightly more elliptical (5 cm) than through the rest of the sedimentary column. These irregular, vertical features are thus either associated with the development of the elliptical shape of the borehole, or to the tangential intersection of N160–N340 vertical fractures.

case, where the borehole is slightly more elliptical (5 cm) than through the rest of the sedimentary column, these irregular, vertical features are either associated with the development of the elliptical shape of the borehole, or to the tangential intersection of N160–N340 vertical fractures. These fractures are more conductive than the surrounding matrix, but not necessarily open to fluid circulation. In the case of drilling induced fractures, a more linear vertical trace would also be expected.

Hole 793B. A consistent hole elongation toward N065 is observed in basement. In the sediment, the same direction is obtained over the lowermost 350 m, with less consistency due to the small magnitude or absence of ellipticity throughout the section.

Hole 792E. No trace of ellipticity was detected in the basement from the analysis of FMS calipers, which negates the possibility of a large stress

anisotropy in the horizontal plane at this site. A steady hole elongation oriented N020 was to the contrary observed over more than 400 m in the sedimentary section, and confirmed by the presence of numerous near-vertical conductive fractures on FMS images.

Additional evidence about the present stress field orientation is (a) given by geological evidence of east–west spreading in the southernmost part of the backarc over the last Ma (Taylor, Fujioka *et al.* 1990), now realized by the Sumisu Rift (Fig. 1b) and (b) by the observation of borehole elongation oriented N040–N220 as revealed by the borehole televiewer (BHTV, Schlumberger®) in Hole 786B. This borehole was drilled within the forearc, 75 km to the west of the Izu–Bonin trench (Fig. 1), during ODP Leg 125 (Fryer, Pearce *et al.* 1990).

A synthesis of the results of these analyses is presented in Fig. 21. The relative velocity vector between the Pacific plate and the Philippine Sea plate (oriented N020) is indicated to the east of

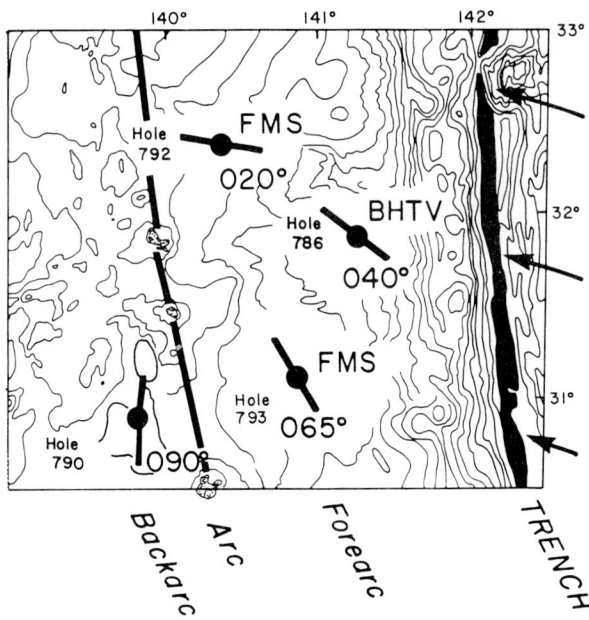

Fig. 21. Direction of the horizontal stress field orientation in the Izu–Bonin Arc, from the analysis of borehole ellipticity with BHTV acoustic images (Hole 786B), FMS electrical images (Holes 792E and 793B) in the forearc, and geological evidence from opening of the Sumisu Rift in the southern-most part of the back-arc region. The relative velocity vector between the Pacific plate and the Philippine Sea plate is indicated to the east of the trench. Whilst the stress field tends to align with this direction in the northern part of the arc (Hole 792E) and near the trench (Hole 786B), it is seen to rotate in the southern region where the back arc is spreading.

the trench. Whilst the stress field tends to align with this direction in the northern part of the arc (Hole 792E), and near the trench (Hole 786B), it is seen to rotate in the southern region where the back arc is spreading. Such an observation is coherent with the model of stress-rotation within a forearc-arc-backarc system of Kanamori & Uyeda (1980). Here also, a more detailed analysis would be needed to determine the exact stress regime active in the different parts of the arc.

Evolution of the stress in the Izu–Bonin Arc

Initiated by rifting of the arc during Oligocene times, basin development was followed by periods characterized by extensional tectonics. Koyama *et al.* (in press) showed from the analysis of cores collected during ODP Leg 126, that the Philippine Sea plate has been rotated 90° clockwise since the Oligocene. It is thus important to take this rotation into account when studying palaeostresses in the arc. For example, a near-vertical mineralized fracture was imaged by the FMS in Hole 793B (Fig. 6). It developed as the result of palaeo-extensional stresses in the conglomeratic base of a turbidite

sequence (from 127.0 to 1272.0 m), aligning with bedding when reaching the more conductive, plastic, and clay-rich beds at 127.1 m. This fracture is sealed, thus related to palaeostresses, and strikes to the west, that is in a direction about 65° away from the present direction of maximum horizontal stress at Site 793. Considering the precision of such techniques (which cannot be better than 20° in azimuth due to analytical variability), the orientation of the stress field in the Izu–Bonin Arc seems to have remained stable with respect to the trench axis since the opening of this fracture.

Conclusions

1. Formation MicroScanner (FMS) electrical images were recorded for the first time in the context of ODP during Leg 126. They permitted the construction of complete bed-by-bed sections through most of the Oligocene succession that mostly filled the Izu–Bonin forearc basin.

2. Bed thicknesses in the forearc turbidite succession are distributed according to a power law with an exponent of about 1.0. Because all beds belong to the same distribution, a single

triggering mechanism for flows of different scales is difficult to isolate from this dataset.

3. The vertical distribution of beds with different thickness is essentially random; there are few, if any, clear examples of thinning- or thickening-upward sequences. Spaced packets of thick and very thick beds may be a response to low stands of global sea level, particularly at 30.0 Ma. The absence of channels, cyclicity, and variable palaeocurrents, point to a basin-plain setting for the Oligocene succession.

4. Palaeocurrents obtained from grain fabric in thicker sandstone beds, and ripple migration directions in thinner beds, suggest a component of transport away from the eastern margin of the basin, and a secondary mode directed northward along the basin axis, for the period 30.2 to 29.5 Ma. Later, from about 28.9 to 27.3 Ma, palaeocurrent data suggest a change to predominantly south-directed axial transport, with a small component of flow away from the western margin of the basin. The direction of slope of the basin floor changed consequently from north to south between 29.0 and 29.5 Ma age. At the same time, the major source of volcaniclastic detritus apparently shifted from the vicinity of the outer arc high to that of the modern volcanic front.

5. Earthquakes are a probable triggering mechanism for sediment gravity flows in the forearc basin. The recurrence interval for earthquakes capable of causing liquefaction of sands is much shorter (a few tens of years) than that of emplacement of sandstone beds greater than 10 m thick (about 300 ka). Instead, the very thickest beds would offer supporting evidence in favour of generation by very large earthquakes (of the order of magnitude 8 or 9) that occurred close to large, unstable accumulation of sand-sized volcaniclastic material.

6. The sharp decline in sedimentation rates and increase in the abundance of mudstones, in the time period from 27.0 to 23.5 Ma, is probably the result of a proposed rise in global sea level of about 100 m at 25 Ma, possibly coupled with subsidence of the arc volcanoes as a result of extension during opening of the Shikoku Basin.

7. Initiated by rifting of the arc during Oligocene times, basin development was followed by continuing extensional tectonics. Post-depositional extensional deformations such as normal microfaults, conjugate high-angle fractures, clastic injections, and dewatering veinlets are identified both in the core and on FMS images. Present-day orientations of the stress field within the arc and forearc obtained at depth in several holes from the analysis of hole ellipticity confirm models of stress distribution in forearc–arc–back arc regions. In particular, a rotation of the maximum horizontal stress trajectory in the overlying plate is observed, in a direction orthogonal to the plate boundary. In spite of a 90° clockwise rotation of the Philippine Sea plate since Oligocene time, the orientation of the stress field seems to have remained stable with respect to the trench axis over this period.

Without the energetic efforts of Roger Anderson, Borehole Research Group of Lamont-Doherty Geological Observatory, the Formation MicroScanner would probably not have been available for Leg 126 of ODP. We heartily thank him for making this research possible. Measurement of features dip and orientation could only be done on the Schlumberger workstation. Sal Gallegos, Schlumberger, Dallas, Texas, provided us with inexpensive workstation time and personal assistance outside normal working hours, for which we are particularly thankful.

References

BOUMA, A., H. 1962. *Sedimentology of some Flysch Deposits: a Graphic Approach to Facies Interpretation.* Elsevier, Amsterdam.

CLOETINGH, S., 1986. Tectonics of passive margins: implications for the stratigraphic record. *Geologie en Mijnbouw*, **65**, 103–117.

EKSTROM, M. P. *et al.* 1986. Formation imaging with microelectrical scanning arrays. *Transactions of SPWLA*, Paper 88.

FRYER, P., PEARCE, J. *et al.* 1990. *Proceedings of the Ocean Drilling Program, Initial Reports.* Ocean Drilling Program, College Station, TX, **125**.

FUJIOKA, K. & SAITO, S., 1992. Heavy minerals and glasses of the sandstones of the Izu–Bonin Arc., *Proceedings of the Ocean Drilling Program, Scientific Results.* Ocean Drilling Program, College Station, TX, **126**.

HAQ, B. U., HARDENBOL, J. & VAIL, P. R. 1987. Chronology of fluctuating sea levels since the Triassic. *Science*, **235**, 1156–1167.

HISCOTT, R. N., PICKERING, K. T. & BEEDEN, D. R., 1986. Progressive filling of a confined Middle Ordovician foreland basin associated with the Taconic Orogeny, Quebec, Canada. *In*: ALLEN, P. A. & HOMEWOOD, P. (eds) *Foreland Basin.* International Association of Sedimentologists, Special Publication, 8. Blackwell, Oxford, 309–325.

—— & GILL, J. 1992. Volcaniclastic sand and sandstone chemistry as a record of magma compositions, Oligocene to Holocene Izu–Bonin intraoceanic arc, *Proceedings of the Ocean Drilling Program, Scientific Results.* Ocean Drilling Program, College Station, TX, **126**.

——, COLLELA, A., PEZARD, P. A. & LOVELL, M. A, 1992. Deep-water volcaniclastic succession and turbidity current pathways, Oligocene Izu–Bonin forearc basin, based on the FMS. *Proceedings of the Ocean Drilling Program, Scientific Results.*

Ocean Drilling Program, College Station, TX, **126**.

HONZA, E. & TAMAZAKI, K. 1985. The Bonin Arc. *In*: NAIRN, A. E. M., STEHLI, F. G. & UYEDA, S. (eds.) *The Ocean Basins and Margins, Volume 7A*. Plenum, New York, 459–502.

HSÜ, K. J. 1983. Actualistic catastrophism: address of the retiring President of the International Association of Sedimentologists. *Sedimentology*, **30**, 3–9.

HUSSONG, D. M. & UYEDA, S., 1981. Tectonic processes and the history of the Mariana Arc: a synthesis of the results of the Deep Sea Drilling Project Leg 60. *In*: HUSSONG, D. M., UYEDA, S. *et al. Initial Reports DSDP, 60*, U.S. Govt. Printing Office, Washington, 909–929.

INGRAM, R. L., 1954. Terminology for the thickness of stratification units and parting units in sedimentary rocks. *Bulletin of the Geological Society of America*, **65**, 937–938.

KAIHO, K. 1992. Oligocene to Quaternary benthic foraminifers and paleobathymetry of the Izu–Bonin Arc, Leg 126, *Proceedings of the Ocean Drilling Program, Scientific Results*. Ocean Drilling Program, College Station, TX, **126**.

KANAMORI, K. & UYEDA, S. 1980. Stress gradient in arc-back-arc regions and plate subduction. *Journal of Geophysical Research*, **85**, 6419–6428.

KARIG, D. E., INGLE, J. C. *et al.* 1975. *Initial Reports DSDP*, 31, U.S. Govt. Printing Office, Washington.

KEEFER, D. K. 1984. Landslides caused by earthquakes. *Bulletin of the Geological Society of America*, **95**, 406–421.

KLAUS, A. & TAYLOR, B. 1992. Submarine canyon development in the Izu–Bonin forearc: a seaMarc II and seismic survey of Aoga Shima canyon, in press.

KOBAYASHI, K. & NAKADA, M. 1979. Magnetic anomalies and tectonic evolution of the Shikoku interarc basin. *In*: UYEDA, S. *et al.* (eds) Geodynamics of the western Pacific. *Advances in Earth and Planetary Sciences, Tokyo (Japan Sci. Soc. Pr.)*, **6**, 391–402.

KOYAMA, M., CISOWSKI, S., HASTON, R. & PEZARD, P. A. 1992. Paleomagnetic evidence for northward drift and clockwise rotation of the Izu–Bonin Arc since the Oligocene. *Proceedings of the Ocean Drilling Program, Scientific Results*. Ocean Drilling Program, College Station, TX, **126**.

Leg 126 Shipboard Scientific Party. 1989. ODP Leg 126 drills the Izu–Bonin Arc. *Geotimes*, **34**, 10, 36–38.

LUTHI, S. M. & BANAVAR, J. R., 1988. Application of borehole images to three-dimensional geometric modeling of eolian sandstone reservoirs, Permian Rotliegende, North Sea. *Bulletin of the Geological Society of America*, **72**, 1074–1089.

MANDELBROT, B. B., 1983. *The Fractal Geometry of Nature*, W. H. Freeman, New York.

MATSUDA, J., 1985. Sr isotopic studies of rocks from the Philippine Sea and some implications for the mantle material. *In*: SHIKI, T. (ed.) *Geology of the Northern Philippine Sea*. Tokai University Press, Tokyo, 63–78.

MOGI, K., 1990. Seismicity before and after large shallow earthquakes around the Japanese islands. *Tectonophysics*, **175**, 1–34.

MOORE, J. G., CLAGUE, D. A., HOLCOMB, R. T., LIPMAN, P. W., NORMARK, W. R., and TORRESAN, M. E., 1989. Prodigious submarine landslides on the Hawaiian Ridge. *Journal of Geophysical Research*, **94**, 17465–17484.

MUTTI, E. 1985. Turbidite systems and their relations to depositional sequences. *In*: ZUFFA, G. G. (ed.) *Provenance of Arenites*. Reidel, Dordrecht, 65–93.

—— & GUIBAUDO, G. 1972. Un esempio di torbiditi di conoide sottomarina esterna: le Arenie di San Salvatore (Formazione di Bobbio, Miocene) nell'Appennino di Piacenza. *Mem. Acc. Sci. Torino Classe Sci. Fis. Nat.*, Series 4, 16.

PARKASH, B. & MIDDLETON, G. V. 1970. Downcurrent textural changes in Ordovician turbidite greywackes. *Sedimentology*, **14**, 259–293.

PEZARD, P. A., LOVELL, M. A., *et al.* 1990. Downhole images: electrical scanning reveals the nature of the ocean crust, *EOS*, 709.

PICKERING, K. T. 1982. The shape of deep-water siliciclastic systems: a discussion. *GeoMarine Letters*, **2**, 41–46.

RICCI LUCCHI, F. & VALMORI, E., 1980. Basin-wide turbidites in a Miocene, over-supplied deep-sea plain: a geometrical analysis. *Sedimentology*, **27**, 241–270.

TAYLOR, B., FUJIOKA, K., *et al.* 1990. *Proceedings of the Ocean Drilling Program, Initial Reports*. Ocean Drilling Program, College Station, TX, **126**.

——, MOORE, G., KLAUSE, A., SYSTROM, M., COOPER, P. & MACKAY, M., 1990. Multichannel seismic survey of the central Izu-Bonin Arc. *In*: TAYLOR., B., FUJIOKA, K. *et al.* (eds). *Proceedings of the Ocean Drilling Program, Scientific Results*, Ocean Drilling Program, College Station, TX, **126**, 51–60.

TAYLOR, R., LAPIERRE, H., VIDAL, P. & CROUDACE, I., 1992. Volcanic stratigraphy of the Izu–Bonin forearc basement. *Proceedings of the Ocean Drilling Program, Scientific Results*, Ocean Drilling Program, College Station, TX, **126**.

TURCOTTE, D. L., 1989. Fractals in geology and geophysics. *Pure and Applied Geophysics*, **13**, 171–196.

WALKER, R. G., 1975. Generalized facies model for re-sedimented conglomerates of turbidite association. *Bulletin of the Geological Society of America*, **86**, 737–748.

WALLMAN, P. C., MAHOOD, G. A., & POLLARD, D. D. 1988. Mechanical models for correlation of ring-fracture eruption at Pantelleria, Strait of Sicily, with glacial sea-level drawdown. *Bulletin of Volcanology*, **50**, 327–339.

YUASA, M., UCHIUMI, S., NISHIMURA, A. & SHIBATA, K. 1988. K/Ar age of a forearc seamount adjacent to the volcanic front of the Izu–Ogasawara Arc (abstract). *Program and Abstracts. Volcanology Society of Japan*, **2**, 63.

Application of FMS images in poorly recovered coring intervals: examples from ODP Leg 129

A. R. M. SALIMULLAH & D. A. V. STOW

Geology Department, The University, Southampton SO9 5NH, UK

Abstract. A thick sedimentary section was penetrated at three sites in the West Central Pacific during the Ocean Driling Program (ODP) Leg 129. Although average core recovery ranged from 17.3 to 29.5%, recovery in some intervals falls as low as 0–5%. Furthermore, the recovered portion of the core is conventionally assigned to the topmost part of the cored interval, although this is now known to be incorrect in most instances in consolidated parts of the section. These drawbacks have led to incomplete and even erroneous sedimentological interpretations. High-resolution electrical (Formation MicroScanner) images of the bore-hole obtained by scanning the borehole wall with arrays of small electrodes pressed against the borehole surface can provide:

(1) detailed sedimentary structure and texture of missing sections of each core, through calibration with visual data obtained from recovered cores;

(2) the original sedimentary features, where these have been disturbed or brecciated during drilling, as well as drilling artifacts on the borehole wall;

(3) recognition of sedimentary facies (e.g. slump units in hole 801B) in non-recovered intervals;

(4) correct location of recovered portions within the cored interval, through matching of specific sedimentary features.

Examples of these applications of the FMS images to the poorly recovered ODP Leg 129 cores are presented. The use of FMS logging has widespread application in any borehole where conventional cores have not been taken or where core recovery is limited. However, log calibration with core samples is essential.

Three sites were drilled during the Ocean Drilling Program (ODP) Leg 129; Sites 800 and 801 in Pigafetta Basin and Site 802 in East Mariana Basin (Fig. 1). A thick sedimentary section was penetrated at each of the sites and the summary lithostratigraphy (Shipboard Scientific Party 1990) is shown in Fig. 2. For this study, the Cretaceous and Mio-Pliocene volcaniclastic Units of sites 801 and 802 are considered. These were deposited in a typical mid-ocean plate

setting in an area surrounded by volcanic sea-mounts that showed intense activity at different periods of ocean development. The full range of resedimentation processes, including slumps,

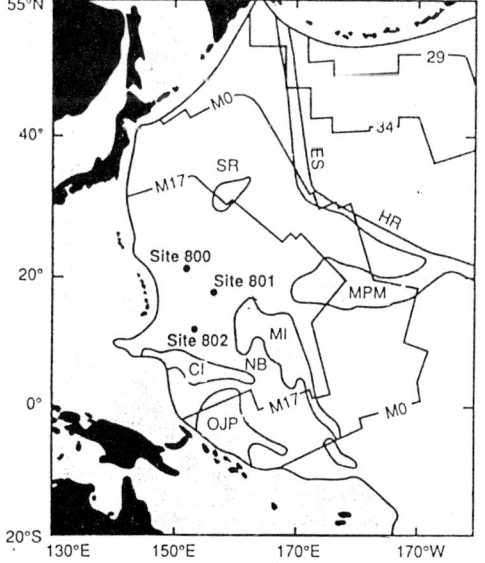

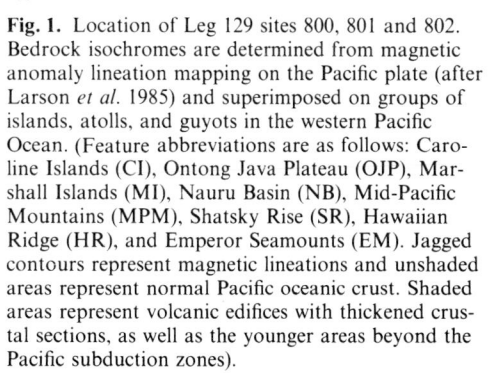

Fig. 1. Location of Leg 129 sites 800, 801 and 802. Bedrock isochromes are determined from magnetic anomaly lineation mapping on the Pacific plate (after Larson *et al.* 1985) and superimposed on groups of islands, atolls, and guyots in the western Pacific Ocean. (Feature abbreviations are as follows: Caroline Islands (CI), Ontong Java Plateau (OJP), Marshall Islands (MI), Nauru Basin (NB), Mid-Pacific Mountains (MPM), Shatsky Rise (SR), Hawaiian Ridge (HR), and Emperor Seamounts (EM). Jagged contours represent magnetic lineations and unshaded areas represent normal Pacific oceanic crust. Shaded areas represent volcanic edifices with thickened crustal sections, as well as the younger areas beyond the Pacific subduction zones).

From Hurst, A., Griffiths, C. M. & Worthington, P. F. (eds), 1992, *Geological Applications of Wireline Logs II.* Geological Society Special Publication No. 65, pp. 71–86.

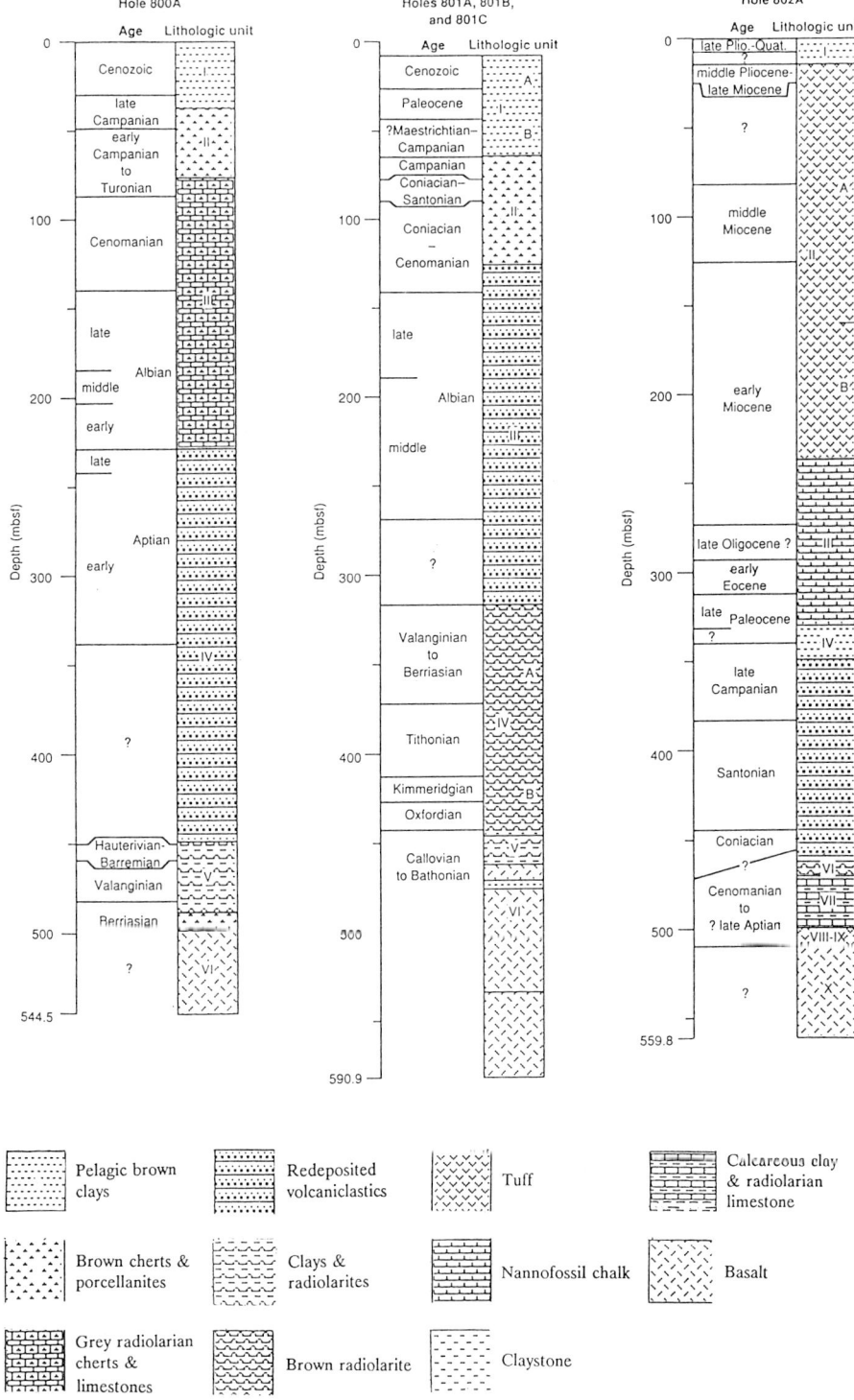

Fig. 2. Summary lithostratigraphic columns for ODP Leg 129 (from Shipboard Scientific Party 1990).

debris flows and turbidity currents (Salimullah & Stow, this volume) were responsible for the reworking of material from the flanks of seamounts and their input into the adjacent basins. Background pelagic sedimentation was mainly radiolarian-rich in the Pigafetta Basin and nannofossil or clay-rich in the East Mariana Basin.

As is commonly the case with unconsolidated cohesionless sediments, core recovery was very poor, as little as 0–5% in some intervals. The borehole walls have most probably suffered disruption as a result of caving and washout, and portions of cores that were recovered are typically abraded, fragmented and brecciated due to drilling disturbance. However, a good suite of logs was taken at two of the sites, and the specific objective of this study is to demonstrate how the FMS images can be used to:

(a) delineate detailed sedimentary features in intervals with poor core recovery;

(b) delineate the original sedimentary features where these have been abraded or brecciated during drilling;

(c) recognize sedimentary facies in non-recovered coring intervals;

(d) assign the correct location of short-length recovered portions within the known cored interval.

FMS tool and image processing

The FMS tool (Fig. 3) is capable of producing high-resolution borehole images from electrical conductivity measurements. Although it has been used in the petroleum industry since 1986, its use in the ODP has been constrained by the narrow internal diameter (10.48 cm) of the drill pipe used on the *Joides Resolution* drillship (Shipboard Scientific Party 1990). Consequently, a modified sensor was developed by Schlumberger for ODP use. This sensor was designed in such a way that four images (of 16 traces each), instead of two images obtained in the original version of Schlumberger (Ekstrom *et al.* 1986) are recorded simultaneously (Shipboard Scientific Party 1989).

The button current intensity (raw data points) is sampled every 2.5 mm (Serra 1989) by the FMS. This contrasts markedly with most conventional downhole measurements that are usually sampled at 150 mm; hence the sampling rate of the FMS is 60 times larger than that of most other logging devices.

Once the data have been acquired in the borehole, the images are processed and enhanced (for details see Serra 1989). Various image enhancement methods can be applied. Different sizes of sliding window are used to improve the local contrast of an image in order to delineate finer details of an event (Serra 1989) such as distribution of clasts (larger than 1 cm) in beds of debrites. For this study, dynamic and 'hilite' normalization images are used. All the 'hilite' images were processed with a 20 cm sliding window and the dynamic images with 10 and 30 cm windows (sliding window size is mentioned for each image in the respective figure caption).

The FMS has an imaging resolution of the order of a few millimetres in both vertical and azimuthal directions (Ekstrom *et al.* 1986). Sedimentary features as small as 1.2 cm can therefore be resolved on the FMS images (Harker *et al.* 1990).

Image interpretation approach and grading

The poor/non-recovery of cores as well as degree of disturbance of recovered-cores and of the borehole wall due to drilling, makes the direct calibration of FMS images difficult. Nevertheless, the recovered material does provide some information on sediment texture, structure, composition, grain density, porosity and water content with which to interpret the images as well as to interpolate and/or extrapolate that interpretation to poorly recovered intervals. In these cases, it is not possible to ascertain the primary control(s) on the scale of grey levels of the images. However, the situation is far better in well-recovered coring intervals where direct calibration of images reveals that the grey scale of the images is largely controlled by grain size. In other words, the image becomes increasingly darker grey with an increase in the amount of clay-size material, as is also observed in siliciclastic rocks (Serra 1989). The change in grain size through a graded bed, for example, is typically accompanied by a slight change in composition (mineralogy), water content (salinity) and permeability, but we are confident that the resistivity contrasts of the images are best interpreted principally as a grain-size phenomenon.

The image interpretation approach adopted here differs in some respects from previously published approaches (e.g. Serra 1989; Harker *et al.* 1990). We have extended Serra's (1989) Grade 1–3 categories of interpretation by adding grades 4 and 5 (Table 1), and shown examples of these in the following section.

Application to ODP Leg 129

The value of the FMS as an aid to the interpretation of poorly-recovered coring intervals is

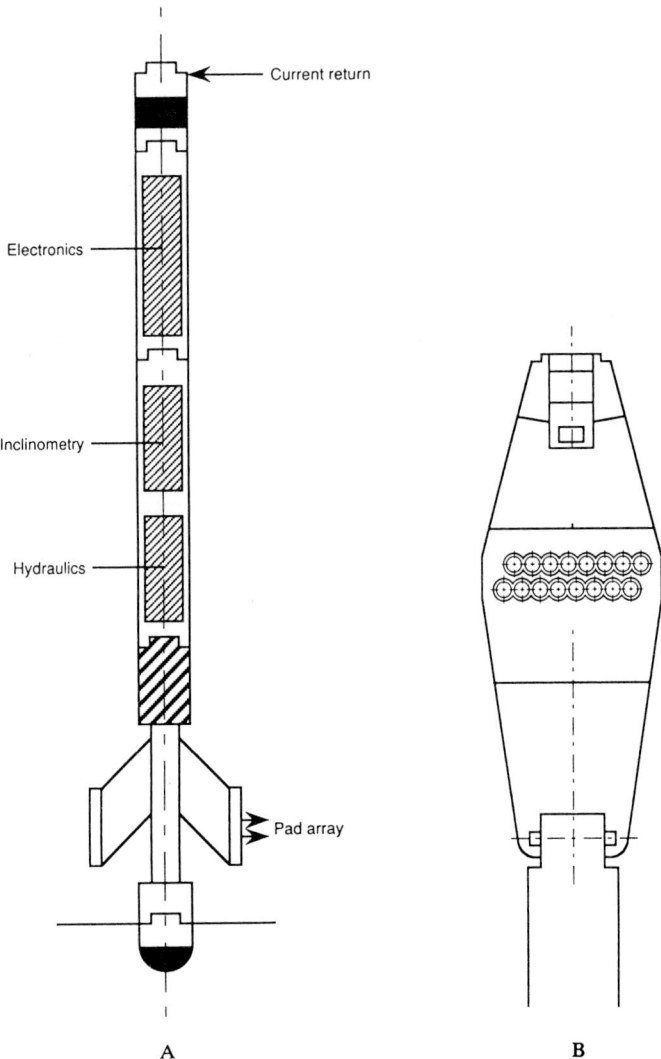

A   B

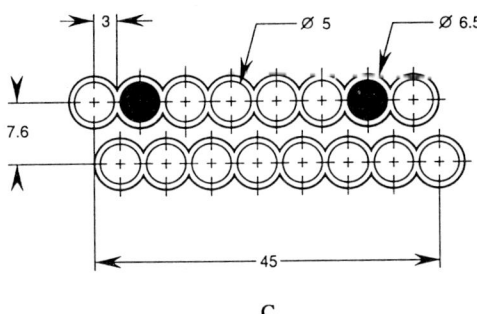

C

Fig. 3. FMS tool (from Shipboard Scientific Party, 1989): A, sketch of the Formation MicroScanner; B, sketch of an FMS pad with a 16-electrode array; C, geometry of the FMS array. Dimensions are in millimetres.

Table 1. *Categories of FMS image interpretation proposed in this study (modified from Serra 1989)*

Grade 1 (self-interpretable)	*Unique features, interpretable by themselves* beds, lamination, cross-lamination, grain-size and shape, grading, erosional surfaces, load clasts, slumps, microfolds, faults, burrows ($\pm$ bioturbation)	*Application in this study* most of these unique features recognized in non-recovered, poorly-recovered and disturbed intervals
Grade 2 (ambiguous)	*Ambiguous features, interpretable with the help of other log data* distinction between conglomerate/breccia pebbles, mud-shale clasts, concretions and bioturbation clarification of indistinct bedding, lamination and grading	*Application in this study* although not discussed in this paper, dipmeter resistivity logs have been found the most useful for identifying different facies, with the aid of FMS images, in Leg 129 sites where no core exists
Grade 3 (unclear)	*Unclear features, interpretable with the help of core data from adjacent intervals* clarification of features listed in Grade 1 above where they are indistinct, weakly developed or on a very fine scale distinction between pebbles, mud clasts, concretions and bioturbation (as for Grade 2 images) identification of limestone textures and stylolites	*Application in this study* interpretation of bioturbation, grading and slumping has been made in non-recovered and disturbed intervals
Grade 4 (unclear)	*Unclear features, interpretable with the help of core data/image calibration from a remote interval* as for Grade 3 above, but interpretation less certain because of lack of core data from adjacent intervals	*Application in this study* interpretation of graded bedding, graded stratification and slumping has been made in disturbed and non-recovered intervals
Grade 5 (interpretable by association)	*Ambiguous features, interpretable by their close association with Grade 1 images* as for Grade 3 above, but interpretation by association with adjacent self-interpretable features	*Application in this study* interpretation of grading, bioturbation, slumping and debris flows has been made in poorly-recovered, disturbed and non-recovered intervals

illustrated with reference to four specific problems from ODP Leg 129.

Poorly-recovered coring interval

This coring interval (core 129-802A-20R, 168.3–177.6 mbsf) is 9.7 m thick but core recovery is only 2.9%. The recovered sediments are mainly composed of fine sand to clay size volcanic glass and smectite/chloritic clays, with an average grain density of 2.82 g/cm^3, 44.6% porosity and 22.4% water content. From these few recovered pieces of bioturbated mud/sandstone (Fig. 4), it is not possible to reconstruct sedimentary facies, features and the type of bedding sequence for the entire coring interval. However, the FMS images over a part of this interval (Fig. 5) show the presence of three more or less distinct beds. The lowermost bed has a sharp, slightly irregular base, and grades upwards from light to very dark grey. A mottled or patchy aspect is evident in the top two-thirds of the bed. These features are most readily interpreted as representing a graded and deeply bioturbated turbidite (Serra 1989; Harker *et al.* 1990).

The middle bed appears to have a sharp planar base (although there seems to be some

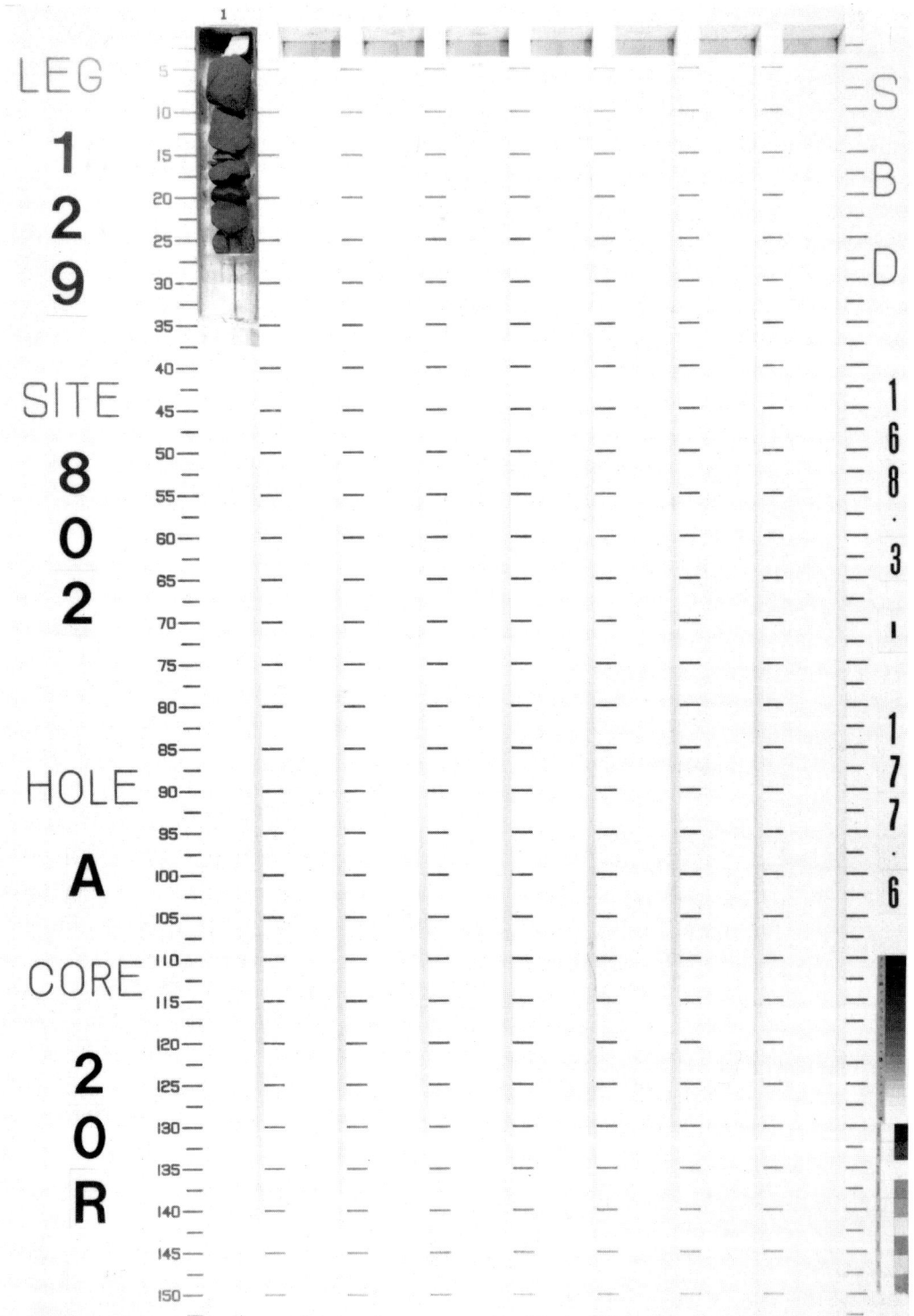

Fig. 4. Core photograph of the poorly-recovered coring interval (core 129-802A-20R).

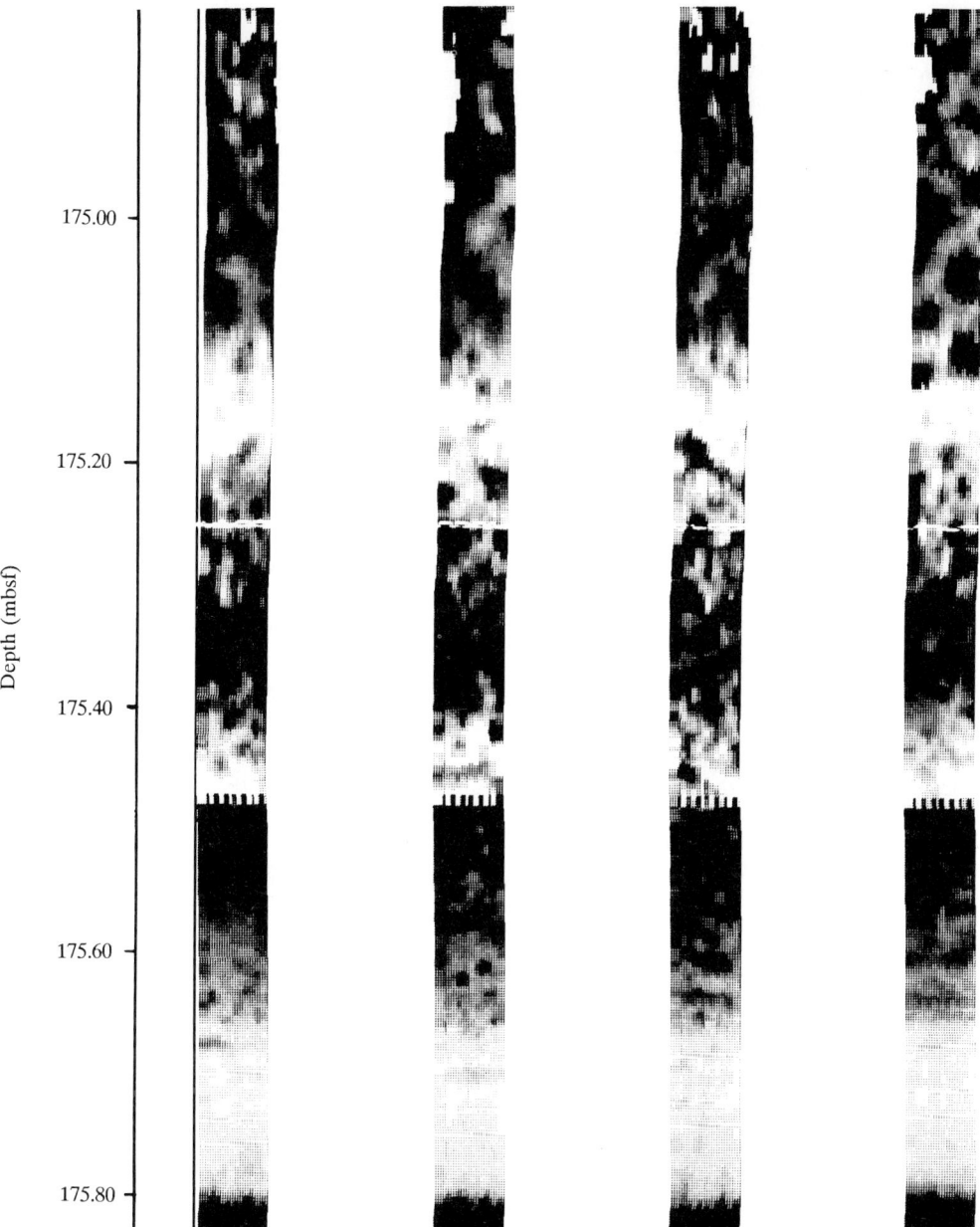

Fig. 5. Characteristic four-pad FMS images (scale 1:5, dynamic normalization, window = 30 cm) of turbidite beds encountered within the coring interval as mentioned in Fig. 4.

image artifact at this point), and then grades upwards from white/light grey to very dark grey. There is then a mottled and/or gradational contact with the overlying 'bed', which again grades upwards from white to very dark grey. The apparent oscillation-type grading and continu-

ous mottling (bioturbation) through at least 1 m of section, superimposed upon episodic (turbiditic) supply, suggests that the uppermost bed and perhaps, the middle bed are best interpreted as hemiturbidites (Stow & Wetzel 1990). These are believed to be characteristic of distal turbidite/

basinal settings where the very slow settling of pelagic and turbiditic material allows for bioturbation to continue throughout deposition. This is periodically interrupted by the arrival of a more distinct turbidite unit.

The distinct graded to bioturbated bed at the base of image in Fig. 5, we believe, is a unique feature interpretable by itself, especially with prior knowledge of the local geological setting, and we would consider it a Grade 1 feature. This bed is one of many similar graded-to-bioturbated beds (17–115 cm thick) throughout the 9.7 m coring interval (Fig. 6). These have also been interpreted directly from the FMS images (Grade 1). The very limited core recovery shows clear bioturbation, so that interpretation of the bioturbated aspect throughout and the hemiturbidite at the top and middle of Fig. 5 would be considered a Grade 3 feature.

Coring interval showing drilling-induced disturbance

The coring interval (129-802A-14R) is 7.7 m thick with a moderately good recovery of 43.77%. However, the recovered portions (Fig. 7) are badly abraded by drilling, so that the sedimentary features are largely obscured. The sediments are mainly medium- to coarse-grained volcaniclastic sands with some granule-size clasts (mudclasts) and having a grain density of 2.65 g/cm³, 31.3% porosity and 14.9% water content. Some sedimentary features are visible including stratification and a sharp bedding contact overlain by subtle graded bedding (Fig. 7), but our interpretation of the non-recovered interval has relied mainly on the calibration of FMS images with cores from other parts of the borehole, in cases where the images are not Grade 1 or 2.

Within this interval we can identify five main types of images or image facies (Fig. 8), which can be interpreted in terms of sedimentary structures. These are as follows:

(a) Light grey becoming progressively darker upwards (the vertical lines are an image artifact) (Fig. 8A). This is a rather subtle or unclear resistivity change which we interpret as graded bedding, at the base of a 1 m thick turbidite, on the basis of core/image calibration from another part of the borehole. The interpretation can therefore be assigned to a Grade 4 degree of confidence.

(b) Interlayered dark and light grey bands showing both gradational and mottled contacts (Fig. 8B). There would appear to be two possible interpretations of this feature: either it represents

a series of thin–medium weakly graded beds with some bioturbation, or it is part of a coarsely-stratified, very thick-bedded turbidite in which the mottling is probably due to granules or cobbles and the stratification to grain size variation. We favour the latter interpretation (Grade 4 degree of confidence), partly because it is possible to recognize the very thick graded bed on the FMS images.

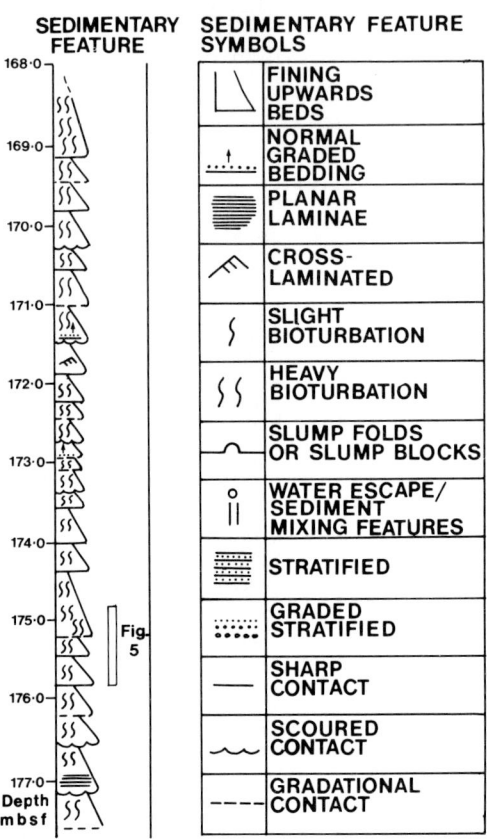

Fig. 6. Vertical sequence of sedimentary features of the coring interval as mentioned in Fig. 4.

(c) Interlayered, distinct dark and light grey bands with irregular contacts; plus some indistinct banding (Fig. 8C). This is best interpreted as a coarse stratification, rather than thin bedding, and is closely comparable to the centimetric stratification observed in the recovered core (Fig. 7, section 1, 62–67 cm), hence giving a Grade 3 degree of confidence to the interpretation. It lies just above the section illustrated in Fig. 8B, forming the upper part of the same thick graded bed.

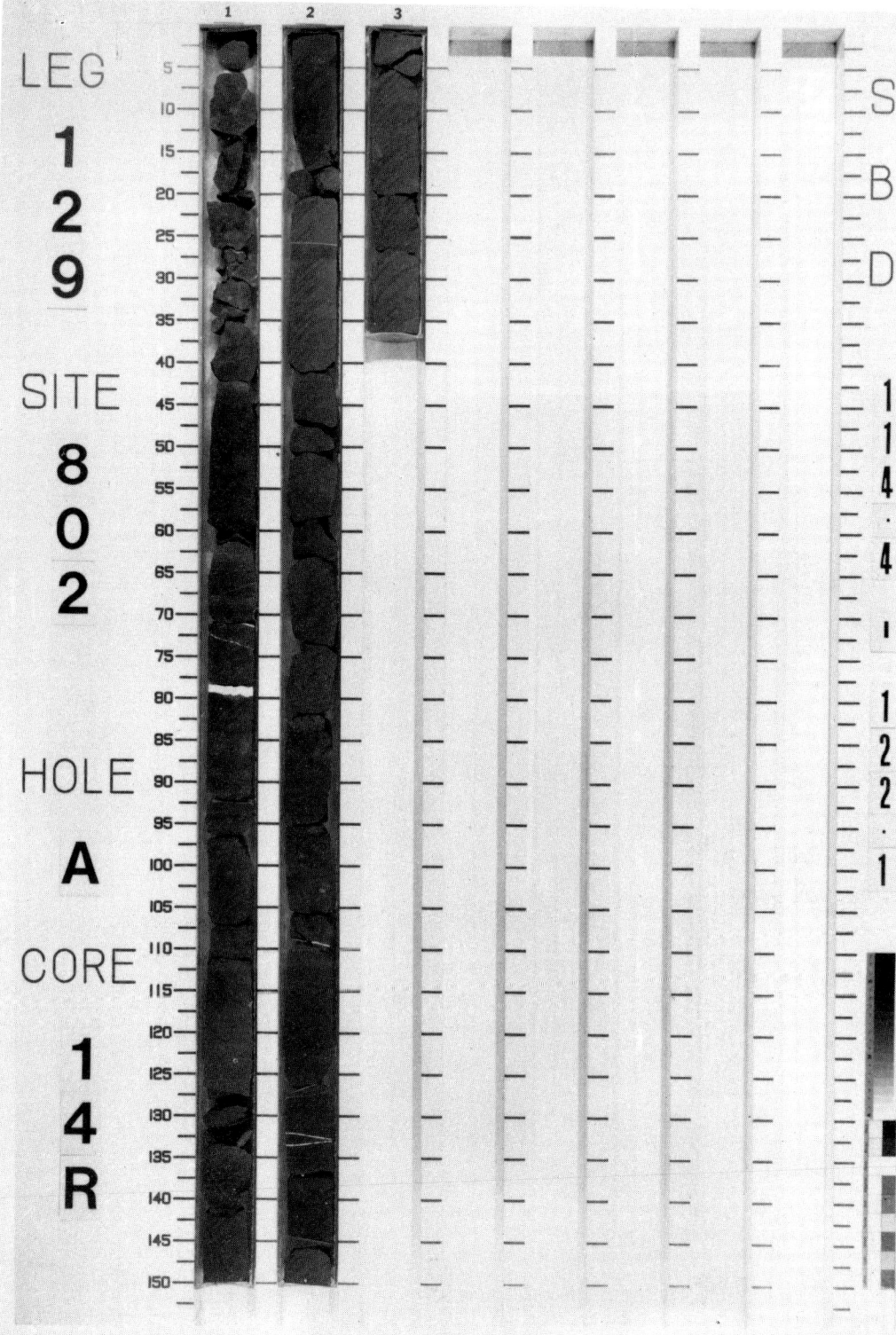

Fig. 7. Core photograph of the disturbed (drilling induced) coring interval (core 129-802A-14R).

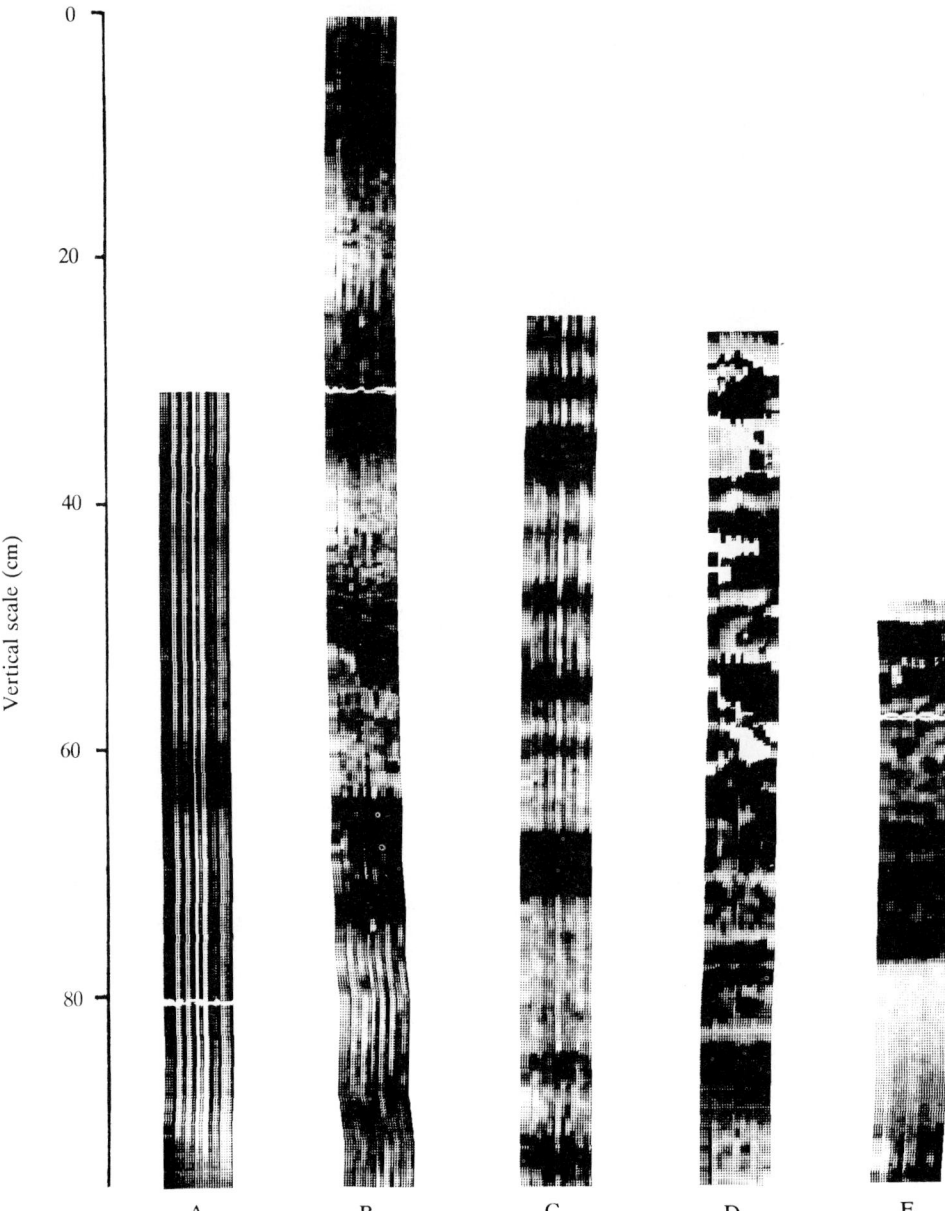

Fig. 8. Characteristic FMS images (scale 1:5) of five main types of image facies, which can be interpreted in terms of sedimentary structures, encountered within the coring interval as mentioned in Fig. 7: 8A ('hilite' normalization, window = 20 cm), light grey becoming progressively darker (graded bedding); 8B (dynamic normalization, window = 30 cm), interlayered dark and light grey bands with gradational and mottled contacts (graded stratification); 8C (dynamic normalization, window = 30 cm), interlayered, distinct dark and light grey bands with irregular contacts (coarse stratification); 8D ('hilite' normalization, window = 20 cm), stratified and cross-stratified interlayering of dark and light grey bands (parallel and cross-stratification); 8E (dynamic normalization, window = 30 cm), mottled aspect near top of section (bioturbation).

(d) Stratified and cross stratified interlayering of dark and light grey bands (Fig. 8D). This is readily interpreted as parallel and cross stratification within a thick bed, and with a Grade 1 degree of confidence.

(e) Mottled aspect near top of section (Fig. 8E). This appears to be part of a graded (lower part) to indistinctly laminated (middle part) to bioturbated (mottled upper part) turbidite bed, so we therefore interpret the mottling as bioturbation from its association with adjacent images (Grade 5 degree of confidence).

The interpreted sequence through the full 7.7 m coring interval is shown in Fig. 9, together with the location of the image facies illustrated in Fig. 8.

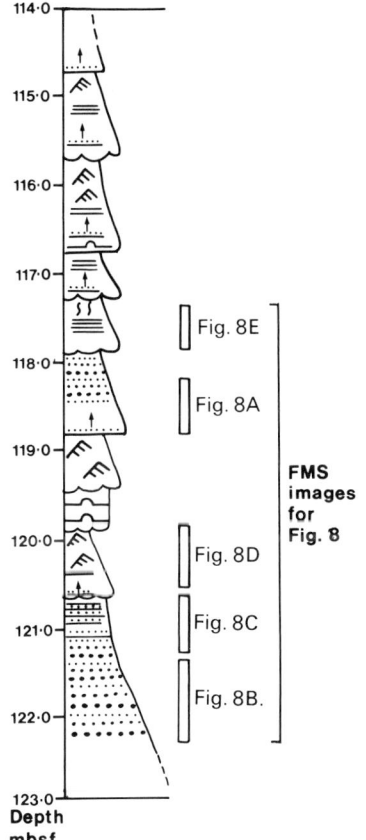

Fig. 9. The interpreted sequence (for key see Fig. 6) of the coring interval as mentioned in Fig. 7, together with the location of the image facies illustrated in Fig. 8.

Non-recovered coring interval

The coring interval 129-801B-9R (262.3–272.0 mbsf) has zero recovery over a section 9.7 m thick. The FMS images from this interval can be divided into three image facies on the basis of the geometry of resistivity events and texture of the images, and these can be interpreted in terms of sedimentary structures representing three different sedimentary facies.

(a) Interlayered light and dark grey bands, variably inclined, lenticular and with evidence of folding (Fig. 10A). The inclined layering and microfolds (recumbent) are clearly interpretable in terms of a slump unit, showing considerable internal contortion (a Grade 1 feature).

(b) Irregular light and dark grey bands, with possible evidence of folding (Fig. 10B). The microfolds together with a somewhat chaotic sediment mixing suggest almost certainly a slump unit, showing less internal contortion than (a) above (Grade 1). However, if we are more cautious in applying a direct image interpretation, we might assign it a Grade 3/4 (comparison with other cored sections) or Grade 5 (association with Grade 1 images) category.

(c) Irregular mottling of light and dark grey (Fig. 10C). Such mottling might be caused by pebbles within a conglomerate, mud clasts, concretions or bioturbation. In this case, however, we favour the complete disruption of interbedded sand and mud units (light and dark greys) and transport/deposition by debris flow, so that the mottling is interpreted as variable lithologies/clasts within a debrite. The larger patches towards the top of the image (Fig. 10C) may represent either large clasts or a gradational transition to an overlying (?) slumped unit. This interpretation is made both by reference to core data from adjacent intervals (Grade 3 feature) and by association with Grade 1 images of slumps (Grade 5 feature).

The full coring interval is shown interpreted in Fig. 11. The gradational contacts between slump and debrite units may indicate the transformation of slumping into debris flow processes.

Location of cores in well-recovered coring intervals

The cored interval 129-801B-1R has a recovery of about 67.73% over 9.5 m of section. From the core photograph (Fig. 12) it can be seen that the interval has encountered two sedimentary facies: one (section 1, 186.0–186.80 mbsf) consists of laminated sandstones and the other (section

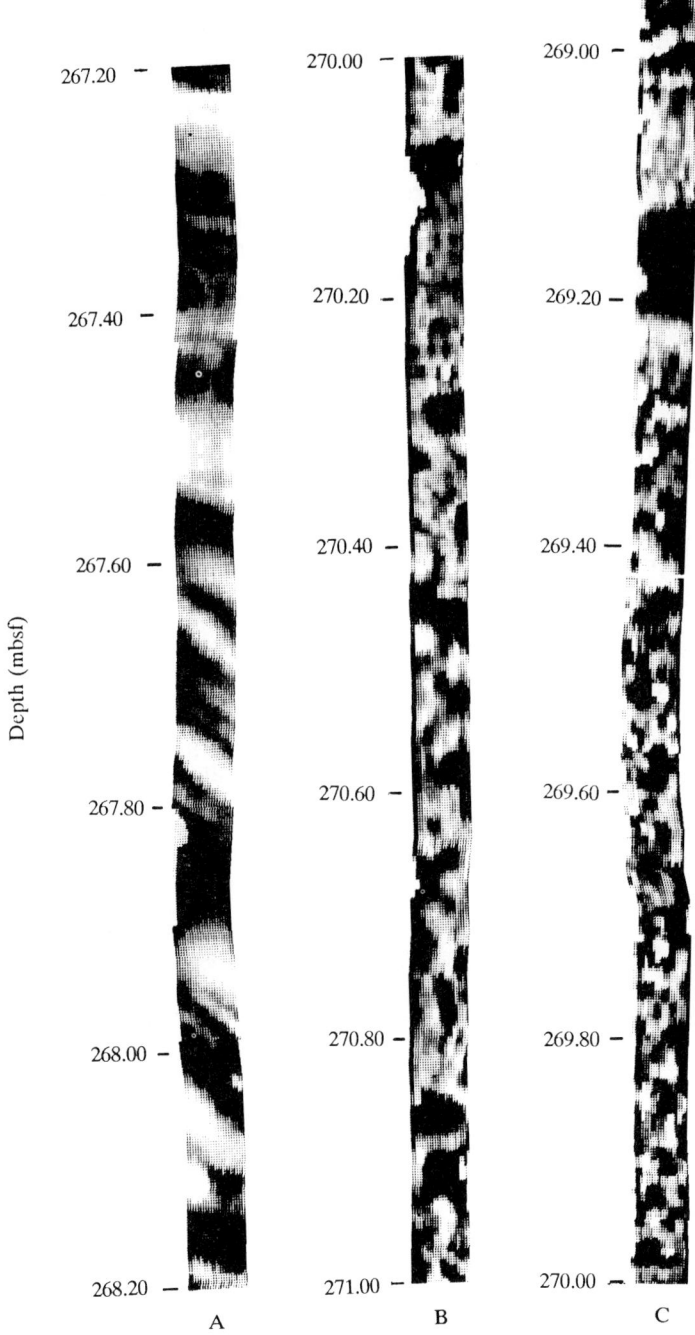

Fig. 10. FMS image facies (scale 1:5) representing three different sedimentary facies within the non-recovered coring interval (core 129-801B-9R): 8A ('hilite' normalization, window = 20 cm), interlayered light and dark grey bands, variably inclined, lenticular and with folding, 8B (dynamic normalization, window = 10 cm), irregular dark and light grey patches, with evidence of possible folding and somewhat chaotic sediment mixing; 8C (dynamic normalization, window = 10 cm), irregular mottling of light and dark grey, which might be caused by variable lithologies/clasts within a debrite.

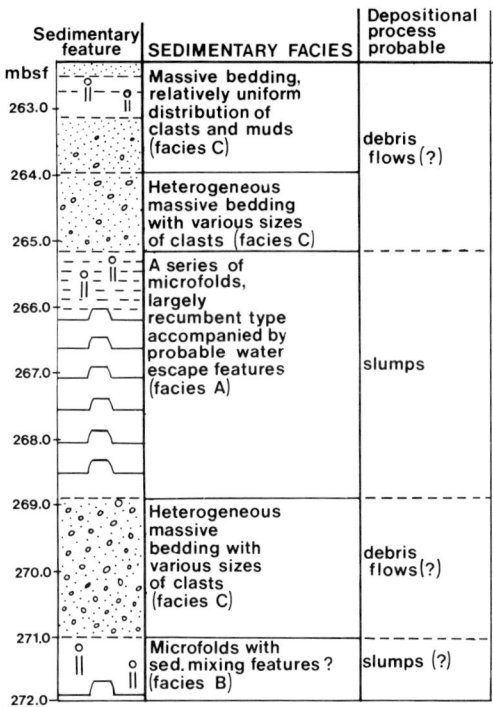

Sedimentary feature	SEDIMENTARY FACIES	Depositional process probable
mbsf	Massive bedding, relatively uniform distribution of clasts and muds (facies C)	debris flows (?)
263.0		
264.0	Heterogeneous massive bedding with various sizes of clasts (facies C)	
265.0		
266.0	A series of microfolds, largely recumbent type accompanied by probable water escape features (facies A)	slumps
267.0		
268.0		
269.0	Heterogeneous massive bedding with various sizes of clasts (facies C)	debris flows (?)
270.0		
271.0	Microfolds with sed. mixing features? (facies B)	slumps (?)
272.0		

Fig. 11. The interpreted sequence (for key see Fig. 6) of the coring interval as mentioned in Fig. 10.

1-cc, 186.80–195.5 mbsf) of massive structureless sandstones. The FMS images also reveal the presence of these two facies within the coring interval precisely and calibrated image facies are shown in Fig. 13. Since recovered sections are conventionally located at the top of the cored interval, the laminated facies has been located in the upper 80 cm of section 1. However, the FMS images show that the laminated interval is, in fact, 2 m thick starting from the topmost part of the cored section, that is from 186 m to 188 m. This implies that part of the missing core is from the top 2 m of the cored interval, and not from the base as convention dictates. The location of the rest of the missing core is not possible to determine as the FMS images show the underlying 7.5 m to be entirely of massive structureless sandstones with varying degree of heterogeneity.

Discussion and conclusion

Core recovery at ODP Leg 129 sites typically ranged from 17.3 to 29.5%, but is 0–5% in a number of cored intervals. In many cases, even the recovered cores are badly disturbed,

abraded, fragmented and brecciated by drilling disturbance. This situation is common in ODP holes, whereas many oil exploration wells are only spot cored if cored at all.

With the aid of a good suite of FMS logs, computer processing, image enhancement, and at least some core recovery to allow calibration, it has been possible to obtain information concerning the sediment from intervals with poor or zero core recovery. The correct location of recovered portions of core within a longer cored interval may also be estimated given that distinctive sedimentary features can be recognized in both core and image. Where parts of cores show signs of disturbance, comparison with the FMS image may allow distinction between drilling-induced effects and those of sedimentary origin.

The interpretation approach used has followed that of previous studies (e.g. Serra 1989; Harker *et al.* 1990), although the sediments described in this study are all of volcaniclastic origin and therefore provide new documentation of FMS images for this class of sediment. Calibration of images with well-recovered coring intervals has shown that resistivity contrasts on images are most closely related to grain-size changes in the sediment.

Furthermore, we have found it helpful to extend Serra's (1989) scheme for the grading of image interpretation, with the addition of two further grades: Grade 4 for unclear features interpretable with the help of core/image data from a remote interval or borehole, and Grade 5 for ambiguous features interpretable by association with Grade 1 images. Harker *et al.*'s (1990) grading scheme does not compare directly with that of Serra (1989), apart from their Grade 1 images, and we have found it to be less helpful.

The sedimentological data gained from this detailed FMS study of ODP Leg 129 volcaniclastics has been extremely beneficial to our understanding of basin development in this part of the Pacific as well as to better characterization of sedimentary structures in volcaniclastic turbidites and associated facies. It has been possible to demonstrate the existence of slumps and of slump–debrite couplets, and to determine a sequence of structures through thick-bedded, coarse-grained, volcaniclastic turbidites. In addition, it may be possible to recognize distinct ichnofacies within these sediments using enhanced FMS images and careful selection of window size during processing.

The primary data were collected during ODP Leg 129 on which one of the authors (ARMS) was a participant. The ODP staff and the captain, officers and crew of *Joides Resolution* are thanked for their part in this

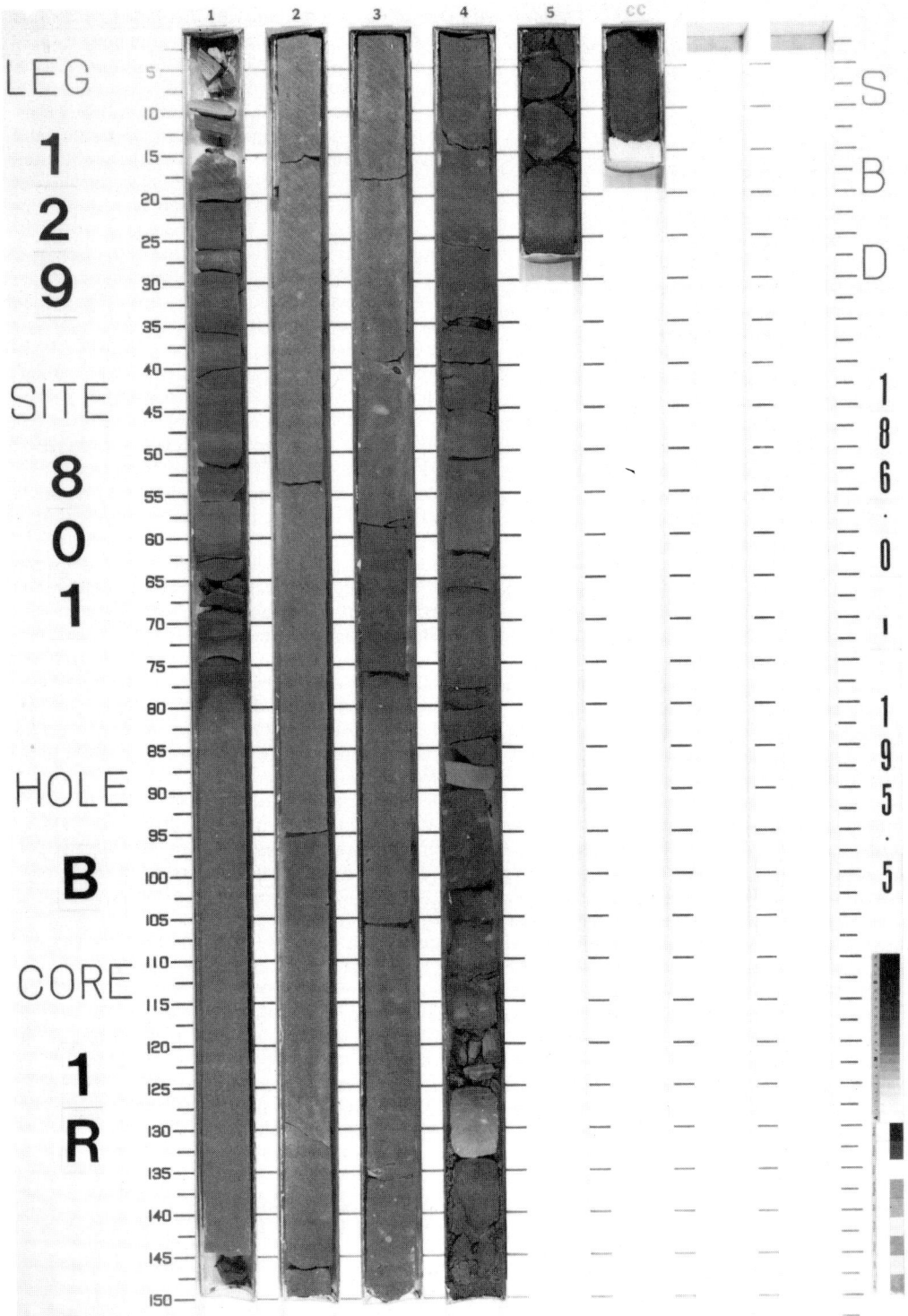

Fig. 12. Core photograph of the well recovered coring interval (core 129-801B-1R) showing the presence of two sedimentary facies: laminated (subtle) sandstones (upper portion of section-1) and structureless massive sandstones with some larger clasts in places.

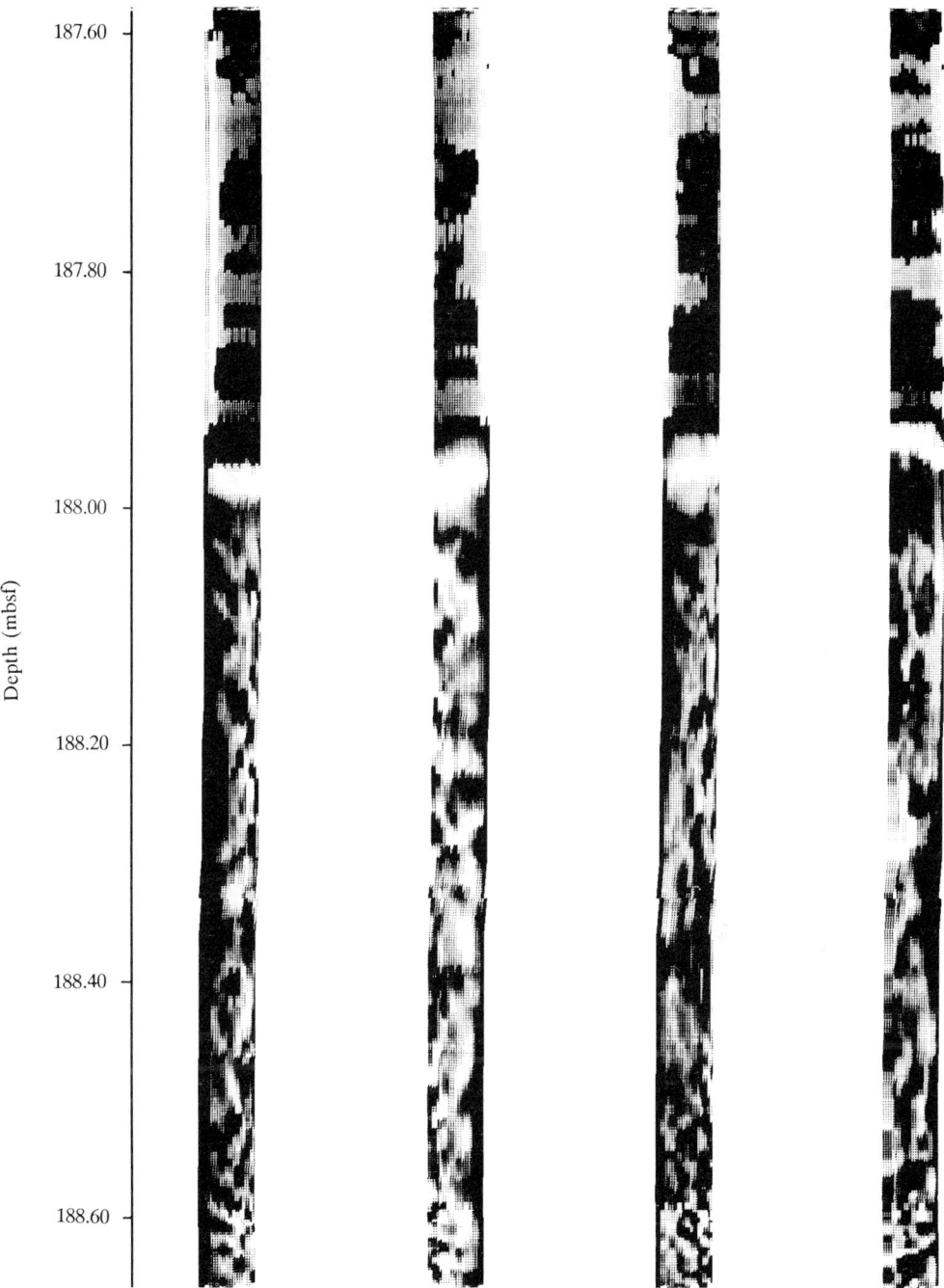

Fig. 13. Four pad FMS images (scale 1:5, 'hilite' normalization, window = 20 cm) showing the presence of two sedimentary facies: laminated (subtle) sandstones (section above 188 mbsf) and structureless massive sandstones (section below 188 mbsf).

cruise. We are also very grateful for the technical help from the Borehole Research Group and especially from Robin Reynolds during the processing of the FMS images at the Lamont Doherty Geological Ob-

servatory. Staff in the Geology Department of Southampton University are thanked for their help in the final preparation of this paper.

References

EKSTROM, M. P., DAHAN, C., CHEN, M.-Y., LLOYD, P. & ROSSI, D. J. 1987. Formation imaging with microelectrical scanning arrays. *The Log Analyst*, **28**, 294–306.

HARKER, S. D. *et al.* 1990. Methodology of formation micro-scanner image interpretation in Claymore and Scapa Fields (North Sea). *In*: HURST, A., LOVELL, M. A. & MORTON, A. C. (eds) *Geological Applications of Wireline Logs*. Geological Society, London, Special Publication, **48**, 13–23.

LARSON, R. L., PITMAN, W. C., GOLOVCHENKO, X., CANDE, S. C., DEWEY, J. F., HAXBY, W. F. & LABRECQUE, J. L. 1985. *The Bedrock Geology of the World*. Freeman, New York.

Shipboard Scientific Party, 1990. *In*: LANCELOT, Y. V. & LARSON, R. *et al. Proceedings of the Ocean Drilling Program, Initial Reports*. Ocean Drilling Program, College Station, TX, **129**, 26, 34.

Shipboard Scientific Party, 1989. Site 787. *In*: TAYLOR, B., FUJIOKA, K. *et al. Proceedings of the Ocean Drilling Program, Initial Reports*. Ocean Drilling Program, College Station, TX, **126**, 39–41.

SERRA, O. 1989. *Schlumberger Formation Microscanner Image Interpretation* Schlumberger Education Services, 16–96.

STOW, D. A. V. & WETZEL, A. 1990. Hemiturbidite: a new type of deep-water sediment. *In*: COCHRAN, J. R., STOW, D. A. V. *et al. Proceedings of the Ocean Drilling Program, Scientific Results*. Ocean Drilling Program, College Station, TX, **116**, 25–34.

Wireline log signatures of resedimented volcaniclastic facies, ODP Leg 129, West Central Pacific

A. R. M. SALIMULLAH & D. A. V. STOW

Geology Department, The University, Southampton SO9 5NH, UK

Abstract. During Ocean Drilling Program Leg 129, two sites (800, 801) were drilled in Pigafetta Basin and one site (802) in East Mariana Basin, West Central Pacific. At all three sites, a thick (192–211 m) Cretaceous succession of dominantly volcaniclastic sediments was encountered, and at site 802 a further 222 m of Miocene–Pliocene volcaniclastics were drilled. These sediments are composed mainly of volcaniclastic material that has been resedimented downslope by various mass-flow processes including slumps, debris flows and turbidity currents. Sedimentary structures and textural variations observed in the recovered section can be used to calibrate the corresponding high-resolution Formation Micro-Scanner (FMS*) images and Dipmeter Micro-resistivity readings. Gamma ray log shapes show grain size evolution within a facies. Fining-upwards and coarsening-upwards grain size trends are observed for some sequences of turbidites and debrites. No obvious change in grain size is observed through massive structureless sandstones. CCA (calcium yield in decimal fraction) and CSI (silicon yield in decimal fraction) readings from the geochemical combination logs, correlated with visual compositional data, are useful to define calcareous and siliceous turbidite facies. This combination of wireline logs has been used in the poorly recovered intervals of volcaniclastic section in the ODP Leg 129 sites to improve significantly our understanding of basinal sedimentation and tectonics in the area.

During Leg 129 of the Ocean Drilling Program (ODP), three sites were drilled in the West Central Pacific. Sites 800 and 801 were drilled in Pigafetta Basin, and site 802 in East Mariana Basin, at present day water depths ranging from 5673.8 to 5968.6 m (Fig. 1). The average core recovery was poor, ranging from 17 to 29%. The lithostratigraphy of these sites, as summarized by the shipboard scientists (Lancelot, Larson *et al.* 1990) is shown in Fig. 2. A thick (192–211 m) succession of Cretaceous mainly volcaniclastic sediments was encountered in all three sites, together with a Miocene–Pliocene (222 m) volcaniclastic succession at site 802. These sediments can be divided into seven process related facies on the basis of texture, structure and composition delineated from recovered portions of cores and/or with information provided by wireline log measurements. The facies are: debrites, slumps, massive sandstones with fluid escape structures, massive structureless sandstones, volcaniclastic turbidites, calcareous volcaniclastic turbidites, and bioclastic (radiolarian) turbidites and pelagites. These facies are typical of deep-water slope and basinal environments (e.g. Stow 1985, 1986), the volcaniclastic and bioclastic composition being common to mid-ocean or active margin settings (e.g. Kelts & Arther 1981).

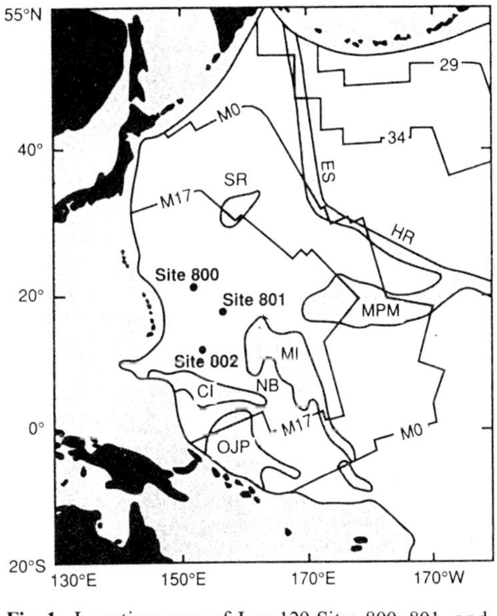

---

* Mark of Schlumberger

Fig. 1. Location map of Leg 129 Sites 800, 801, and 802. Bedrock isochrons are determined from the magnetic anomaly lineation mapping on the Pacific plate (after Larson *et al.* 1985) and superimposed on groups of islands, attols and guyots in the western Pacific Ocean. (Feature abbreviations are as follows: Caroline Islands (CI), Ontong Java Plateau (OJP), Marshall Islands (MI), Nauru Basin (NB), Mid-Pacific Mountains (MPM), Shatsky Rise (SR), Hawaiian Ridge (HR) and Emperor Seamounts (ES).

From HURST, A., GRIFFITHS, C. M. & WORTHINGTON, P. F. (eds), 1992,
Geological Applications of Wireline Logs II. Geological Society Special Publication No. 65, pp. 87–97.

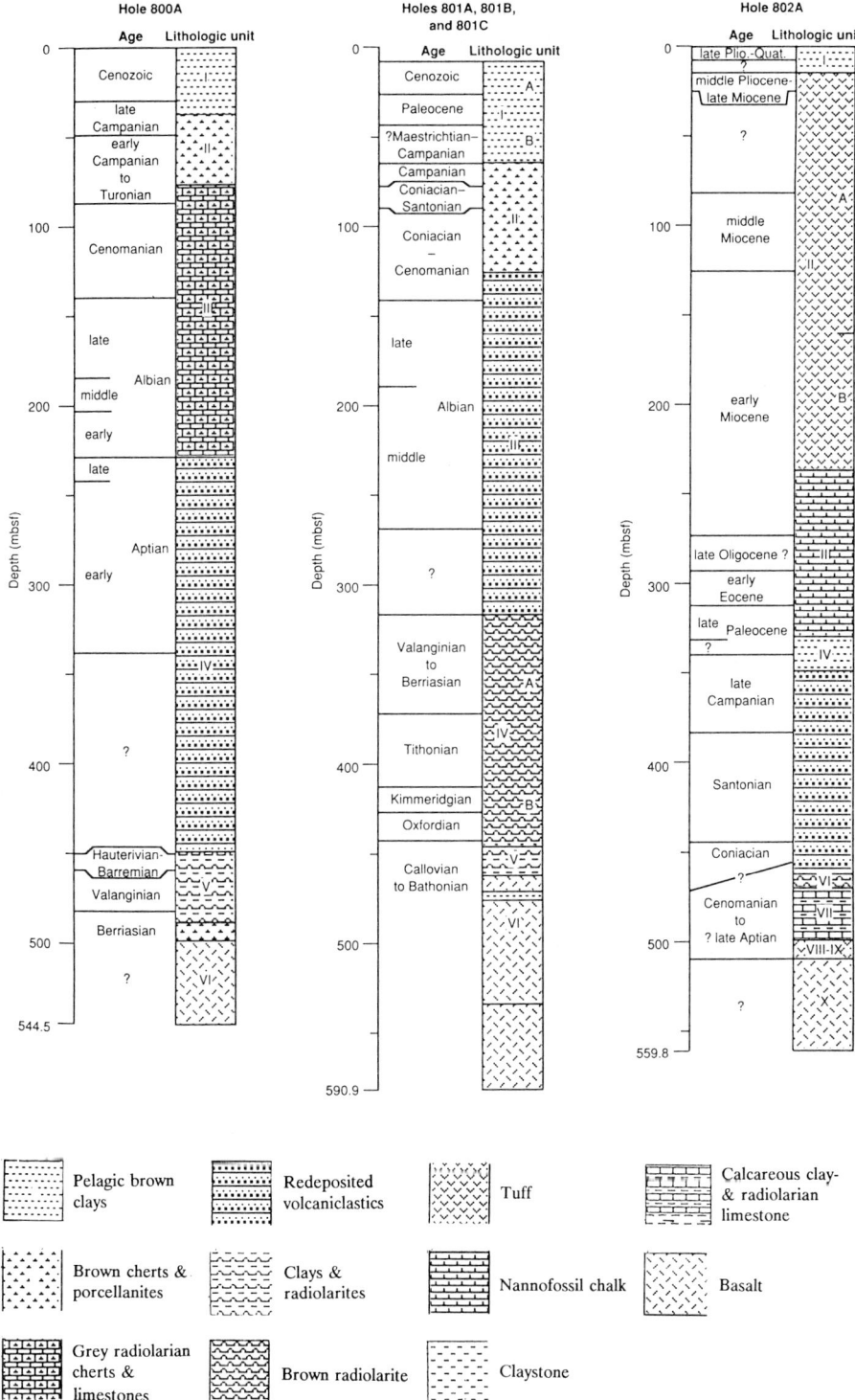

Fig. 2. Summary lithostratigraphy of ODP sites 800, 801 and 802.

Following the normal ODP continuous coring operation, an extensive suite of wireline logs was run at each site in open hole conditions, using small diameter (3.625 in^3/9.2 cm^3) borehole tools (Lancelot, Larson *et al.* 1990). The following data set was therefore available for use in this study:

Core and laboratory data

(A) visual core descriptions (V C D)
(B) core photos
(C) thin section data

Wireline log data

(D) Formation MicroScanner (FMS) images (scale: 1/5)
(E) Dipmeter Micro-resistivity logs
(F) Gamma ray logs
(G) Geochemical logging tool
    (i) CSI (silicon yield in decimal fraction)
    (ii) CCI (calcium yield in decimal fraction)
    (iii) LIR (lithology indicator ratio–silica/ silica + calcium).

The FMS, Dipmeter-Microresistivity and gamma ray logs are used in this study to provide information on the texture and structure of the sediments, whereas the geochemical logs are used to give information on their composition. It must be recognized that the generally very high porosities of volcaniclastic sediments serve to degrade the geochemical signatures. There have been very few published studies of wireline log traces through volcaniclastic successions (e.g. Stow 1984).

The primary objectives of this study, are therefore: (a) to document the signatures of various wireline logs in the different facies present within the volcaniclastic units, having first compared these carefully with the recovered cores, (b) to show how various combinations of logs can be used to enhance the delineation of facies within the volcaniclastic sediments.

Core–log correlation methodology

The FMS tool provides the high resolution and continuous downhole borehole coverage necessary to distinguish fine bedding, internal bed structures and subtle changes in rock properties (Ekstrom *et al.* 1986). However, it is still crucial to integrate core data with open hole logs (Harker *et al.* 1990) to confidently establish lithological changes related to log trends. The general core–log correlation and interpretation approach adopted was as follows.

(a) Common sedimentary features, such as graded bedding, lamination, bioturbation etc. within the various facies observed in recovered cores were correlated with the FMS images. These images were also compared with previously published FMS interpretations (e.g. Serra 1986, 1989; Luthi 1990). In this way we were able to build up a database, for this suite of sediments, of FMS images and corresponding sedimentary interpretations.

(b) In the intervals where there was no recovery of cores, the FMS images were compared with the database developed in (a), hence allowing an interpretation of the non-recovered section.

(c) Dipmeter arrow plots (angle and azimuth) were examined for each facies, especially noting the vertical evolution and distribution of dips. Different facies show different dipmeter patterns, so it is possible to distinguish between facies. Distinctive trends may also be related to changes in rock properties (e.g. grain size) within and between individual beds.

(d) Microresistivity log readings were examined, especially to identify homogeneity and heterogeneity in terms of grain size for the massive structureless sediments recognized from core and images.

(e) Gamma ray log responses were correlated with the various facies, noting especially any vertical trends in terms of grain size.

(f) Higher CCA (calcium yield in decimal fraction) log readings were correlated with the core and thin section data in order to locate carbonate-rich intervals and to help identify possible calcareous volcaniclastic turbidites.

(g) Higher CSI (silica yield in decimal fraction) log readings were correlated with the core and thin section data with a view to locating siliceous intervals (e.g. clayey radiolarites/radiolarian claystones) and to help distinguish pelagites from pelagic/bioclastic turbidites in terms of composition.

(h) LIR (lithology indicator ratio) log readings were used to show the bulk compositional evolution of certain facies and to help establish the composition of the background sediments, in order to provide information on biological (production) and chemical (CCD/SCD) aspects of the oceanic environments during volcaniclastic sedimentation.

Sediment facies and wireline log signatures

Debrites

This facies occurs in relatively thick (70–150 cm), structureless beds that are poorly sorted

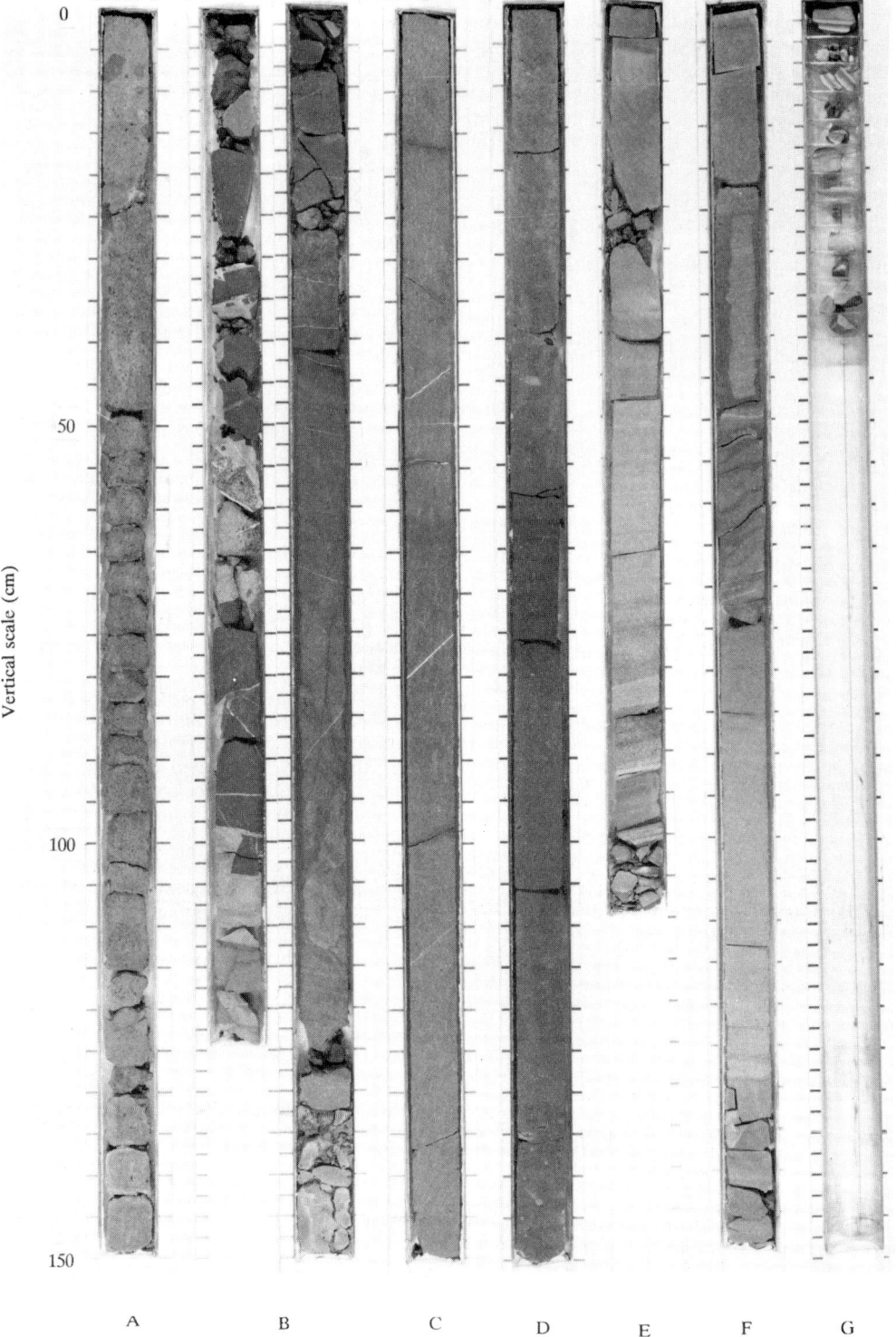

Fig. 3. Core photographs of volcaniclastic facies: A, debrites; B, slumps; C, massive sandstones with fluid escape structures; D, massive structureless sandstones; E, volcaniclastic turbidites; F, calcareous volcaniclastic turbidites; G, pelagites.

Fig. 4. Characteristic FMS images of volcaniclastic facies: A, debrites, B, slumps; B1 microfolds (recumbent); B2, microfaults; C, massive structureless sandstones; D, volcaniclastic turbidites; E, calcareous volcaniclastic turbidites; F, pelagites; G, bioclastic turbidites.

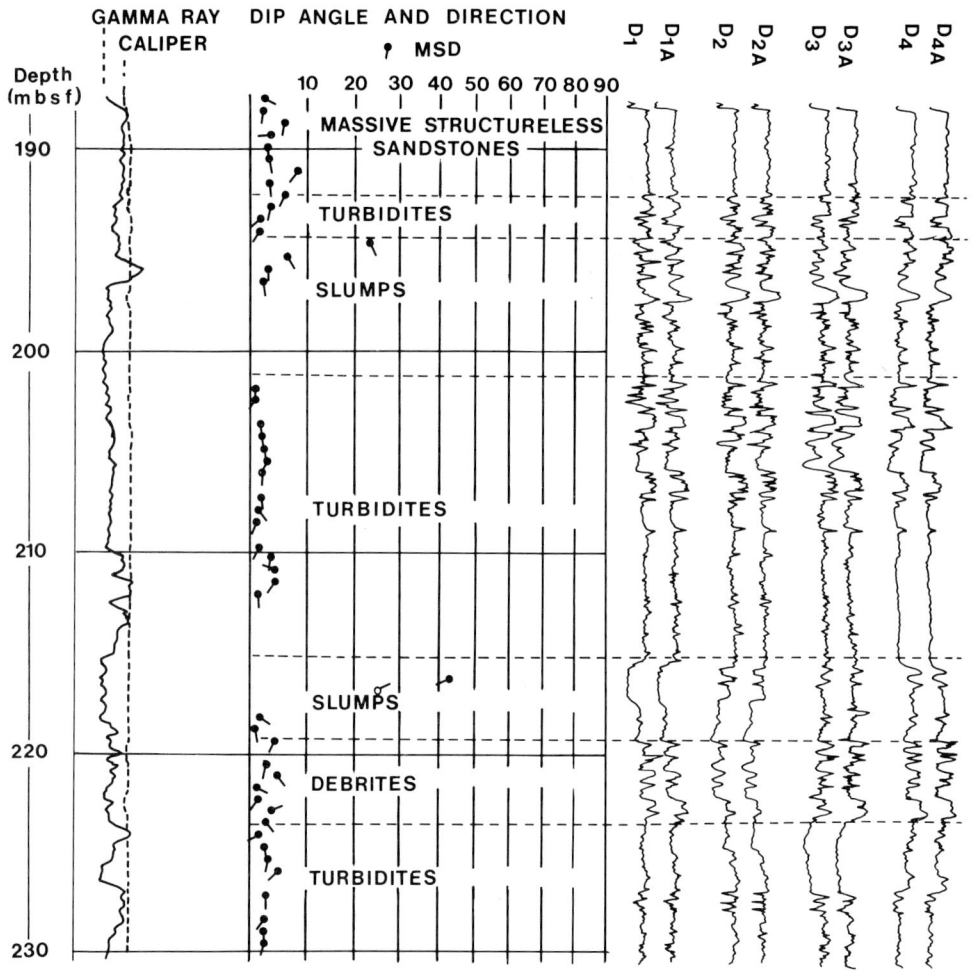

Fig. 5. Dipmeter plots and micro-resistivity log readings of volcaniclastic facies from ODP Hole 801B.

with random fabrics and poor grading (Fig. 3A). The matrix is clay to sand size and is composed of smectite/chlorite clays and volcanic glasses. The clasts are very coarse sand to pebble size, sub-rounded to well rounded and are mainly composed of muds (mud clasts) and igneous rock fragments.

FMS images show an uneven distribution of white (more resistive) irregular patches or dots (clasts/pebbles) in a dark/black (more conductive) medium or matrix, and confirm the heterogeneous and structureless natures of the beds. The shape of these white patches (dots, subrounded to rounded or angular) helps distinguish between conglomerate and breccia and also helps to distinguish conglomerate/breccia

from bioturbational mottles which can create a similar FMS image. Figure 4A shows that the white patches (dots) are subrounded to rounded in shape which indicates the presence of conglomerates. Where there is a relatively higher abundance of closely spaced white patches (dots), this indicates clast-supported conglomerates, whereas a lower abundance is indicative of matrix-supported conglomerates.

For the very thick-bedded debrites, these features can also be seen on other logs. Dipmeter plots reveal variable dips and azimuths throughout the debrite intervals (Fig. 5). Micro-resistivity curves show the heterogeneous nature of debrite beds with an absence of correlation between the different traces in some instances

(Serra 1985). Where clasts/mudclasts are close to each other (clast-supported conglomerates), the peaks are relatively close (Fig. 4A); but where the clasts are isolated in a sandy or silty/clayey matrix, the peaks are isolated and/or absent. Gamma ray logs (Fig. 6) show both serrated and relatively smooth parts of different debrite beds, again indicating variable grain/clast size and both clast-supported and matrix-supported conglomerates, respectively. The responses also reveal general coarsening-upward sequences of debrites in some intervals of Hole 801B. LIR (lithology indicator ratio) log readings reveal distinct fluctuations in silica content within these sequences.

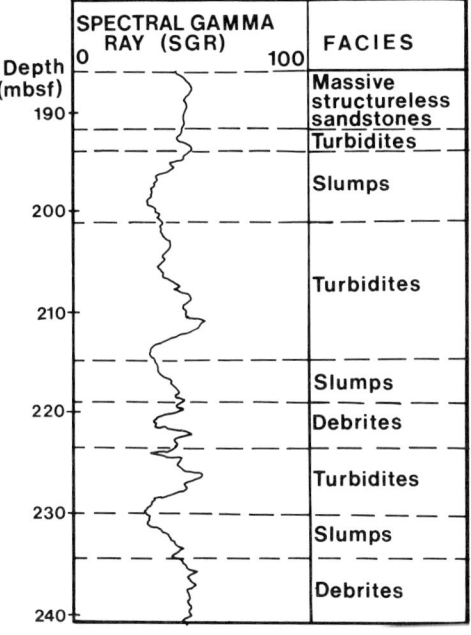

Fig. 6. Gamma ray log readings of volcaniclastic facies from ODP Hole 801B.

Slumps

There was very little recovery of the sediments in those intervals which are thought to be related to sliding or slumping on the basis of their wireline log responses. Two possible intervals from which cores had been recovered are shown in Fig. 3B. However, the FMS images reveal unique features of slumps including microfolds (Luthi 1990) and microfaults. The geometry and shape of the images provide evidence of recumbent slump folds (Fig. 4B1), accompanied by prob-

able water escape features, steeply inclined lamination and possible chevron folding. On the other hand, where images on the pads have a different texture or aspect, or display a loss of continuity between the two sides, they may indicate faulting (Serra 1989). Small depth shifts between similar features on each pad can indicate a microfault. In Fig. 4B2 a small displacement (depth shift) is observed between the two sides of series of microfaults, and the throw can be measured at around 5 cm. The faults make a steep angle (80–90 degrees) with the stratification and appear to show normal displacement.

A continuous change of dips, both in magnitude and azimuth, indicates recumbent folding (Serra 1989), which is commonly present in various slump intervals (Hole 801B). Moreover, a relatively high degree of apparent dips is observed in some intervals (e.g. 42 degrees around depth 252 m, Hole 801B). CSI (silicon yield in decimal fraction) logs (Fig. 9) reveal that the slumping also affects some siliceous sediments (clayey radiolarites/radiolarian claystones) within the volcaniclastic interval, because the log readings correspond to the underlying lithostratigraphic unit which is radiolarian rich (Brown Radiolarites, Figure 2, site 801).

Massive sandstones with fluid escape structures

The sediments of this facies are dominantly medium to fine sand size with a minor proportion of very fine sand to silt size. Individual 'beds' or units (typically 1–8 m thick) are homogeneous with a lack of grading, well to very well sorted and contain many fluid escape structures among which vertical pipes (Fig. 3C) and flame-like pipes are common and locally dominant. Load structures and convolute laminae are rare. The sediments are mainly composed of volcanic glass, igneous rock fragments, pyroxene and (secondary) calcite, with a minor proportion of smectite clays, radiolarians, and red algae.

Unfortunately, wireline logs were not generally obtained for the interval(s) in which this facies occurs. However, in one case we do have a full suite of logs over an interval (292.0–295.0 mbsf, site 800) including massive sandstones with flame-like pipes. This facies is not readily distinguished on the basis of log characteristics from the massive sandstones described below.

Massive structureless sandstones

The sediments of this facies are dominantly fine to medium sand size with some mud size mat-

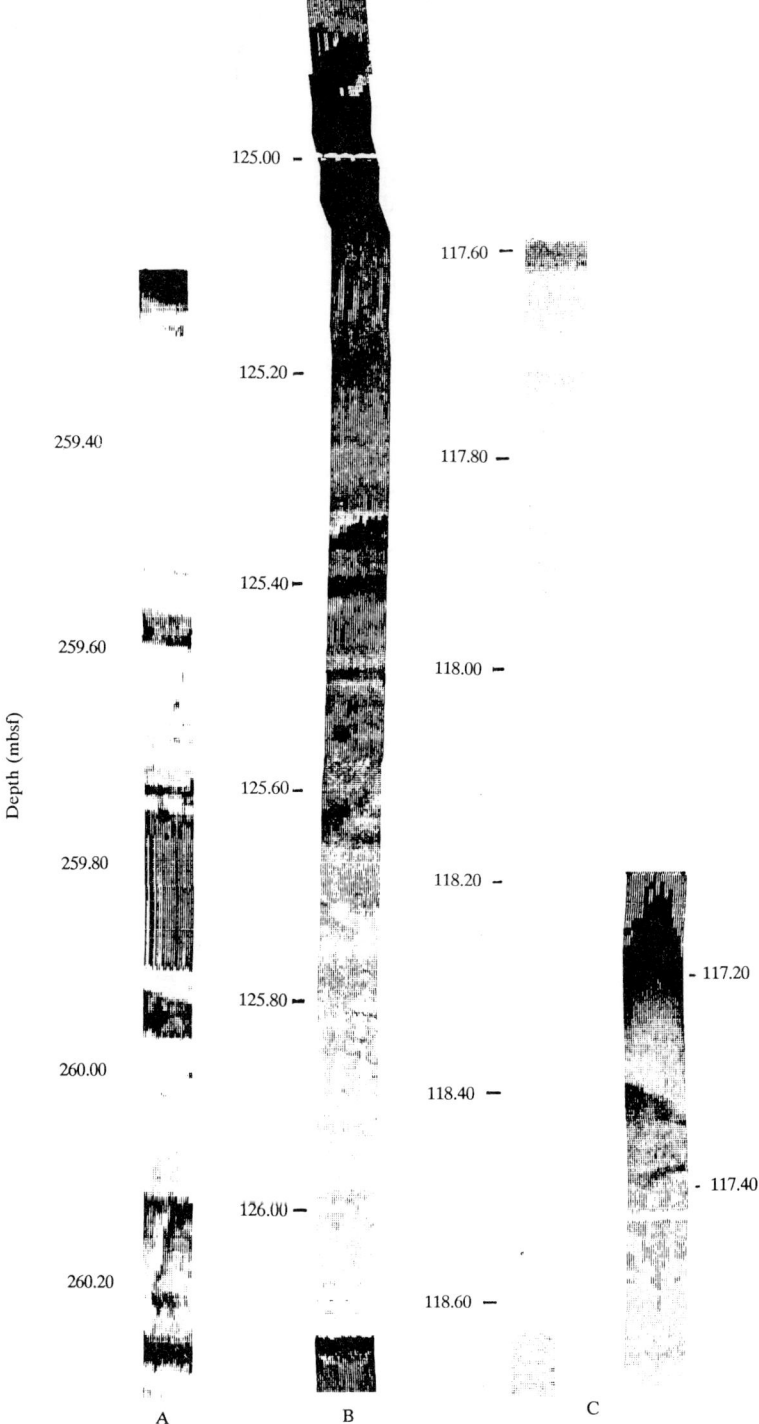

Fig. 7. FMS images of various types of volcaniclastic turbidite beds: A, thin-bedded, medium-grained turbidites; B, thick-bedded, medium-grained turbidites; C, thick-bedded, coarse-grained turbidites.

erial. Individual units or beds (Fig. 3D) appear massive with a general lack of grading, apart from reverse grading in places, and have a typical range of 1–5 m. The sands are moderate to well sorted, and contain rare isolated large clasts (coarse sand to granule size). They are mainly composed of volcanic glass, igneous rock fragments, (secondary) calcite, clays and radiolarians with a minor proportion of calcareous nannofossils, red algae, feldspar and pyroxene.

FMS images reveal a uniform distribution of grey tone within the interval (Fig. 4C) which indicates the structureless nature of the beds and uniform distribution of grain size. Moreover, the presence of the isolated white patches (dots) (more resistive), represents the presence of large clasts as observed in the cores. Where this facies is sufficiently well developed in vertical extent, some of the other logs show similar features. There are consistent dip readings (magnitude and azimuth) in dipmeter logs (Fig. 5) and a very smooth homogeneous aspect of the microresistivity logs (Fig. 5). Gamma ray responses (Fig. 6) are also relatively very smooth and do not show any significant trends in terms of grain size. LIR (lithology indicator ratio) log readings in this facies reveal that the parts adjacent to beds or unit boundaries contain higher silica contents in comparison with the middle portion of the units.

Volcaniclastic turbidites

This facies displays a range of grain sizes from sand to mud grade with granule to pebble size particles at the base of some beds. Individual beds range from 2–400 cm in thickness, show normal grading (Fig. 3E) and contain features typical of coarse-, medium-, and fine-grained turbidites (e.g. Stow 1985, 1986). The sediments are largely composed of volcanic glass, igneous rock fragments, palagonite, pyroxene and (secondary) calcite, with a variable proportion of smectite/chloritic clay, radiolarians, zeolites, feldspar and quartz.

The FMS images clearly document the fining upwards nature of turbidite beds. Graded bedding is indicated by the progressive vertical change in image density, such that an upward increase in grey tone indicates progressive increase in less resistive clay-size particles and hence defines normal graded bedding (Fig. 4D). This is the most common sedimentary structure present within the turbidites, whereas other structures (primary, biogenic and secondary) can be recognized more rarely. These subtle structures are more difficult to resolve through the static normalization image (static pass),

although it may be possible to resolve those structures having vertical continuity (resolution) greater than one centimetre using dynamically normalized images. Various types of turbidite beds can be documented on FMS images (Fig. 7).

Dipmeter log readings vary from highly consistent to sequential increase and decrease in magnitude (Fig. 5), which are most probably related to vertical evolution of the turbiditic intervals in terms of grain size and/or sedimentary structure. Moreover, variable dip readings are also observed at the base of some coarse-grained turbidite beds. Microresistivity readings (Fig. 5) support the dip readings in terms of grain size variation. Gamma ray responses (Fig. 6) show a serrated nature reflecting the presence of a series of turbidite beds and, in places, they also indicate fining-upward intervals of turbidite packages.

Calcareous volcaniclastic turbidites

As the name implies, this facies is similar to the volcaniclastic turbidites in terms of sedimentary textures and structures but differs in composition. These turbidites contain 10 to 55% calcareous material, composed of shallow water carbonates (ooids, coralline limestone clasts, red algae), calcareous nannofossils and foraminifers, along with the volcaniclastic and other components that are commonly present in the volcaniclastic turbidite.

FMS, dipmeter, microresistivity, and gamma ray log signatures are similar to those described above for the volcaniclastic turbidite facies. Higher CCA (calcium yield in decimal fraction) log readings (Fig. 8) can be calibrated with the compositional data of turbidite beds and reveal the presence of calcareous turbidite interval(s) in sections with no core recovery.

Pelagites and bioclastic turbidites

Poor core recovery did not allow the clear definition of the sedimentary features within this facies. The sediments are largely composed of clayey radiolarites and radiolarian claystones with a variable proportion of volcaniclastic components.

The FMS images allow us to clearly distinguish pelagic intervals (pelagites, Fig. 4G1) from bioclastic turbidite beds (Fig. 4G2). Erosive lower bedding contacts, normal grading of beds and the fining-upwards nature of turbidite sequences, substantially differ from the horizontally bedded/banded aspect of pelagites (Fig. 3G). Higher CSI (silicon yield in decimal frac-

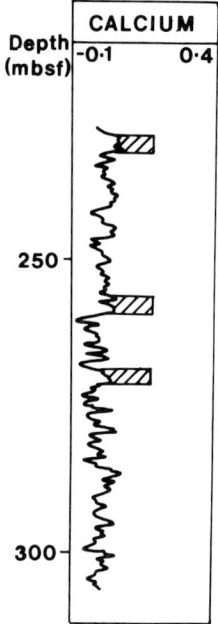

Fig. 8. CCA (calcium yield in decimal fraction) log readings (hatched intervals) revealing the presence of calcareous volcaniclastic turbidites in ODP Hole 800.

tion) log readings (Fig. 9) calibrated with the compositional data and compared with signatures of radiolarian-rich lithostratigraphic unit(s) (Fig. 2, see site 801) reveal the presence of the facies where there is no recovery.

Discussion and conclusions

All the three sites drilled during the ODP Leg 129 have encountered thick volcaniclastic sediments. These sediments show many classical features of the resedimentation processes responsible for their deposition (e.g. Stow 1985, 1986), and their log characteristics, we believe, are applicable to other resedimented successions. Sedimentary features delineated from the recovered cores are initially calibrated with the corresponding FMS images because they have the capability of fine bed resolution. Moreover, the FMS images are the wireline log data which can be compared more closely with the core photographs. If the facies/sedimentary features have sufficient vertical resolution, the other open hole logs including dipmeter–microresistivity and gamma ray readings can be further used to confirm the textural and structural evolution as observed in the FMS images. This study reveals that the combined use of the FMS image (cali-

brated/non-calibrated), dipmeter–microresistivity and gamma ray logs have made it possible to delineate debrites, slumps, tubidites and massive structureless sandstones as well as to distinguish the one from the other. It also shows that the specific elemental signature(s) of the geochemical combination logs can be used to enhance the compositional aspects of these facies. Higher concentrations of calcium and silica delineated from thin sections are calibrated with the corresponding log readings which are then combined with the image interpretations to distinguish calcareous volcaniclastic turbidites from volcaniclastic turbidites and bioclastic turbidites from pelagites. This combination of wireline logs has been very useful to delineate the vertical extent of facies as well as vertical sequences of facies throughout the volcaniclastic sedimentation and tectonics in the area (Salimullah, in press).

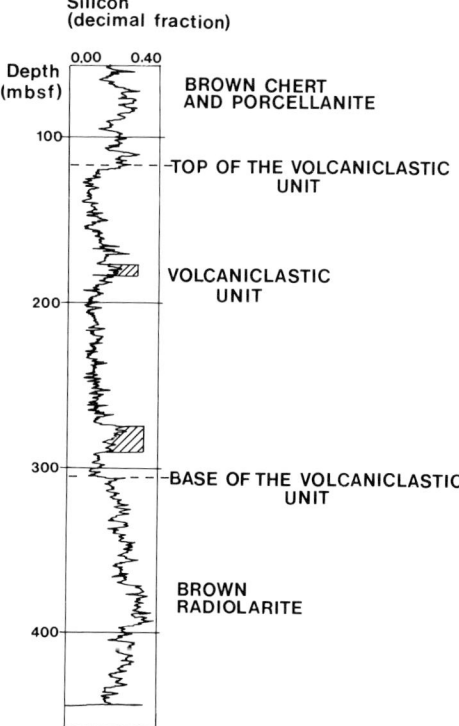

Fig. 9. CSI (silicon yield in decimal fraction) log readings (hatched intervals) revealing the presence of siliceous sediments within the Volcaniclastic Lithologic Unit. These readings largely resemble radiolarian-rich sediments (Brown radiolarites and Brown chert and porcellanites, Fig. 2, site 801) which are underlying and overlying respectively the Volcaniclastic Unit in site 801.

References

EKSTROM, M. P., DAHAN, C., CHEN, M.-Y., LLOYD, P. & ROSSI, D. J. 1987. Formation imaging with microelectrical scanning arrays. *The Log Analyst*, **28**, 306.

HARKER, S. D., McGANN, G. J., BOURKE, L. T., & ADAMS, J. T. 1990. Methodology of formation micro-scanner image interpretation in Claymore and Scapa Fields (North Sea). *In*: HURST, A., LOVELL, M. A. & MORTON, A. C. (eds) *Geological Applications of Wireline Logs*. Geological Society, London, Special Publication, **48**, 13–23.

KELTS, K. & ARTHER, M. A. 1981. Turbidites after ten years of Deep Sea Drilling—wringing out the mop? *In*: WARME, J. E., DOUGLAS, R. G. & WINTERER, E. L. (eds) *The Deep Sea Drilling Project: A Decade of Progress*. S.E.P.M. Special Publication, **32**, 91–127.

LANCELOT, Y. V., LARSON, R. *et al.* 1990. *Proceedings of the Ocean Drilling Program, Initial Reports*. Ocean Drilling Program, College Station, TX, **129**, 34–39, 91–100, 171–177, 283–321 and 412–435.

LUTHI, S. M. 1990. Sedimentary structures of clastic rocks identified from electrical borehole images. *In*: HURST, A., LOVELL, M. A. & MORTON, A. C. (eds) *Geological Applications of Wireline Logs*.

Geological Society, London, Special Publication, **48**, 4–6.

SERRA, O. 1985. *Sedimentary Environments from Wireline Logs*. Schlumberger, 28–35, 160–168.

—— 1986. Information on sedimentary structures. *In*: *Fundamentals of Well-Log Interpretation, Volume 2*. Elsevier, Amsterdam, 137–152.

—— 1989. *Schlumberger Formation Microscanner Image Interpretation*. Schlumberger Education Services, 16–19, 32–72, 96.

SALIMULLAH, A. R. M. Volcaniclastic facies and sequences, ODP Leg 129 (in press).

STOW, D. A. V. 1984. Cretaceous to recent submarine fans in the southeast Angola Basin. *In*: HAY, W. W., SIBUET, J.-C. *et al. Initial Reports of the D.S.D.P.*, Volume LXXV, 771–784, U.S. Government Printing Office, Washington.

—— 1985. Deep-sea clastics: where we are and where we are going? *In*: BRENCHLEY, P. J. & WILLIAMS, B. P. J. (eds) *Sedimentology: Recent Developments and Applied Aspects*. Geological Society, London, Special Publication, **18**, 67–93.

—— 1986. Deep Clastic Seas. *In*: READING, H. G. (ed.) *Sedimentary Environments and Facies*. Blackwell, Oxford, 399–444.

Evaluating thinly laminated reservoirs using logs with different vertical resolution

N. RUHOVETS, R. RAU, M. SAMUEL, H. SMITH, Jr. & M. SMITH

Halliburton Logging Services, Inc., P.O. Box 42800 Houston, Texas 77242-2800, USA

Abstract. Reservoirs with thin laminations can be more accurately evaluated by using logging tools with inherently better vertical resolution, by employing enhanced vertical resolution input processing methods, and by incorporating interpretation models that properly handle log inputs with different vertical resolutions and reconstruct all outputs with high vertical resolution. The paper discusses a specific high-resolution interpretation model and provides comparative analyses of how model outputs are affected by the vertical resolution of its input logs and by the reservoir type. Three field examples are provided. Increases in predicted hydrocarbons were noted in two of these examples when the input log resolution was increased. In these two examples, the observed increases were confined to a number of isolated thin beds. In the third field example, significant decreases in predicted hydrocarbons were observed when high-resolution input data were used; the reservoir in this case appears to illustrate thin sand–shale laminae extended over a 27 metre interval.

Reservoirs composed of thin laminations can be more accurately evaluated by using input logs with high vertical resolution. In the past, logging tool designers have compromised between such features as vertical resolution, depth of investigation, and log repeatability. For example, conventional deep induction logging tools have good depth of investigation, but poor vertical resolution. Newer logging tool designs feature higher vertical resolution measurements and, in certain cases, excellent depth of investigation. For example, a new dual induction logging tool has an improved vertical resolution (60–90 cm) and a greater depth of investigation than conventional induction tools (Strickland 1987). Appendix 1 contains a more comprehensive discussion of logging tool vertical resolution improvements, including reflectivity measurements from a new high frequency dielectric tool and Pe measurements from newer generation spectral density tools.

The vertical resolution of logging measurements can also be improved by using more sophisticated processing techniques (e.g. deconvolution) on the raw data signals before further processing occurs. For example, Smith (1990) applied deconvolution to a dual-spaced neutron logging tool and vertical resolution improved from about 75 cm to 40 cm. Appendix 2 contains a more complete discussion of some newer methods for enhancing a vertical resolution prior to application of log analysis programs.

Thus, improved tool designs and more sophisticated processing methods can enhance the vertical resolution of raw logging data. Two problems still remain before these data can be used to improve the evaluation of reservoirs containing thin laminations. First, the enhanced vertical resolution of the induction and porosity logs is still not always adequate to evaluate certain reservoirs. Second, logs with enhanced vertical resolution may differ in their inherent resolutions. High-resolution log analysis models must be specifically designed to handle these differences.

To overcome these above mentioned problems, several high-resolution models have been developed (Allen 1984; Bateman 1990; Quinn & Sinha 1985; Raiga-Clemenceau 1988; Ruhovets 1990). Most of these methods reconstruct the conductivity and porosity logs to the high vertical resolution level of a shale indicator such as a dipmeter or a high-frequency dielectric log.

Main objectives

The present paper documents the use of one high-resolution interpretation technique called Laminated Reservoir Analysis (LARA; Ruhovets 1990) developed to analyse logs with improved vertical resolution relative to similar processing with standard logs. This model assumes that the sand and shale laminae have different resistivities and porosities, but these

From HURST, A., GRIFFITHS, C. M. & WORTHINGTON, P. F. (eds), 1992,
Geological Applications of Wireline Logs II. Geological Society Special Publication No. 65, pp. 99–121.

Table 1. *Vertical resolution levels, physical properties, and associated logging tools. Three vertical resolution levels are currently used by the high-resolution reconstructive model LARA.*

Vertical resolution level	Standard value (cm)	Enhanced value (cm)	Physical property	Logging tools
High	5 (H)	5 (H)	Shale volume	Dipmeter, microlog, high-frequency dielectric, unfiltered Pe
Medium	75 (M)	40 (Me)	Shale volume, porosity	Density and neutron porosity, gamma ray
Low	180 (L)	90 (Le)	Conductivity, resistivity	Induction, laterolog, high resolution induction

properties remain fixed for the laminae contained within the vertical resolution distances of their respective logging tools. A key objective of the paper is to investigate to what extent this assumption is reduced in importance as higher vertical resolution input logs are used in place of conventional logs with poor resolution.

The effects of reservoir type are examined. The paper also illustrates the effectiveness of several new high-resolution logging measurements as shale indicators, including a high-frequency dielectric log and an enhanced gamma ray log.

Laminated reservoir analysis model (LARA)

Table 1 summarizes the vertical resolution levels, physical properties and logging tools involved in the high-resolution interpretation model. Calculations are performed at three different levels of vertical resolution. Each of these levels has an adjustable resolution value that is set according to the actual tool or datum used. For example, if

a high-resolution dual induction log is used, its resolution level is set to a value of 90 cm (Le). The medium level has a resolution of 75 cm (M) for standard processing of the gamma ray, density, and neutron raw data; a value of 40 cm (Me) is used if enhanced vertical resolution processing of these logs is performed. See Appendix 2 for more details.

Conductivity and porosity are reconstructed to the high vertical resolution level by performing the basic operations in Table 2. While LARA is analogous to other models, several of its features are unique. First, the total porosities in the sand and shale laminae are not forced to be equal (or approximately equal). Instead LARA merely assumes that these porosities do not change significantly within the vertical resolution of the porosity logging tools. Second, all sand and shale laminae within the vertical resolution of the density, neutron and shale indicator logging tools contribute equally to the measured signals independently of their distances from the

Table 2.

Vertical resolution level	Laminated reservoir analysis model operation
High	(1) obtain high resolution shale volume;
Medium	(2) integrate shale volume to medium level and reconcile it with the shale volume from the neutron-density crossplot;
Medium	(3) conventionally determine total and effective porosities;
Medium Low High	(4) determine dispersed and laminated shale volumes, reconstruct them to the high-resolution level, and integrate them to the low-resolution level;
Medium Low	(5) separate sand and shale components of total porosity and conductivity by using the corresponding laminated shale volumes;
High	(6) reconstruct total and effective porosities and conductivity to the high-resolution level;
High	(7) compute water saturation at the high-resolution level from Waxman–Smits model modified to handle both dispersed clay and laminated shale.

Table 3. *Cumulative hydrocarbon feet (HCFT) and porosity feet (PEFT) of selected zones as functions of logging tool type and input vertical resolution level*

Text figure	Shale indicator		IND	NEU	DEN	PEFT	HCFT
				Well 1 (X152–X182)			
2	GR	M	L	M	M	2.88	0.89
3	GR	M	L	Me	Me	2.86	0.92
4	GR	Me	L	M	M	2.88	0.98
5	GR	Me	L	Me	Me	2.88	1.03
6	DIEL.	H	L	M	M	2.52	0.83
7	DIEL.	H	L	Me	Me	2.51	0.84
				Well 2 (X100–X165)			
9	GT	M	Le	M	M	18.60	14.43
10	DIP.	H	Le	M	M	16.72	12.38
				Well 3 (X500–X530)			
12	GR	M	L	M	M	4.35	0.77
13	GR	Me	L	M	M	4.39	0.80
14	GR	M	Le	Me	Me	4.26	0.78
15	GR	Me	Le	Me	Me	4.30	0.82

measure point. For the induction and laterolog tools, the total conductivity uses the standard geometrical factors as weights for the sand and shale laminae. Third, a modified Waxman–Smits formula is used to compute water saturation. Originally, this formula was developed for reservoirs containing only dispersed clay; later it was modified (Ruhovets 1990) to describe reservoirs containing both dispersed clay and laminated shale. This was achieved by changing the formulae for computing the effective concentration of exchange ions Q_v', the formation resistivity factor F^*, the compound water resistivity R_{wc}, and the equivalent conductance of clay exchange cations B_c.

LARA is primarily designed to evaluate shaly sand reservoirs in which effective porosity depends mainly on shale volume and mode of shale distribution, i.e. laminated and dispersed. Another present limitation is that the vertical resolutions of the input neutron and density logs must be the same.

The following log examples illustrate the effects of changing the vertical resolution of selected individual input logs on the water saturation, porosity and lithology estimates output by the model. Table 3 summarizes the cumulative hydrocarbon feet (HCFT) and porosity feet (PEFT) of a single zone selected from each well. However, the full impact of varying input log

vertical resolution is best judged by a detailed study of Figs 1–15.

Field examples

Well 1

This well is in a reservoir in southeast Texas representative of the fine grained laminated sandstone of the Wilcox formation. Effective porosity is 15–20% and permeability is 5–80 mD. Hydrocarbon bearing zones contain medium gravity oil, often with condensate or dissolved gas.

Logs recorded in this well include a standard dual induction, dual-spaced density and neutron, high-frequency dielectric attenuation, and gamma ray (Fig. 1). Other input logs, with enhanced vertical resolution processing applied, are labelled as the EVR gamma, neutron and density curves.

Figure 2 represents the standard output interpretation for this well; standard vertical resolutions for all inputs were used. Two prospective oil-bearing shaly sands are indicated between (X152–X182'), with porosities 13–17% and water saturation about 30–35%. In this case, the high-resolution interpretation model is functioning simply as a conventional log analysis program.

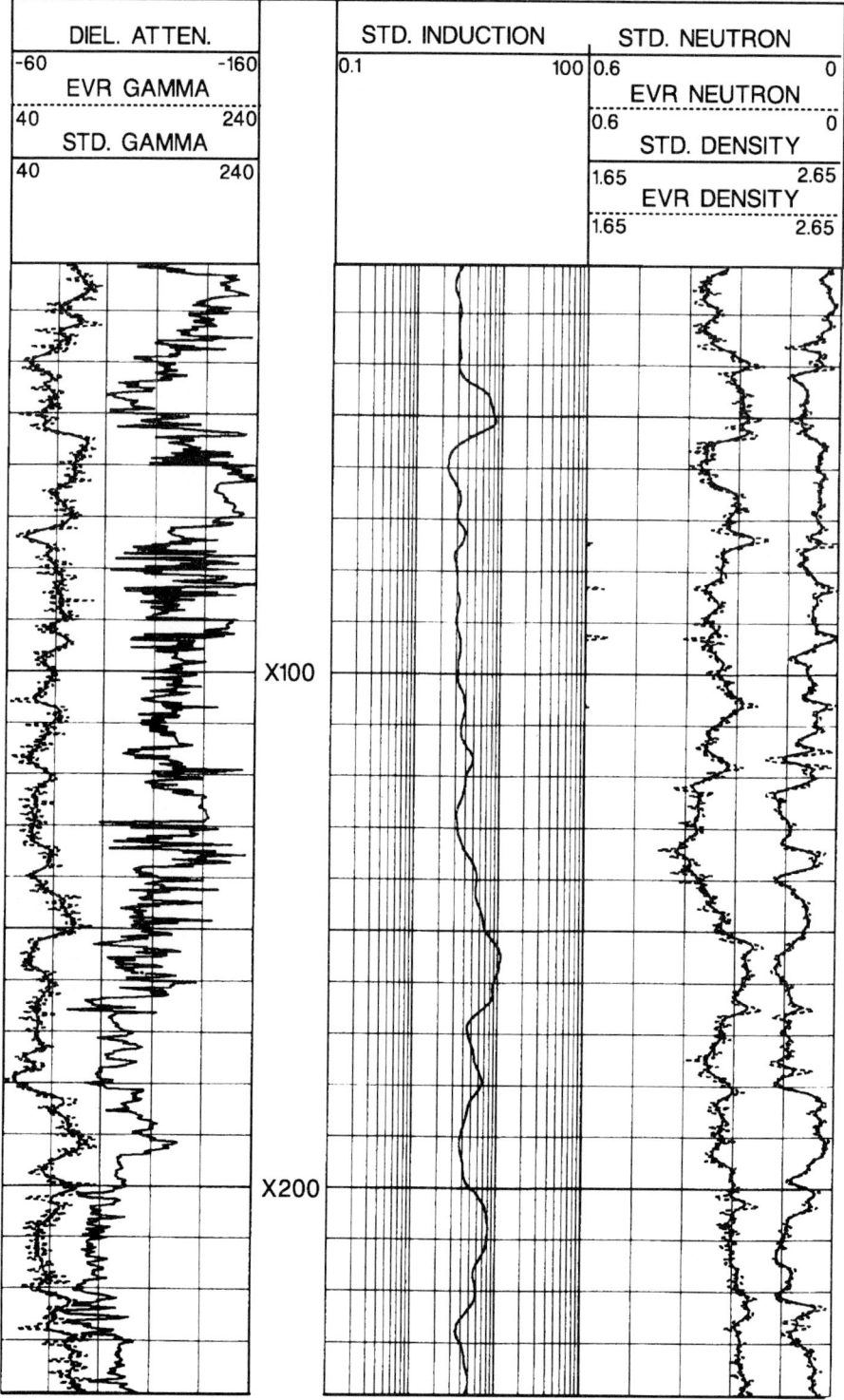

Fig. 1. Well 1: Logs used for interpretation. (Vertical scale in feet.)

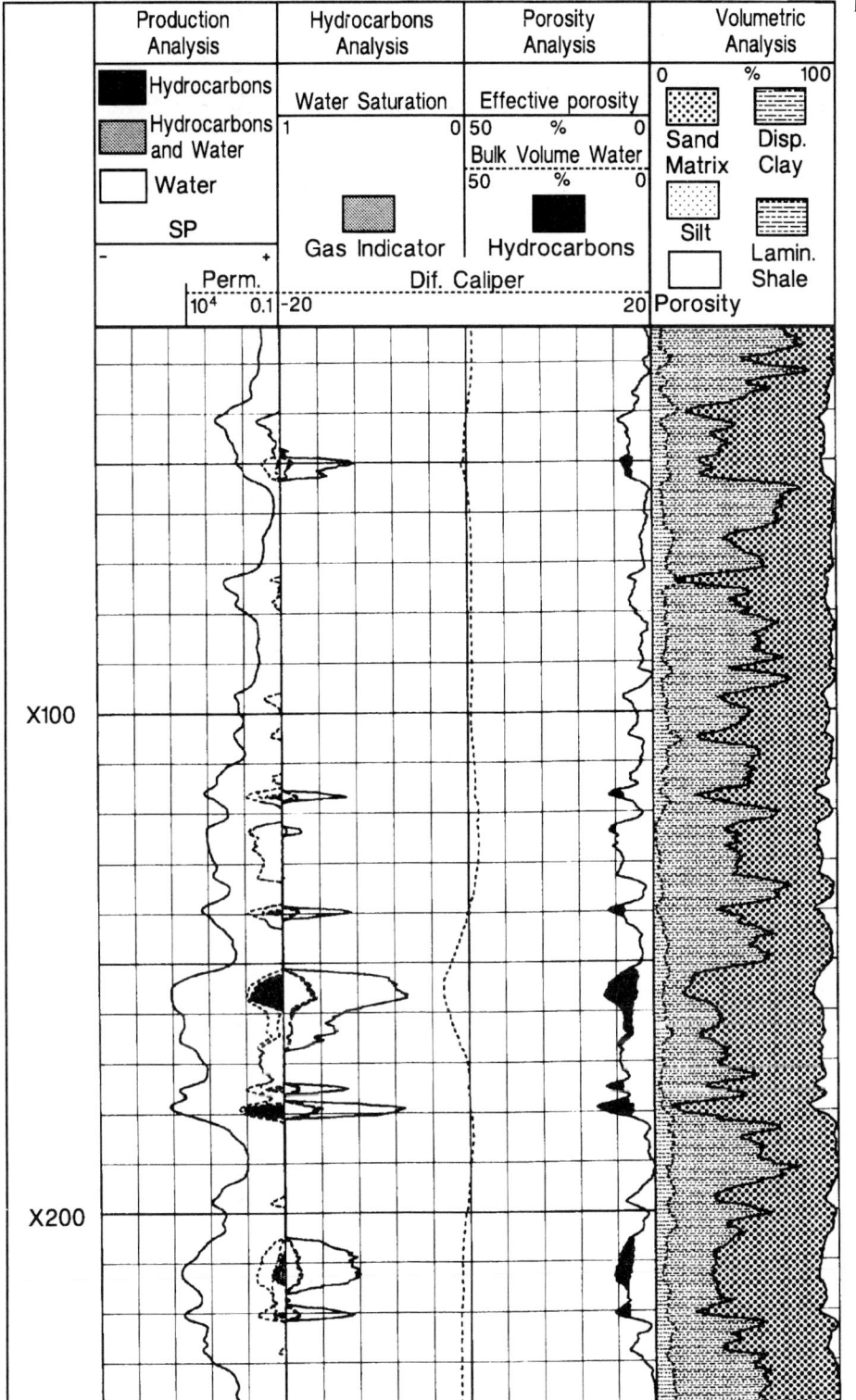

Fig. 2. Well 1: Interpretation results with standard GR, density, neutron, and induction logs. (Vertical scale in feet.)

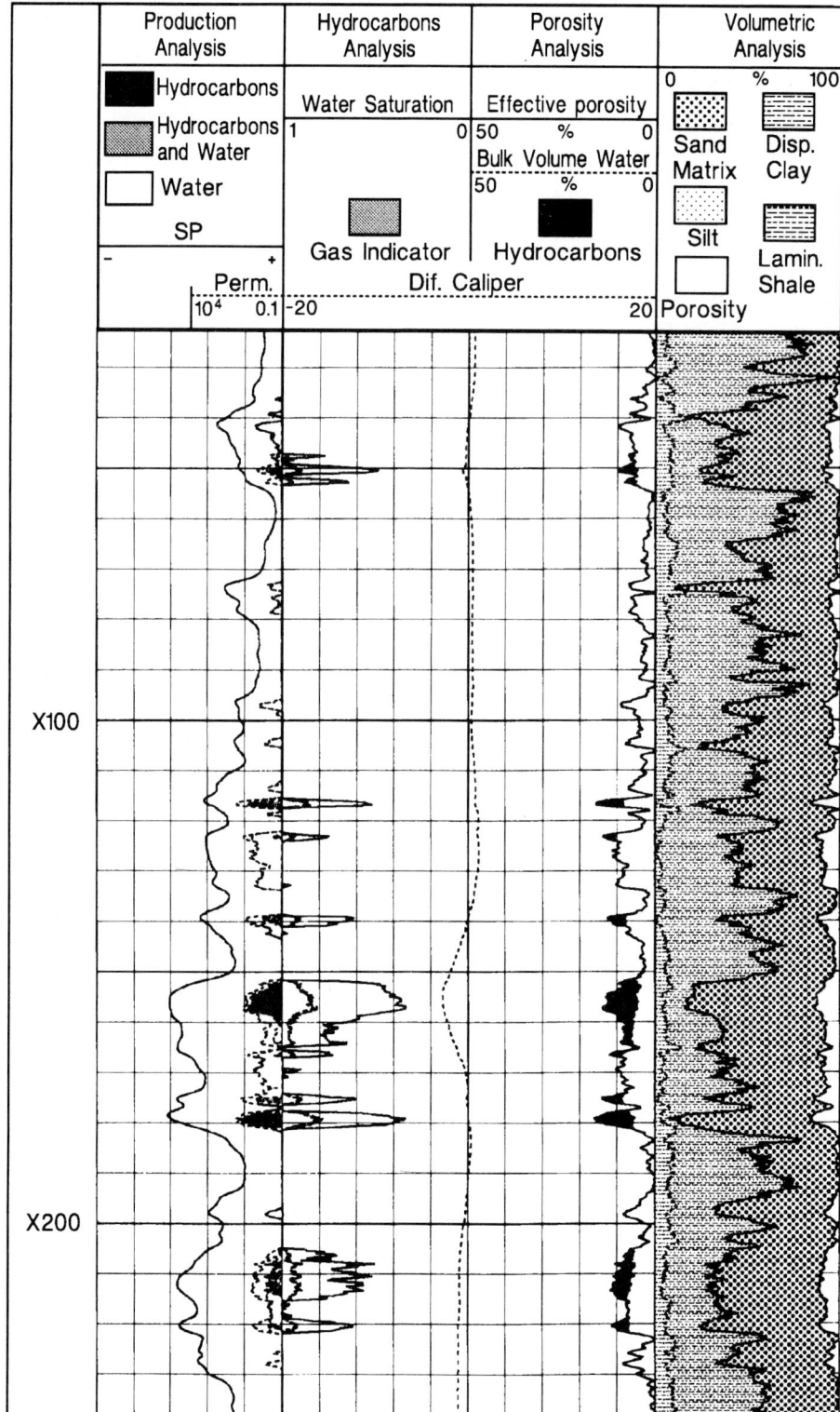

Fig. 3. Well 1: Interpretation results with standard GR and induction logs, and enhanced resolution density and neutron logs. (Vertical scale in feet.)

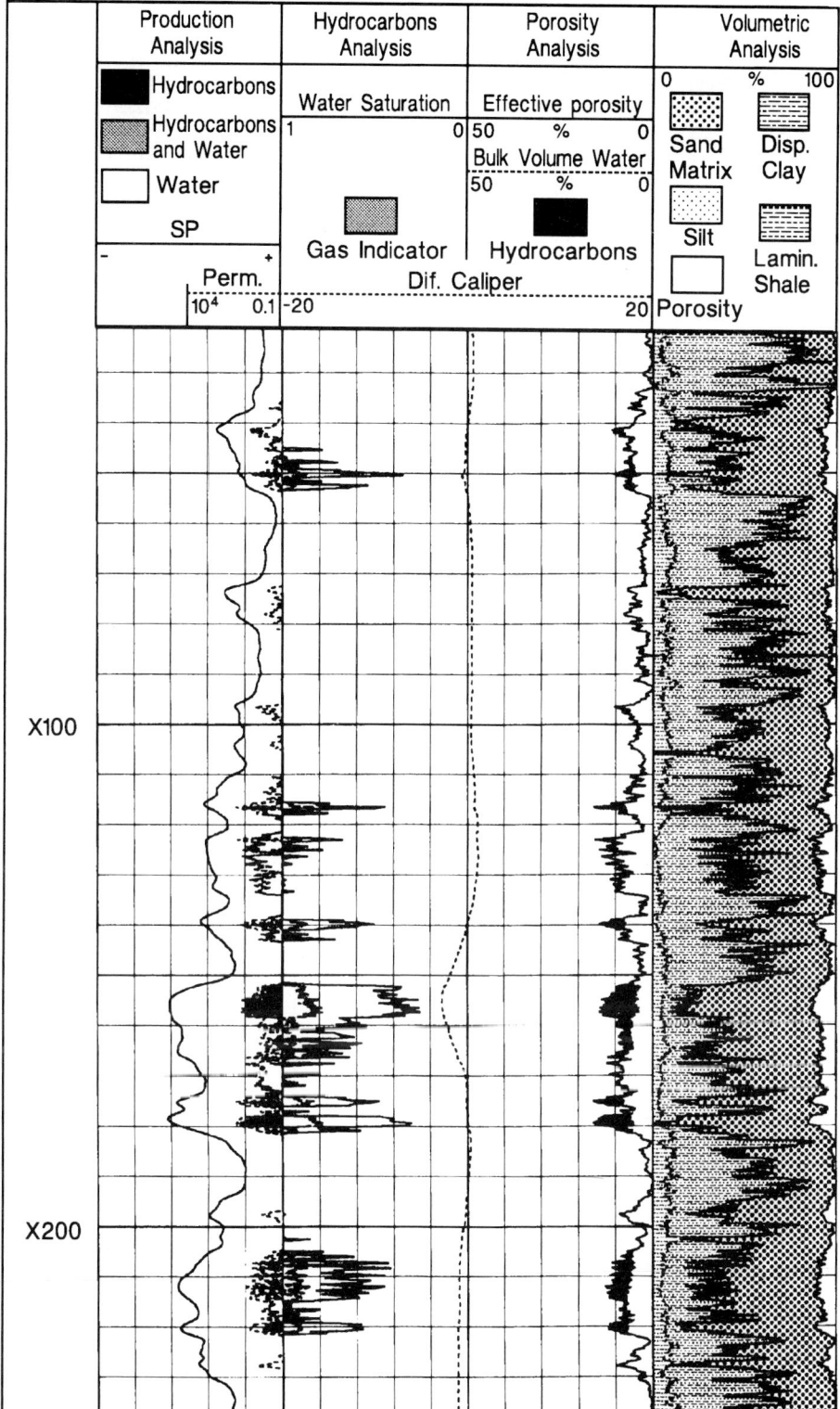

Fig. 4. Well 1: Interpretation results with enhanced resolution GR, standard density, neutron, and induction logs. (Vertical scale in feet.)

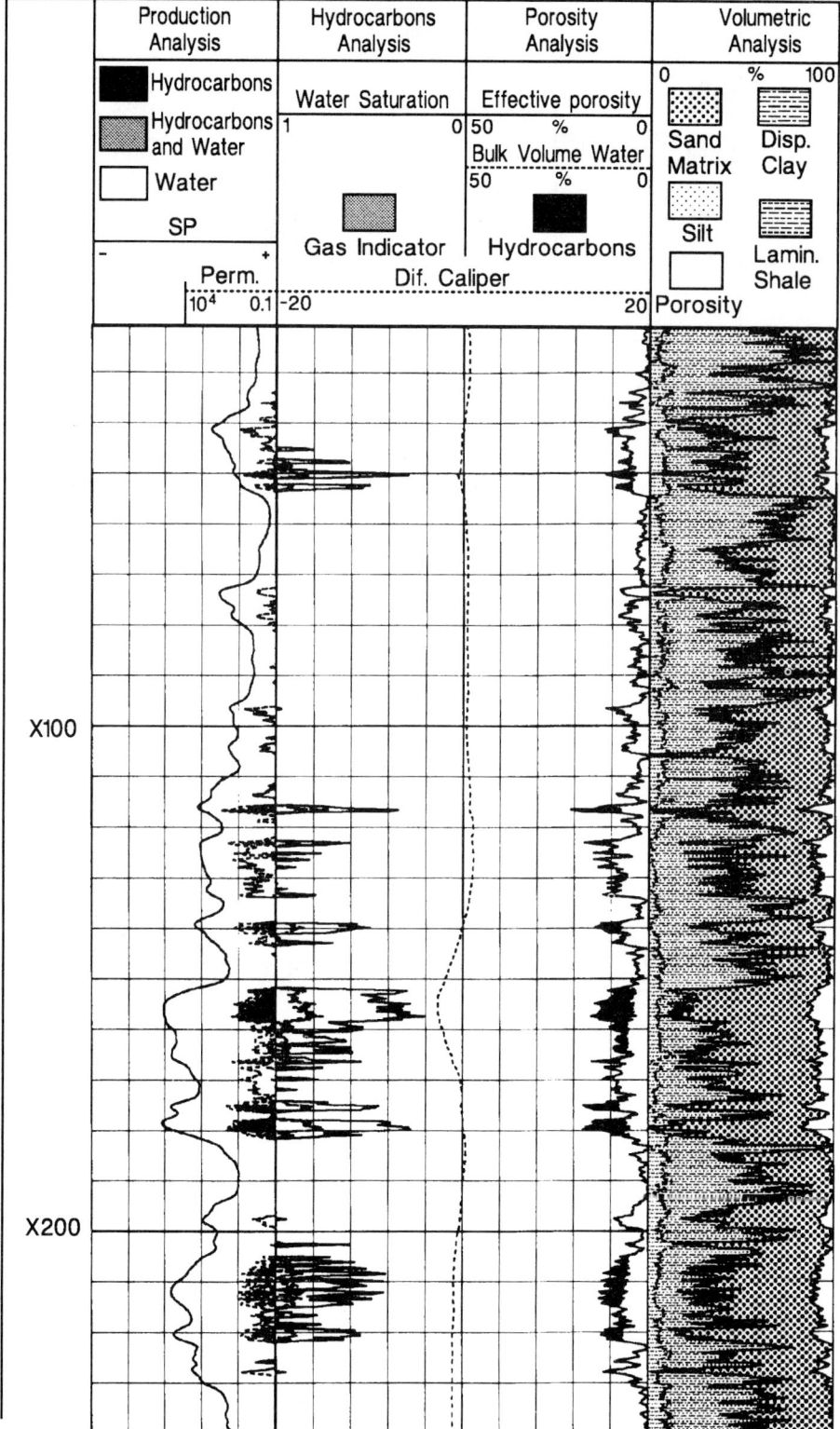

Fig. 5. Well 1: Interpretation results with enhanced resolution GR, density and neutron logs, and standard induction log. (Vertical scale in feet.)

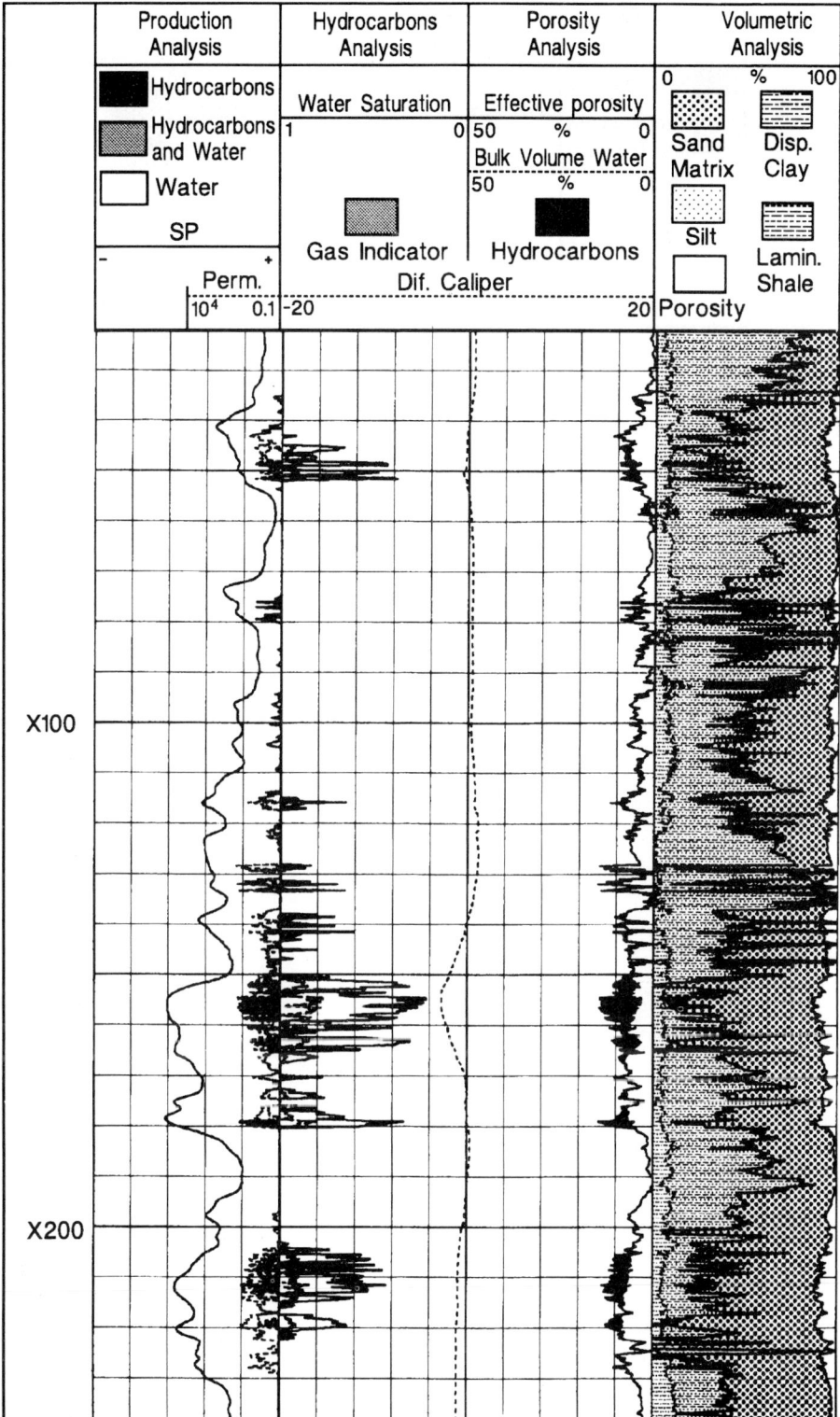

Fig. 6. Well 1: Interpretation results with high frequency dielectric attenuation, standard density, neutron, and induction logs. (Vertical scale in feet.)

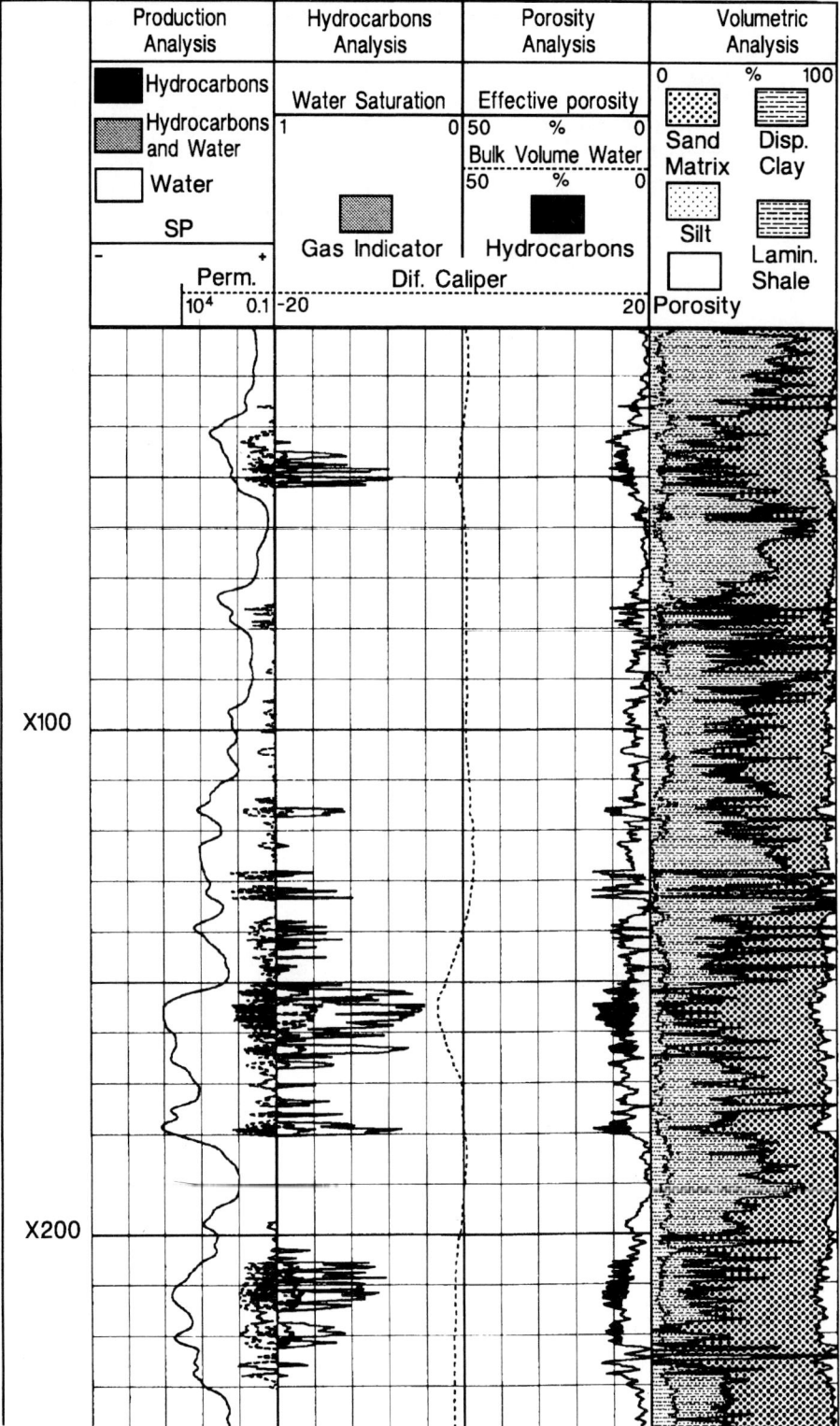

Fig. 7. Well 1: Interpretation results with high frequency dielectric attenuation, enhanced resolution density and neutron logs, and standard induction log. (Vertical scale in feet.)

Figures 3 to 7 show the model outputs as different vertical resolution inputs were selected. For this well, only the standard dual induction log was available, with its low (L) vertical resolution. Two shale indicators, with three different vertical resolutions, were processed: gamma ray (M), enhanced gamma ray (Me), and dielectric attenuation (H). The density and neutron porosity logs were resolution matched, and had either standard (M), or enhanced (Me), vertical resolution.

Table 3 shows the cumulative hydrocarbon feet and cumulative porosity feet for the single zone (X152–X182') as various input/resolutions were selected. Cumulative hydrocarbon feet increased by about 11% when the enhanced gamma ray was used. Smaller increases were observed when enhanced resolution neutron-density porosities were used. A 16% overall increase in hydrocarbon feet, with no change in porosity feet, occurred when the enhanced vertical resolution processed gamma ray, density and neutron porosities were used in place of their standard resolution values. These increases are confined to the lower section of this interval which consists of a number of isolated thin beds.

In the case shown in Fig. 3, for which the standard gamma ray and induction logs were combined with the enhanced density and neutron logs, the porosities and water saturations of the relatively thick zones (up to 6 feet) were virtually unchanged, but several thin beds above and below this interval gained porosity and hydrocarbon saturation.

Considerably greater increases in porosity and hydrocarbon saturation for virtually all thin beds occurred in the case when the enhanced resolution gamma ray was used as a shale indicator, even though standard density, neutron, and induction logs were used (Fig. 4). A further slight increase in computed porosity and hydrocarbon saturation occurred when enhanced resolution gamma ray, density, and neutron logs were used together with the standard induction log (Fig. 5). Especially noticeable increases are seen in the interval (X040–X150') in the very thin shaly sand beds.

No improvement was seen when the standard/enhanced density-neutron porosities were used with the dielectric attenuation as a shale indicator (Figs 6 & 7; Table 3).

The enhanced resolution gamma ray shale indicator (Figs 4 & 5) delineated thin beds about as well as the high-frequency dielectric attenuation measurements (Figs 6 & 7).

Two intervals have been perforated in this well to date. One of them (X205–X222') did not produce any hydrocarbons. The second interval (X042–X054') has produced 200 mcf of gas, 12 barrels of condensate and 3–5 barrels of water with a flowing test pressure of 600 psi. According to the baseline model output (Fig. 2) both of these intervals have water saturations of about 60%. Based on the model outputs maximizing the use of enhanced resolution logs (Figs 5 & 7), water saturations in the first interval are 40–50%, and in the second interval 30–50%.

Well 2

This well is located in the Louisiana Gulf Coast and is characteristic of the highly porous and permeable sandstone of the Frio Formation. Maximum clean porosity is 33%. The available logging suite included a high resolution induction (Le), dual-spaced density and neutron logs (M), dipmeter (H), and gamma ray (M) (Fig. 8).

Enhanced vertical resolution processing of the gamma ray, density, and neutron logs was not available, so the only contrast occurred between the gamma ray and the dipmeter as shale indicators. Figure 9 shows the interpretation output with the standard gamma ray log as the shale indicator. Two major oil deposits occur in the interval (X104–X204'), with water saturations between 10 and 20%, and shale content of only 5–10%.

Figure 10 shows the interpretation results using the high-resolution dipmeter as the shale indicator. A classical thinly laminated sand–shale reservoir is revealed. This is a good example of the original design objectives of a high-resolution interpretation model. Clean sands are interbedded with shaly ones, with shale content of 30 to 50%. Bed thicknesses vary from less than one foot up to approximately 3 feet. Water saturation changes from 10% in clean beds to 40% in shaly sands. From Table 3, the cumulative porosity feet drops 10% and the cumulative hydrocarbon feet drops 14%, when the dipmeter is used as the shale indicator. Obviously, the gamma ray indicates a much cleaner reservoir than the dipmeter.

Well 3

This well is located in South Texas. The reservoir is a thinly laminated sand–shale sequence of the Yegua Formation. Maximum porosity encountered in the clean parts of the reservoir reaches about 32%. Standard (L) and high resolution (Le) induction logs were recorded, as were dual-spaced density and neutron logs and gamma ray logs (M). Enhanced vertical resolution processing (Me) of all the nuclear logs was performed. All input logs are shown in Fig. 11.

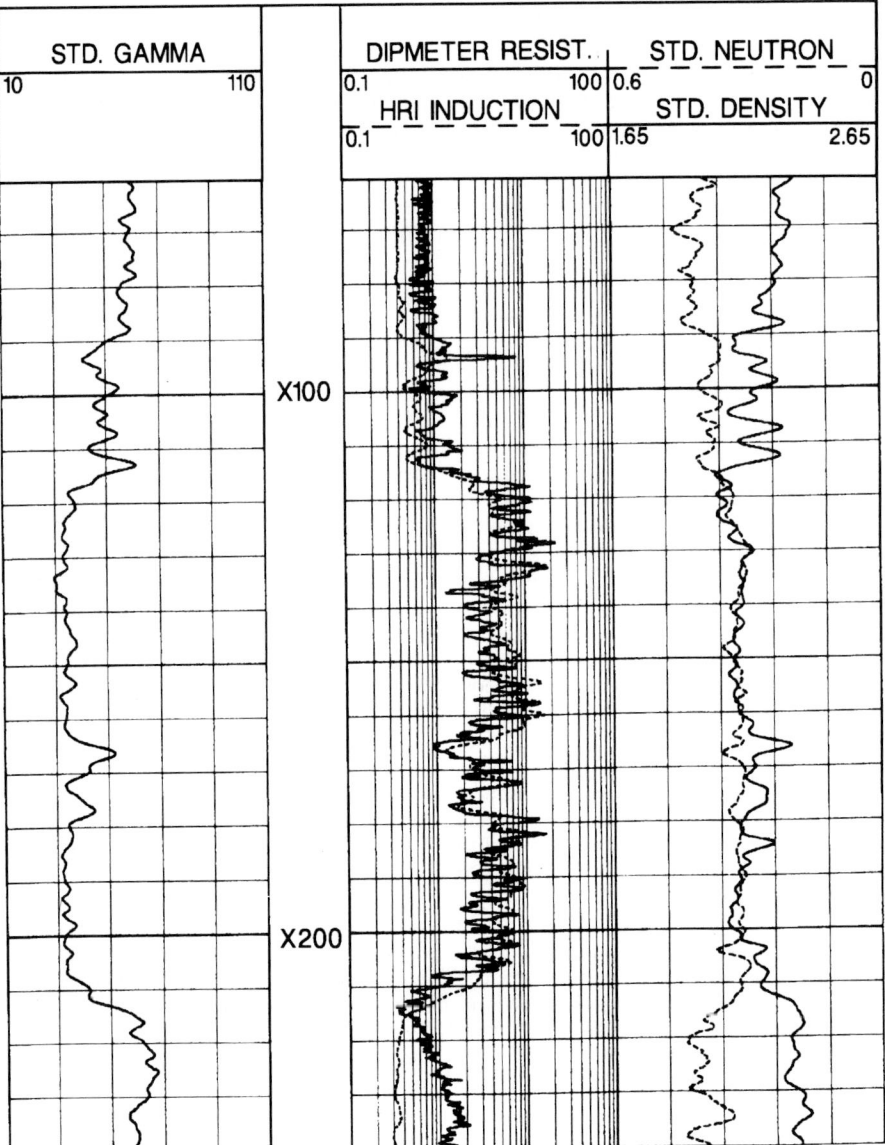

Fig. 8. Well 2: Logs used for interpretation. (Vertical scale in feet.)

The standard interpretation output is shown in Fig. 12, with the gamma ray as the shale indicator. Two oil-bearing zones are indicated in (X518–X524′) and (X596–X600′), with corresponding thicknesses 6 and 4 feet. Several thin beds with residual oil can also be seen in the illustrated section of the well.

Figure 13 displays the model output using the enhanced vertical resolution gamma ray (Me) as a shale indicator, in combination with the stan- dard density, neutron, and induction logs. Small increases in porosity and hydrocarbon satu- ration in several isolated thin sands in the inter- val (X544–X564′) may be seen. The porosity and water saturation of two thicker beds mentioned above remained virtually unchanged.

When enhanced resolution density and neu- tron logs and a high-resolution induction log were used with the standard resolution gamma ray, model output is as shown in Fig. 14. Some-

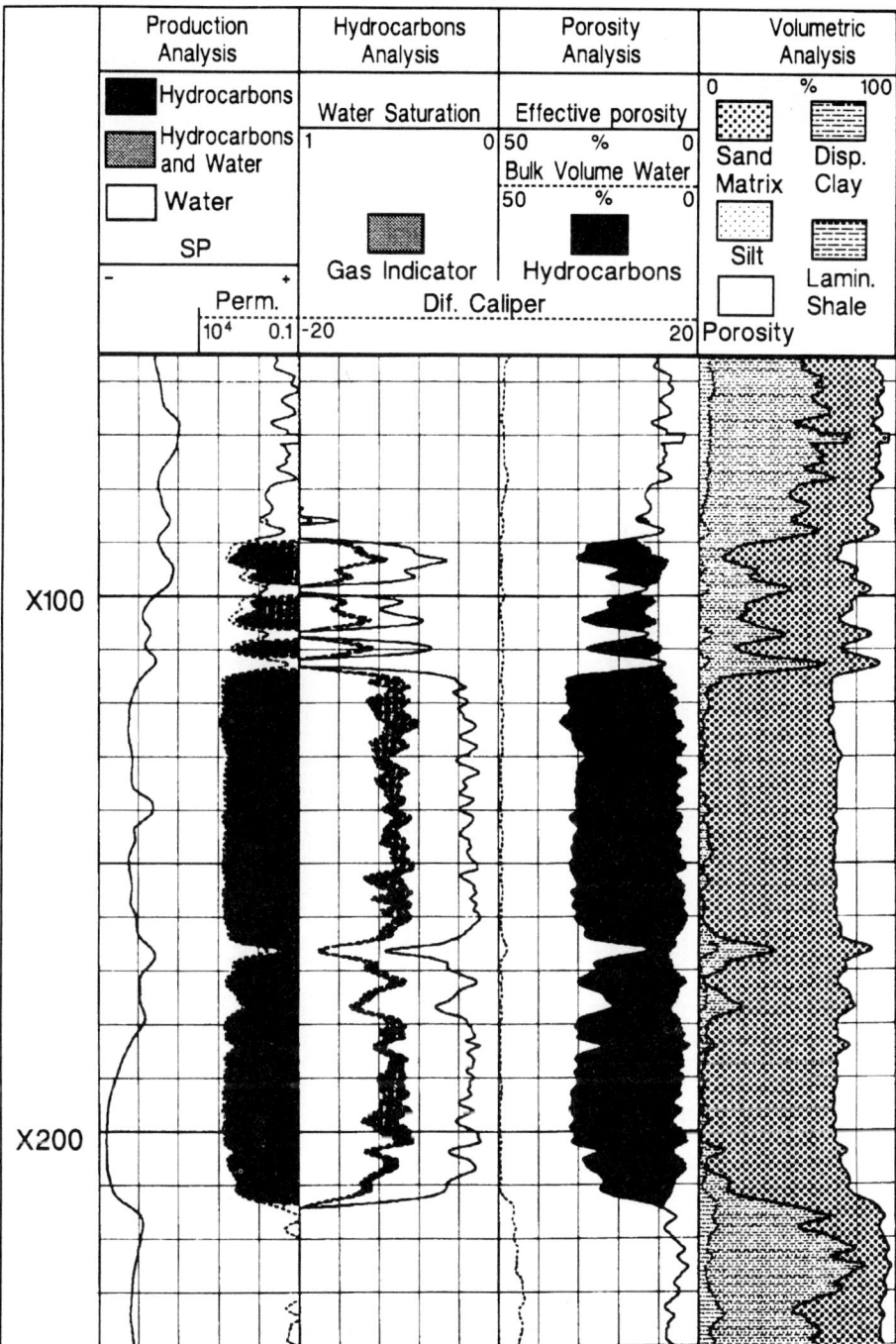

Fig. 9. Well 2: Interpretation results with standard GR, density and neutron logs, and high resolution induction log. (Vertical scale in feet.)

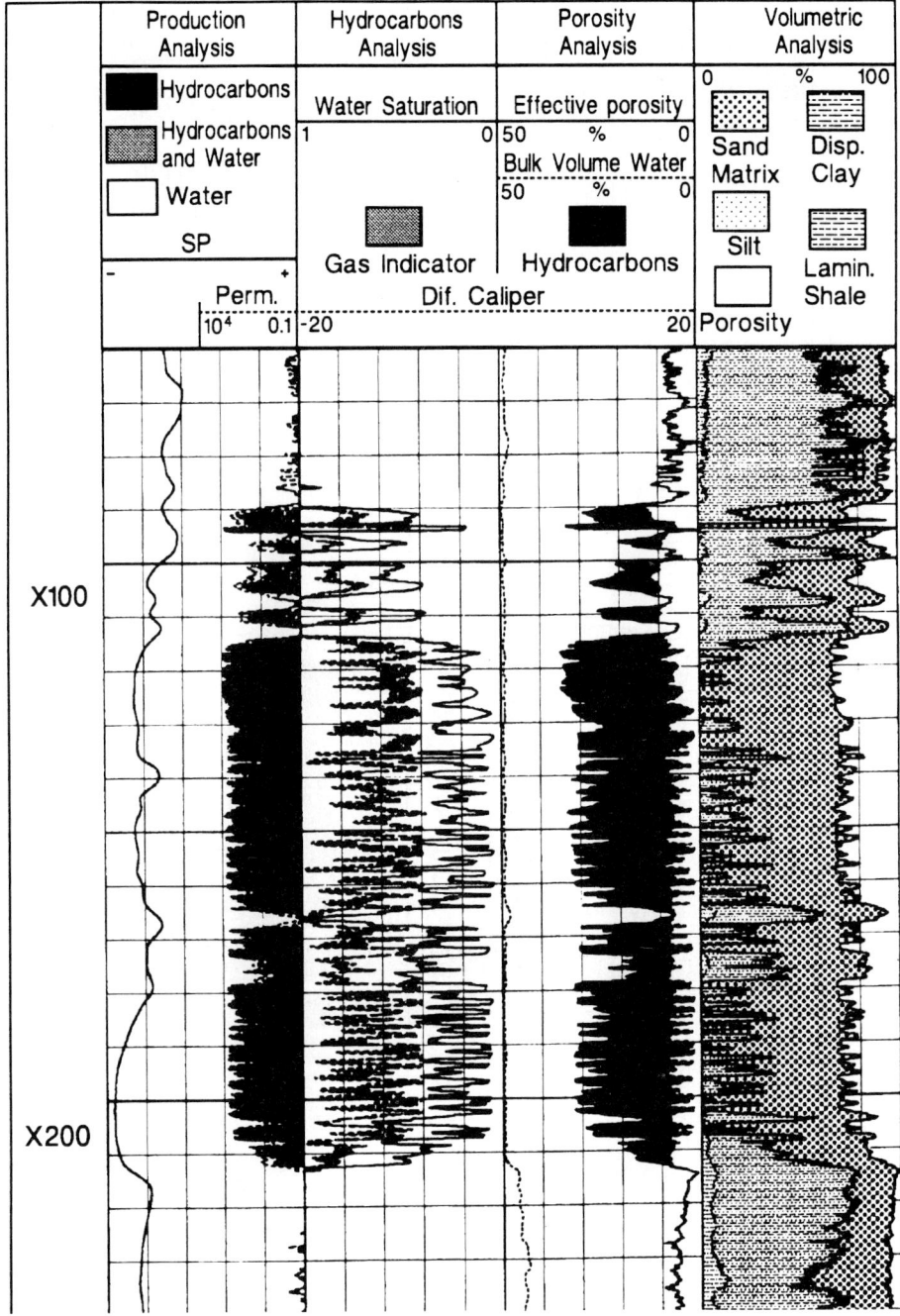

Fig. 10. Well 2: Interpretation results with dipmeter, standard density and neutron log, and high resolution induction log. (Vertical scale in feet.)

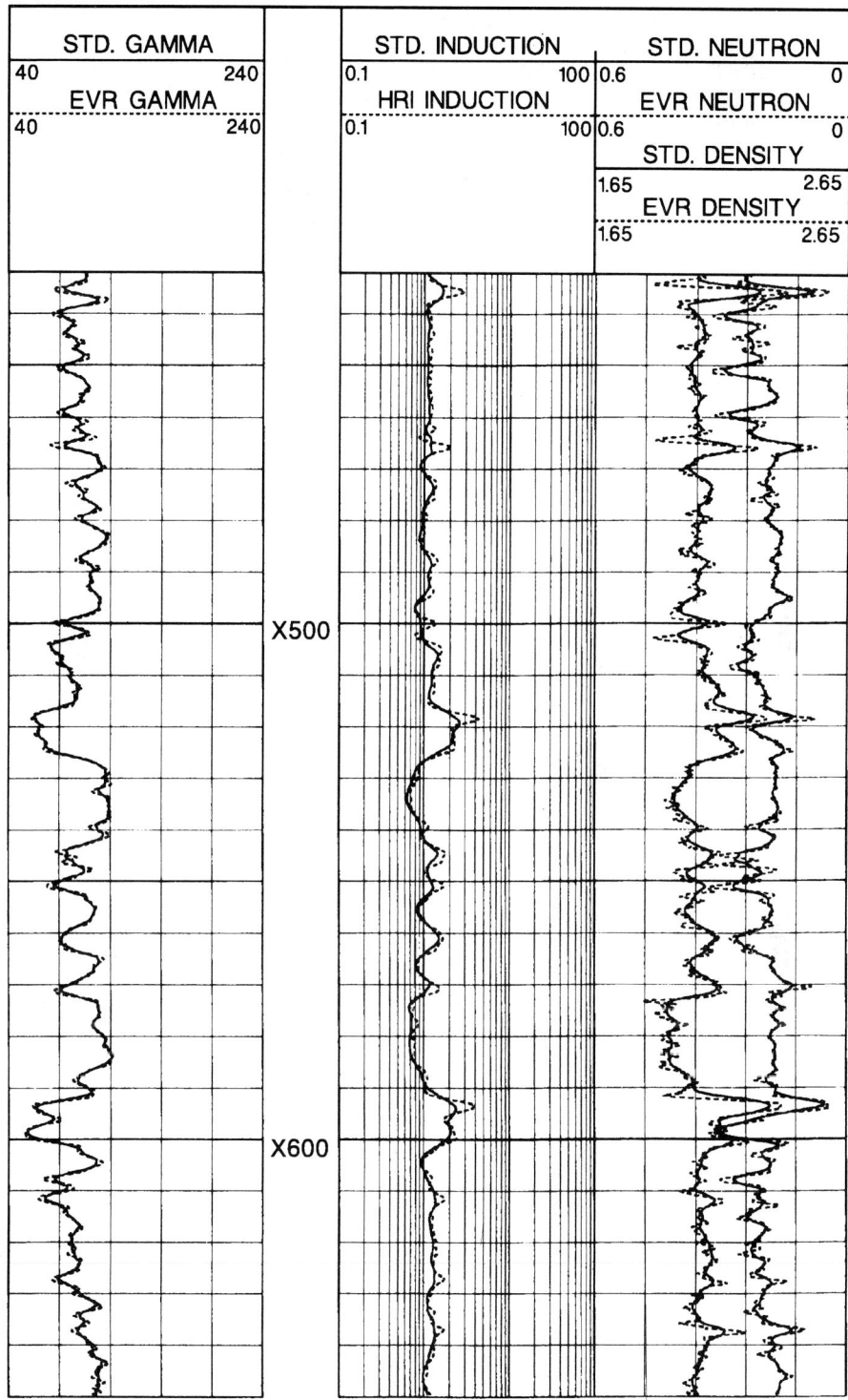

Fig. 11. Well 3: Logs used for interpretation. (Vertical scale in feet.)

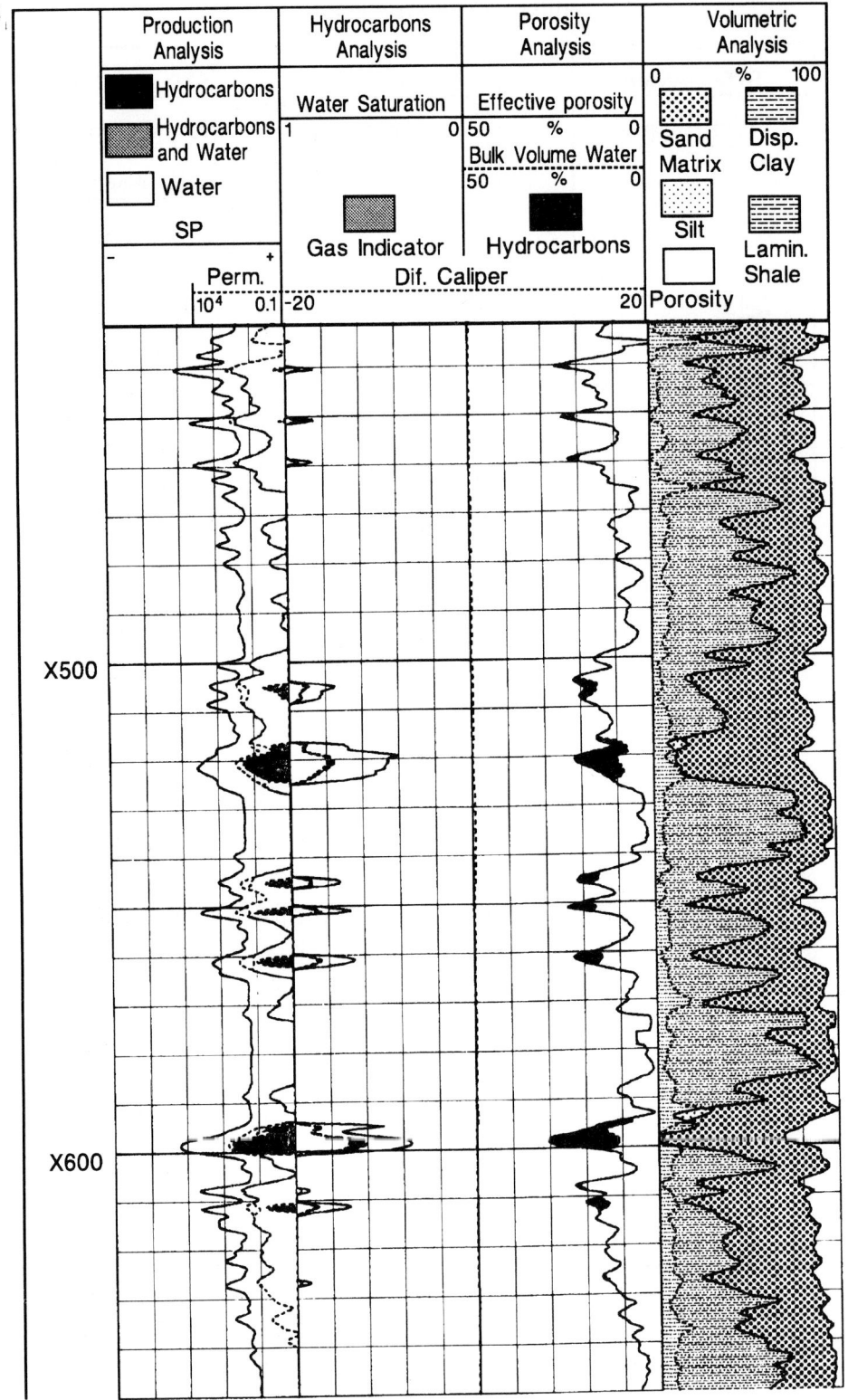

Fig. 12. Well 3: Interpretation results with standard GR, density, neutron and induction logs. (Vertical scale in feet.)

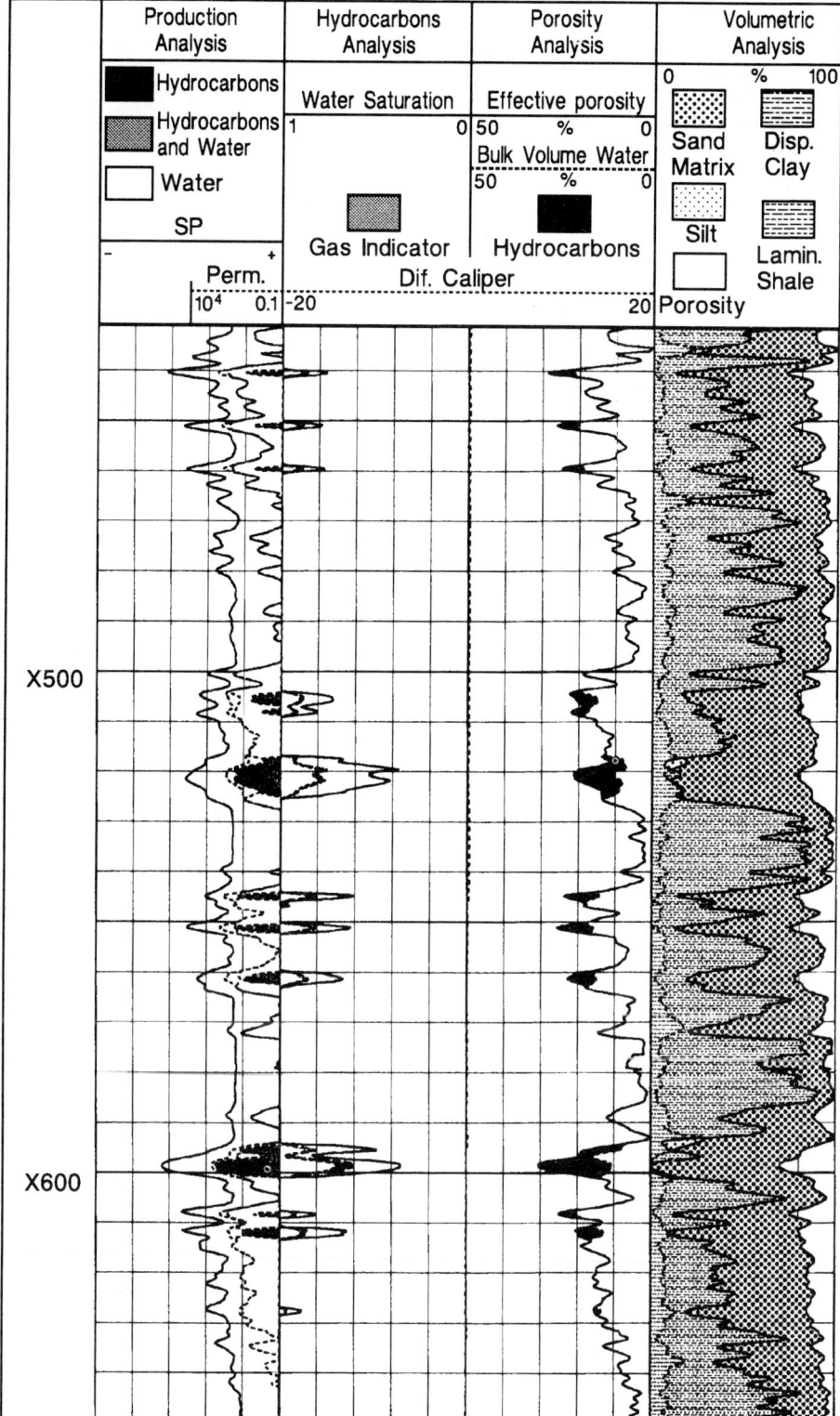

Fig. 13. Well 3: Interpretation results with enhanced resolution GR and standard density, neutron, and induction logs. (Vertical scale in feet.)

116

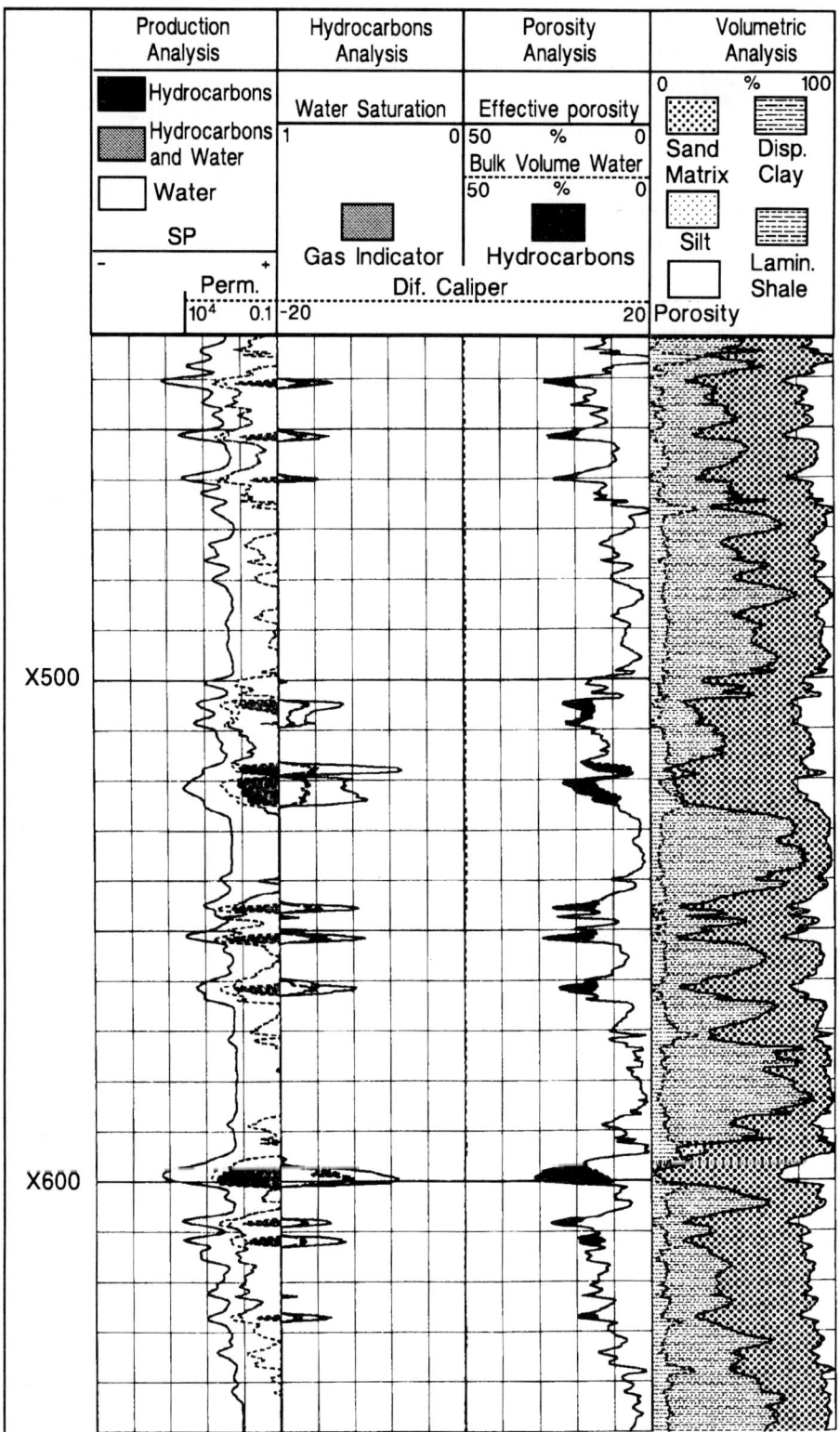

Fig. 14. Well 3: Interpretation results with standard GR, enhanced resolution density and neutron logs, and high resolution induction log. (Vertical scale in feet.)

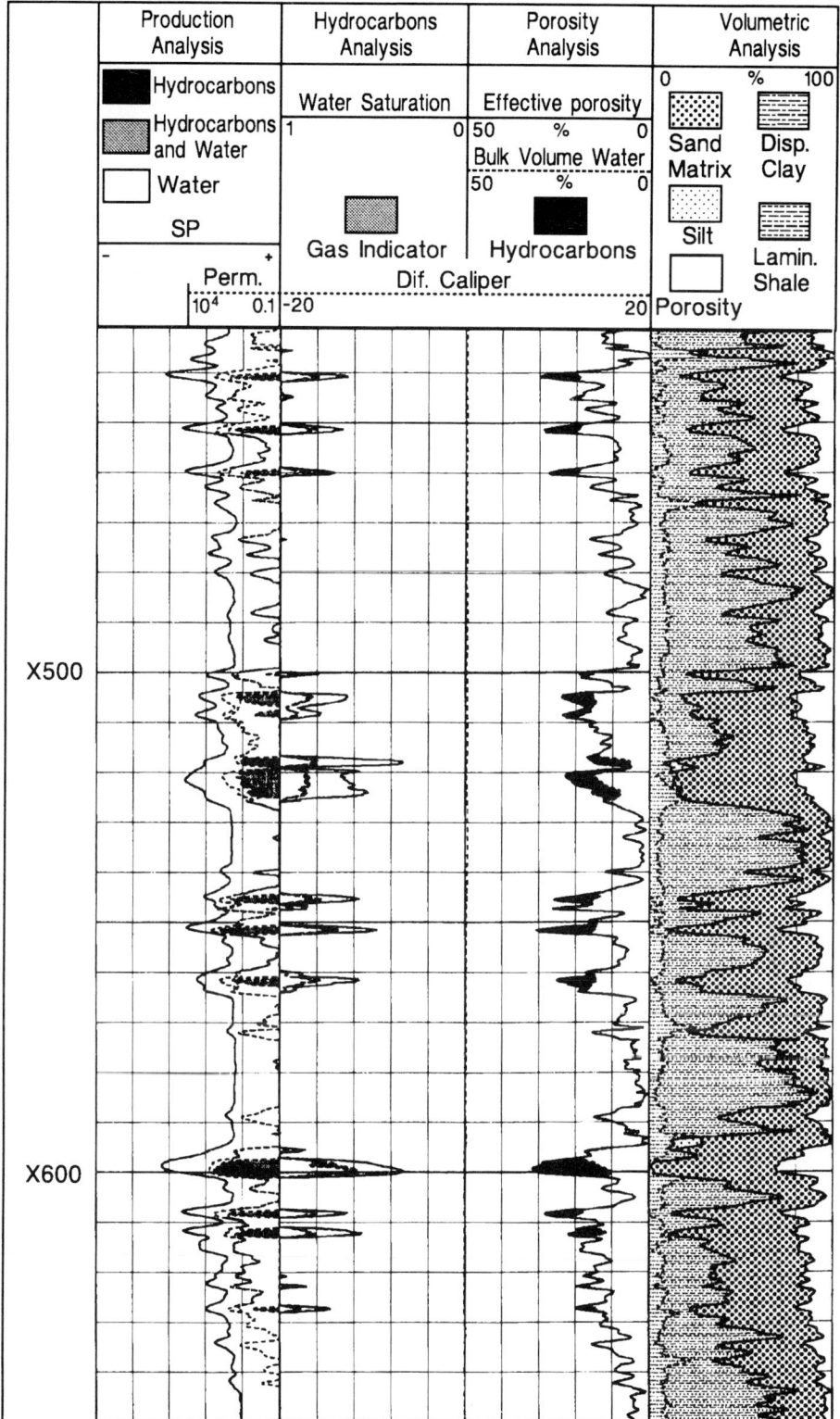

Fig. 15. Well 3: Interpretation results with enhanced resolution GR, density and neutron logs, and high resolution induction log. (Vertical scale in feet.)

what greater increases in porosity and water saturation in the same thin beds were observed. These increases are even more apparent when all the enhanced resolution logs, including the gamma ray, are input to the interpretation model (Fig. 15). One of these thin beds in the interval (X550–X552′), with water saturation less than 50%, can now be considered as potentially productive.

The increases indicated for well 3 apply to several isolated thin beds; these increases are quite significant for each individual bed. However, as indicated in Table 3, the impact on cumulative hydrocarbons for the larger interval (X500–X530′) is not as significant.

Discussion

Examination of the interpretation model outputs for the three wells in the paper, clearly illustrates that the vertical resolution of the input logs produces significant changes to the model output values, and dramatic increases in apparent vertical resolution of output porosity, water saturation, and lithology. Production testing for two intervals of well 1 support the conclusion that interpretation results are more accurate when higher vertical resolution logging data are input to the model.

For well 2, substantial decreases in cumulative porosity feet and cumulative hydrocarbons were noted when the dipmeter was used as a shale indicator. Hence, higher vertical resolution input data and high resolution interpretation models improve accuracy but do not guarantee more optimistic results.

In many cases, output differences were further increased when enhanced vertical resolution porosity logs were used with a high-resolution shale indicator. These results support the concept that interpretation accuracy improves as additional enhanced inputs are used. Two good reasons can be offered to explain such accuracy improvements. First, the model that was used in this study generally assumes the absence of changes in porosity and resistivity within the vertical resolution distances of the individual logging measurements. This assumption is better satisfied for input logs with enhanced vertical resolution because the distances involved all become shorter. Second, logs with enhanced vertical resolution require smaller corrections for shale. Consequently, smaller errors will be introduced into corrected porosity or conductivity values as a result of inaccurate shale volume calculations or unnoticed changes in shale properties.

The dipmeter may be the best high-resolution shale indicator for most applications since it has the best vertical resolution of all the logging tools. This does not mean that other high-resolution shale indicators will always provide results inferior to the dipmeter. For example, in well 2 in the interval (X100–X210′), the deep resistivity of the high-resolution induction tool matches the dipmeter pad resistivity very well because most of the thin beds in this interval have a thickness between 1 and 2 feet. Under such conditions, the high frequency dielectric log or unfiltered Pe log might do as good a job as the dipmeter for a shale indicator.

In well 1, the enhanced resolution gamma ray shale indicator delineated thin beds as well as the high frequency dielectric attenuation measurement. This is very significant, since a gamma ray log is run in nearly every borehole. Thus, enhanced resolution processing of the gamma ray log can provide, in many cases, very valuable additional information about thin bedding conditions for little or no additional cost. In fact, the data in this paper strongly support the concept that enhanced vertical resolution processing of gamma ray, density and neutron logs should become a standard procedure.

Conclusions

(1) For the high-resolution log analysis model discussed in this paper, the vertical resolution of the input logs produces significant changes to model output values and dramatic increases in apparent vertical resolution of output porosity, water saturation and lithology.

(2) Limited production testing supports the idea that interpretation modelling results are more accurate when higher vertical resolution logging data are used. These results are not necessarily more optimistic.

(3) For the model used in this paper, it is not necessary to assume that the total porosities in the sand and shale laminae are equal. It is sufficient to assume that these porosities do not change significantly within the vertical resolution distances of the porosity logging tools. This assumption is further strengthened if enhanced vertical resolution processing is applied to the density and neutron input logging data.

(4) A modified Waxman–Smits saturation formula can be used when both dispersed clay and laminated shale are present in reservoirs.

(5) An enhanced vertical resolution gamma ray shale indicator can sometimes delineate thin beds as well as a high frequency dielectric attenuation measurement.

(6) Enhanced vertical resolution processing of gamma ray, density and neutron logs should

become a standard procedure, even when being used in conventional interpretation models.

(7) If the occurrence of thinly laminated reservoirs is possible, at least one high-resolution measurement such as a dipmeter, high-frequency dielectric log, or unfiltered Pe log should be included in the logging suite. The most appropriate one of these measurements should be utilized as the shale indicator in a high-resolution log analysis model.

The authors thank Brayton Oil Corp., Shell Offshore, Inc. and Ultramar Oil and Gas, Ltd for permission to use log and production data for this paper. The authors also thank Michael Manning, Ronald Menzel, Ronald Sarrat, and Carlos Silva for help in preparing the paper.

Appendix 1: Improvements in vertical resolution by recent tool designs

High-resolution induction tool

Conventional dual induction tools are based on a coil array known as the 6FF40, the industry standard for twenty years. The vertical resolution of the deep induction measurements from this tool is approximately 5 to 6 feet, depending on borehole conditions and side bed conductivities. Thus, thinly bedded reservoirs having zones of a foot or less in extent can easily be hidden. Attempts to improve the 6FF40 vertical resolution (Anderson 1986) using digital signal processing techniques have met with limited success.

A recent induction tool development has greatly improved this situation. In 1987, Strickland *et al.* reported on a new high-resolution induction tool (HRI*) for which the vertical and radial coil arrays are optimized independently. It is a dual induction tool since it has two coil arrays having matched vertical responses. However, each array has a different depth of investigation. A new processing scheme ensures that the HRd (deep measurement) and HRm (medium measurement) will overlay in the absence of invasion. Also, the tool is designed to be less sensitive to eccentring. HRd can resolve beds as thin as two feet and is accurate in beds thicker than three feet, while reading 40% deeper than the conventional deep induction measurement.

High-frequency dielectric tool

A new 1 GHz high-frequency dielectric logging tool (HFD*) is also capable of improving logging accuracy in thinly laminated reservoirs.

This tool also has distinctive features to aid in understanding the invaded zone and in minimizing the effects of borehole washouts and rugosity. Multiple antennas and long-spaced receivers provide deeper investigation depths than earlier tools. Tool sensors are deployed on an independently articulated pad instead of a mandrel configuration. A back-up arm opposite the main pad deploys a microlog sensor.

Unlike earlier tool designs in which the complex propagation constant measurement was a differential measurement made between a pair of receivers, the measurements in this tool are made between the transmitter and each individual receiver. In addition to providing multiple depths of investigation and multiple vertical resolutions, this technique provides a tremendous increase in the dynamic range of the measured signals. Unlike earlier designs, the transmitter and receiver channels continuously sample the formation, thereby improving the measurement of phase and amplitude of weak signals. This feature permits the deployment of a receiver at a longer transmitter-to-receiver spacing and provides a greater depth of investigation.

Another novel feature of this tool is the measurement of incident and reflected signals at the transmitter; dielectric constant and resistivity measurements are obtained after the effects of the transmitter antenna are removed through calibration. This is equivalent to a zero spacing receiver and so maximizes the vertical resolution of the tool. A very high resolution shale indicator, quite insensitive to borehole effects, is thereby provided. For a detailed theoretical description of this tool's measurements, see (Rau *et al.* 1991).

During logging, the phase and amplitude from the incident, reflected, and receiver signals are digitized and recorded at 0.2 inch intervals. In order to maximize vertical resolution and thin bed accuracy further, the tool also provides a Z-axis accelerometer measurement that is used to correct for depth inaccuracies.

Dielectric constant and 1 GHz resistivity values are computed in real time from calibrated and averaged phase and amplitude data, using algorithms developed from mathematical modelling. Microlog resistivity and borehole radii from the two caliper arms are also provided. In addition to providing an independent verification of mudcake thickness, the microlog measurement also provides a second high-resolution resistivity measurement which can be used to verify or supplement the dielectric and resistivity measurements from the dielectric pad.

When sand porosity shows small variations, either dielectric attenuation or propagation time

---

* A mark of Halliburton Logging Services

measurements may be used separately as high-resolution shale indicators. Conversely, when sand porosity shows wider variations, attenuation and propagation time measurements must be used together to obtain a high-resolution shale indicator independent of porosity. We are continuing to search for the best method for determining shaliness from dielectric measurements. In the main part of this paper, a simple linear shale index based on dielectric attenuation alone was used to indicate shaliness.

Pe measurements from spectral density tools

The unfiltered Pe measurement from a spectral density tool can also provide an excellent high-resolution shale indicator for laminated shaly sand evaluation, since its inherent vertical resolution is only 2–3 inches (Moake & Schultz 1987). (Of course, the depth sampling rate should be 4 to 6 times per foot.)

Recent spectral density logging developments have made this measurement even better. In some tools, Pe is being measured by the short-spaced detector, which has both improved vertical resolution and reduced statistical errors relative to Pe, obtained using long-spaced detector measurements.

Appendix 2: Enhanced vertical resolution processing for the gamma ray and dual-spaced density and neutron logging tool raw data

In the past, the vertical resolution of the standard nuclear logging tools had been limited to three or more feet, mainly because of limitations imposed by the vertical response of the far neutron detector measurement and the need to keep all measurements from all tools both depth-aligned and vertically matched.

More sophisticated processing of the raw data from the standard nuclear logging tools can improve vertical resolution. Early efforts (Galford *et al.* 1986; Flaum *et al.* 1987) often referred to as 'alpha processing' or 'near detector profiling' focused on the use of the near detector measurement to improve the dual-detector response of density and neutron tools. These methods assumed that borehole conditions remain constant over a distance of about 3 feet. This assumption is invalid if washouts, caves, or general rugose borehole are present. Such methods interfere with the basic shop calibrations and dual-detector compensation methods originally developed for both the density and neutron logging tools. These methods

cause an imbalance between the near and far detector vertical response intervals, i.e. they disturb the vertical matching of these tools.

In this paper, as well as many other logging applications, it is vital to ensure that the logging measurements provided by all the nuclear tools, both individually and collectively, remain vertically matched to the same depth interval. This means that, at least vertically, all the tool measurements are sensing the same formation material. In this way, the shale volume and porosity computations will be accurately made; spurious beds (sometimes called horns) will not appear at the transitions from one bed to another; and false gas shows and mineralogical changes will be minimized. These remarks are true whether the logs are processed conventionally or by a reconstructive model (such as the one used in this paper) in which calculations are performed on several different vertical resolution levels.

A new method has recently been developed for obtaining dual-spaced neutron measurements with a vertical resolution of about 40 cm by applying a deconvolution method to the far detector count rate log (Smith 1990). A special smoothing technique is also used on the near count rate log to keep both neutron detector logs vertically matched. This smoothed near rate and deconvolved far rate provide a deconvolved near/far ratio with enhanced vertical resolution. The standard shop calibration procedures and borehole compensation methods can be used with this enhanced ratio to compute a neutron porosity with substantially improved vertical resolution. The density log does not require deconvolution to achieve the same vertical resolution of this enhanced neutron measurement, but significantly less filtering is applied than in standard density processing. Like the neutron tool, conventional shop calibrations and borehole compensation methods are retained for the density tool. Thus the full accuracy of the dual-spaced method is retained (separately) for both the neutron and density tool measurements under all logging conditions, including rugose boreholes. Log repeatability also remains very good. These factors combine to produce a much improved enhanced vertical resolution neutron/density log for use in laminated shaly sand thin bed analysis.

Enhanced processing of the gamma ray tool data (sometimes called H6 or matched filtering) may also be performed (Jacobson *et al.* 1990). The combined new processing of dual-spaced neutron and density data, and the gamma ray data, is called enhanced vertical resolution processing (EVR). Using the notation of Table 1 in the main text, this processing improves the vertical resolution of the combined gamma ray, density, and neutron tool measurements from the medium level (M) to the enhanced level (Me), i.e. from 75 cm to about 40 cm.

As mentioned above, enhanced vertical resolution processing centres around deconvolution of the far neutron detector count rate. This is a new and controversial subject, but several key issues can briefly be addressed. Van Cittert deconvolution is used. This type of deconvolution is not extremely aggressive, but a statistical pre-filter operation is applied to the raw data before the actual deconvolution process. This pre-filter is heavy in high porosity formations where count rates are low and statistical pre-

cision and repeatability are poor; this ensures that the deconvolved far detector count rate remains well-behaved during the logging process.

Another important issue concerns repeatability more directly. Although vertical resolution improves, and exactly the same shop calibration and borehole compensation procedures are retained, repeatability is somewhat reduced. The EVR processing reduces repeatability by about a factor of two. Experience shows that this is quite acceptable, i.e. that a new compromise between vertical resolution and repeatability has been established. This point is covered in detail by Smith (1990).

References

ALLEN, D. F. 1984. Laminated Sand Analysis. *Transactions of the SPWLA 25th Annual Logging Symposium*, Paper **XX**.

ANDERSON, B. 1986. The Analysis of Some Unsolved Induction Interpretation Problems Using Computer Modelling. *Transactions of SPWLA 27th Annual Logging Symposium*, Paper **II**.

BATEMAN, R. M. 1990. Thin Bed Analysis with Conventional Log Suites. *Transactions of SPWLA 31st Annual Logging Symposium*, Paper **II**.

FLAUM, C., GALFORD, J. E. & HASTINGS, A. 1987. Enhanced Vertical Resolution of Dual Detector Gamma-Gamma Density Logs. *Transactions of SPWLA 28th Annual Logging Symposium*, Paper **M**.

GALFORD, J. E., FLAUM, C., GILCHRIST, Jr., W. A. & DUCKETT, S. W. 1986. Enhanced Resolution Processing of Compensated Neutron Logs. *61st SPE Annual Technical Conference and Exhibition*, Paper 15541.

JACOBSON, L. A., GADEKEN, L. L., MERCHANT, G. A. & WYATT, Jr., D. F. 1990. Enhancement of Nuclear Logging Measurements Through Deconvolution. *Transactions of SPWLA 31st Annual Logging Symposium*, Paper **TT**.

MOAKE, G. L. & SCHULTZ, W. E. 1987. Improved Density Log Lithology Identification Using A Borehole-Compensated Photoelectric Factor. *Transactions of SPWLA 28th Annual Logging Symposium*, Paper **FF**.

QUINN, T. H. & SINHA, A. K. 1985. Comparative Results of Quantitative Laminated Sand/Shale Analysis in Gulf Coast Wells Using Maximum Diplog Microresistivity Information. *Transactions of SPWLA 26th Annual Logging Symposium*, Paper **QQ**.

RAIGA-CLEMENCEAU, J. 1988. Taking into Account the Conductivity Contribution of Shale Laminations When Evaluating Closely Interlaminated Sand–Shale Hydrocarbon Bearing Reservoirs. *Transactions of SPWLA 29th Annual Logging Symposium*, Paper **DD**.

RAU, R., DAVIES, R., FINKE, M. & MANNING, M. 1991. Advances in High Frequency Dielectric Logging. *Transactions of SPWLA 32nd Annual Logging Symposium*.

RUHOVETS, N. 1990. A Log Analysis Technique for Evaluating Laminated Reservoirs in the Gulf Coast Area. *The Log Analyst*, **31**, 294–303.

SMITH, M. P. 1990. Enhanced Vertical Resolution Processing of Dual-Spaced Neutron and Density Tools Using Standard Shop Calibration and Borehole Compensation Procedures. *Transactions of SPWLA 31st Annual Logging Symposium*, Paper **SS**.

STRICKLAND, R., SINCLAIR, P., HARBER, J. & DEBRECHT, J. 1987. Introduction to the High Resolution Induction Tool. *Transactions of SPWLA 28th Annual Logging Symposium*, Paper **E**.

Automated prediction of sedimentary facies from wireline logs

ERIK BØLVIKEN,[1] GEIR STORVIK,[2] DAG ERIK NILSEN,[3] ERLING SIRING[4]
& DIRK VAN DER WEL[3]

[1] *University of Oslo and The Norwegian Computing Center, P.O. Box 114 Blindern, N-0314
Oslo 3, Norway*
[2] *The Norwegian Computing Center, P.O. Box 114 Blindern, N-0314 Oslo 3, Norway*
[3] *Norsk Hydro a/s, P.O. Box 200, N-1321 Stabekk, Norway*
[4] *Statoil a.s, P.O. Box 300, N-4001 Stavanger, Norway*

Abstract. The problem addressed is whether a computer can be programmed to identify depositional facies from a set of wireline logs. The basic approach is to let the computer learn by itself the patterns to search for by feeding it log signatures that have already been assigned facies labels. Having gone through this training phase, it can make sedimentary predictions from new data. The underlying model is a mathematical formalization of the idea that sedimentary processes have deposited lithological sequences which influence the observed log traces. Stochastic descriptions are used for these relationships. Markov chains link the lithology to the underlying sedimentary facies. The upward transition probabilities of the Markov chain are the main features which discriminate sedimentary facies.

An efficient reconstruction algorithm permits probabilistic restoration of both lithology and sedimentology. This allows the uncertainty of the conclusions to be quantified, and more than one interpretation can be put forward where appropriate. Results of the tests are promising.

Can a computer be programmed to produce sensible guesses concerning depositional facies from a set of wireline log traces? Such a method could be routinely applied to any logged well. Its verdict might not be the final one, but it would provide a quick and easy way to inspect many wells. The output would be objective and reproducible and could be coupled with other programs. In partially cored wells, conventionally obtained sedimentological interpretations could be interpolated across gaps.

The paper reports on an attempt to set up an automatic, computerized method of this kind. The ultimate aim is to develop an easy-to-use procedure which can make useful sedimentological summaries of many wells with little human intervention. The method may not outperform experienced log analysts, but it will be a work saver. It is important that a high degree of automation is maintained. Figure 1 illustrates the general approach. The idea is to let the computer teach itself the log shapes to search for by feeding it log signatures that have already been assigned facies labels. If successful, this holds the big advantage that the procedure can easily be readapted for use in other fields. The method takes as its starting point one or more training wells where a sedimentological 'ground truth' has been set up in some way, in practice by conventional sedimentological studies

in cored regions of these wells. The computer uses the training sample to calibrate a preprogrammed, probabilistic model which defines log shapes mathematically. When this task has been completed so that the characteristic log signatures have been formalized, the information is used to search for similar patterns in other wells. The computer will then suggest sedimentary classifications in the new wells. This kind of approach is standard in image processing. Kerzner (1986) applies such ideas to well logs.

The success of this ambitious scheme hinges on several factors. Firstly, there is the big question as to whether log signatures of depositional facies are sufficiently stable and sufficiently discriminating among themselves for sedimentary interpretations to be possible at all. This varies, and some facies will be more easily confused than others. For example, log signatures of mouthbars and tidal sandflats may look quite similar, while those of coastal swamps or marine facies are very different. The human log analyst encounters the same problem, however. Secondly, it is obvious that the log shapes used to tune the method must be, in some sense, 'representative' or 'typical' for the facies to be identified. Moreover, facies not present in the training wells will by definition not be picked up elsewhere. (Information could be supplied in other ways, but this would reduce the automa-

From HURST, A., GRIFFITHS, C. M. & WORTHINGTON, P. F. (eds), 1992,
Geological Applications of Wireline Logs II. Geological Society Special Publication No. 65, pp. 123–140.

123

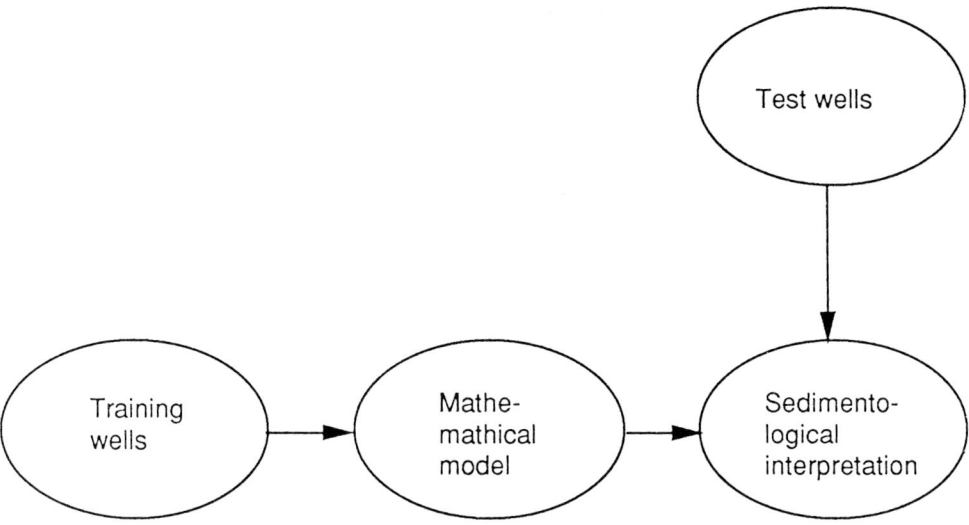

Fig. 1. The main elements of the approach.

tion). Finally, the mathematical model must be able to describe complex and fluctuating patterns in several log variables simultaneously. It is also desirable that it should handle additional sources of information such as prior expectations of facies thickness and sedimentological context. We shall see later that all these discriminating factors are balanced against each other in one large computation. A single piece of information is rarely decisive. The sedimentological interpretation must be based on an overall assessment.

The mathematical model, presented in the next section, is a formalization of the idea that sedimentary processes produce stacked lithofacies which in turn influence the log responses. Writing L for lithofacies, the log traces X and the sedimentary facies S are thus linked through the scheme

$$S \to L \to X \qquad (1)$$

which expresses the idea that sedimentary environments influence the logs indirectly through sequences of rocks deposited. Both relationships in (1) are described by stochastic mechanisms. The computer is given X. Its job is to recover L, and having done so, use the L-sequences with other sources of information to make sedimentological interpretations. Actually, as we shall see, the restoration of L and S is done in a joint operation which is a summation of a large number of probabilities. This also provides a framework for evaluation of the uncertainty of

the conclusions. More than one sedimentary interpretation can be put forward when there is doubt.

Model adaptation is an important part of the approach. It will be demonstrated how key parameters in (1) can be adjusted to a given field. This includes attempts to construct lithofacies statistically by an advanced type of clustering procedure specifically designed for the present purpose.

A final word on some of the key concepts is in order. In the present paper, sedimentary facies are considered to be a set of labels assigned in the training wells, thereby defining log patterns to search for later. Of course for the eventual classification to make sense sedimentologically *these labels must correspond to depositional facies*. However, this level of interpretation is irrelevant for the discussion of the methodology presented here. Similarly, we envisage lithofacies as groups of rocks with homogeneous properties. But our treatment lacks any deeper meaning since lithofacies will be defined statistically to be used as tools to describe log shapes mathematically. Our lithofacies correspond to geological entities only to the extent that these entities produce stable log responses. Finally, the mathematical models with which we shall be concerned are not genetic. They say little about how reservoirs have been created, and they are not even approximately 'true'. They should be judged for their predictive ability rather than for their proximity to the 'truth'. The data analysis presented here therefore concentrates on discussing the actual sedimentological interpretation. The methodology will be tested on three wells from an off-shore sandstone reservoir. One well is used for training and the others for testing. Detailed discussion of the results will indicate the potential of the method.

The model

Each recorded log depth, indexed t, is assigned a sedimentary facies s_t and a lithofacies l_t. The latter are envisaged as discrete classes such as coarse sand, medium sand, shale, coal and so on. It may be that lithofacies in reality are on a continuous scale. Discretization is justified, however, both because it leads to an operational mathematical model and because the method, in sticking to geological tradition, is expressed in a language familiar to users. A continuous phenomenon can in any case be approximated by a discrete model provided the discretization is sufficiently fine. In the present situation this means a large number of lithofacies. The technology has been developed with this possibility in mind (see further comments in the final section).

The model belongs to the so-called Bayesian class where the pair (s_t, l_t) is assumed to be the outcome of an unobservable, stochastic process (S_t, L_t). What is observed is a suite of log variables at each t. The first part of the model deals with the hidden or latent process (S_t, L_t), known as the *prior* in Bayesian statistics. The most important part of this submodel is the assumption that a lithological sequence within a sedimentary facies is the product of a Markov chain with transition probabilities characteristic of that facies. Thus we are introducing probabilities of the form

$$p_s(l'|l) = \Pr\{L(t-1) = l'|L(t) = l, \\ S(t-1) = S(t) = s\} \qquad (2)$$

which quantifies the likelihood that lithofacies l' (at $t-1$) succeeds lithofacies l one depth unit below (at t) given s as the sedimentary facies. (We use throughout the paper the convention that the index t increases with depth, so that $t = 1$ at the top and $t = n$ at the bottom. All transition probabilities are defined upwards.) (2) is a so-called conditional probability, the statements to the right of the vertical bar defining the condition. (Similar notation is used with (5) and (6) below.) Note that this is a way to express that different lithological patterns are expected under different sedimentological regimes. Thus, facies are, within the model, discriminated by their transition probabilities. This is not in itself a new

idea (see Read & Merriam 1971) and Markov chains have, of course, been used repeatedly in quantitative sedimentology (for example Schwarzacher 1975). These are not in any way genetic models. They are simply a convenient way to state prior anticipations about certain lithofacies sequences being more likely than others.

Most geological applications of Markov chains employ the so-called homogeneous version where the transition probabilities are the same everywhere within a sedimentary bed. We believe this form to be inadequate for the present purpose. It is obviously not true with upfining or upcoarsing sequences where the likelihood of passage to another lithofacies depends on position within the current bed. Neither are homogeneous Markov chains able to express the important discriminating feature that some sedimentary facies typically start or terminate with abrupt changes in lithology. It follows from these observations that the transition probabilities should be related to location within the underlying bed. This may appear to be a problem since the bed boundary is to be reconstructed by the technique. Our solution is to divide a facies into distinctive zones called subfacies. The hierarchic scheme (1) is thus expanded to

$$S \rightarrow B \rightarrow L \rightarrow X \qquad (3)$$

where B stands for subfacies. This means that each sampled point t is assigned a triple

$$\omega_t = (s_t, b_t, l_t) \qquad (4)$$

which will be called the *state* at t. Here b_t is a subfacies number. A sedimentary facies is entered at subfacies one, then after a while it is succeeded by subfacies 2 and so on until the maximum subfacies specified for the facies is reached. By this technique the model keeps track of the approximate position within the facies, and it becomes possible to link the transition probabilities (2) to both s and b.

The states (4) are regarded as the outcome of a stochastic state process $\Omega_t = (S_t, B_t, L_t)$. It was assumed above that the sequence of lithofacies l_t was the outcome of a Markov chain working within the current sedimentary facies. This assumption is now extended to the whole state vector so that Ω_t itself is a homogeneous Markov chain with transition probabilities

$$p(\omega'|\omega) = \Pr\{\Omega_{t-1} = \omega'|\Omega_t = \omega\} \qquad (5)$$

There are two reasons for imposing this restriction on the Ω_t process. Firstly, it is actually possible to recover it from the data and keep the technicalities under control, as is demonstrated in the next section. Secondly, Markov models respresent a versatile class which enables us to enter several discriminating features known to influence log analysts making sedimentary interpretations from well logs. Indeed, a formal decomposition of the right hand side of (5) yields

$$\Pr\{\Omega_{t-1} = (s',b',l')|\Omega_t = (s,b,l)\} = \quad (6)$$
$$\Pr\{S_{t-1} = s'|\Omega_t = (s,b,l)\}$$
$$\times \Pr\{B_{t-1} = b'|S_{t-1} = s',$$
$$\Omega_t = (s,b,l)\}$$
$$\times \Pr\{L_{t-1} = l'|B_{t-1} = b',S_{t-1} = s',$$
$$\Omega_t = (s,b,l)\}.$$

The detailed modelling is done in terms of the three conditional probabilities on the right. Many of these probabilities are trivially zero or one, but it is nevertheless important to define them carefully. The first of the factors on the right in (6) respresents the likelihood of finding s' at $t-1$ given (s,b,l) as the state immediately below. If b is less than the maximum subfacies number for s, the process is not allowed to leave s and transitions to other facies s' have zero probability. Prior ideas on sedimentological order can be fed into the system through this factor. For example, if it is considered unlikely, although not impossible, for a marine facies to succeed a coastal one, the corresponding transition $s \to s'$ is registered with low probability. The second factor specifies the probability for subfacies b at $t-1$ given s' as the facies there and (s,b,l) as the state at the preceding point. Note that if $s' \neq s$, then $b' = 1$ by definition, all other possibilities being forbidden, i.e. having zero probability. On the other hand, if $s' = s$, either $b' = b$ or $b' = b + 1$. Typically short facies would have higher values for transitions $b \to b + 1$ than would thicker ones, making it possible to build prior information about thickness into the model. The final factor is the most important one as it defines the lithofacies transition probabilities. Note that it is an extension of (2) in that it depends on both facies s' and subfacies b'. The other conditions to the right of the vertical bar are immaterial.

The second part of the model is a link between L_t and X_t, that is a mathematical description of the second relationship in (3). We have used the specification

$$X_t = \mu(L_t) + \varepsilon_t \quad (7)$$

where $\mu(l)$ are mean log responses at lithofacies l, $l = 1, 2, \ldots$, and ε_t random disturbance terms

which we shall call 'noise', 'error process' or 'residual process'. Note that (7) is a vector equation with one relationship for each log variable. The mean log responses $\mu(l)$ are regarded as fixed parameters to be determined later by statistical estimation. The noise terms ε_t have zero means and their standard deviations could depend on the underlying lithofacies l. We must also assume some specific probability distribution for them, for example Gaussian or Student-type distributions. It is unlikely that ε_t is independent from one point t to another. The degree of dependency is a function of the sampling rate and the number of lithofacies used. The higher the number of lithofacies, the more structure is removed by first term on the right in (7), and the lower influence of the residual process ε_t. Error processes that are autocorrelated along the borehole are known as *coloured noise* in engineering as opposed to white noise when there is independence. We shall use these phrases. It will emerge at the end of the present section that our test model with six lithofacies has a residual process that is highly autocorrelated.

Figure 2, where a model identified from a real well has been simulated, illustrates how the model works. The three columns to the far left show sedimentary facies, subfacies and lithofacies, the latter varying more rapidly. The model producing this synthetic well works from base to top. When one sedimentary bed is finished, a successor facies is chosen according to the specified probabilities using a pseudo-random number generator. The new bed is then entered at subfacies 1, which after a while is replaced by subfacies 2, then by the third one and so on until the last subfacies is left, and the layer is completed. (This is sometimes called a Monte Carlo procedure.) The thickness of the subfacies is also drawn randomly, from probability distributions carefully specified to reflect prior notions about the size of sedimentary beds. Observe that there are in Fig. 2 considerable variations in subfacies proportions within the same facies. This is due to a combination of model specification and randomness. For example, very thin zones created by chance could be regarded as an effect of erosion, while sharp transitional zones would always be thin. The third round of Monte Carlo simulation selects the lithofacies. There are six of them, the same as in Table 1. Four ranged from medium coarse sand to shale, one represented coal and one carbonates. Their dependence on the underlying sedimentary facies is evident in Fig. 2. Upcoarsening patterns were produced in some instances (i.e. for the proximal delta facies,

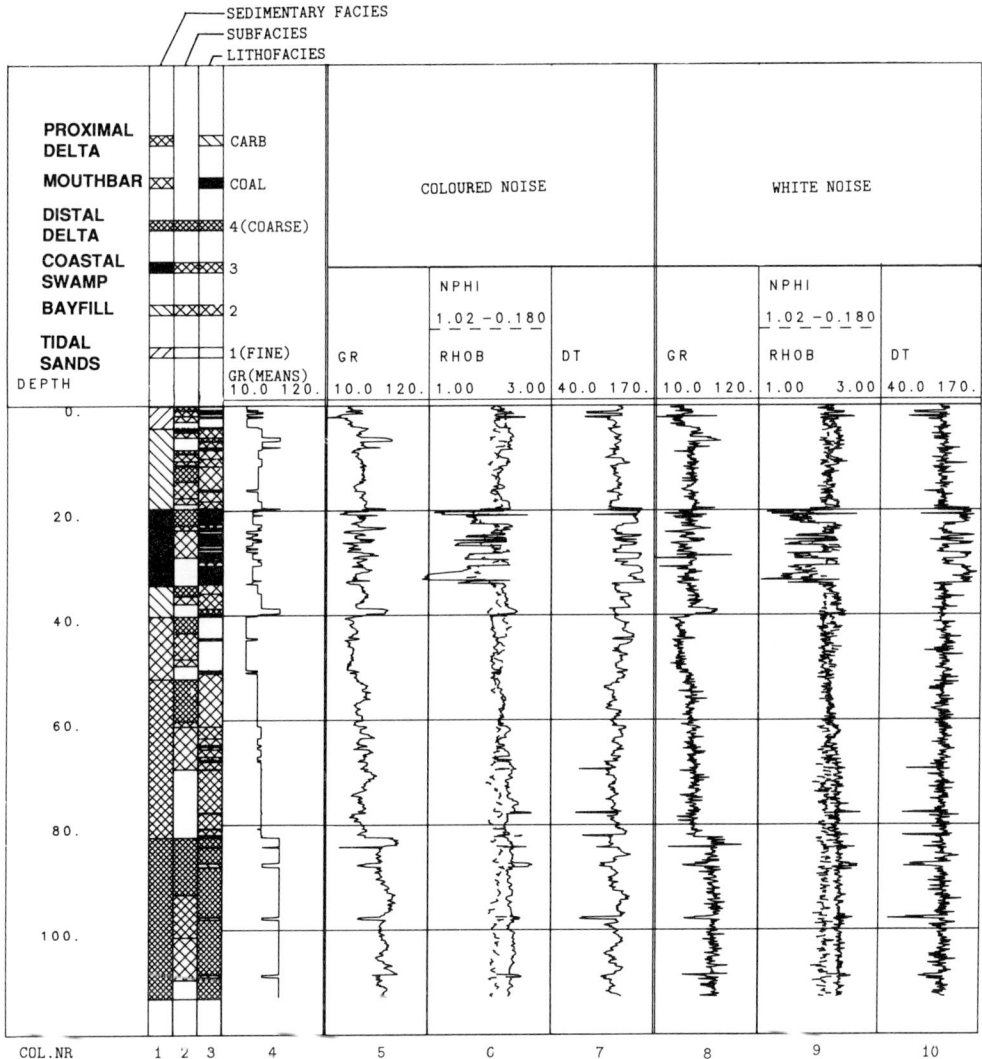

Fig. 2. Reservoir and log traces simulated according to a model adapted from a real well. The sedimentary facies are: tidal sandflats, bayfills, coastal swamps, mouthbars, proximal deltafront and distal deltafronts. Facies, subfacies, lithofacies and GR means to the left, log traces based on coloured noise in the middle and white noise to the right.

second from bottom, and for the bayfills, one at about $t = 35$ and some stacked ones higher up). In these cases the lithofacies transition probabilities varied with subfacies.

The other parts of the model, connecting log traces and lithology, are shown in Fig. 2 from column 4 and rightwards. Take the gamma log (GR) as an example. The mean gamma response of the selected lithofacies has been plotted against depth in column 4 using values estimated empirically from one of the wells below. This step function, jumping up and down at the

lithofacies boundaries, would have been the exact log trace had there been no noise and had the discrete lithofacies model been an accurate description of reality. With the addition of coloured noise, the gamma trace becomes the curve in column 5. Other logs have been simulated in the same manner in columns 6 and 7. White noise was supplied for the three remaining columns which are exact reproductions of columns 5–7, except for the different model of random errors.

These synthetic log signatures may be com-

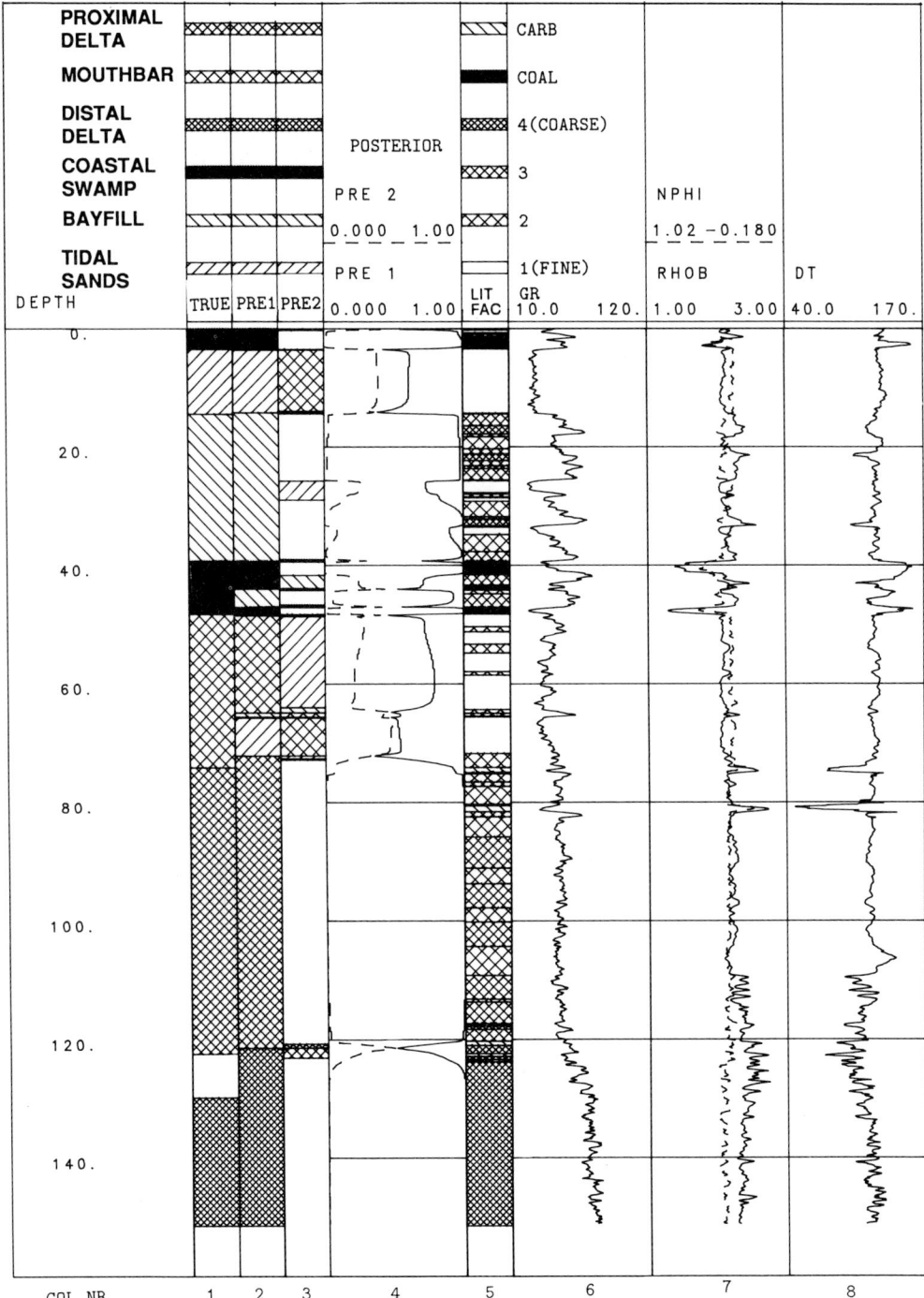

Fig. 3. Sedimentological interpretation of well 1 based on a model adapted from well 2. TRUE is the classification given by the sedimentologists (coded blank if none). PRE1 and 2 are due to the computer with PRE1 the primary interpretation. The posteriors estimate the uncertainties of PRE1 and 2. Lithofacies reconstructions (LITFAC) are also given.

pared with observed ones in Figs 3–5 (the log-scales are identical, while the depth scales differ slightly). The general impression is that there is considerable similarity. The real data are much smoother than portrayed by white noise, but the very simple model used for coloured noise in Fig. 2 (a so-called AR(1) with coefficient 0.90, see Priestly (1981)) seems to imitate the observed traces fairly well. One of the facies (bayfills) has been repeated three times. These four log shapes are quite different in appearance. A general shortcoming with this model is that it produces unnaturally sharp transitions across lithofacies boundaries. This is due both to the crude discretization of the lithofacies and to the fact that the effect of instrument smoothing has been neglected completely in case of white noise, and partially when using coloured noise.

The classification algorithm

The model introduced in the preceding section was defined as a Markov chain buried in noisy data. These so-called hidden Markov models have attracted interest in many fields during the past few years. They have for a considerable period been used in speech recognition (see Rabiner (1989) for a tutorial). Other areas are symbol recognition (Kundu *et al.* 1989) and image restoration (Devijver & Dekesel (1988)). A model belonging to this class, although different from the one employed here in other respects, was applied to the dipmeter log in Tjøstheim & Karlsen (1990). There are not many references in theoretical statistics. Baum *et al.* (1970) and Lindgren (1978) are two relevant ones.

Excluding ad hoc methods, there is a choice between two approaches for the restoration of the latent process ω. The so-called maximum *a posteriori* procedure (MAP) is to take the sequence of ω's as the single most probable sequence given the data. This criterion is often used in image restoration, see, for example, Dubes & Jain (1989). The opposite is to deal with one depth t at a time, select the most likely ω and then go on to the next depth. Both rules utilize all available information and both can be computed precisely because an appropriate statistical model has been established. Both rely on Bayes' rule (found in any textbook on statistics and probability) to invert, so to speak, the arrows in (3), and both require careful development of an algorithm to avoid computational explosion. We have chosen the second alternative for no very good reason. It is cheaper computationally and it permits quantification of

the uncertainty of the conclusions in a more direct way.

We introduce

$$\pi_t(\omega) = \Pr\{\Omega_t = \omega | \text{all data}\} \qquad (8)$$

as the posterior probability for state ω at t given all the data. Clearly (8) yields a similar posterior for $S_t = s$ by summation, i.e.

$$\pi_t^s(s) = \sum_b \sum_l \pi_t(s,b,l) \qquad (9)$$

Note that uncertainties in the restoration of subfacies b are averaged out. Hence, if (8) has been computed, the sedimentary facies s at t can be predicted as the maximizers of (9), i.e. as the most likely possibility, taking into consideration data and prior information. This is how the classifications in Figs 3–5 have been obtained. Obviously, alternative solutions can be put forward when there is doubt, i.e. when (9) fails to identify one convincing candidate. The numerical values of the posterior probability distribution provide a basis for doing this. In the examples shown in Figs 3 & 4 the two largest values of $\pi_t^s(s)$ have been plotted against t as PRE1 and 2.

The computation of (8) remains. It is extemely important that this problem is included in an algorithm for hidden Markov models in general as we would otherwise, with so complicated a model, be rapidly bogged down with endless details. There are in the literature two competing procedures. The original one is due to Baum *et al.* (1970), and most of the papers cited above use their technique. However, we prefer the adaptation of an algorithm coming originally from control engineering (Askar & Derin 1981), which has the advantage over Baum's that it avoids numerical underflow. Suppose $x_n, \ldots, x_1$ are the observed log recordings such that $t - n$ is at the bottom and $t = 1$ at the top. Let

$$\pi_t(\omega | m) = \Pr\{\Omega_t = \omega | x_n, \ldots, x_m\} \qquad (10)$$

be the conditional (posterior) probability distribution for Ω_t given the data from the bottom up to m. In particular, $\pi_t(\omega | 1) = \pi_t(\omega)$ are the quantities sought. Let $f(x_t | \omega)$ be the probability density function for the suite of logs X_t given $\Omega_t = \omega$ as the depth at t. It follows from (7) that we could have written $f(x_t | \omega) = f(x_t | l)$ since the values of b and s are immaterial. Then by adapting the algorithm in Askar & Derin (1981) (see also Kitagawa 1987) we have

$$\pi_{t-1}(\omega | t) = \sum_{\omega'} p(\omega | \omega') \pi_t(\omega' | t) \qquad (11)$$

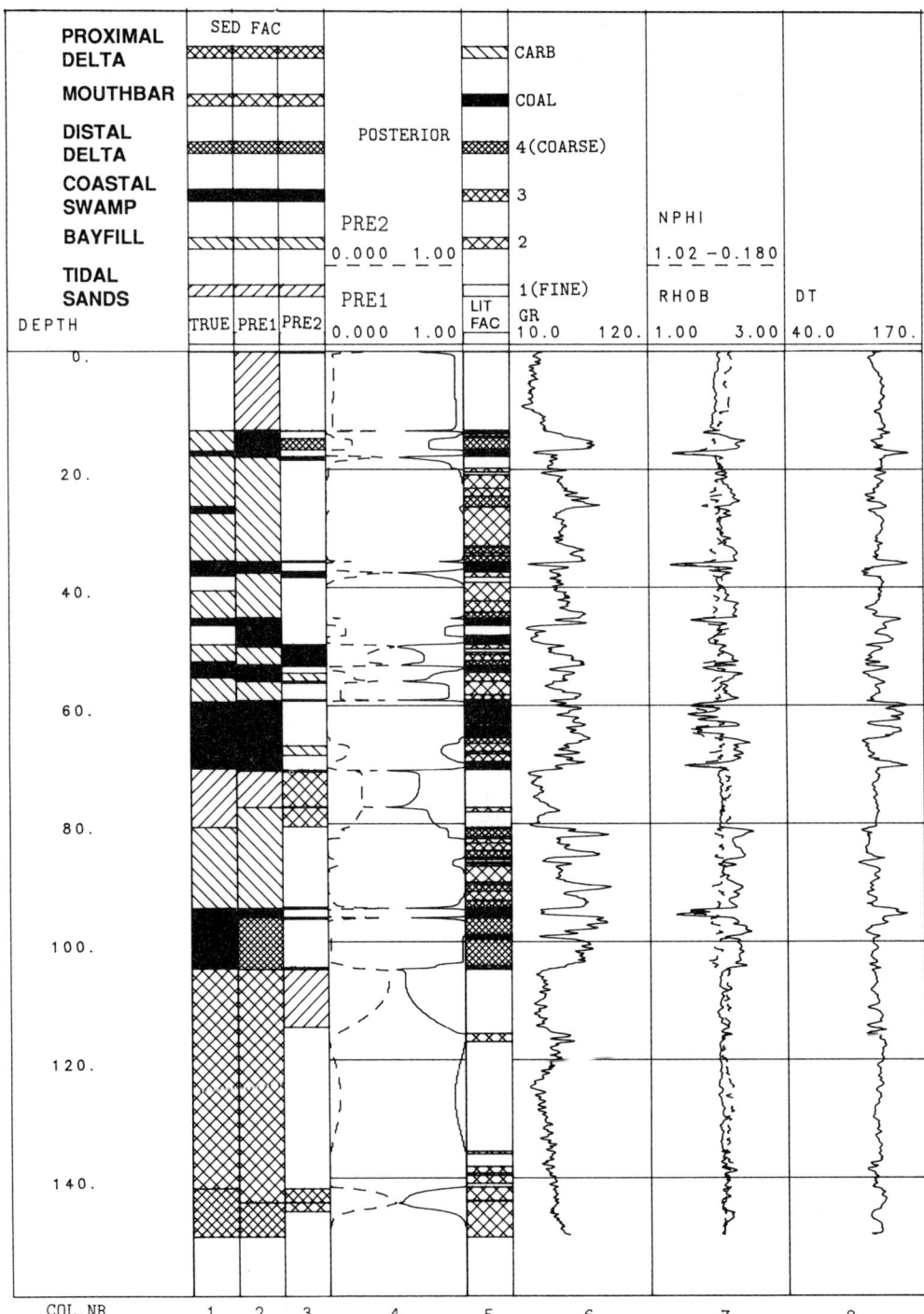

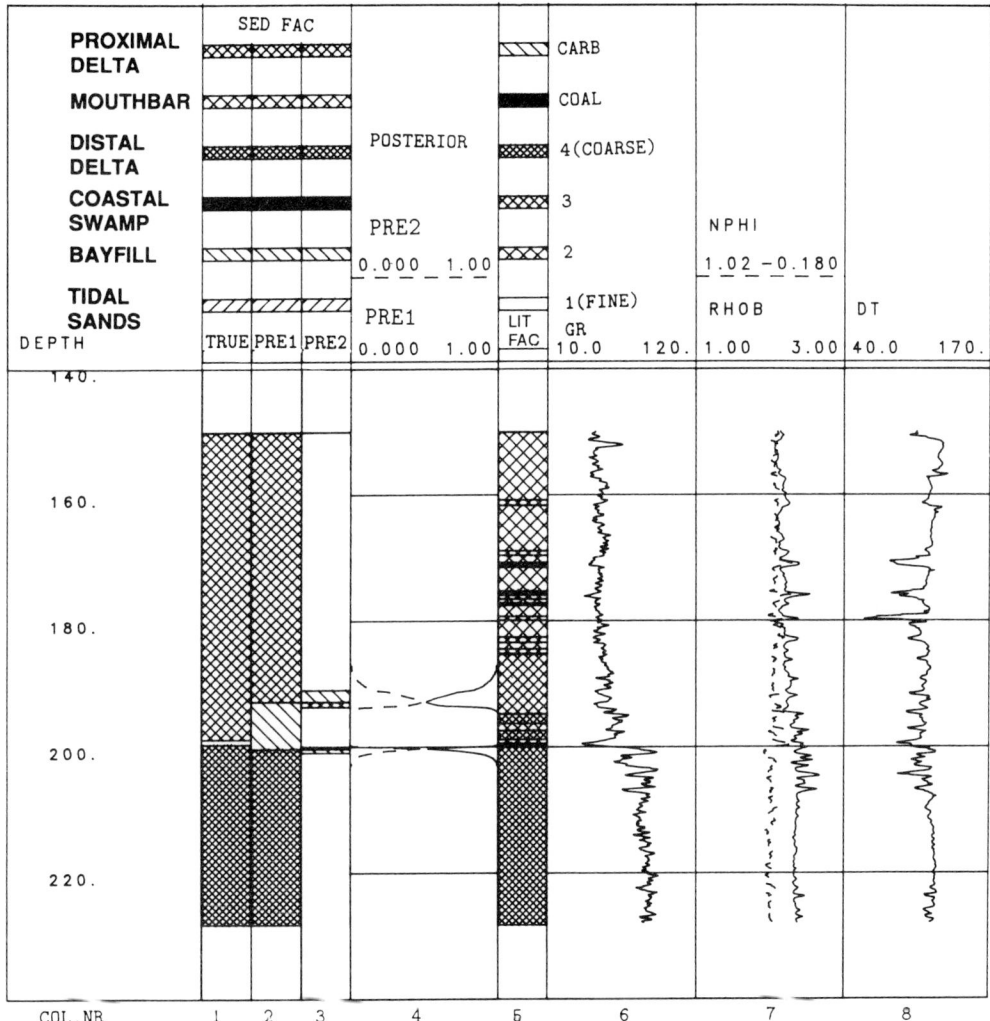

Fig. 4. Sedimentological interpretation of well 2 based on a model from well 1. TRUE is the classification given by the sedimentologists (coded blank if none). PRE1 and 2 are due to the computer with PRE1 the primary interpretation. The posteriors estimate the uncertainties of PRE1 and 2. Lithofacies reconstructions (LITFAC) are also given.

$$\pi_{t-1}(\omega|t-1) = \frac{f(x_{t-1}|\omega)\pi_{t-1}(\omega|t)}{\sum_{\omega'} f(x_{t-1}|\omega')\pi_{t-1}(\omega'|t)} \quad (12)$$

and

$$\pi_n(\omega|1) = \pi_t(\omega|t) \left\{ \sum_{\omega'} p(\omega|\omega') \frac{\pi_{t-1}(\omega'|n)}{\pi_{t-1}(\omega'|t)} \right\} \quad (13)$$

(11, 12) is an upwards recursion starting at the bottom at $t = n$ and storing on the way up the left-hand sides. The latter are in turn used for the downwards recursion (13), which produces the required posteriors for ω.

During implementation we took advantage of the observation that many of the transitions $\omega \rightarrow \omega'$ are forbidden. This reduces the number of terms in the sums (11) and (12). Both computation and storage increase linearly with the size of the well. It took about 30 CPU seconds on an IBM 3090 to produce Fig. 3. The storage was about 300 000 reals.

Model adaptation

All the discriminating features presented above have to be quantified and adapted to specific applications. This is done partly by subjective judgement, partly by computer assisted learning. The lithofacies and their transition probabilities are found by statistical estimation, trying to make the model mimic observed log shapes. This requires a training sample. The minimum requirement is a sedimentological 'ground truth' *s* which can be matched against the log traces *x*. The *s*, which may come from conventional sedimentological interpretations, should be read as a collection of layers, each identified with some well defined sedimentary environment. Gaps are allowed. Each sedimentary facies must occur at least once. Useful supplements to *s* and *x* would be core measurements, but to utilize such data effectively with logs, local depth shift errors must be overcome. In the present study, cores have only been used to establish *s*.

The statistical estimation of the lithofacies and their upward transition probabilities involves the following stages: The lithofacies must first be characterized in terms of mean log responses and variances. A lithological sequence must then be identified, and linked to the sedimentological beds to establish the transition probabilities within sedimentary facies. The latter is in its simplest form a question of counting passages from one lithofacies to another, but see further comments in the Appendix.

During the first phase of development our approach was sequential. The lithofacies were defined from clusters in *x*, *s* being entered afterwards for the purpose of finding the transition probabilities. Although this did work reasonably well, it was felt that the set-up would be more robust if the lithofacies determination was related to the eventual use of the model. In the present context lithofacies play a supporting role. They are primarily tools to discriminate log signatures and were constructed according to their effectiveness in this respect. Our implementation of this idea has been to estimate the parameters as maximizers of the so-called conditional likelihood $Pr(x|s)$ of the log data given *s*. This is further explained in the Appendix. The statistical model presented earlier makes $Pr(x|s)$ a well defined function of the unknown parameters. With some care it can be programmed and fed into a numerical optimizer. Alternatively, rather simple expressions for the derivatives can be used to devise an iterative scheme, similar, although more involved, to the standard expectation-maximization algorithm for mixture data (see, for example, Titterington *et al.* (1985)

paragraph 4.3.2, or Campbell (1984)). We have followed the latter course. All the unknown parameters, the mean log responses, the variances and the lithofacies transition probabilities are estimated in one operation.

The procedure works from an initial, possibly very inaccurate, guess on mean log responses and standard deviations for the lithofacies. Using the logs *x*, the lithological sequence corresponding to these lithofacies is reconstructed. This yields, when matched to the sedimentological 'ground truth' *s*, estimates of transition probabilities for all members *s* of the library of sedimentary facies. It also produces a revised set of parameters for the lithofacies, which in turn updates the lithological sequence extracted from the logs, and so on. After a while, the process stabilizes so that no change is observed from one cylce to the next one. It follows from general results in Dempster *et al.* (1977) that we have then reached a local optimum on the conditional likelihood surface $Pr(x|s)$. The process is then terminated, and the current values of parameters taken as estimates.

The estimation procedure has been tested on two wells 1 and 2 from a sandstone reservoir in the North Sea. The sedimentary environment was from lower to upper shoreface (about 200 metres) and the facies consisted of distal and proximal deltas, mouthbars, coastal swamps, bayfills and tidal sandflats. The logs, sampled at 0.125 m, were GR, RHOB, NPHI and DT. The lithology was predominantly sand, silt and shale with some coal (in the coastal swamps) and a tiny percentage of carbonates. The computer was asked to construct six lithofacies. Due to scarcity of carbonates it proved unable to set up a pure carbonate facies (it was mixed with shales). Such effects are common with clustering. Mean log responses for carbonates were therefore specified in advance and for coal too (although the computer did manage to define a pure coal class in other tests). These simplifications mean that there are four remaining lithofacies to determine. It is believed that this is a course that can be followed in practice. Carbonates and coal have characteristic log responses which are fairly stable over many sandstone reservoirs. It should therefore be possible to use the same parameters (i.e. mean values) for carbonates and coal in many applications.

Table 1 shows the lithofacies obtained by separate computations from each of the two wells 1 and 2. The lithofacies labelled 1–4 were identified by means of GR, RHOB and the difference (called DIFF) between the RHOB and NPHI–traces (DIFF = RHOB −(2.7–NPHI/ 0.6)). NPHI and DT were additional variables to separate coal and carbonates, but they were not used to discriminate the sand, silt, shale group. Note the similarity of the solutions from the two wells. It is tempting to interpret facies 1 as medium coarse sand and facies 4 as shale, with facies 2 and 3 something in between.

Table 1 also displays the lithofacies distributions for different sedimentary facies. Again there is a strong resemblance between the results from the two wells. This is also true for Table 2 where the lithofacies distributions have been allocated to subfacies. Notice the stability for tidal sandflats, mouthbars and the distal deltafront as opposed to the upcoarsening patterns for bayfill and the proximal deltafront. Recall that these descriptions are produced by the computer on its own.

Experimental results

The tests reported below were conducted on wells 1 and 2 from the preceding section and on a third well from the same field. The latter had not been cored. The mathematical model was developed from wells 1 and 2 in turn, in the manner explained earlier. Recall that this means that the lithofacies (coal and carbonate cemented sand excepted) and their transition probabilities have been constructed automatically by the computer from given log signatures. The actual models are those described in Tables 1 & 2. For all the experiments, the model was calibrated on a different well from that used for testing.

Additional experimental conditions were as follows. Six sedimentary facies were to be identified. The thickness of each was characterized by a mean value which was 10 metres for tidal sandflats, 6 metres for bayfills and coastal swamps and 30 metres for mouthbars, distal and proximal deltafronts. By adjusting these specifications the method can be made more or less inclined to select short sequences as sedimentary facies. The option of using sharp transitions in sediments as a special discriminating feature for facies boundaries had not at the time of writing been fully implemented and information of this kind has not been utilized. Preconceived ideas concerning sedimentological order were only taken into account for the last round of experiments illustrated in Fig. 5.

The lithofacies used were those listed in Table 1. The logs used were GR, RHOB, NPHI and DT. GR, RHOB and DIFF (i.e. the difference between the RHOB and NPHI traces as explained earlier) were responsible for differentiating between sands, silt and shale. NPHI and DT gave important supplementary information for the identification of coal and carbonates. Random errors, the term used for everything not captured by the step function model in Fig. 2, were assumed to be stochastically independent from one depth to another. Autocorrelated models would have produced a much better approximation to reality, as we have seen, but this was not implemented. As to error distributions, the results did not seem to be unduly sensitive. It seems reasonable on empirical

Table 1. *Lithofacies (mean log response) and their distribution (in percent) on six sedimentary facies. Construction from well 1 and well 2. Estimated standard deviations for lithofacies 1–4: For GR 5.5 (well 1), 7.8 (well 2), for DIFF 0.052, 0.063 and for RHOB, 0.048, 0.051. (Abbreviations: TS for tidal sandflat, BF for bayfill, CS for coastal swamp, MB for mouthbars, PD for proximal deltafront, DD for distal deltafront.)*

	Lithofacies	GR	Mean DIFF	RHOB	TS	BF	CS	MB	PD	DD
Well 1	1 (coarse)	35.1	−0.135	2.08	100	18	9	81	0	0
	2	47.9	0.027	2.18	0	33	18	17	62	0
	3	52.0	0.183	2.30	0	34	31	1	33	0
	4(fine)	72.1	0.273	2.34	0	15	2	0	0	100
	coal	42.0	−0.270	1.59	0	0	40	0	0	0
	carb	53.1	0.034	2.55	0	0	0	0	5	0
Well 2	1(coarse)	35.9	−0.093	2.02	89	7	3	85	0	0
	2	47.4	0.043	2.09	11	32	8	15	44	0
	3	52.9	0.176	2.21	0	37	22	0	48	1
	4(fine)	76.2	0.360	2.34	0	21	33	0	6	98
	coal	42.0	−0.270	1.59	0	0	33	0	0	0
	carb	53.1	0.034	2.55	0	2	0	0	2	2

Table 2. *Lithofacies distribution (in percent) broken down on subfacies for wells 1 and 2. 4 subfacies used. Abbreviations as in Table 1.*

		Well 1						Well 2					
		Lithofacies						Lithofacies					
		coarse			fine			coarse			fine		
	subfacies	1	2	3	4	coal	carb	1	2	3	4	coal	carb
	4(top)	100	0	0	0	0	0	89	11	0	0	0	0
Tidal	3	100	0	0	0	0	0	93	6	0	0	0	0
sandflat	2	100	0	0	0	0	0	86	14	0	0	0	0
	1(bottom)	100	0	0	0	0	0	89	11	0	0	0	0
	4(top)	39	49	12	0	0	0	48	42	10	0	0	0
Bayfill	3	1	74	25	0	0	0	0	71	29	0	0	0
	2	0	43	50	7	0	0	0	32	60	8	0	0
	1(bottom)	0	0	47	47	1	5	0	0	47	53	0	0
	4(top)	6	9	14	1	70	0	1	8	11	19	61	0
Coastal	3	11	20	32	3	33	0	4	9	19	33	36	0
swamp	2	11	22	43	3	21	0	4	8	29	42	17	0
	1(bottom)	6	17	32	2	43	0	3	7	33	41	16	0
	4(top)	87	13	0	0	0	0	92	8	0	0	0	0
Mouth-	3	83	16	1	0	0	0	92	8	0	0	0	0
bar	2	84	14	2	1	0	0	90	10	0	0	0	0
	1(bottom)	70	28	1	1	0	0	66	34	0	0	0	0
	4(top)	0	75	7	0	0	18	0	77	22	0	0	1
Proximal	3	0	73	25	0	0	3	0	33	62	0	0	5
delta-	2	0	58	42	0	0	0	0	0	94	5	0	0
front	1(bottom)	0	4	67	24	0	5	0	0	32	68	0	0
	4(top)	0	0	0	100	0	0	0	0	2	94	0	5
Distal	3	0	0	0	100	0	0	0	0	0	100	0	0
delta-	2	0	0	0	100	0	0	0	0	0	100	0	0
front	1(bottom)	0	0	0	100	0	0	0	0	0	100	0	0

grounds to apply models that would make gross deviations from the average more likely than under a Gaussian family. Our choice was one of the so-called Student *t*-distributions which are bell-shaped like the Gaussian, but, in the language of statisticians, with heavier tails. The t_5 distribution was used in the examples.

Under the classification, coal was assumed to be present in the coastal swamps only, while carbonate cemented sands were awarded no importance for sedimentological interpretation at all. The latter type of deposits still had to be identified, of course, but only to avoid confusions with other lithofacies.

Figure 3 shows sedimentary classifications from well 1 based on model identification from well 2, and Fig. 4 is the opposite, well 2 being the test well and well 1 the training well. A sedimentological ground truth was known for both wells and is plotted to the far left. The prediction (PRE1) regarded by the computer to be the one

most likely is displayed next and then the second most likely interpretation (called PRE2), provided the likelihood is larger than 0.15. The posterior probabilities for PRE1 and 2 are given in column 4 to indicate the degree of uncertainty (but see warnings in the closing section about this), and the lithofacies reconstruction in column 5. The three remaining columns to the far right are for the observed log traces. Keep in mind that the sedimentological interpretations mirror log signatures found in the training well. Non-representative patterns in the test wells will lead to errors. The experiments reported in Figs 3 & 4, where facies thickness is the only information used in addition to the log signatures, can be regarded as an examination of log shape similarity between wells. The discussion will try to refer discrepancies between the given sedimentary ground truth and the computer-made predictions back to specifics in the training well.

There are two or three errors in the well-1

classification (Fig. 3). Firstly, a faulty bayfill has been inserted within the coastal swamp close to depth 45. This is a reflection of configurations in the training well (well 2) where coastal swamp/ bayfill sequences are associated with similar log responses (consider the interval from 35 to 60 in Fig. 4). Secondly, the bottom part of the mouth-bar further down in Fig. 3 has mistakenly been taken for a tidal sandflat. Log signatures from this pair of facies are often difficult to distinguish from one another (recall the summary descriptions in Tables 1 and 2). However, the computer has marked 'mouthbar' as a strong secondary solution almost equal in likelihood for the two alternatives.

Turning to the predictions for well 2 (in Fig. 4), there are again two main errors, disregarding the tiny coastal swamp not picked up a little below depth 25. At about depth 100 a persistent layer of shale within the coastal swamp there has erroneously been classified as a distal deltafront. The interpretation of a marine facies in between two coastal ones may be regarded as unreasonable, but it should be recalled that the computer has received no warning against such a solution. On the contrary, it has been taught to associate thick beds of shale with off-shore facies. The interpretation is made all the more plausible (for the computer) by the two coastal swamps in the training sample (well 1) containing comparatively little shale. The other major error in well 2 (Fig. 4) is the incorrect handling of the transition zone between the distal and proximal deltafront at depth 200 where the upcoarsening has been confused with a bayfill. Again no rule disallowing such a configuration was included in the model. The error may be partly due to the information loss caused by lithofacies discretization. It may be easier to distinguish this pattern from the real bayfills if the number of lithofacies was increased.

The final round of experiments compared alternative predictions on the same, new well (well 3). The objective was (i) to see what can be gained by introducing knowledge about order between sedimentary facies and (ii) to examine the importance of different training wells. The two columns to the far left in Fig. 5 show classifications from well 1 training while the next columns two are based on a model adapted from well 2. Columns 1 and 3 are without prior assumptions on stratigraphic succession whereas columns 2 and 4 impose low probabilities for distal and proximal deltafronts adjacent to tidal sandflats, bayfills and coastal swamps (in other respects all configurations of sedimentary facies are equally likely). The general impression after inspecting the alternative interpretations in Fig.

5 is that the method is stabilized by taking prior ideas concerning sedimentological order into account. Differences between model adaptations from well 1 and well 2 have caused some minor classification discrepancies. Several of the differences have similar causes to those given for the errors in Figs 3 & 4.

Some specific comments on the predictions follow. (i) As remarked earlier, it is difficult to distinguish the log signatures of tidal sandflats and mouthbars. The well-2 calibrated methods result in solutions that are different from the others in this respect (see the interval from 0 to 30 in Fig. 5). The uncertainty quantifications (not shown) expressed great doubt between the two alternative environments. (ii) The well-1 trained model is not inclined to accept thick beds of shales within coastal swamps, whereas the other version is (see examples below depth 40 and at 145). This is a direct consequence of conditions varying between the training wells. Note that the interpretation below 40 is the same (see column 2) even when the unlikelihood of a marine environment between two coastal environments is taken into account. The low *a priori* probability has been beaten by the high probability of associating thick shale beds with offshore facies. The effect of introducing sedimentological order is different at 145 where the shale bed inside the coastal swamp now is interpreted as a bayfill (it is, perhaps, slightly upcoarsening). Again the coastal swamp has been rejected by the well-1 method because of the low shale content of well-1 realizations of this facies. (iii) The well-1 model returns a tiny coastal swamp at depth 60 while the others do not. (iv) Some upcoarsening shapes (around 60 and 120) have been identified as proximal deltafronts rather than bayfills by the well-2 method in column 3 (with, however, large uncertainty). This classification is revised when prior expectations about stratigraphic sequences are entered. (v) Several interpretations are put forward for the interval from 150 to 175. Well-1 training produces a mouthbar complex, believed to be correct while the well-2 method first picks a tidal sandflat and then further down a seemingly incorrect bayfill (the shoulders of the bed of carbonate cemented sand close to depth 170 are partly responsible for this as they were confused with shales). (vi) Finally, there is an error around 215 where in some of the predictions a bayfill has been inserted between the distal deltafront and the succeeding proximal deltafront on top of it.

Concluding remarks

The results are in our view promising, but the

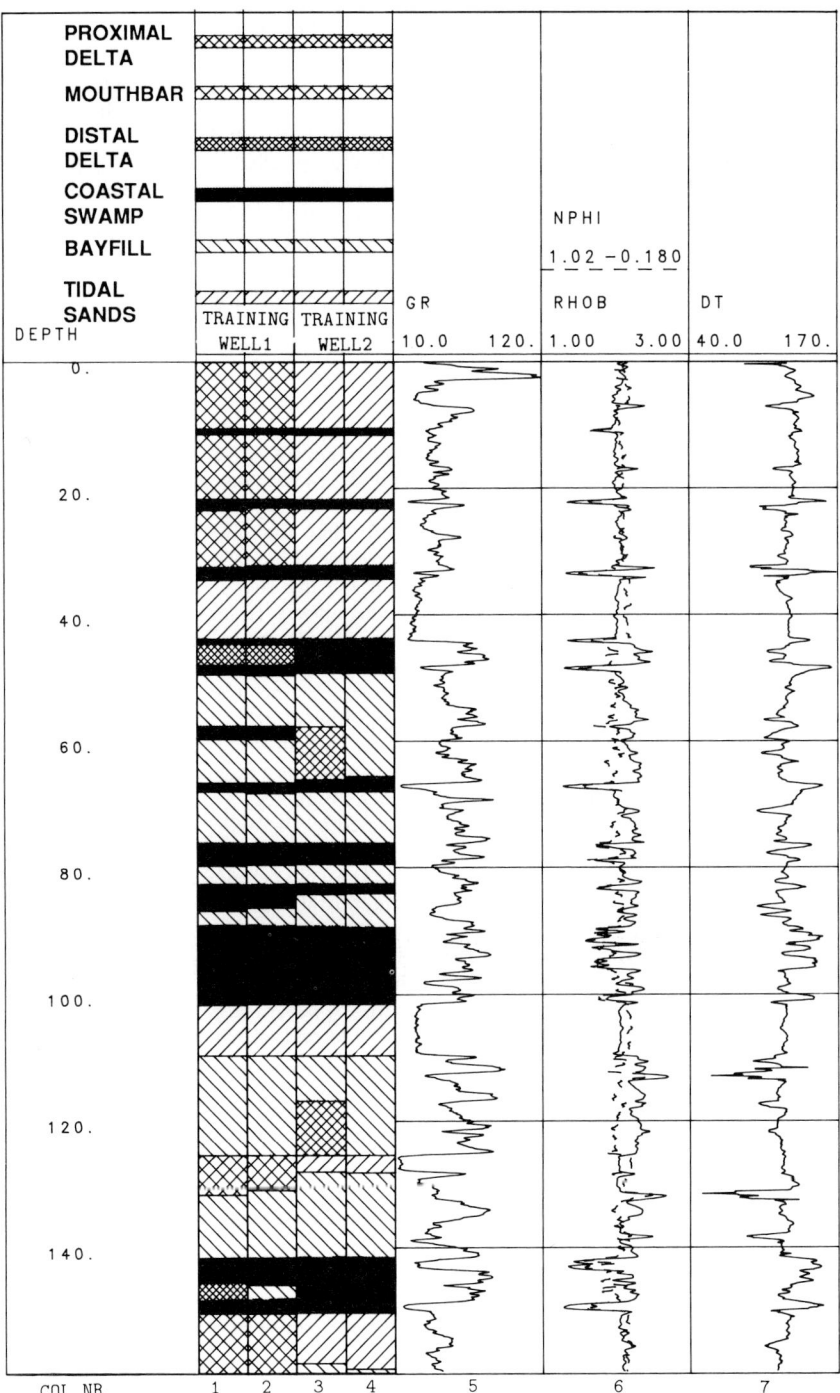

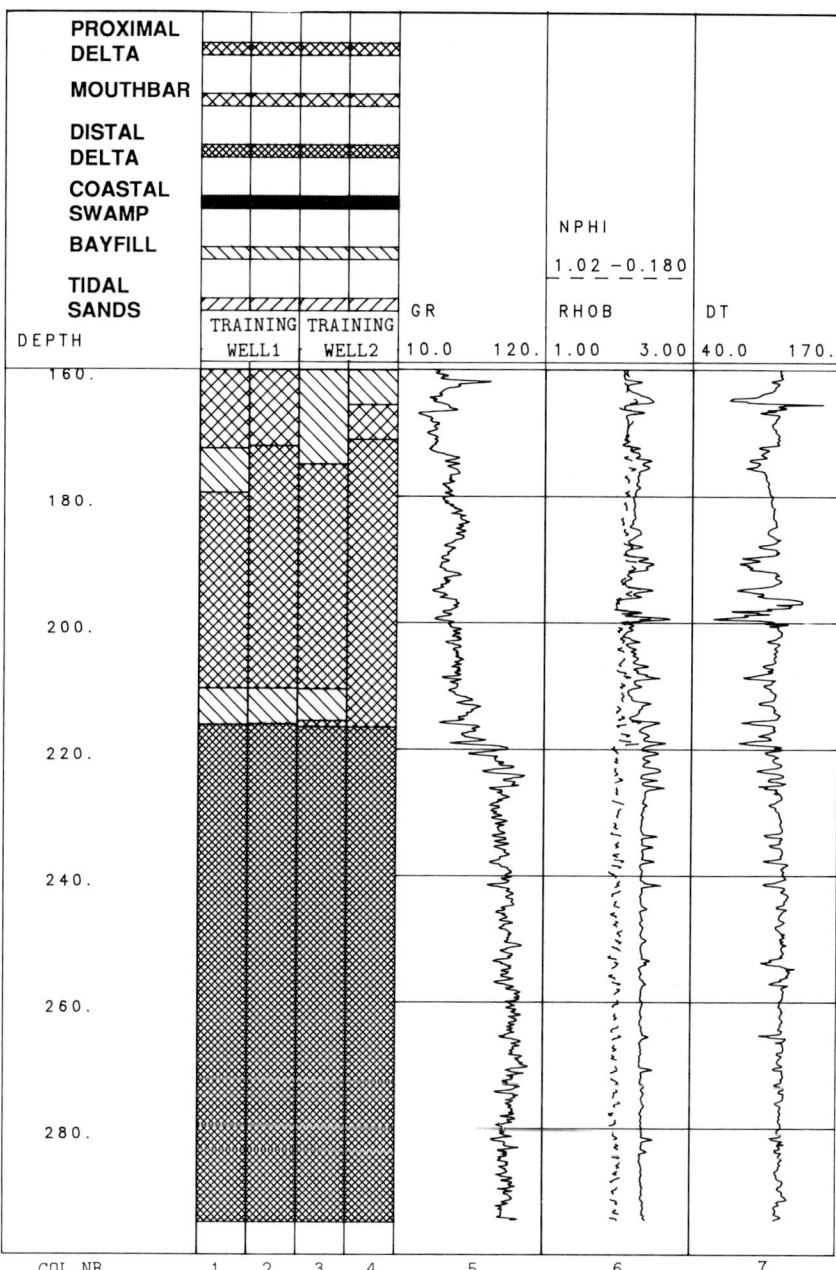

Fig. 5. Sedimentological interpretation of well 3 based on models from both well 1 and 2. The left (right) columns for wells 1 and 2 are without (with) prior expectations on facies order as explained in the text.

final judgement can only come through use by stratigraphers. There is certainly scope for improvement. It is important to allow as many discriminating features as possible. In that way the computer can balance many factors. The decisive ones will vary, but the more of them there are, the better are the chances that some of them will have sufficient discriminating power to identify the 'correct' solution. We shall in later applications introduce information about certain facies, typically starting or ending with abrupt changes in deposits. The subfacies idea can cope with this.

Due to errors in the transition probability estimates (which are, in effect, multiplied many times and have considerable impact) and partly to high auto-correlations in the random errors having been neglected, the *a posteriori* probabilities do not give uncertainty assessments that show the real doubt. High *a posteriori* probabilities are probably overestimated, and low *a posteriori* probabilities underestimated.

Coloured noise should be allowed (this is not difficult technically). It produced better descriptions of the log traces which could perhaps lead to more accurate predictions. An immediate effect of including coloured noise is that *a priori* assumptions will be more influential. This may stabilize the method. It will also imply more realistic uncertainty assessments.

Automated handling of different fluid zones might be handled by the same basic technology.

We shall in future work test the effectiveness of using many lithofacies for more accurate descriptions of log traces. That will inflate the number of transition probabilities enormously, but the smoothing procedure, briefly mentioned in the Appendix, effectively keeps down the number of independent parameters to be estimated and prevents the training data from being overfitted. It is an open question whether this will result in any improvements, and both CPU use and storage will go up.

Finally, this paper has attempted to tackle a very complicated problem through the use of hidden Markov chains. We believe that these models might be utilized for other applications in quantitative geology.

We are grateful to Eivind Damsleth and Jan Evensen at Norsk Hydro for going through an earlier version of the manuscript. Helpful suggestions were received from Dirk Kassenaar and Brian Moss. Finally, Cedric M. Griffiths, as editor, scrutinized the various drafts carefully, and his critical comments were most valuable.

Appendix. Iterative equations for model adaptation

A detailed presentation of the equations used to adapt the model is beyond the scope of the paper. The following discussion is intended to give the flavour of the approach.

The training sample presupposes that a sedimentary interpretation *s* is given. This attaches to the observed data *x* a conditional likelihood $\Pr(x|s)$, which is a function of the parameters of the model. Consider, in particular $\mu(l)$ and $\sigma(l)$ as the mean log responses and the standard deviation of log responses for the lithofacies; let $p_s(l'|l)$ or $p_{s,b}(l'|l)$ be the probabilities for transitions from lithofacies *l* to lithofacies *l'*, given *s* as the sedimentary facies or (s,b) as the facies–subfacies pair. The model adaptation algorithm maximizes $\Pr(x|s)$ over all these parameters. By differentiation, the following equations can be derived. Consider first $\mu(l)$ and $\sigma(l)$. Let

$$\tilde{\pi}_t(l) = \Pr(L_t = l | x,s) \qquad (14)$$

be the probability that lithofacies *l* is found at depth *t* given the data and the sedimentological ground truth *s*. These quantities can be computed by the Askar–Derin algorithm as presented earlier, the only difference being that the transitions for the Ω_t process are modified by incorporating the fact that the *s* vector is known. Under the assumption of the data being Gaussian, it can be shown that $\mu(l)$ and $\sigma^2(l)$ satisfy

$$\mu(l) = \frac{\sum_t \tilde{\pi}_t(l) x_t}{\sum_t \tilde{\pi}_t(l)} \qquad (15)$$

$$\sigma^2(l) = \frac{\sum_t \tilde{\pi}_t(l)\,(x_t - \mu(l))^2}{\sum_t \tilde{\pi}_t(l)} \qquad (16)$$

For other distributions, (15) and (16) must be modified as in Campbell (1984). Note that (15) defines $\mu(l)$ as weighted averages, with high weights for those *t* for which the data suggests *l* as a likely lithofacies. There is a problem, however, in that the weights $\tilde{\pi}_t(l)$ depend on the quantities $\mu(l)$ and $\sigma^2(l)$ to be estimated. This apparent circularity is broken by iteration. From $\mu(l)$ and $\sigma^2(l)$, $\tilde{\pi}_t(l)$ are computed. The parameters can be revised according to (15) and (16). New values of the weights $\tilde{\pi}_t(l)$ can be found, and so on until the procedure stabilizes. Convergence is certain (Dempster *et al.* 1977).

To estimate the transition probabilities $p_s(l'|l)$ or $p_{s,b}(l'|l)$ a similar procedure is available, although more involved. We refer the reader to Bølviken *et al.* (1992). What makes this problem more intractable, is the high number of different probabilities, which causes estimation instabilities if care is not exercised. We have implemented a smoothing procedure to circumvent this problem.

References

ASKAR, M. & DERIN, H. 1981. A recursive algorithm for the Bayes solution of the smoothing problem. *IEEE. Transaction on Automatic Control*, **26**, 558–650.

BAUM, L. E., PETRIE, T., SOULES, G. & WEISS, N. 1970. A maximisation technique occurring in the statistical analysis of probabilistic function of Markov chains. *Annals of Mathematical Statistics*, **41**, 164–171.

BØLVIKEN, E., STORVIK, G., NILSEN, E., SIRING, E. & VAN DER WEL, D. 1992. Identifying sedimentary layers from patterns in well log traces.

CAMPBELL, N. A. 1984. Mixture models and atypical values. *Journal of the International Association for Mathematical Geology*, **16**, 465–578.

DEMPSTER, A. P., LAIRD, N. M. & RUBIN, D. B. 1977. Maximum likelihood from incomplete data via the EM algorithm (with discussion). *Journal of the Royal Statistical Society*, Ser. B., **39**, 1–38.

DEVIJVER, P. A. & DEKESEL, M. M. 1988. Cluster analysis under Markovian dependence with application to image segmentation. *In*: BOCK, H. H. (ed.) *Classification and Related Methods of Data Analysis*. North-Holland. Amsterdam, 2203–2217.

DUBES, R. C. & JAIN, D. C. 1989. Random field models in image analysis. *Applied Statistics*, **16**, 131–164.

KERZNER, M. C. 1986. *Image processing in well log analysis*. Reidel. Boston.

KITAGAWA, C. 1987. Non-gaussian state-space modeling of nonstationary time series (with discussion). *Journal of American Statistical Association*, **82**, 1033–1063.

KUNDU, A., HE, Y. & BAHL, P. 1989. Recognition of handwritten words: First and second order hidden Markov model based approach. *Pattern Recognition*, **22**, 283–297.

LINDGREN, G. 1978. Markov regime models for mixed distributions and switching regressions. *Scandinavian Journal of Statistics*, **5**, 81–91.

PRIESTLEY, M. B. 1981. *Spectral Analysis and Time Series*. Academic, London.

RABINER, L. R. 1989. A tutorial on hidden Markov models and selected applications in speech recognition. *Proceedings of the IEEE*, **77**, 257–285.

READ, W. A. & MERRIAM, D. F. 1971. A simple quantitative technique for comparing cyclically deposited succession. *In*: MERRIAM, D. F. (ed.) *Mathematical Models of Sedimentary Processes*. Plenum, New York, 203–232.

SCHWARZACHER, W. 1975. *Sedimentation Models and Quantitative Stratigraphy*. Elsevier, Amsterdam.

TITTERINGTON, M., SMITH, A. F. M. & MAKOV, U. E. 1985. *Statistical Analysis of Finite Mixture Analysis*. Wiley, New York.

TJØSTHEIM, D. & KARLSEN, H. 1990. Autoregressive segmentation of signal traces with application to geological dipmeter measurements. *IEEE on Geophysics and Remote Sensing*, **28**, 2, 171–181.

Analysis of dipmeter data for sedimentary orientation

GAVIN I. F. CAMERON

Rider-French Computing Ltd, 153 Cambridge Science Park, Milton Road, Cambridge CB4 4GG, UK

Abstract. Dipmeter data are logged as a vertical traverse through formations of interest. By analogy to concepts of dip analysis at surface outcrops, not all of the dip data collected by a logging run are of direct relevance to specific interpretation aims. A methodology is required whereby only relevant dip magnitude and azimuth data can be extracted from the mass of logged results. One approach involves the deductive use of log zonation and dip filtering techniques to isolate the dip results that most closely match the geological requirements. Statistical analysis for preferred orientation in the selected dataset gives a result to which clear geological significance can be attached. The principles described can be applied to a variety of sedimentary dipmeter interpretation problems, such as the determination of palaeocurrent directions, considerations of reservoir geometry and preferred bedding orientation.

The dipmeter log has long been considered as something apart from the mainstream of wireline logs, and the interpretation of dipmeter data as something of a specialist area. It is, however, the primary source of detailed structural and sedimentary orientation data from wells, and has additional spin-offs for questions of detailed reservoir geology, such as facies definition and thin-bed analysis. The standard technique of presenting dipmeter results as dip and azimuth pairs on a dip arrow plot, creates a clear contrast with the more familiar curve traces used to present other wireline data, and is to some extent a discouragement to their integration. However, the geological interest in the results and the uniqueness of dip and azimuth amongst other wireline parameters is incentive enough to spend time on interpretation. This paper considers questions of sedimentary dipmeter interpretation and how best to extract the maximum of useful sedimentary orientation data from the results of a logging run in an accurate and efficient manner.

Changes in emphasis of dipmeter usage

Dipmeter logging has its origins in the early 1940s. A review of the early evolution of dipmeter logging technique reveals a constant desire for an improvement in resolution (Allaud & Ringot 1969). Since the first of the modern dipmeter tools, the High Resolution Dipmeter Tool* (HDT), was introduced in 1969, there have been important increases in resolution achieved such that the tools today can record up to a resolution of one microresistivity value

_____
* Mark of Schlumberger

for every 0.25 cm of borehole section. The comparison, shown in Fig. 1, of a set of dipmeter microresistivity curves to the gamma ray curve recorded over the same interval illustrates the strength of this achievement.

With the increases in dipmeter tool resolution has come a change in expectation as to the type of geological interpretation that can be made from the logging results. Initially the dipmeter was used just to provide structural dip information. Eventually it became clear that the dipmeter could also give useful sedimentary orientation data (Campbell 1968). The cumulative effect of tool improvements has greatly enhanced the accuracy of the microresistivity data and the reliability of the sedimentary dip data that are obtained.

Recent developments in dipmeter interpretation divide into two strands, one dealing with understanding and exploiting the orientation data, the other with uses of dipmeter microresistivity curves for textural and lithology recognition.

On the first count, the problem for geologists is to devise techniques for analysing the mass of orientation data given on processed dipmeter logs and to interpret something useful, such as the preferred bedding orientation within a reservoir, or directions of sand body elongation, or a palaeocurrent/sediment transport scheme. There is a general lack of documented methodologies for achieving this aim and this is responsible for the situation in which dipmeter data, though expensive to acquire, are not widely exploited, and are sometimes regarded as an optional extra in the log evaluation process. However, detailed dip measurements at outcrop have provided valuable models of dip distribution at surface

From HURST, A., GRIFFITHS, C. M. & WORTHINGTON, P. F. (eds), 1992, *Geological Applications of Wireline Logs II.* Geological Society Special Publication No. 65, pp. 141–154.

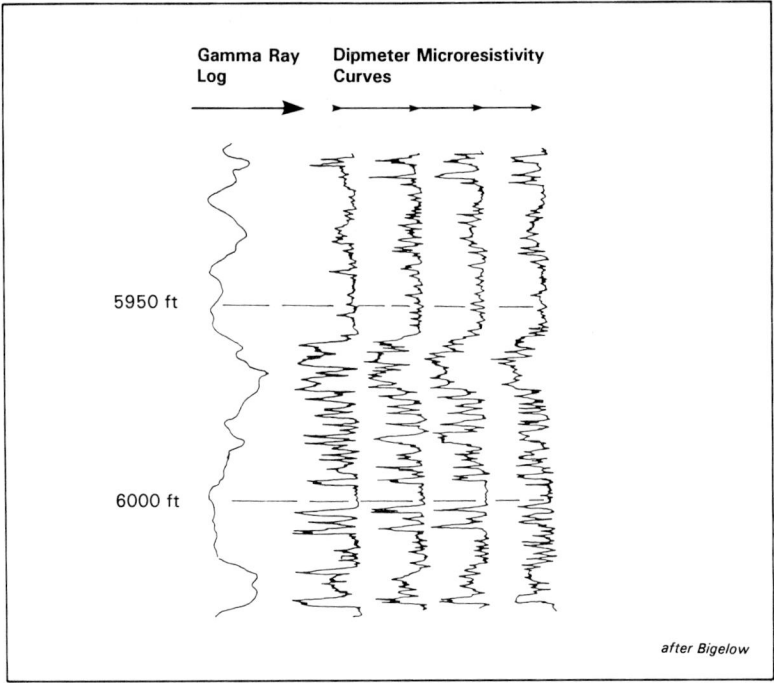

Fig. 1. Comparison of dipmeter microresistivity curve resolution to a standard gamma ray log (after Bigelow 1982).

examples and have given some useful pointers as to how the problem of dip interpretation should be approached in the subsurface (Perrin 1975; Cameron *et al.* in press; Williams & Soek in press).

On the second count, there has been an increasing realization that the dipmeter micro-resistivity curves in themselves are a valuable source of detailed textural information about subsurface formations. A number of papers have been published on quantitative and qualitative uses of dipmeter microresistivity curves in reservoir applications. Dipmeter microresistivity data have been used to analyse evaporite formations and distinguish between salt facies (Curial 1988), for thin bed analyses (Sallee & Wood 1984) and in the generation of core-like descriptions from open-hole logs (Anxionnaz *et al.* 1990). This style of approach is set to be further developed by the increasing use of borehole imaging techniques such as the Formation MicroScanner* (FMS), Formation MicroImager* (FMI) and Borehole Televiewer* (BHTV) for detailed formation evaluation.

---

* Marks of Schlumberger

The need to be selective with subsurface orientation data

Concepts of the use of sedimentary structures, such as cross-bedding and other directional laminae, in contributing to palaeogeographic reconstructions of sediment transport directions and depositional geometries, have been developed largely from regional studies at surface outcrop. To extend the use of these sedimentological concepts into the subsurface with dipmeter data, it is necessary to appreciate certain differences between the well log data and the data gathered at the surface.

First, there is the question of validity of the dipmeter data. This centres around the extent to which planes correlated between anomalies in dipmeter microresistivity curves are a true expression of geological bedding dip. Detailed core to dipmeter comparisons can be used to provide an empirical demonstration that the method is valid (Cameron 1986).

Second, an important, but often overlooked point is a required change of attitude towards the dip data when using dipmeter results compared to the use of data collected at the surface. For example, a geologist at outcrop is normally very selective about where to take readings of dip and strike. The geologist makes a discerning choice between different types of dipping surface, and collects data only from those surfaces that are relevant to the aim of the study at hand. So for a palaeocurrent study, the geologist measures dip at foreset laminae, while parallel bedding and bed boundary surfaces are largely ignored. When it comes to analysis and interpretation of the outcrop data, the geologist has a selected dataset which bears only on the problem in hand. A dipmeter log, by contrast, gives a regularly sampled single traverse through formations of interest, with a resulting dataset in which *no pre-selection* has occurred. Therefore, when dealing with subsurface data, a greater emphasis must be placed on the extraction of relevant details from the mass of undiscriminated data.

Dipmeter processing and noise effects

There are also good data processing reasons to be selective with dipmeter data. On the earliest dipmeter logs, the correlation of surfaces in the dipmeter curves was done manually. However, with the advent of digitally recorded data and the ever-increasing resolution of the tools, computer programs were developed that performed the correlation and dip computation process from the dipmeter curves automatically (Moran *et al.* 1962).

The majority of present-day dipmeter processing packages employ some variant of the cross-correlation technique which is controlled by the setting of three correlation parameters, namely correlation length, step distance and search angle (Fig. 2, Schlumberger 1981; Rudman & Lankston 1973). A minority of packages use other techniques of pattern recognition and frequency analysis for dipmeter curve correlation. However, for reasons of reliability and general robustness of the method, the fixed interval cross-correlation technique remains by far the most widely used.

The correlation length parameter controls the length of microresistivity curve compared at each correlation attempt. The longer the correlation length, the greater the averaging by the

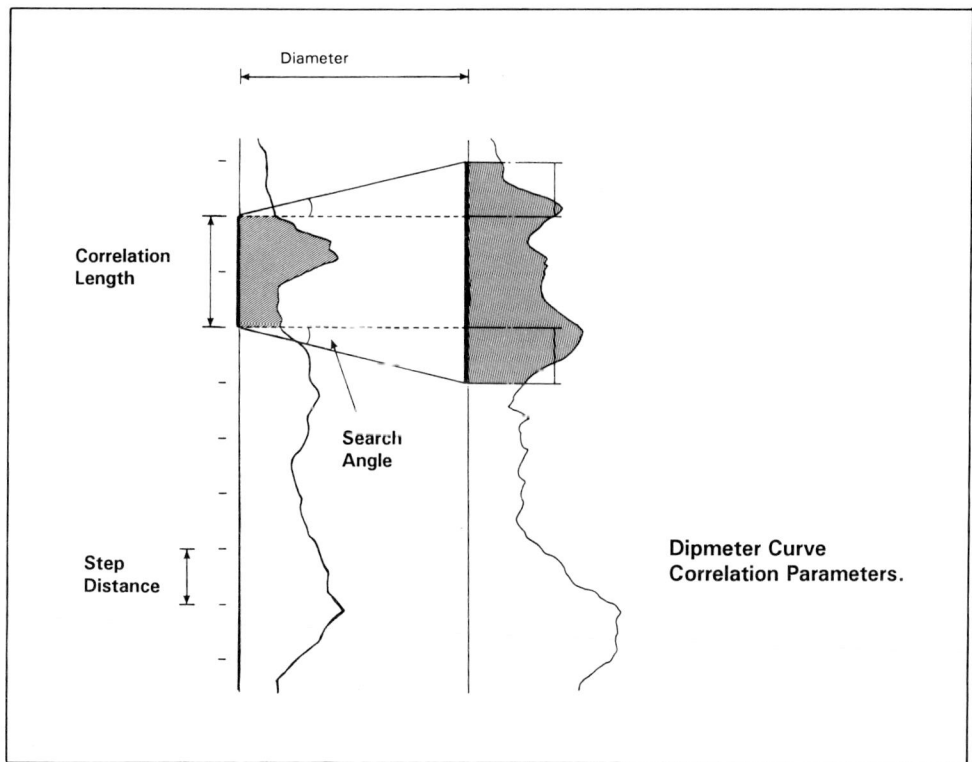

Fig. 2. Diagram to illustrate the principle of fixed-interval cross correlation used for dipmeter processing (see text for explanation).

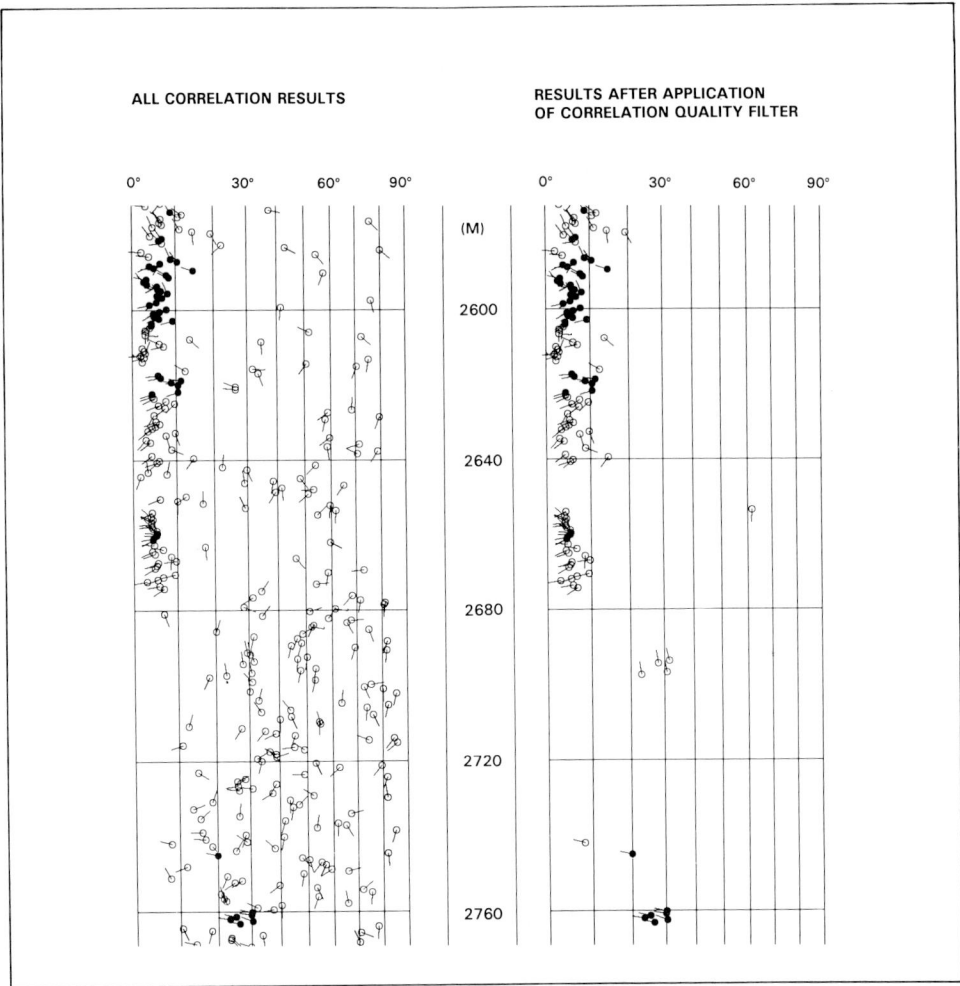

Fig. 3. Comparison of dipmeter arrow plots showing all dip correlation results, and the same results after application of a correlation coefficient cut-off to reject all poor quality results.

correlation method, the more likely the computed dip results will reflect structural dip, and the less likely it is that sedimentary dips will be correlated. However, it is the step distance which controls the number of correlation attempts and, therefore, the number of dip results that are computed by a dipmeter processing run. For example, setting a step distance of 0.5 m means that the correlation programme will attempt to correlate a dip surface every 0.5 m along the length of the dipmeter microresistivity curves. A step distance of 0.2 m will result in five dip correlations for every one metre of borehole section. It is in the nature of the correlation process that a result is always obtained for every correlation attempt. The left-hand track in Fig. 3 shows all of the dipmeter results computed for a given 200 m interval by a cross-correlation dipmeter processing package. The track shows a well developed stable dip down to 2680 m, but from there downwards the computed results show dips in all directions and with all possible dip magnitudes. Clearly the microresistivity curves over this lower interval contain little dip significance and yet the correlation process is obliged to seek some kind of result. It is normal to remove this type of 'noise' on output logs, (Fig. 3, right-hand track), using the correlation coefficients calculated during curve correlation as a means of rejecting any dips that do not come up to a specified level of confidence.

The level at which to apply a cut-off is un-

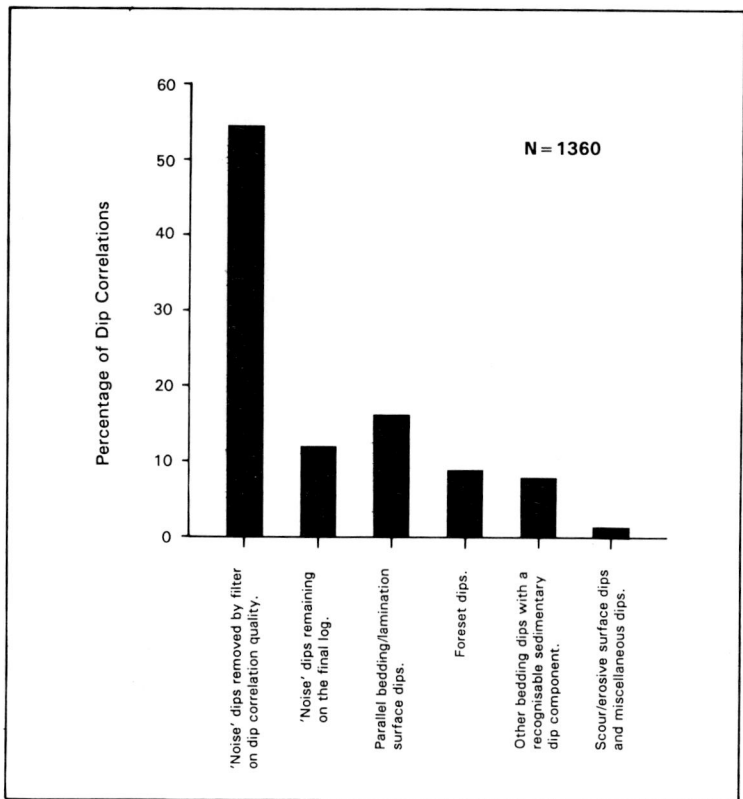

Fig. 4. Histogram showing an analysis of dipmeter dip results and core data from a 136 m test interval. From a total of 1360 dips, each dipmeter result was compared to dip surfaces seen at the same depth on the core, and classified according to type. Dips for which no corresponding surface was seen on the core were classified as 'noise'.

certain. There is something to be said for applying a very strict cut-off so that only the very best results are retained, and yet this may ignore other geologically more important, but texturally less distinct, surfaces which are correlated with lower coefficients. Similarly, the optimum cut-off over one lithology may be undesirable over another. With interactive computers and appropriate software it is possible to leave this decision to the geologist during interpretation. However, a large majority of dipmeter data are still interpreted from paper logs on which this decision is to all practical purposes unalterable.

The histogram in Fig. 4 shows the results from one of a series of exercises carried out on detailed dipmeter results. In this case, from a 136 m borehole interval for which cores were also available. The dipmeter was processed using a fixed interval cross-correlation programme with a correlation interval of 0.2 m, step distance of 0.1 m and search angle of $35° \times 2$ (giving a

maximum search range of 70°). The correlation process gave 1360 dip results, of which only 619 (45%) were presented on the final log as reliable dips.

The data acquisition over the interval was good. The results on the final log were compared to the core and a breakdown of how each dip related to dip surfaces on the cores was prepared. Any dip for which no corresponding dip surface could be found on the core was regarded as a 'noise' dip. In this case, although the data were regarded as good, some 26% of the dips of the final log did not appear to relate directly to surfaces seen on the core. Moreover, the 'noise' dips were not just confined to the lower end of the correlation coefficient spectrum but were more evenly distributed. Some of the dips may come from true geological features that were not seen by eye, but it is probable that a significant proportion of the dips have no geological meaning and are spurious data generated

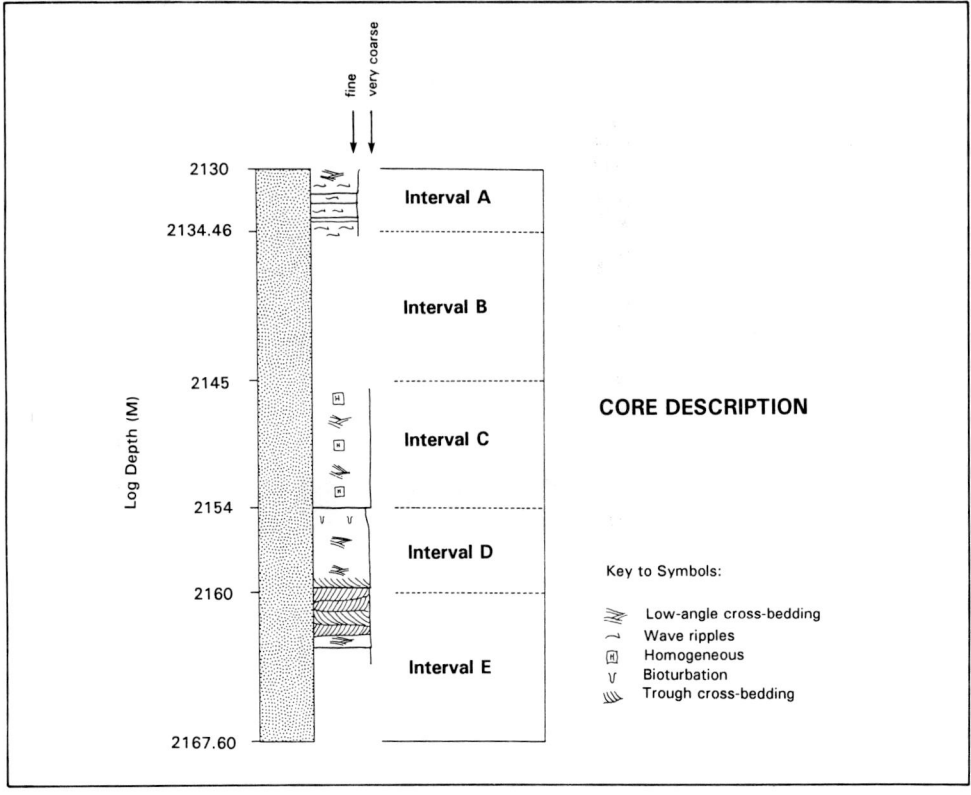

Fig. 5. Core description through an example sequence used for a test of the dipmeter processing method and its sensitivity to changing correlation parameters.

during the acquisition and computer correlation process.

This example illustrates that, even with carefully selected correlation coefficient cut-offs, a certain amount of 'noise', perhaps up to the level of about 25% of the final dips, is endemic on dipmeter logs. This points to the need for robust methods of dip results analysis which can cope with a certain amount of inherent noise.

Processing strategies

The importance of using dipmeter curve correlation parameters which are appropriate to the interpretation aim is already documented (Bigelow 1985). For example, it is widely appreciated that a long correlation interval of 1 m or more will favour the correlation of bed boundary surfaces and so offer structural dip information, while shorter correlation lengths will promote the correlation of texturally more subtle sedimentary dip surfaces. It may be, however, that outside of the broad distinction between struc-

tural and sedimentary dip, the choice of correlation parameters in the dip correlation process is not as critical as it is sometimes held to be.

Figure 5 shows the core description of an example sequence for which the dipmeter data were processed eight times, each time with a different set of correlation parameters. The aim was to assess how changes in correlation parameters affected the dip correlations that were found in the microresistivity curves during fixed interval dipmeter processing. The sequence shows a good range of sedimentary structures, from trough cross-bedding at the base, through low-angle cross-stratification and homogeneous/burrowed intervals, to wave ripple-bedding at the top. For purposes of the trial, the core interval was divided into five sections A–E. Readings of apparent dip were taken from the core slabs at prominent sedimentary surfaces to give an independent record of dip in the interval. No azimuth data were acquired from the core.

The raw dipmeter data for the interval were processed eight times using a spread of different

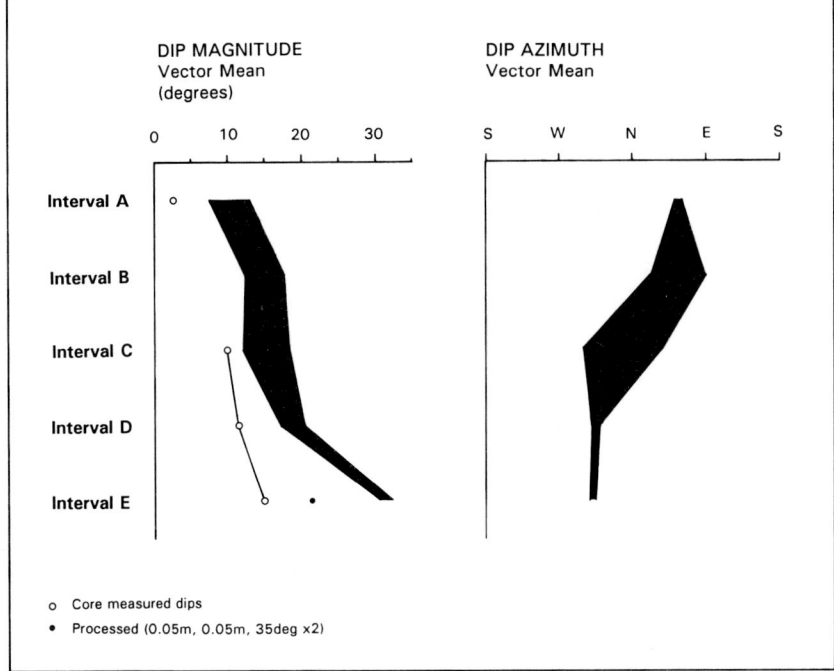

Fig. 6. Graphs summarizing the results of eight separate dipmeter processing runs over the same sequence, each time using a different set of correlation parameters (as detailed in Table 1). Mean dip magnitude and vector mean azimuth are shown calculated for each sedimentary interval A–E (see Fig. 5) over each of the eight processing runs. There is good agreement between the results of each run. The spread of results reaches its widest in the more homogeneous intervals B and C.

Table 1. *Dipmeter processing parameters used in eight separate processing runs over the same test sequence (see Fig. 5)*

Processing run number	Correlation interval	Step distance	Search angle
Run No. 1	0.05 m	0.05 m	35° × 2
Run No. 2	0.1 m	0.05 m	35° × 2
Run No. 3	0.1 m	0.09 m	35° × 2
Run No. 4	0.2 m	0.1 m	35° × 2
Run No. 5	0.2 m	0.18 m	35° × 2
Run No. 6	0.35 m	0.175 m	35° × 2
Run No. 7	0.35 m	0.315 m	35° × 2
Run No. 8	0.5 m	0.375 m	35° × 2

correlation parameters (Table 1) and the dip results from each processing run were subjected to a simple statistical analysis for vector mean dip azimuth and magnitude. Structural dip was low (2°–3°) so no correction was applied. The results of this analysis are shown in two X–Y graphs (Fig. 6). The azimuth graph shows the envelope of vector mean values calculated for

the processing runs over each of the core sedimentary intervals A–E. The graph shows very clear agreement of vector means between all of the processed data. This is particularly marked in core intervals, A, D and E where the core has good sedimentary structures. The agreement between processing runs is not so good in core intervals B and C, where a maximum variance of

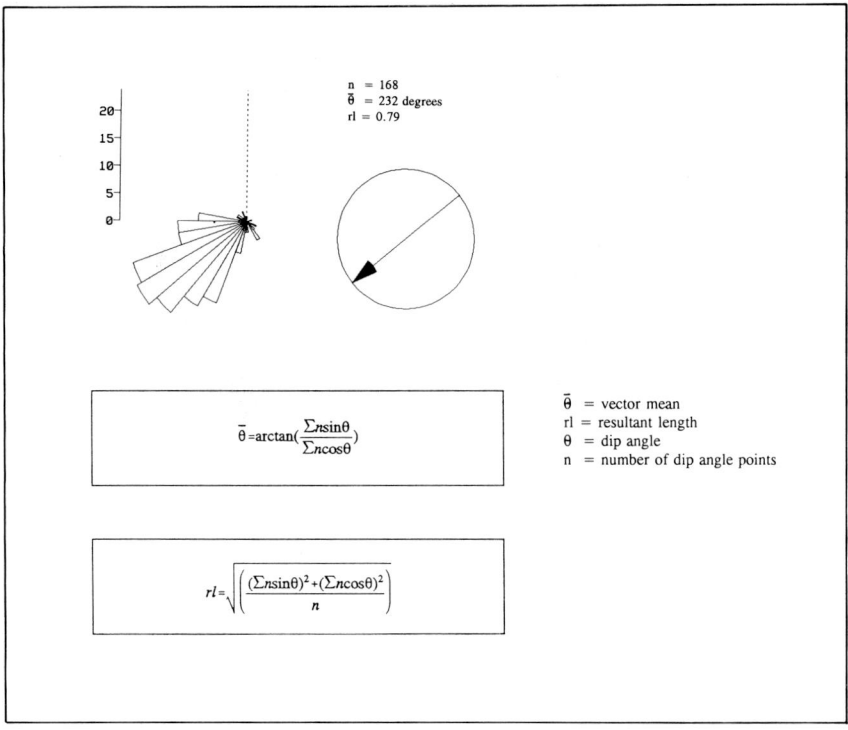

Fig. 7. Formulae for calculation of vector mean from circular orientation data (Curray 1956). The rose diagram and accompanying vector mean arrow illustrate the usefulness of this simple statistic as a method of summarizing dipmeter azimuth or magnitude data.

90° is observed, but this is attributed to the homogeneity which can be observed in the core sediments.

The dip magnitude graph, with vector mean dip, shows good agreement within a five degree spread between all sets of processed data. An anomalous response occurs with Run 1 data in core interval E, but this is the only source of disagreement and, moreover, this is not reflected by any corresponding anomaly in the azimuth data of this interval. The core-measured data, shown on the graph, form a matching parallel trend to the processed data but are consistently five degrees or more below the processed mean dips. This difference is explained mainly by the fact that the core data were measured from slabbed core surfaces giving a record of apparent dip only.

The graphs show that in spite of wide differences in the correlation parameters used, the computed dips give very similar results in terms of vector mean azimuth and mean magnitude. From this it can be inferred that in sedimentary work, the choice of correlation parameters in

processing raw dipmeter data is not nearly so critical to the final result as is sometimes believed. There is a useful degree of robustness in the correlation method which allows the same basic information to emerge from variously processed data.

Summary statistics for orientation data

It has become clear that, rather than trying to assign geological significance to every dip magnitude and azimuth value on a dipmeter log, it is more rewarding to analyse the data for trends. There is no statistic more useful in this respect for analysing orientation data than the vector mean (Fig. 7). The calculation and use of this statistic is well described in the reference given, but some additional comments must be made regarding its use for dipmeter interpretation.

The vector mean calculation gives valid results on dip data with a unimodal distribution. Where more than one mode exists (e.g. bimodal distribution) the vector mean calculation may give a

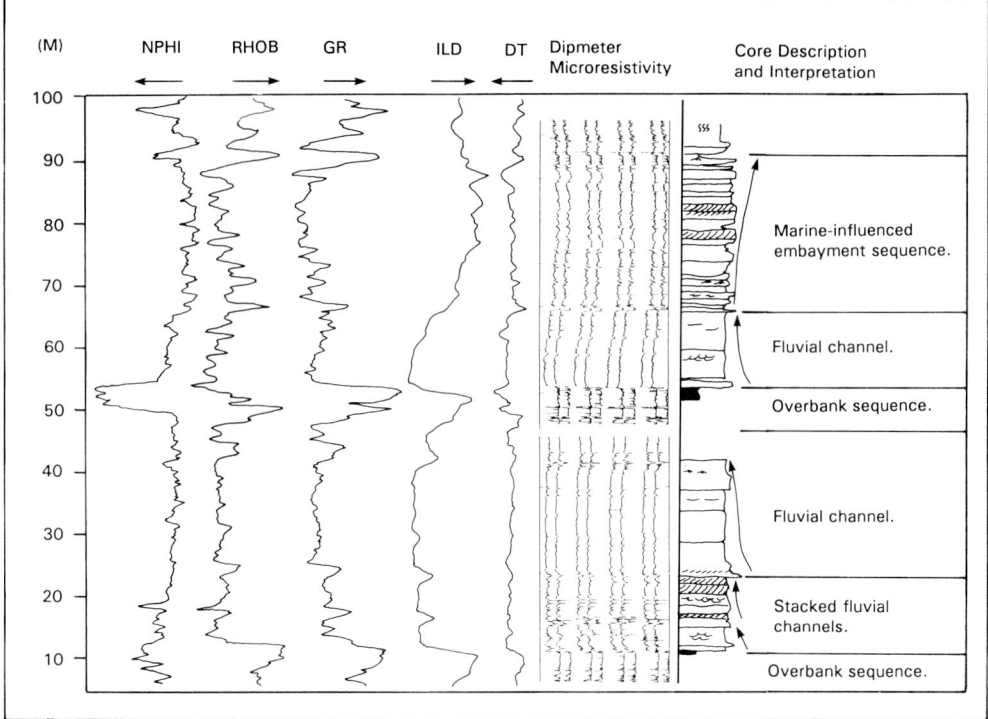

Fig. 8. Dipmeter microresistivity curves, open hole logs and core descriptions compared. This example illustrates how textural information expressed by dipmeter curves can be used in a qualitative way to provide excellent definition of sedimentary facies boundaries, as a supplement to the other open hole logs.

misleading result. The accompanying resultant length statistic should always be calculated when using vector means. This gives a useful indication of the spread of the results about the vector mean, which can be used to gauge the strength of a given preferred orientation. In addition, some test of significance of preferred orientation, for example the Rayleigh test (Curray 1956), gives a further check on the calculated mean.

Fortunately for many geological purposes, the assumption that a distribution is basically unimodal (i.e. that some preferred orientation exists), or at least that some dominant mode exists, is a valid one. This is after all one of the main reasons for running the dipmeter in the first place.

The vector mean calculation can be adapted for use on dip magnitude values which only range from 0°–90° rather than the full 0°–360°. All dip magnitude values are multiplied by four and the vector mean calculated in the normal way over the 0°–360° range. The result is then divided by four to give the vector mean dip

magnitude. This gives a more reliable average than the arithmetic mean.

Log zonation

The geological value of statistics and summary graphics such as histograms and rose diagrams in dipmeter interpretation, depends very much on how samples are selected from among the data. In sedimentary dipmeter studies, the borehole succession can be considered as an arrangement of more or less discrete packages of sediment deposited one on top of another. Each package is distinct from the one above and below by virtue of different gross lithologies or by the presence of different sedimentary structures. Part of the interpretation of the sedimentary succession depends on where boundaries (whether sharp or gradational) are recognized between individual packages.

The standard open hole logs, such as Gamma Ray, FDC/CNL, resistivity, provide an excellent record from which to set up log zones based on

**Uncored
Interval**

**Cored
Interval**

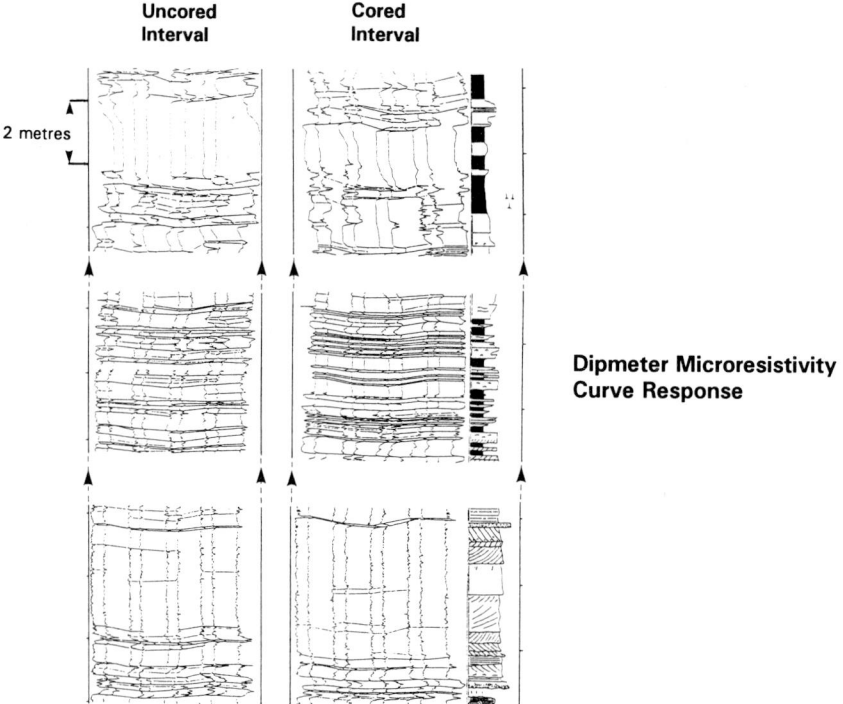

2 metres

**Dipmeter Microresistivity
Curve Response**

Fig. 9. Dipmeter microresistivity curves through fining-upwards sequences, one cored and the other uncored, from the same well. The curves from each sequence show detailed agreement of curve character, illustrating a well-developed facies cyclicity, and more generally the principle that dipmeter curve responses calibrated against core can be used to extend interpretations into uncored areas.

downhole lithology and facies variations. In addition to this, qualitative use of dipmeter microresistivity curves, calibrated against core intervals, has shown them to be a valuable source of textural information for accurate log zonation. Figure 8 shows the log responses, including dipmeter, through an interpreted cored clastic succession and illustrates how closely the combined logs are a reflection of sedimentary facies.

The dipmeter microresistivity curves bring out, with unrivalled detail, such subtle sedimentary features as the boundaries between stacked channels and the boundary between the highest channel sequence and the marine influenced embayment sequence above.

Figure 9 is another example of the detailed appreciation of sedimentary facies that can be gathered from dipmeter microresistivity curves. This figure shows dipmeter curves through two separate fining-upwards sequences from the same well through a reservoir interval. The right-hand sequence has been calibrated against a set of cores, while the left-hand sequence is interpreted from an uncored section. The cored

sequence occurs approximately 10 m above the uncored sequence in the well.

The curves of the upper sequence match up very well to the core. A large resistivity kick marks the base of the channel. The clean channel sands show fine 'sawtooth' resistivity variations, with minor peaks caused by coarser beds. Above this, an inter-bedded overbank sequence of mudstones and siltstones results in very regular resistivity contrasts. Finally a coal and seatearth sequence marks the conclusions of the fining-upwards sequence, and is reflected by a mixture of saturated dipmeter curves (flat spots indicating loss of tool pad contact) and a raggedness of curve response which seems to be characteristic of the presence of organic material.

The comparison between the dipmeter curves of the cored and uncored sequences shows a striking degree of similarity. This is a clear indication of sedimentary cyclicity in the reservoir formation and is also good evidence that dipmeter interpretations validated by core material can be extended with a degree of confidence into uncored sequences.

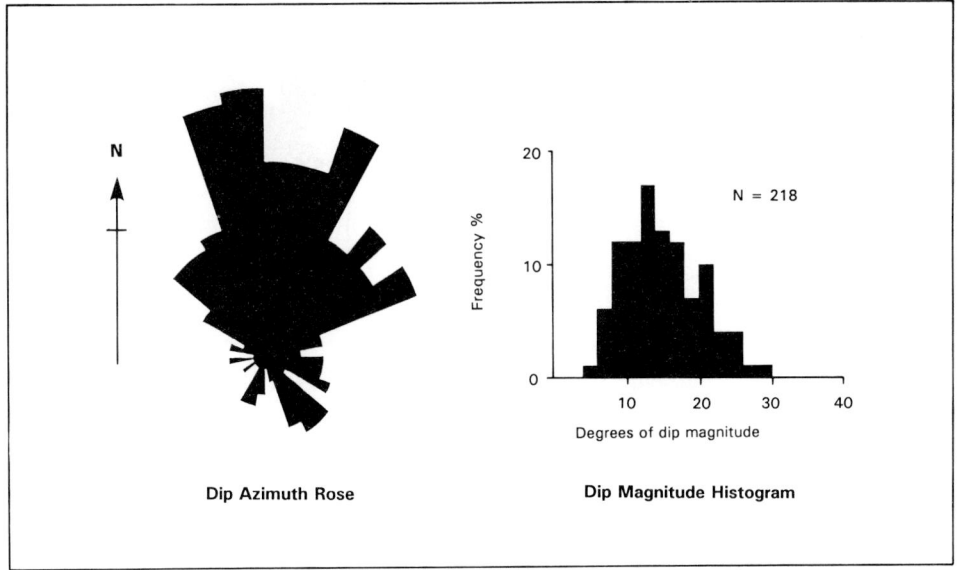

Dip Azimuth Rose   Dip Magnitude Histogram

Fig. 10. Dip azimuth rose and dip magnitude histogram summarizing the distribution of core-validated foreset information on the dipmeter log of a case study well. These core-validated results give a set of basic dip information from which hypotheses about the likely foreset dip distribution in uncored parts of the same well or adjacent wells can be made.

Dip filters

Statistics and summary graphics for dipmeter interpretation are enhanced still further by the use of dip filters to enable the selection of geologically reasonable samples. For example, in a palaeocurrent study of foreset dips, the aim is to restrict the analysis to dip results that come from foreset laminae only. A sample of dipmeter data taken wholesale from a given interval may contain dips not just from foreset laminae but from a variety of structures and bedding surfaces, such as cosets and planar bedding surfaces etc. (Rider 1978). A filter on allowable dip magnitudes, for example restricting the analysis to dips that fall between a typical foreset dip magnitude range of 10°–30°, has the effect of focusing the analysis on those dips in a sequence that are most likely to come from foresets (Cameron *et al.* in press; Williams & Soek in press).

A case example

The following discussion gives an example of how the principles discussed above can be applied to a practical well study. The example concerns a palaeocurrent study of dipmeter data from a well through a fluvial sand and shale sequence. Channels were interpreted as low-sinuosity braided type and so it was anticipated

that a study of foreset dip directions from the dipmeter would reveal a reliable down-current orientation.

The fluvial formation covered 522 m of borehole section. Cores were available for most of the upper 100 m of the formation, the remaining section being uncored. A detailed dipmeter processing, with parameters of 0.2 m correlation interval, 0.1 m step distance and 35° × 2 search angle was available. Structural dip in the well was negligible.

The study began with a phase of detailed dipmeter/core comparison in the uppermost 100 m of section. The aim was to detect all dips on the dipmeter log that could be related to foreset laminae on the core, and so establish a set of baseline core-validated orientation data. The results of this comparison are presented in Fig. 10. The azimuth rose shows a primary spread of north to east facing foreset dips, with a secondary spread developing towards the southeast. The dip magnitude histogram indicates a slightly negative skewed distribution of foreset dips spread around a modal class of 12°–14°.

The second element of the study was to devise a set of conditions whereby likely foreset information from the uncored intervals could be extracted from the dipmeter results and analysed for preferred orientation. First, it was assumed

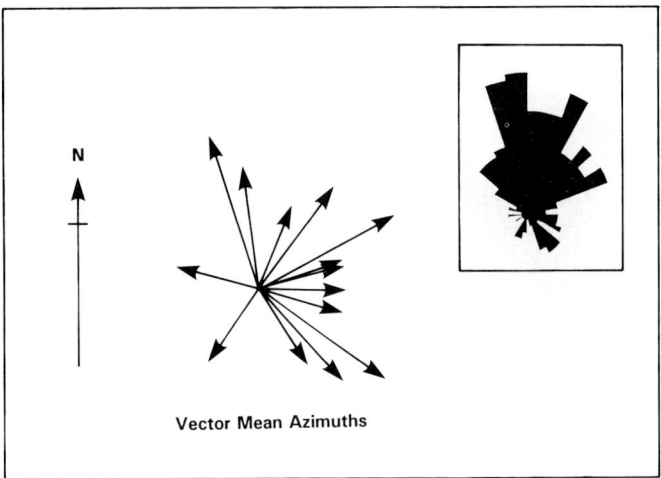

Fig. 11. Results of a preferred orientation analysis for foreset dip data in dipmeter results of a 522 m fluvial formation. Dipmeter results from suspected channel intervals were isolated from the main body of results. A dip magnitude filter of 12°–30° was then applied to exclude all dips outside of the likely foreset dip range. The resulting channel datasets were tested for preferred orientation using the vector mean statistic. Fourteen of the suspected channel intervals gave a significant indication of preferred orientation. These mean azimuths are shown in the figure, the longer the arrow the stronger the preferred orientation. The results of this analysis of foreset orientations can be compared with the dip azimuth rose of core-verified foreset data from the upper 100 m of formation (inset).

that foreset dips would occur only within channel sand intervals, so the dipmeter log was zoned to identify likely channel intervals. A broad lithologic zonation was achieved using the standard open hole logs. The dipmeter microresistivity data of each sand zone were then compared to microresistivity data of proven channels from the cores. This comparison led to rejection of some intervals as channels. Each of the remaining sand intervals was regarded as a likely channel interval in which there was the potential for foreset dips to occur. From the dip magnitude histogram of core-verified dip data (Fig. 10), it was deduced that a dip filter of 12°–30° would be sufficient to bracket the majority of foreset dips in channel intervals while rejecting other low-angle bedding dips and higher angle noise. A statistical analysis for preferred orientation was performed on the dipmeter data of all possible channel intervals with the dip filter in place. A vector mean azimuth was calculated for each interval, together with the resultant length, and tested for significance using the Rayleigh test (see Curray (1956) for method). All vector means indicating a statistically valid preferred orientation were retained, and the others rejected. The result of the whole well analysis was that fourteen channel sand intervals were

found to show a significant preferred orientation of azimuth data among the dip results between 12°–30° dip magnitude (Fig. 11). All vector means, except for two of lesser significance, lie within a north to east facing spread. This spread of results obtained by analysis over the entire 522 m of formation, shows a very marked similarity to the spread obtained from core-verified foreset dips in the top 100 m of the formation. It was concluded for the well study that foresets within the fluvial channel sands have a northeast bisected 160° spread of azimuths. This is the downstream orientation of foreset laminae and indicates an overall northeast palaeoflow. Closer inspection of the mean azimuths and their relationships in depth indicates wandering of the flow direction in time and thus accounts for the relatively wide spread of azimuths.

An interesting sideline to this interpretation is the extent to which this method of analysis reduces a mass of dip results into a set of significant vector means that relates directly to a geological interpretation aim. The process of deduction using a number of geologically reasonable assumptions was able to extract 14 statistically significant vector means, as the summary of 1110 probable foreset dips, out of an original number of 5220 computed dips for

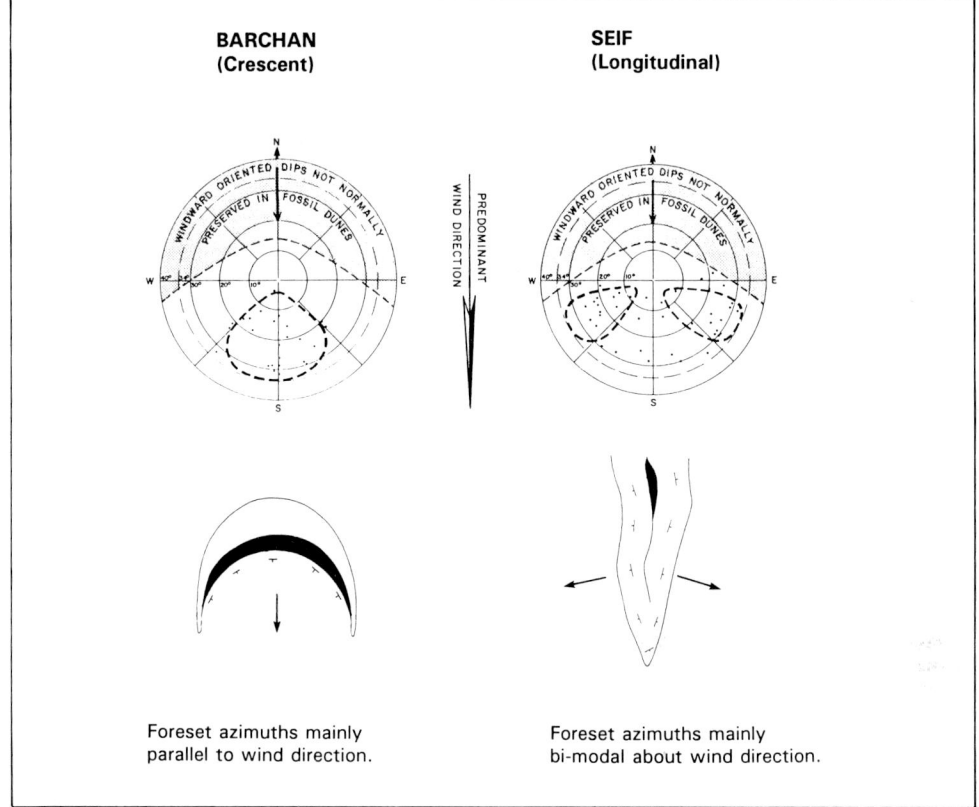

Fig. 12. Contrasting relationships between preferred orientation data, prevailing wind direction and sand body geometry. Crescent and longitudinal dunes after Glennie (1970).

the entire formation interval. The results are valuable orientation data for integration into the reservoir model.

Would the same result have been achieved without the support of the core data? Certainly the same type of analysis could have been attempted. The core data provided useful input for the calibration of microresistivity curve responses and for setting the dip magnitude filter. However, the microresistivity response can be calibrated almost as well against the other open hole logs, and the dip filter can be established by trial and error by a number of iterations in the analysis.

Interpretation of analysed orientation data

Having completed an analysis of dipmeter data for preferred orientation, there still remains the task of understanding the geological significance of any results that emerge. This is more a ques-

tion of sedimentological models than of dipmeter interpretation. For example, in the case discussed above it was reasonably obvious that the preferred orientation established from dipmeter was derived from downstream-oriented structures, and therefore the palaeocurrent direction. From this, and by comparison with established sedimentological models, it is interpreted that channel sand bodies in the formation were oriented along the palaeocurrent axis and may show rapid lateral thinning orthogonal to this axis. Here is an example where the relationship between the dipmeter derived preferred orientation and such factors as palaeocurrent and direction of reservoir elongation/thinning is fairly obvious. However, there are many sedimentary environments where these relationships are not so clearly defined.

One such example can be found in the interpretation of preferred orientation within aeolian dune sands. Given the discussion above, it is not difficult to imagine how a sedimentary dip-meter analysis of a dune sand sequence might proceed. However, if a preferred dune foreset orientation is established, what is its relationship to the prevailing palaeowind direction and to the direction of sand body elongation? This depends very much on the type of dune-form that is interpreted. Figure 12 illustrates the typical relationship between the prevailing wind direction and dipping surfaces within dune-forms of longitudinal and barchan type. Preserved dunes of longitudinal (seif) type will show a bimodal spread of azimuths orthogonal to the predominant wind direction but with a component in the palaeowind direction. The palaeowind axis will be parallel to the direction of dune elongation. The barchan or crescent dune, by contrast, will give dip azimuths that are mainly oriented in the downwind direction and which are orthogonal to the front of sand migration. In this aeolian example, the sedimentological model that is applied to the interpretation of dipmeter azimuth spreads and statistics is obviously critical to achieving a correct understanding of likely sand-body geometry and palaeogeographical context.

Conclusions

Reasoned use of log zonation and dip filtering techniques, combined with statistical analysis for preferred orientation, provide an efficient methodology for extracting useful sedimentary orientation data from dipmeter logging results. The approach overcomes two of the main obstacles to integration of dipmeter results with the other data from the open-hole logging suite. First, the mass of undiscriminated dip results produced by a normal dipmeter run can be summarized by a set of well focused essentials, and thus problems of data-handling and assimilation are minimized. Second, the statistical approach can largely override the random effects of 'noise' on dipmeter logs giving positive results where good data are in the majority.

I would like to thank Malcolm Rider and the reviewers for their comments on drafts of this paper.

References

ALLAUD, L. A. & RINGOT, J. 1969. The high resolution dipmeter tool. *The Log Analyst*, May–June 1969, 3–11.

ANXIONNAZ, H., DELFINER, P. & DELHOMME, J. P. 1990. Computer-generated corelike descriptions from open-hole logs. *AAPG Bulletin*, **74**, 375–393.

BIGELOW, E. L. 1982. Application of dip related measurements to a complex carbonate-clastic depositional environment. *The Log Analyst*, March–April 1985.

—— 1985. Making more intelligent use of log derived dip measurements. Part III: Computer processing considerations. *The Log Analyst*, May–June 1985, 18–31.

CAMPBELL, R. L. 1968. Stratigraphic applications of dipmeter data in mid-continent. *AAPG Bulletin*, **52**, 1700–1719.

CAMERON, G. I. F. 1986. Confidence and the identification of foresets in stratigraphic dipmeter surveys. *Transactions of the Tenth European Formation Evaluation Symposium, April 1986*.

——, COLLINSON, J. D., RIDER, M. H. & XU, L. 1992. Analogue dipmeter logs through a prograding deltaic sandbody. *In*: ASHTON, M. (ed.) *Advances in Reservoir Geology*. Geological Society, London, Special Publication, in press.

CURIAL, A. 1988. Multiple applications of dipmeter curves for analysing evaporite formations: examples from Paleogene of Bresse Trough, France. *AAPG Bulletin*, **72**, 1323–1333.

CURRAY, J. R. 1956. The analysis of two-dimensional orientation data. *Journal of Geology*, **64**, 117–131.

GLENNIE, K. W. 1970. Desert sedimentary environments. *Developments in Sedimentology*, 14, 222.

MORAN, J. H., COUFLEAU, M. A., MILLER, G. K. & TIMMON, J. P. 1962. Automatic computation of dipmeter logs digitally recorded on magnetic tapes. *Journal of Petroleum Technology*, July, 771–782.

PERRIN, G. 1975. Comparaison entre des structures sedimentaires a l'affleurement et les pendagemetries de sondage. *Bulletin Centre Recherche Pau-S.N.P.A.*, **9**, 147–181.

RIDER, M. H. 1978. Dipmeter log analysis—an essay. *SPWLA Ninth Annual Logging Symposium, June 1978*.

RUDMAN, A. J. & LANKSTON, R. W. 1973. Stratigraphic correlation of well logs by computer techniques. *AAPG Bulletin*, **57**, 577–588.

SALLEE, J. E. & WOOD, B. R. 1984. Use of microresistivity from the dipmeter to improve formation evaluation in thin sands, Northeast Kalimantan, Indonesia. *Journal of Petroleum Technology*, **36**, 1535–1544.

SCHLUMBERGER (Ltd.), 1981. *Dipmeter Interpretation—Fundamentals*. Schlumberger, Paris.

WILLIAMS, H. & SOEK, H. F. in press. Predicting reservoir sand body orientation from dipmeter data: the use of sedimentary dip profiles from outcrop studies. *In*: BRYANT, I. D. & FLINT, S. (eds) *Quantitative clastic reservoir modelling*. International Association of Sedimentologists, Special Publication.

Fractures and stress

Borehole breakouts and stress analysis in the Timor Sea

R. R. HILLIS[1] & A. F. WILLIAMS[2]

[1] School of Earth Sciences, The Flinders University of South Australia, GPO Box 2100, Adelaide, SA 5001, Australia (Present address: Department of Geology and Geophysics, University of Adelaide, GPO Box 498, Adelaide, SA 5001, Australia)

[2] CSIRO Division of Geomechanics, PO Box 54, Mt Waverley, VIC 3149, Australia

Abstract. Boreholes drilled in the search for oil in the Vulcan Sub-basin (Timor Sea, Northwest Shelf, Australia) commonly exhibit an elliptical cross section believed to be the result of wellbore failure known as borehole breakout. The azimuths of the long axes of breakouts identified in 13 wells in the Vulcan Sub-basin show a reasonably consistent 130–170°N trend implying that maximum horizontal compressive stress (SH_{max}) is oriented 040–080°N.

This NE–ENE SH_{max} orientation in the Vulcan Sub-basin does not coincide with the direction of absolute plate velocity, nor is it consistent with compression transmitted from the nearby Australia/Banda Arc collision zone. However, it is in agreement with theoretical models of stress distribution based on the plate-driving forces throughout the Indo-Australian plate.

The breakout analyses, in conjunction with structural interpretations, suggest that the instabilities observed in vertical wells will be exacerbated by orienting horizontal wells 040–080°N, and that drilling in the 130–170°N direction may enhance both stability and production rates.

The Vulcan Sub-basin (hereafter referred to as the Vulcan) underlies the Timor Sea on the North West Australian Continental Shelf. The Australian North West Shelf is a passive margin formed by the final, late Jurassic–early Cretaceous break-up of eastern Gondwanaland (Audley-Charles 1988). The Vulcan is located at its northeastern end, immediately south of the present-day zone of collision between the Australian Continent and the Indonesian Banda Island Arc (Fig. 1).

The Banda Arc contains the classic elements of an island arc (volcanic arc, forearc basin, forearc ridge, and trench), and was formed by northward subduction of oceanic crust to the north of the Australian continent (Hamilton 1977, 1988). However, both geochemical (Whitford & Jezek 1982) and geophysical data (Chamalaun et al. 1976; Jacobson et al. 1978) show that continental crust is continuous between the Australian continent and the Banda Arc. The relative aseismicity of the Timor Trough (Cardwell & Isacks 1978), the slowing of convergent motion in the Trough since the mid Pliocene (Johnston & Bowin 1981), and the cessation of volcanism at 3 Ma in the Alor/Wetar area north of Timor (Abbott & Chamalaun 1981) all suggest that the arrival of buoyant Australian continental crust has 'choked' the Banda Arc subduction system. In this paper, the inferred in situ stress in the Vulcan, as determined from

analysis of four-arm dipmeter logs, is interpreted in terms of collision in the area.

The collision between Australia and the Banda Arc is believed to be responsible for a major phase of late Miocene–Recent tectonism in the Vulcan (e.g. Pattillo & Nicholls 1990; Nelson 1989). The resulting structures have been interpreted both as predominantly extensional and as predominantly strike-slip (Woods 1988; Nelson 1989). Determination of the stress regime in the Vulcan may help unravel the structural complexities of the region. Furthermore, although the Vulcan contributes some 20% of Australia's current oil production, field size in the area is relatively small (of the order of tens of millions of barrels). Collision-related, late Miocene–Recent tectonism may have been responsible for the reduction in size, and even destruction, of existing hydrocarbon traps in the Vulcan. The relatively small field size in the Vulcan requires that novel technologies be used in their commercial exploitation, hence horizontal drilling is likely to play an important role. This paper discusses the use of horizontal principal stress data in predicting those horizontal well orientations least likely to give rise to extensive wellbore caving or spalling.

Borehole breakouts

Subsequent to drilling, the cross sectional shape

From HURST, A., GRIFFITHS, C. M. & WORTHINGTON, P. F. (eds), 1992,
Geological Applications of Wireline Logs II. Geological Society Special Publication No. 65, pp. 157–168.

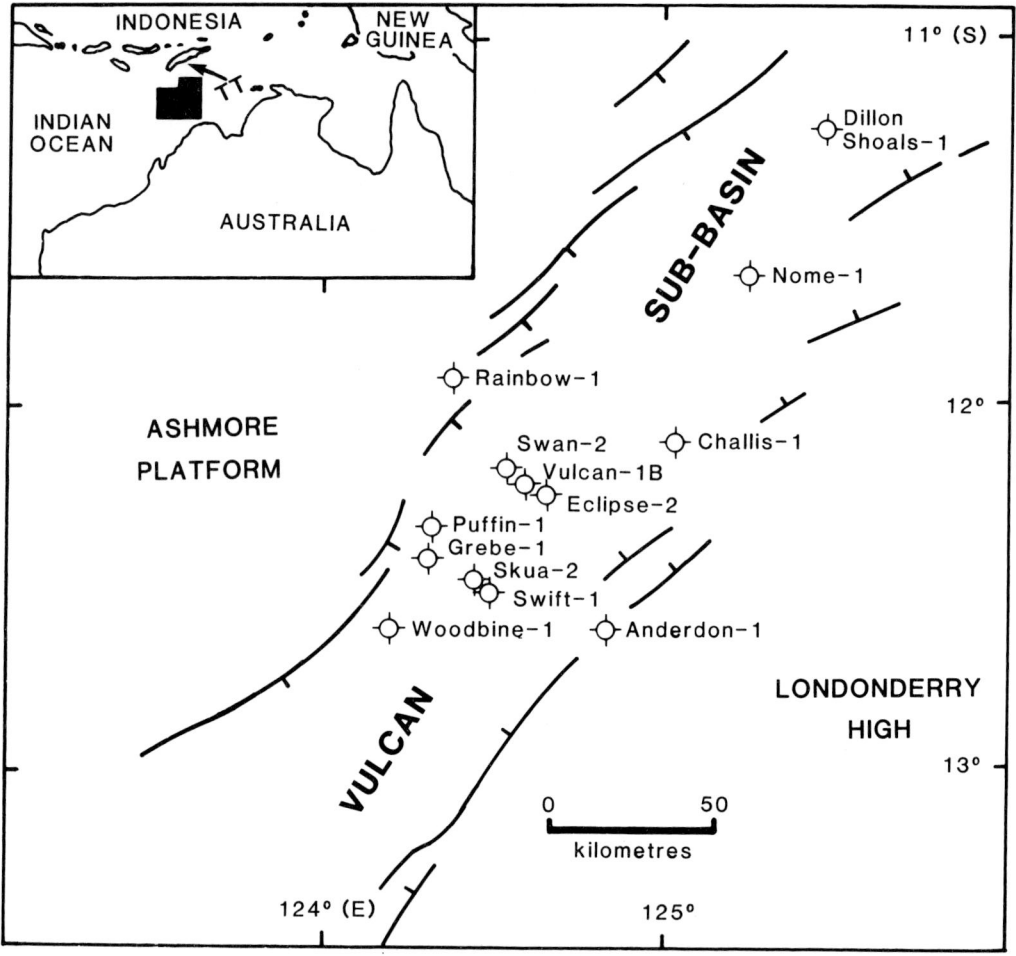

Fig. 1. Location of wells in the Vulcan analysed for borehole breakouts. Marked faults are normal with a possible component of strike-slip movement. The shaded area on the inset map shows the location of the detailed map. TT marks the location of the Timor Trough, immediately south of the island of Timor.

of many boreholes is deformed, usually over limited, discrete vertical intervals, into an ellipse, known as a breakout (Fig. 2A). Breakouts commonly show a consistent direction of elongation throughout the hole (e.g. Dart & Zoback 1989). The empirical relation that borehole breakouts are aligned parallel to the direction of minimum in situ horizontal stress is demonstrated by their orthogonality to the direction of maximum horizontal compression (SH_{max}), as determined by earthquake focal mechanisms and hydraulic (induced) fracturing (e.g. Plumb & Hickman 1985; Zoback et al. 1985). This relation has also been demonstrated in mechanical theory (Bell & Gough 1979; Zoback et al. 1985). Failure of the borehole wall is due to a local concentration of

the ambient stress field near the free surface of the hole wall (Fig. 2A).

The dipmeter tool and breakout recognition

Boreholes drilled in the course of oil exploration in the Vulcan have been logged by four-arm dipmeter tools (Schlumberger 1981). As the four-arm dipmeter tool is raised up the borehole, cable torque causes it to rotate about a semi-vertical axis. The tool records its orientation and inclination. Its four pads, spaced at 90° intervals and hydraulically pressed against the borehole walls, measure formation resistivity. The caliper arms also measure the width of the hole (Fig. 3).

Successful breakout recognition requires the

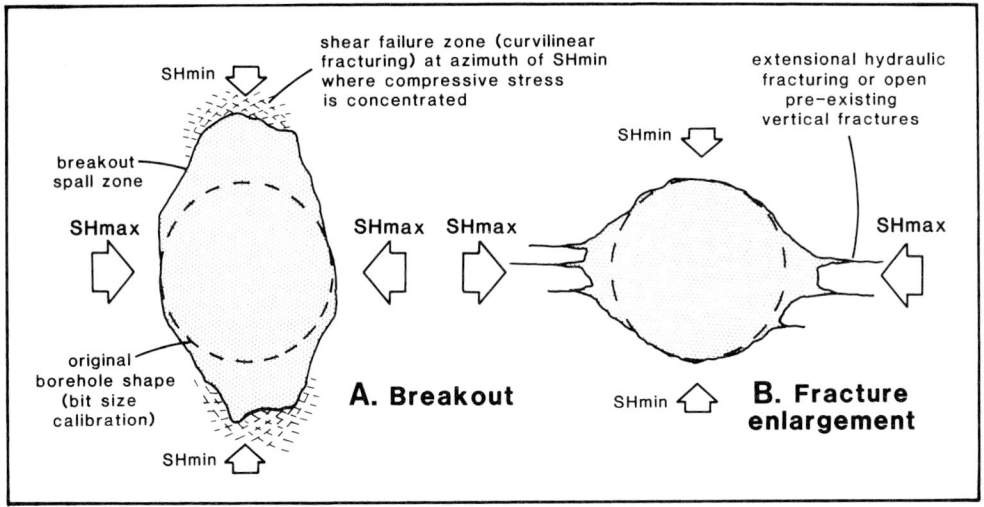

Fig. 2. Stress-induced wellbore cross sectional elongations. A, Breakouts are elliptically shaped zones of borehole–wall spalling over discrete vertical intervals, formed primarily by compressional shear failure due to the concentration of horizontal compressive stress at the azimuth of SH_{min}. B, Drilling-induced, extensional hydraulic fractures elongate the wellbore cross section at the azimuth of SH_{max}. Inferred SH_{max} direction in the Vulcan parallels the structural trend of the basin (Fig. 1), and the intersection of wellbores with open, pre-existing fractures may also generate NE–SW trending cross sectional elongations. Modified from Dart & Zoback (1989).

Table 1. *Criteria used in the recognition of stress-induced borehole cross-section elongation on four-arm dipmeter logs. See Plumb & Hickman (1985) for a more detailed description of these criteria.*

1. The tool rotation stops in the zone of elongation. Ideally the tool should rotate before and after the breakout, but in a zone containing several discrete breakouts, rotation may stop altogether.

2. The difference between the calipers is greater than 6 mm and the length of the elongation zone greater than 1.5 m.

3. The larger caliper should be greater than (drill) bit size.

4. The smaller of the caliper readings should not be significantly greater or smaller than bit size and its trace should be straight.

5. The direction of elongation should not coincide with the high side of the tool if deviation is greater than 5°.

distinction between stress-induced elongation of the borehole cross-section and that induced by the effects of drilling itself, such as mudcakes, washouts and key-seats (asymmetric abrasions due to wear by the drill string on the side of a hole which deviates from the vertical) (Fig. 3). Table 1 lists the criteria used for recognition of stress-induced cross sectional elongation in this

study. They are derived from Plumb & Hickman (1985) who give a more detailed description of the breakout identification procedure.

The length of each zone of borehole cross sectional elongation that satisfied the above criteria, and, at the point of maximum eccentricity, its depth, caliper width readings and long axis azimuth were recorded. Cross sectional elongation azimuths were corrected for magnetic declination. Finally, since cross sectional elongation detected on the four-arm dipmeter log must be assumed to be symmetrical, all azimuths were assigned to the range 0 to 180°. The Appendix outlines the statistical analysis applied to the resultant azimuths of borehole cross sectional elongations.

Breakout orientations in the Vulcan

One hundred and seven discrete vertical intervals of borehole cross sectional elongation were recorded in 13 wells in the Vulcan (Table 2 and Figs 4–6). The summary rose diagrams for unweighted, and length- and eccentricity-weighted azimuths of cross sectional elongation, show a pronounced SE–SSE trend (Fig. 6). The mean azimuths for the entire unweighted and length- and eccentricity-weighted datasets in the Vulcan lie in the range 142–158°N (Table 2). The mean azimuths of all the wells except Vulcan-1B

Table 2. *Summary of borehole breakout data from the Vulcan Sub-basin. Lat and Lon are the latitude and longitude of the well locations respectively. N is the total number, L/ N the mean length, and E/N the mean eccentricity of breakouts in the well. Mean and sd are the mean azimuth (0–360° N) of breakouts in the well and their standard deviation in degrees (see Appendix). Q is the quality rank in the World Stress Map scheme (Zoback & Zoback 1989), and ecc is the eccentricity of the breakout, i.e. the difference between the two caliper readings.*

Well	Lat (N)	Lon (E)	N	Unweighted			Length-weighted			L/N (m)	Ecc-weighted			E/N (cm)
				mean	sd	Q	mean	sd	Q		mean	sd	Q	
Anderdon-1	12°39'	124°48'	7	139	36	D	133	30	D	9.0	127	30	D	1.7
Challis-1	12°07'	125°00'	6	146	31	D	131	19	C	11.9	145	19	C	3.4
Dillon Shoals-1	11°14'	125°27'	12	138	23	C	139	12	B	13.0	140	20	B	2.0
Eclipse-2	12°14'	124°39'	15	172	45	D	173	37	D	24.6	158	29	D	6.7
Grebe-1	12°27'	124°15'	1	110		E	110		E	60.4	110		E	2.5
Nome-1	11°39'	125°13'	8	143	27	D	117	54	E	6.3	142	22	C	2.0
Puffin-1	12°19'	124°20'	5	132	29	D	132	34	D	7.4	131	31	D	2.8
Rainbow-1	11°56'	124°20'	18	140	31	D	148	22	C	17.0	148	22	C	5.8
Skua-2	12°31'	124°24'	10	131	49	D	103	56	E	15.1	110	65	E	2.6
Swan-2	12°12'	124°30'	9	132	43	D	167	46	D	35.1	141	46	D	5.3
Swift-1	12°32'	124°27'	2	110	36	E	107	36	E	2.6	101	35	E	1.5
Vulcan-1B	12°15'	124°33'	9	048	42	D	030	40	D	74.6	038	51	E	4.9
Woodbine-1	12°39'	124°09'	5	168	40	D	158	13	C	52.9	161	24	C	6.2
Vulcan Sub-basin			107	142	41		158	44		23.6	149	34		4.2

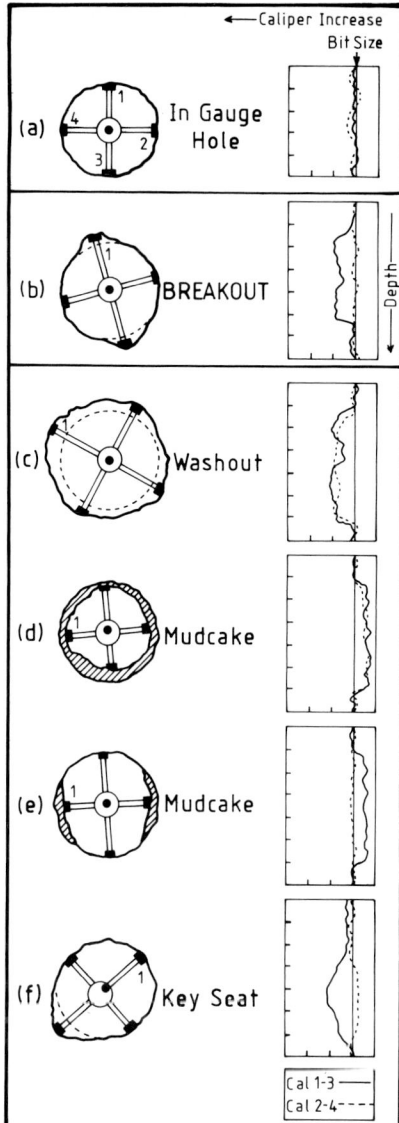

Fig. 3. Cross sectional schematics of possible wellbore conditions and their appearance on the four-arm dipmeter caliper log. Dashed circle on schematics represents bit size, solid line hole shape and shading mudcake. The orientation of the four-arm dipmeter in the hole is shown. Solid dot indicates centre of the in-gauge hole. Cal 1–3 and cal 2–4 indicate borehole diameter as measured between opposing caliper arms. (a) An in-gauge hole; (b) a stress-induced breakout; (c) a washout; (d) and (e) mudcaked holes; (f) key-seat induced by eccentric wobble of the drill string abrading wall of borehole which deviates from the vertical. Note that only (b) satisfies criteria 3 and 4 of Table 1. However, in (f) the smaller bit size may be only slightly less than caliper size, hence rule 5 of Table 1 is also applied. Modified from Plumb & Hickman (1985).

lie in the SE quadrant (Table 2 and Fig. 5). If Grebe-1 and Swift-1, where only one and two cross sectional elongations were observed respectively, are also ignored, the range of mean unweighted azimuths of borehole cross sectional elongation is 131–172°N (Table 2). A NE–ENE SH_{max} orientation in the Vulcan is inferred from the SE–SSE trend of cross sectional elongation in the Vulcan.

Elongated borehole cross sections in the Vulcan were recorded over a depth range of 1235–3997 m (below drilling rotary table). There is no systematic variation in the mean azimuth of cross sectional elongation with depth. However, cross sectional elongation azimuths occurring at depth show a tighter cluster around the mean azimuth than those occurring in the shallower section. Considering cross sectional elongations occurring above and below 2.4 km depth, unweighted and length- and eccentricity-weighted mean azimuths are similar in the two groups (Table 3). However, the 63 cross sectional borehole elongations observed above 2.4 km show a standard deviation about the mean of 39–57°, whereas the 44 deeper cross sectional elongations show a standard deviation of 29–34°. The data were also divided into those occurring in formations younger, and those in formations older than the Callovian rift onset unconformity in the Vulcan (Hillis 1990). Considering unweighted and length- and eccentricity-weighted mean azimuths, the 72 cross sectional elongations developed in formations post-dating rift onset show a standard deviation about the mean of 36–48°, whereas the 35 pre-rift cross sectional elongations show a standard deviation of 20–27° (Table 3). It is considered that with increasing lithification, as witnessed by burial depth and/or age, formations are less likely to develop cross-sectional elongations with a 'spurious' azimuth (i.e. one that deviates from the mean direction, and hence the inferred direction of least horizontal stress).

Although statistical analysis shows a clear SE–SSE cross sectional elongation azimuth trend, Eclipse-2 and Vulcan-1B appear to show bimodal, orthogonally-oriented azimuth distributions (Fig. 4). Indeed the mean cross sectional elongation azimuth trend in Vulcan-1B is 048°N (Table 2 and Fig. 5). The rose diagram summarizing all length-weighted data from the Vulcan also shows a subsidiary cross sectional elongation trend which is orthogonal to the principal SE–SSE trend (Fig. 6).

Morin et al. (1989) outlined a technique for determining whether presumed bimodality is statistically significant, or whether it is simply a projection of substantial data scatter extending

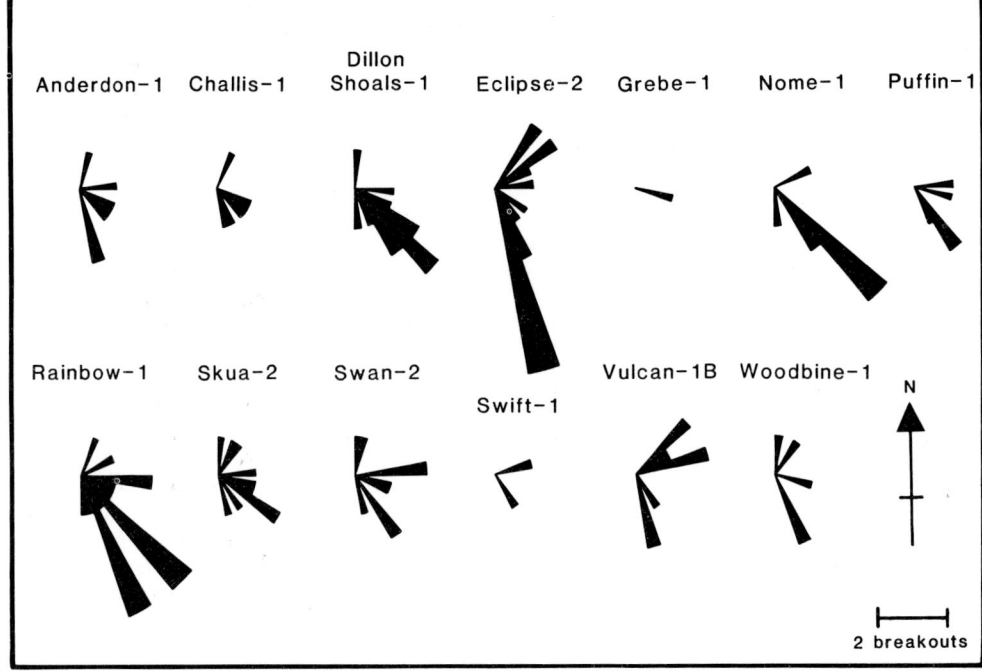

Fig. 4. Rose diagrams illustrating the distribution of unweighted breakout azimuths for wells in the Vulcan.

Table 3. *Comparison of borehole elongation azimuths recorded above and below 2.4 km depth, and in formations younger and older than the Callovian rift onset unconformity. Analysis and abbreviations as Table 2.*

Well	N	Unweighted mean	sd	Length-weighted mean	sd	L/N (m)	Ecc-weighted mean	sd	E/N (cm)
above 2.4 km	63	138	46	162	57	24.7	148	39	4.1
below 2.4 km	44	144	34	157	29	21.8	150	29	4.2
post-rift onset	72	147	48	165	46	31.0	152	36	5.2
pre-rift onset	35	137	27	138	20	8.2	140	24	2.1

from a single predominant azimuthal mode. However, the possible subsidiary, orthogonal mode observed in the Timor Sea comprises only 10–20 azimuthal readings. The paucity of data precludes meaningful statistical analysis of this subsidiary mode (but see Appendix for details of the statistical analysis applied).

Orthogonal cross sectional elongation trends have been ascribed to the generation of true breakouts and extensional, drilling-induced hydraulic fractures in the same well (Dart & Zoback 1989; Morin *et al.* 1989; Figure 2B). It is not possible to differentiate conclusively between the two types of stress-induced borehole cross

sectional elongations using only wellbore-enlargement log data. However, SH_{min} direction is generally determined from the primary cross sectional elongation mode (e.g. Morin *et al.* 1989). The assumption that the primary mode of azimuths of cross sectional elongations represents the true breakout trend is supported in this study because the subsidiary orthogonal mode is only well developed in the post-rift onset sequences (all cross sectional elongations in Eclipse-2 and Vulcan-1B occur within the post-rift onset sequence). Weaker (shallower/younger) rocks are more likely to undergo hydraulic fracturing for a given excess mud weight.

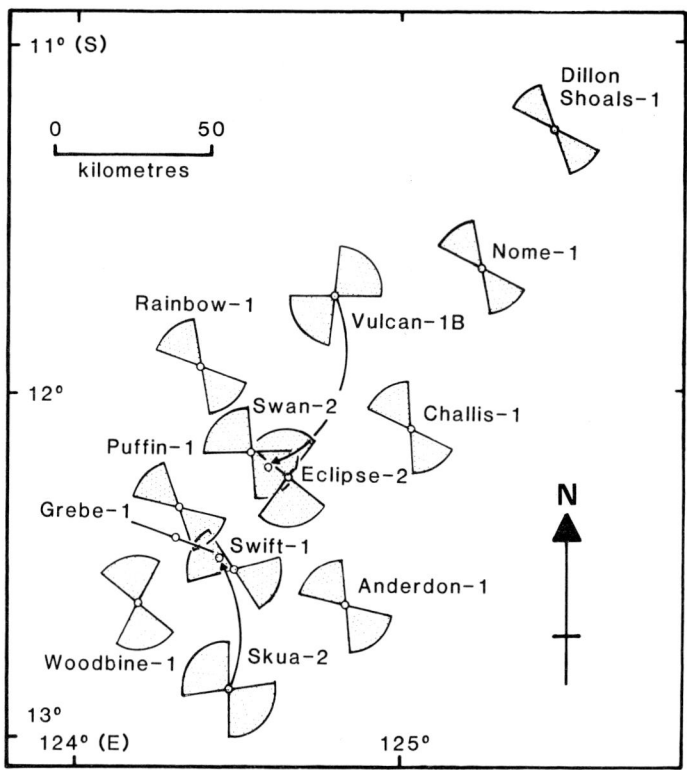

Fig. 5. Mean unweighted breakout azimuths for wells in the Vulcan. Small circles indicate well locations. The dotted 'bow-tie' for each well illustrates mean breakout azimuth ± one standard deviation.

Furthermore, it is likely that drilling engineers paid less attention to mudweight in the upper hole. It is planned to analyse further wells in order to determine whether the orthogonal, bimodal cross sectional elongation trend is significant. The authors are also currently collaborating on modifying standard oil company leak-off (formation integrity) tests in wells on the North West Shelf to simulate hydraulic fracture experiments (Bell 1990).

The subsidiary, orthogonal group of cross-sectional elongations may be due to the wellbore intersecting existing steeply dipping fractures (Babcock 1978; Bell & Gough 1979; Fig. 2B). The Vulcan exhibits a pronounced NE–SW structural grain (Fig. 1), and boreholes intersecting steeply dipping fractures with this orientation would encounter a strength anisotropy favourable to the development of NE-trending wellbore cross sectional elongation.

World stress map project

A global compilation of in situ lithospheric stress data is being carried out under the auspices of the World Stress Map Project (Zoback et al. 1989). In situ stress indicators are quality ranked (A > B > C > D > E) on the reliability of the method used and the internal consistency of the data (Zoback & Zoback 1989; M. L. Zoback, pers. comm. 1990). Data ranked above D has been included in the World Stress Map of Zoback et al. (1989). Most of the data derived in this study are assigned a rank of D or E because the standard deviation of cross sectional elongation azimuths in individual wells exceeds 25° (Table 2). However, if the unweighted and the length- and eccentricity-weighted data are considered, five wells attain at least one rank better than D. The non-D ranked datasets indicate mean SH_{max} orientations in the range 041–071°N, close to the means for the entire dataset.

The SH_{max} orientations inferred herein are in reasonable agreement with the rather sparse World Stress Map data for the Timor Sea and adjacent regions. Extant borehole breakouts from the Vulcan (Swan-2), Browse (Scott Reef-2A), the onshore Canning Basins, and an earth-

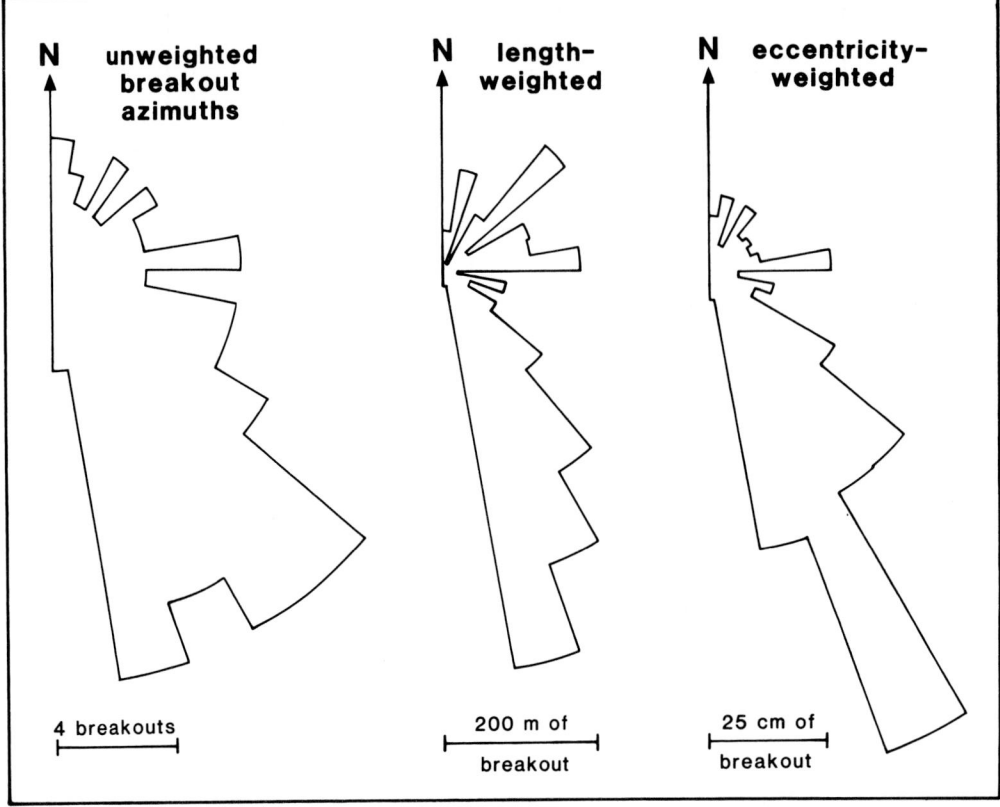

Fig. 6. Composite rose diagrams of unweighted, length-weighted and eccentricity-weighted breakout data for the entire Vulcan.

quake focal mechanism (Friedrich *et al.* 1988) from the offshore Canning suggest that SH_{max} is oriented NE–E.

Interpretation of the in situ stress data

SH_{max} orientation in some lithospheric plates shows a strong correlation with the direction of absolute plate motion (Zoback *et al.* 1989). However, mean SH_{max} orientations inferred by the present study are rotated some 20–40° east of the NNE direction of absolute plate velocity determined by Minster & Jordan (1978) for the Indo-Australian plate (IAP).

The northeastern boundary of the IAP is defined by a complex and laterally variable system of plate convergence. From northwest to southeast this includes Himalayan continent–continent collision, subduction of oceanic IAP under the Sunda Island Arc, and diachronous continental IAP–Banda Island Arc collision. Variation continues through the more mature collisional belt of New Guinea. At the Solomon–New Hebrides Trench, IAP of mixed continental/oceanic affinity is being subducted under the Pacific Plate, and at the Tonga–Kermadec Trench, the IAP overrides the Pacific Plate.

This complex convergence scheme imposes a variety of collision and trench-resistive, and subducting slab pull forces on the IAP (Richardson *et al.* 1979; Cloetingh & Wortel 1986). Hence SH_{max} orientation would not be expected to simply parallel the direction of absolute plate motion. However, the data presented here do show good agreement with theoretical models of horizontal stress trajectories in the IAP based upon numerical modelling of the effect of plate-driving forces. Both the most sophisticated global stress model (E31) of Richardson *et al.* (1979) and the IAP-specific model of Cloetingh & Wortel (1986) predict a NE SH_{max} orientation in the Timor Sea area.

The data presented herein, combined with extant World Stress Map data, show that the stress regime of the Australian North West Shelf is complex. At the southwestern end of the North West Shelf, SH_{max} is oriented NW–SE, orthogonal to that in the Timor Sea. Modelling the effect of far-field stresses (e.g. Cloetingh & Wortel 1986) only partially accounts for the observed complexity. Second-order effects such as topography, sedimentary loading and the density contrast between continental and oceanic crust probably complicate the stress regime at the scale of the margin. In order to further investigate the variability of the stress regime within the North West Shelf the authors plan to investigate wellbore cross sectional elongation in more wells in the area.

Implications for the controls on and structural style of Miocene–Recent tectonism in the Vulcan

Late Miocene–Recent tectonism in the Vulcan was associated with a period of rapid subsidence (e.g. Pattillo & Nicholls 1990; Hillis 1990). Although the onset of subsidence pre-dates mid Pliocene formation of the present Timor Trough (Veevers 1974) and the synchronous uplift of Timor (Audley-Charles 1986), this phase of tectonism has been widely ascribed to Australia/Banda collision. Since the Vulcan lies outboard (on the foreland side) of the tectonically-loaded foredeep of the Timor Trough, and the onset of late Miocene–Recent tectonism and subsidence preceded formation of the Trough, late Miocene–Recent tectonism cannot be attributed to foreland basin-style loading from the Timor orogenic belt. Furthermore, the 050–060°N SH_{max} orientation determined in this study, is aligned parallel to the Trough. Hence, in spite of its proximity, compression from the Australia/Banda collision zone does not control SH_{max} orientation in the Vulcan.

The inferred SH_{max} orientation is consistent with extension developed in the outer part of a lithospheric plate as it is downwarped into a subduction zone, or, as in the case of the Timor region, a young collision zone. At trench depths, extension is developed downdip, into the trench/trough as the plate is bent (e.g. Isacks & Molnar 1969). Hence SH_{max} is parallel to the trough. However, geometrical reconstructions of the Australia/Banda collision zone suggest that the Vulcan is south of the region of plate bending associated with the Timor Trough (Hamilton 1977, Fig. 78; Price & Audley-Charles 1987). Thus the Vulcan should not be subject to bending-related extension. The inferred SH_{max} orientation is also consistent with compression trans-

mitted from the more mature collisional belt in New Guinea, and with slab-pull-related extension transmitted from the nearby Java Trench where oceanic IAP continues to subduct under the Sunda Arc. Extensional forces may persist from the pull of the dense slab of now-subducted oceanic IAP beneath Timor, if the now-subducted oceanic lithosphere is still attached to the continental Australian IAP. However, both McCaffrey et al. (1985) and Price & Audley-Charles (1987) suggest, from seismological and mechanical/regional geological considerations respectively, that the now-subducted, oceanic part of the IAP is in the process of becoming, or has already become, detached from the overriding Australian Continent. Nonetheless, while the Vulcan may no longer be subject to 'pull' from the subducted oceanic lithosphere, the inferred SH_{max} orientation implies that there is not yet significant 'push' from the young continent/island arc collision zone.

The Vulcan exhibits a pronounced NE structural grain that developed during the late Jurassic–early Cretaceous breakup of the continental margin and also during Miocene–Recent tectonism (Woods 1988; Pattillo & Nicholls 1990). The observed NE–ENE SH_{max} orientation would tend to cause wrench movements on the NE-trending faults if the overall stress field is compressional, or orthogonal extensional movements if it is extensional (the borehole breakout method can determine the orientation of SH_{max} but not whether the stress field is extensional or compressional). Further work determining the magnitudes of SH_{max}, SH_{min}, and S_v (vertical stress) from modified leak-off (formation integrity) tests and density log data (Bell 1990) could be used to resolve the sense of movement implied by the in situ stress field.

Implications for horizontal drilling in the Vulcan

Companies operating in the North West Shelf/Timor Sea region, like operators elsewhere in the world, are planning to take advantage of the economics and efficiencies offered by horizontal drilling technology. The feasibility of horizontal drilling plays a significant role in assessing the viability of many smaller fields, and it forms part of the strategy for several field developments in the Vulcan Sub-basin.

The in situ stress environment impacts on two broad issues which need to be addressed in the design of horizontal wells. The first concerns the stability of the well, the second concerns any anisotropy in reservoir permeability. From a stability point of view, there is a need to use mud weights which will not cause hydraulic fracturing or fluid loss, but which ensure stability and minimize compressional shear failure. These issues depend on stress magnitudes, the stress ratios and on the orientation of the well with respect to the in situ stresses. From a permeability point of view, the orientation of the in situ stresses, in conjunction with the structural data, may indicate the orientation of any anisotropy in reservoir permeability, and suggest a preferred well direction.

Evidence of distress, such as borehole breakouts, in vertical exploration wells such as those analysed herein has alerted operators to the potential of greater difficulties when highly deviated or horizontal wells are attempted. Of particular concern is the observation that breakouts frequently occur in the shales which directly overlie the reservoir sands in strata where it is necessary to build a high-deviation angle.

If it is assumed that the maximum principal stress is vertical, and neglecting formation strength anisotropy, then the argument follows two important steps. Firstly, the occurrence of breakouts suggests that the magnitudes of horizontal stresses are high relative to the material strength, and that the stress ratios are also high, i.e. $SH_{max}/SH_{min} \gg 1$. Secondly, it is clear that wells drilled horizontally in the direction of SH_{max} will be subjected to high vertical stress SV and a low horizontal stress SH_{min}, with $SV/SH_{min} > SH_{max}/SH_{min}$, implying greater instability for the horizontal well than the vertical case. On the other hand, wells drilled in the direction of SH_{min} will be subjected to the same vertical stress but a higher horizontal stress so that $SV/SH_{max} < SV/SH_{min}$, implying improved stability over the SH_{max} direction, and probably improved stability over the vertical case.

If the major Miocene–Recent structures are, as interpreted, predominantly extensional and strike-slip (Woods 1988; Nelson 1989), not only would they have lowered horizontal stresses in the direction of extension, but they imply that the maximum permeability may be oriented 040–080°N (Fig. 1). Wells drilled normal to this direction should maximize intersection of high permeability features.

It is suggested that the breakout analysis of vertical wells in the Vulcan Sub-basin, in conjunction with other geological data, indicates that the instability of horizontal wells will be exacerbated by drilling towards 040–080°N, and that both stability and production rates will be enhanced by drilling towards 130–170°N, or 310–350°N. However, this is a preliminary interpretation and further work determining the magnitudes of SH_{max}, SH_{min} and SV from

modified leak-off (formation integrity) tests and density log data will be used to confirm the relative magnitudes of the principal stresses.

RRH has been funded by a research grant from the The Leverhulme Trust (UK) and BP Australia Ltd. Additional support for the project has come from the Energy Research and Development Corporation (Australia), CSIRO (Australia) and Flinders University. Wiltshire Geological Services, Ampol Exploration Ltd, BHP Petroleum, BP Exploration Ltd, Norcen International Ltd, SANTOS Ltd, TCPL Resources Ltd and WMC Ltd are all thanked for providing well data. Mary Lou Zoback and David Denham kindly provided information on, and results of, in situ stress analysis.

Appendix

Intuitively, it is clear that 'normal', linear statistics are not appropriate for the analysis of directional data. For example, the directional mean of two breakout azimuths of 010°N and 170°N is not 090°N (Table A1). The statistical analysis of borehole cross sectional elongation azimuths followed Mardia's (1972) techniques for the analysis of directional data.

Mean azimuth (radians) is given by:

$$\cos^{-1}(C/R) \text{ or } \sin^{-1}(S/R) \qquad (1)$$

where $C = 1/n\sum f_i\cos\theta_i$ (average of the cosines of the azimuths)

$S = 1/n\sum f_i\sin\theta_i$ (average of the sines of the azimuths)

$R = (C^2 + S^2)^{1/2}$.

In the above equations θ is the azimuth of the observation, f is the frequency of a given azimuth and n is the total number of observations in the population. Note that for the unweighted analyses the frequency of each cross sectional elongation azimuth was taken as one. In the length- and eccentricity-weighted analyses length and eccentricity were, respectively, taken as the frequency.

The standard deviation about the mean azimuth (radians) is given by:

$$[(-2\ln R)^{1/2}]/l \qquad (2)$$

where azimuths are of range (periodicity) $(0,2\pi/l)$. Since cross sectional elongations determined from the four-arm dipmeter must be assumed to be symmetrical, their azimuthal range is $(0,\pi)$, hence $l = 2$. Where range is $(0,\pi)$, azimuths (θ) should be doubled before application of equation (1). The mean azimuth given by equation (1) must then be halved to give the correct mean

Table A1. *Comparison of directional statistical analysis for data of ranges $(0,\pi)$, and $(0,2\pi)$, and linear statistical analysis. The first three datasets are test data, the fourth is the azimuths of cross sectional elongations in Vulcan-1B. Results quoted are mean azimuth $(0-360°N)$/standard deviation in degrees.*

Azimuthal data (°N)	Directional statistics		Linear statistics
	range $(0,\pi)$	range $(0,2\pi)$	
010, 170	180/10	090/107	090/113
140, 145, 150, 155, 160	150/7	150/7	150/8
140, 145, 150, 155, 160, 055, 065	150/38	128/42	124/44
041, 043, 054, 063, 074, 074, 142, 167, 169	048/42	087/52	092/52

azimuth. Similarly, the value R in equation (2) is that derived from the doubled angles.

Bell (1990) quotes equations appropriate for the analysis of data of range $(0,360°)$ i.e. $(0,2\pi)$. These are not suitable for the statistical analysis of borehole cross sectional elongation data determined from the four-arm dipmeter (see also Table A1).

Extensional, drilling-induced, hydraulic fracturing may generate a cross sectional elongation direction orthogonal to the 'true' breakout trend, as inferred at Eclipse-2 and Vulcan-1B. If Mardia's (1972) equations are applied, as appropriate for data of range $(0,\pi)$, the directional mean of a bimodal, orthogonal distribution of azimuths is representative of the dominant mode. The mean will not be biased towards a subsidiary, orthogonal mode. However, the standard deviation will be high. This is illus-

trated in Table A1, where the third and fourth datasets comprise bimodal, orthogonal distributions with one mode dominant. This property of Mardia's (1972) equations may account for the relatively consistent azimuthal means but high standard deviations exhibited by the cross sectional elongation data from the Vulcan.

If Mardia's (1972) equations as appropriate for data of range $(0,2\pi)$, or linear statistics, are applied to a bimodal, orthogonal distribution, then the resultant mean is biased towards the subsidiary, orthogonal mode (Table A1). However, *contra* Morin *et al.* (1989), provided that the azimuths are treated as being of range $(0,\pi)$, it is not necessary to reject directional means in favour of modes in wells exhibiting possible bimodal, orthogonal azimuth distributions of borehole cross sectional elongations.

References

ABBOTT, M. & CHAMALAUN, F. 1981. Geochronology of some Banda Arc volcanics. *In*: BARBER, A. & WIRYOSUJONO, S, (eds) *The Geology and Tectonics of Eastern Indonesia* Geological Research and Development Centre, Bandung, Indonesia, Special Publication, **2**, 253–268.

AUDLEY-CHARLES, M. G. 1986. Rates of Neogene and Quaternary tectonic movements in the Southern Banda Arc based on micropalaeontology. *Journal of the Geological Society, London*, **143**, 161–175.

—— 1988. Evolution of the southern margin of Tethys (North Australian region) from early Permian to late Cretaceous. *In*: AUDLEY-CHARLES, M. G. & HALLAM, A. (eds) *Gondwana and Tethys*. Geological Society, London, Special Publication, **37**, 79–100.

BABCOCK, E. A. 1978. Measurement of subsurface fractures from dipmeter logs. *AAPG Bulletin*, **62**, 1111–1126.

BELL, J. S. 1990. Investigating stress regimes in sedimentary basins using information from oil industry wireline logs and drilling records. *In*: HURST, A., LOVELL, M. A. & MORTON, A. C. (eds) *Geological Applications of Wireline Logs*. Geological

Society, London, Special Publication, **48**, 305 325

—— & GOUGH, D. I. 1979. Northeast–southwest compressive stress in Alberta: evidence from oil wells. *Earth and Planetary Science Letters*, **45**, 475–482.

CARDWELL, R. K. & ISACKS, B. L. 1978. Geometry of the subducted lithosphere beneath the Banda Sea in eastern Indonesia from seismicity and fault plane solutions. *Journal of Geophysical Research*, **B83**, 2825–2838.

CHAMALAUN, F., LOCKWOOD, K. & WHITE, A. 1976. The Bouguer gravity field of eastern Timor. *Tectonophysics*, **30**, 241–259.

CLOETINGH, S. & WORTEL, R. 1986. Stress in the Indo-Australian plate. *Tectonophysics*, **132**, 49–67.

DART, R. L. & ZOBACK, M. L. 1989. Wellbore breakout stress analysis within the central and eastern continental United States. *Log Analyst*, **30**, 1, 12–25.

FRIEDRICH, J., MCCAFFREY, R. & DENHAM, D. 1988. Source parameters of seven large Australian earthquakes determined by body waveform inversion. *Geophysical Journal*, **95**, 1–13.

HAMILTON, W. B. 1977. Subduction in the Indonesian

region. *American Geophysical Union Maurice Ewing Series*, **1**, 15–31.

—— 1988. Plate tectonics and island arcs. *Geological Society of America Bulletin,*, **100**, 1503–1527.

HILLIS, R. R. 1990. Post-Permian subsidence and tectonics, Vulcan Sub-basin, North West Shelf, Australia. *Australasian Institute of Mining and Metallurgy, Pacific Rim Congress 90 Proceedings*, **2**, 203–211.

ISACKS, B. & MOLNAR, P. 1969. Mantle earthquake mechanisms and the sinking of the lithosphere. *Nature*, **223**, 1121–1124.

JACOBSON, R. S., SHOR, JR., G. G., KIECKHEFER, R. M. & PURDY, G. M. 1978. Seismic refraction and reflection studies in the Timor–Aru Trough system and Australian continental shelf. *American Association of Petroleum Geologists Memoir*, **29**, 209–222.

JOHNSTON, C. R. & BOWIN, C. O. 1981. Crustal reactions resulting from the mid Pliocene–Recent continent-island arc collision in the Timor region. *BMR Journal of Australian Geology and Geophysics*, **6**, 223–243.

MCCAFFREY, R., MOLNAR, P., ROECKER, S. W. & JOYODIWIRYO, Y. S. 1985. Microearthquake seismicity and fault plane solutions related to arc-continent collision in the eastern Sunda Arc, Indonesia. *Journal of Geophysical Research*, **B90**, 4511–4528.

MARDIA, K. V. 1972. *Statistics of Directional Data*. Academic, London, New York.

MINSTER, J. B. & JORDAN, T. H. 1978. Present-day plate motions. *Journal of Geophysical Research*, **B83**, 5531–5534.

MORIN, R. H., ANDERSON, R. N. & BARTON, C. A. 1989. Analysis and interpretation of the borehole televiewer log: information on the state of stress and the lithostratigraphy at hole 504B. *In*: BECKER, K., SAKAI, H., *et al.* (eds) *Proceedings of the Ocean Drilling Program, Scientific Results*. Ocean Drilling Program, College Station, TX, **111**, 109–118.

NELSON, A. W. 1989. Jabiru field—horst, sub-horst or inverted graben? *Australian Petroleum Exploration Association Journal*, **29**, 176–194.

PATTILLO, J. & NICHOLLS, P. J. 1990. A Tectonostratigraphic framework for the Vulcan Graben, Timor Sea region. *Australian Petroleum Exploration Association Journal*, **30**, 27–51.

PLUMB, R. A. & HICKMAN, S. H. 1985. Stress-induced borehole elongation: a comparison between the four-arm dipmeter and the borehole televiewer in the Auburn geothermal well. *Journal of Geophysical Research*, **B90**, 5513–5521.

PRICE, N. J. & AUDLEY-CHARLES, M. G. 1987. Tectonic collision processes after plate rupture. *Tectonophysics*, **140**, 121–129.

RICHARDSON, M., SOLOMON, S. C. & SLEEP, N. H. 1979. Tectonic stress in the plates. *Reviews of Geophysics and Space Physics*, **17**, 981–1019.

SCHLUMBERGER 1981. *Dipmeter Interpretation*. Schlumberger Educational Services, Vol. 1.

VEEVERS, J. J. 1974. Sedimentary sequences of the Timor Trough, Timor, and the Sahul Shelf. *In*: VEEVERS, J. J., HEIRTZLER, J. R. *et al.* (eds) *Initial Reports of the Deep Sea Drilling Project*, **27**, 567–569.

WHITFORD, D. J. & JEZEK, P. A. 1982. Isotopic constraints on the role of subducted sialic material in Indonesian island-arc magmatism. *Geological Society of America Bulletin*, **93**, 504–513.

WOODS, E. P. 1988. Extensional structures of the Jabiru Terrace, Vulcan Sub-basin. *In*: PURCELL, P. G. & PURCELL, R. R. (eds) *The North West Shelf, Australia. Proceedings Petroleum Exploration Society Australia Symposium*, 311–330.

ZOBACK, M. D., MOOS, D., MASTIN, L. & ANDERSON, R. N. 1985. Well bore breakout and in situ stress. *Journal of Geophysical Research*, **B90**, 5523–5530.

ZOBACK, M. L. & ZOBACK, M. D. 1989. Tectonic stress field of the continental United States. *In*: PAKISER, L. C. & MOONEY, W. D. (eds) *Geophysical Framework of the Continental United States, Geological Society of America Memoir*, **172**, 523–539.

—— *et al.* 1989. Global patterns of tectonic stress. *Nature*, **341**, 291–298.

Stress trajectory determinations in southwestern Ontario from borehole logs

NAJWA A. YASSIR & MAURICE B. DUSSEAULT

Department of Earth Sciences, University of Waterloo, Waterloo, Ontario N2L 3G1, Canada

Abstract. This paper reviews the benefits and limitations of the use of four-arm dipmeter logs to detect breakouts, and to estimate horizontal stress trajectories. An example is given from southwestern Ontario, a low seismic risk area, yet one with unusually high horizontal stresses, causing contemporaneous deformations such as pop-ups and tunnel closures. The preliminary results show that the relatively few breakouts detected follow two main trends, approximately N80°E and S80°E. The breakouts often have different orientations within the same well, but do not seem to 'prefer' a particular lithology. The data, when compared to other stress measurements, do show a general NE trend for compressive stress, as popularly believed. However, they are as yet too few to make statistically reliable assumptions on the stress state in the area.

Borehole breakouts are elongations in wellbores (Fig. 1). They were first detected on four-arm dipmeter logs run on Alberta oil wells by Cox (1970). The NW–SE trending orientation was unrelated to the structural dip in the area and was therefore interpreted by Babcock (1978) to be caused by NW-striking vertical fractures. Bell & Gough (1979), however, postulated a probable origin for the breakouts that was to make them a useful tool in detecting horizontal stress directions: they pointed out that the well elongations were probably related to compressive failure of the rock in the borehole wall, and that the maximum compressive tangential stress was caused by the regional NE–SW compressive stresses in the Alberta area. These regional stresses would clearly be linked to the tectonics associated with the formation of the Rocky Mountains.

Their conclusions were confirmed by extensive work in Eastern British Columbia (Gough & Bell 1981, 1982). Since that time, the consistency of their proposed interpretation has been confirmed all around the world by correlation with other stress measurement methods (Plumb & Hickman 1985). Examples include eastern Canada and northeastern USA (Plumb & Cox 1987), northeastern Mexico (Suter 1987), Britain (Brereton & Evans 1987) and western Europe (Janot *et al.* 1988).

This paper reviews some of the research done to date on borehole breakouts and describes the use of four-arm dipmeter logs to detect breakout orientations in southwestern Ontario.

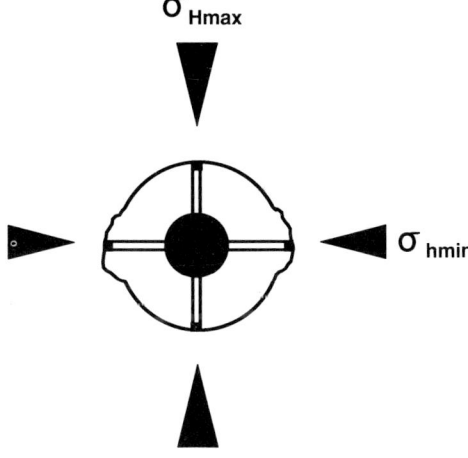

Fig. 1. Schematic illustration of a four-arm dipmeter in a breakout zone (plan view).

Breakout detection

Breakouts are most commonly detected by the study of four-arm dipmeter geophysical logs. This device consists of four orthogonal spring-loaded pads that contact the borehole wall directly and measure the resistivity of the rock over a small vertical distance. The four-arm dipmeter log is drawn up the hole using a standard geophysical cable that is shielded by a helically wound high-strength-steel sheath. The tension on this cable diminishes as the tool is

From HURST, A., GRIFFITHS, C. M. & WORTHINGTON, P. F. (eds), 1992, *Geological Applications of Wireline Logs II.* Geological Society Special Publication No. 65, pp. 169–177.

drawn up the hole; there is therefore a tendency for the tool to spiral upwards slowly in response to the steel sheathing re-coiling from the extension at depth.

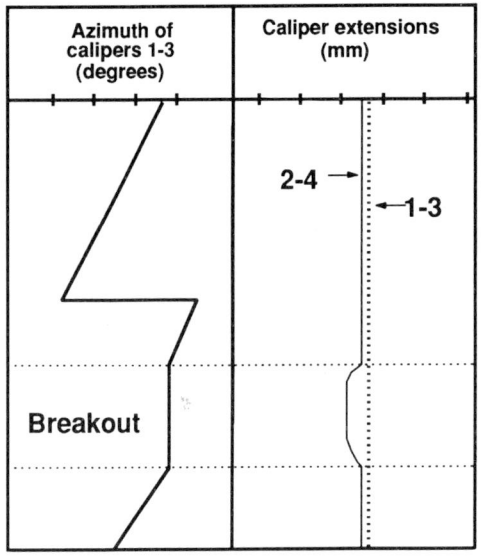

Azimuth of calipers 1-3 (degrees)	Caliper extensions (mm)

2-4 →

← 1-3

Breakout

Fig. 2. Schematic illustration of a breakout on a dipmeter log.

The two opposing pairs of calipers in the four-arm dipmeter log extend differently when they encounter a breakout section in a well, recording different hole diameters. The tool stops rotating at the same time because the pads become temporarily 'stuck' in the breakout grooves, so that a constant azimuth reading is obtained (Fig. 2). This is used to measure the azimuth of the elongation which is assumed to correspond to the direction of σ_{HMIN}, following the consistent interpretation scheme developed by previous users. σ_{HMAX} is perpendicular to the direction of σ_{HMIN}. This interpretation assumes that the principal stresses are oriented in the vertical and horizontal planes—a reasonable assumption at depth and in areas of flat topography and simple structural framework. In order to ensure that the readings taken are not of mechanically induced elongations, for example of key seating or wash-out features (Cox 1983), certain criteria have to be met (Plumb & Hickman 1985):

(a) the length of breakout has to exceed 30 cm;
(b) the smaller caliper reading has to be close to bit size;
(c) the logging tool has to stop completely, and
(d) the direction of elongation must not consistently coincide with the azimuth of the high side

of the borehole when the borehole deviates from vertical.

Bell (1990) adds,

(e) the larger diameter of hole elongation should not coincide with the azimuth of hole deviation;
(f) significant changes in conductivity should be recorded by all four pads and
(g) there should be no conductivity anomaly recorded by one or two pads.

Brereton & Evans (1987) developed a method that was independent of breakout 'relief'. They used resistivity traces to detect 'incipient' break-outs, or breakouts which have fractured but not yet collapsed, and find that their results correspond with readings taken from the caliper readings. These incipient breakouts may be interpreted in a consistent manner because it is the same mechanism that gives rise to them: large compressive stresses in the borehole wall lead to dilation and microcracking, and the invasion of drilling fluid of a different conductivity from that of the formation fluids will give an anisotropic resistivity which echoes the stress directions.

More sophisticated equipment can now be used to measure or examine breakouts, for example, televiewer cameras or formation micro-scanners (Plumb & Hickman 1985; Paillet & Kim 1987; Laubach et al. 1987; Bell & Mayers 1990), but these are considerably more expensive to run, and the images thus obtained require sophisticated analysis. Bass et al. (1986) describe the use of holography to measure borehole deformations.

There are several potential sources of error in breakout interpretation. For example, apart from often being operator-dependent, log interpretation itself may be difficult: anomalous readings may be caused by the existence of a strong rock fabric or of intersecting fractures, or by wellbore deviation from vertical, yet still meet the above criteria.

Controls on breakout formation

Breakouts are generally believed to develop as a result of tangential stress concentration perpendicular to the maximum stress causing failure of the borehole wall (Gough & Bell 1979), (Fig. 1). Some of the controls on their shape and mode of formation will be briefly discussed here.

Lithology, fabric and strength

Lithology and strength are obvious controls on breakout formation, as the strength of the rock has to be exceeded by the maximum hoop stress before the breakout can form. Guenot (1990) listed clay (shale), coal, carbonates, and metamorphic rocks at the higher pressures as being the most susceptible rock types to wellbore failure. Different rock types, or rocks of different strengths, also seem to produce different breakout shapes, as shown by experimental work (e.g. Santarelli & Brown 1989; Addis *et al.* 1990). It is generally agreed, however, that if the stresses are high enough, the breakout should not discriminate between rock types, as suggested by numerous observations such as in Alberta (Bell & Gough 1979), Nevada test site (Springer *et al.* 1984), and Britain (Brereton & Evans 1987).

Some observations do show anomalous orientations associated with changes in lithology (Allison & Nielson 1987), and cementation (Laubach *et al.* 1987).

Stress regime

The ratios between the three principal regional stresses acting on a borehole depend on the geological setting of the area. Breakouts are known to occur in different stress regimes, varying from compressional, such as throughout Alberta (Gough & Bell 1979), to extensional, such as the continental margin basins on the Scotian Shelf (Bell 1989).

The role of the vertical stress, or depth, is as yet not well understood. Geological observations show that breakouts seem to increase in frequency with depth, e.g. Springer *et al.* (1984) and Zoback *et al.* (1986). This is supported by laboratory experiments, e.g. Maloney & Kaiser (1989) and theoretical studies, e.g. Zoback *et al.* (1986), both relating the higher stress magnitudes to depth. Depth is probably not independent of the horizontal stresses in influencing breakout formation. For example, Zoback *et al.* (1986) expect breakouts to occur at deeper levels in a strike-slip regime as compared to a reverse faulting regime. Zoback *et al.* (1985), who model breakouts using a Mohr Coulomb yield criterion, suggest that horizontal stress contrast is an important control on their shape and size. This hypothesis was experimentally confirmed by Haimson & Herrick (1986) and Maloney & Kaiser (1989), and has been analysed more theoretically by Guenot (1987) among many others.

Wellbore deviation and orientation

Stresses acting on the wellbore can vary tremendously depending on whether or not the borehole deviates from vertical. If it does, several additional factors have to be taken into account, such as angle of deviation from vertical and orientation of the wellbore with respect to the horizontal stresses (if anisotropic). Also, stress anisotropy (between vertical and horizontal, and the two horizontal stresses) discussed by Guenot (1987) may play a more important role when dealing with deviated wells.

Bradley (1979) showed that deviation from vertical reduces the stability of the wellbore, and that, when the horizontal stresses were anisotropic, orientation had a notable effect on the hole angle which can be safely used. Aadnoy (1987) found that if the well was parallel to the direction of σ_{HMIN}, it is more stable than if it is in the σ_{HMAX} direction. Also, if the horizontal stresses are equal, an angle of 10–35° causes collapse for laminated rocks. McLean (1988) theoretically demonstrated a similar result: deviated wells were more stable if drilled towards the σ_{HMIN} direction; in fact, their greatest stability would seemingly be at an angle where the stresses would be closest to isotropic. A similar result was obtained experimentally (polyaxial testing) by Addis *et al.* (1990).

Of particular interest to this study is the effect wellbore deviation has on stress orientation determination. Mastin (1988) found wellbore deviation from vertical and from the σ_{HMIN} planes to affect the horizontal projection of σ_{HMIN}. This effect was more pronounced in the normal and reverse faulting regimes than in the strike-slip faulting regime. Guenot (1989) reviews the validity of obtaining stress orientation from deviated wells, because of the complexity of the stress field and the greater likelihood of drilling-related artifacts which may be interpreted as breakouts. Many workers (e.g. Brereton & Evans 1987; Plumb & Cox 1987) prefer to ignore data from wells deviated at >5–$10°$.

Study area

Southwestern Ontario is the region bound by lakes Huron, Erie and Ontario (Fig. 3). The geology of the area is dominated by thick Lower Palaeozoic shales and carbonates dipping gently southwestwards and overlying the Precambrian basement which outcrops further north. The structure of southwestern Ontario, although subtle, is complex and the subject of much controversy. Seismically, the region is regarded as 'low risk' (Whitham & Hasegawa 1975). However, there is a great deal of evidence to suggest that the horizontal stresses are very high, such as the occurrence of pop-ups (sediment upheavals forming ridges), quarry floor buckles and tunnel closures (White *et al.* 1973; Lo 1978; White & Russel 1982; Lee & White 1986). A number of

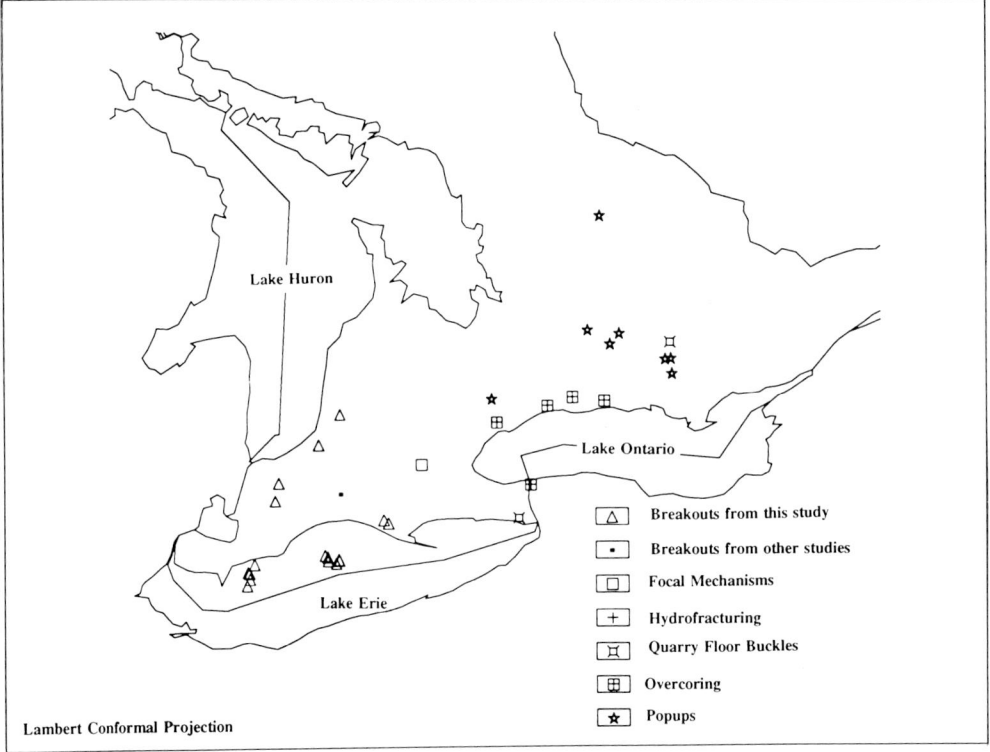

Fig. 3. Map of southwestern Ontario showing locations of stress orientation measurements.

oil and gas wells have been drilled in south-western Ontario, and it was decided to investigate the occurrence of wellbore elongation as an indication of horizontal anisotropy in the region.

Results

All the available files for wells completed in 1969 to 1987 have been accessed. Of the wells on which dipmeter logs were run, 17 have yielded usable elongation data (less than one third of accessed well data, Fig. 3). The breakouts were analysed and grouped into five classes (1–5) in order of reliability:

High reliability: Class 1. Follows criteria listed above: Perfect agreement between tool extension and constant azimuth. Large breakout. No well deviation. Smaller caliper reading remains constant.

Low reliability: Class 5. Does not follow some of the above-mentioned criteria: Imperfect agreement between tool extension and azimuth. Some tool rotation. Small breakout. Some well deviation ($<9°$).

The breakouts considered reliable are classes 1–3; however, all the results were included for the sake of comparison (Fig. 4, Table 1).

The formations showing breakouts range in age from Cambrian to Lower Devonian (nearly all formations in the region). The greatest concentration falls in the Middle Silurian to Lower Devonian. The breakouts occur mostly in dolostone (17), shale (15) and limestone (9), the three predominant rock types in the area. No strong trends were established between the lithology and the length of the breakouts, which varies between a few metres to over 100 metres. The longer breakouts tend to occur at greater depths; however, they are also associated with hole deviation ($<9°$), observed in three wells. (Two of these were assigned class 4 status, despite their apparent reliability, because they corresponded strongly with the onset of deviation.)

The individual breakout orientations are presented in Table 1, and in Fig. 4 in cumulative order of reliability. Initially, with the class 1 breakouts, an ESE and a N trend can be seen. The latter becomes weaker with the addition of lower quality breakouts and an ENE trend emerges (Fig. 4).

Table 1. *Summary of breakout locations and quality*

Well name	Latitude (N)	Longitude (W)	Breakout number	Azimuth	Quality
SR 4-7-XI	43-28-51	81-38-52	1	172	5
R76 2-26-VI	42-39-01	82-16-50	2	173	4
R80 1-5-IV	42-48-37	82-16-16	3	003	1
C 308L	41-57-10	82-26-43	4	164	3
			5	021	4
C 353G	41-35-37	82-28-04	6	128	3
			7	156	4
C 56R	42-36-19	80-52-29	8	040	4
			9	003	1
			10	104	3
			11	085	1
			12	129	4
			13	141	4
C 222M	42-12-47	81-32-42	14	157	5
			15	108	2
			16	082	5
C 222D	42-14-30	81-33-11	17	008	4
C 220E	42-14-15	81-24-56	18	001	5
C 220F	42-13-50	81-24-23	19	011	4
			20	077	5
			21	172	4
C 221L	42-12-05	81-26-24	22	028	5
			23	140	5
			24	102	4
			25	012	5
			26	095	4
C 55L	42-37-39	80-56-08	27	107	5
			28	076	5
			29	072	2
			30	084	4
C 176U	42-15-11	81-35-24	31	079	5
			32	113	1
			33	096	1
			34	098	1
			35	107	4
			36	077	3
P 289 W	42-00-02	82-27-41	37	169	5
			38	114	5
P 289X	42-00-29	82-28-55	39	025	4
			40	120	5
			41	096	5
P 289A	42-04-57	82-25-01	42	078	4
			43	144	4
			44	057	4
SB 273	43-11-20	81-51-22	45	130	5
			46	081	4

Discussion and conclusions

It is difficult to draw strong conclusions from the above preliminary results because the data are few which makes them statistically unreliable. The fact that less than one third of the accessed wells showed apparent elongation suggests that the lithology of the region is generally strong and not too affected by potential stress aniso-tropy at depth. Also, from the above discussion, it can be seen that interpretation can be subjective, depending largely on the judgement of the operator. Some comparisons were made between the above results and stress orientation measurements (both direct and indirect, surficial and at depth) taken in the region. These include data from pop-ups and quarry floor buckles (from

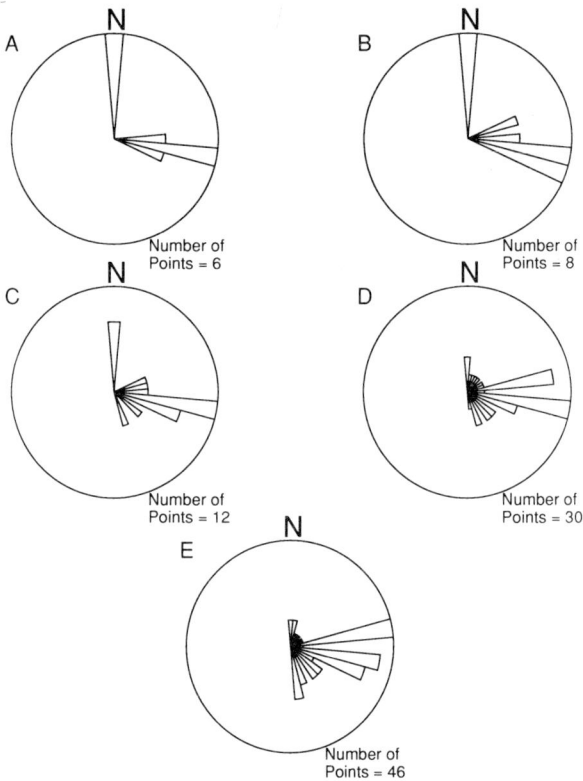

Fig. 4. Orientation of borehole elongations in southwestern Ontario. A (class 1) to E (class 5).

White & Russel 1982), overcoring and hydraulic fracture tests (e.g. Haimson & Lee 1980), earthquake focal mechanisms (Mereu 1986) and one breakout studied previously (Cox 1983), Fig. 3. The results, corrected by 90° to show maximum horizontal stress where necessary, are shown in Fig. 5. It can be seen that the maximum horizontal stress seems to fall in the NE quadrant.

The origin of the stresses in southwestern Ontario is, as yet, not well understood. Some workers attribute high surface horizontal stresses to glacial rebound (see review by Adams 1981). The theory suggests that when the thickness of ice (up to 3 km) receded from southwestern Ontario, the rocks unloaded vertically, but the horizontal stresses remained 'locked in'. Many workers, on the other hand, attribute these stresses to tectonic movements (e.g. Sbar & Sykes 1973; Sanford et al. 1985). The directionality (approximately NE) of the maximum stress orientations (Fig. 5) argues against a purely glacial origin. It could, in fact, correspond to a regional trend observed by various methods in eastern Canada (Herget & Arjang 1990; Hase-

gawa et al. 1985; Linder & Halpern 1978), New York (Engelder & Geiser 1980) and Michigan (Haimson 1978). Geological evidence also seems to point to this trend. For example, two major fracture sets are observed in the northern Michigan Basin (Holst 1982) and the Appalachian Plateau, New York (Engelder & Geiser 1980), striking roughly NE and SW in both areas. Engelder and Geiser correlate the NE set with hydraulic fracture tests, indicating present stresses, and the NW set (which is associated with deformed fossils) with older, Appalachian compression. Holst explains his results in a similar fashion.

But what of southwestern Ontario? Sanford et al. (1985) demonstrated an intricate fracture network in the area related to compressional and extensional deformations in the region. The fracture-bound blocks may control the stress regime locally, in which case closer attention should be paid to stress orientations in the region. Indeed, despite the apparent correlation between different stress data in the region, there is sufficient scatter that a definite stress determination can-

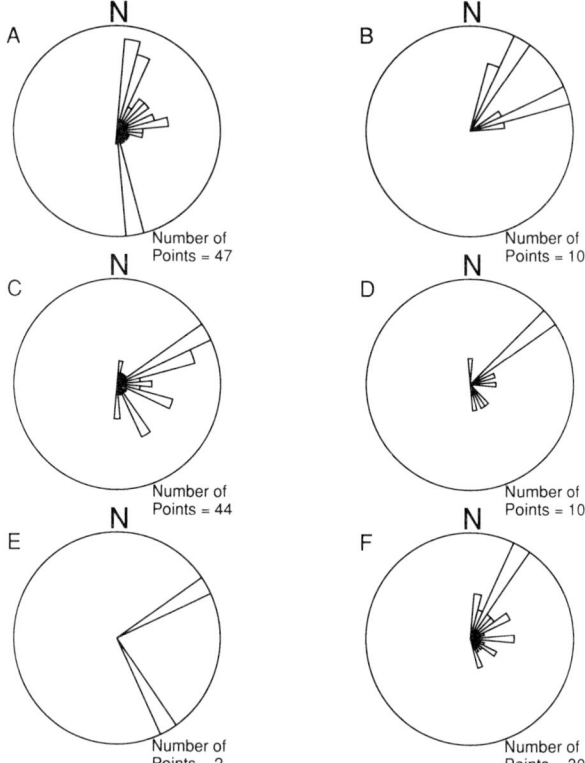

Fig. 5. Maximum horizontal stress orientations in southwestern Ontario. A, breakouts from this study, including data from Cox (1983); B, hydraulic fracture; C, overcoring; D, quarry floor buckles; E, focal mechanisms; F, pop-ups—references in text.

not be made at the moment. The data were collected at different depths, and stress controls could vary with depth, as observed at Darlington, Ontario, between Precambrian and Palaeozoic rocks (Haimson & Lee 1980). Also, the data sets were largely taken from different parts of southwestern Ontario. Overcoring data, for example, were concentrated along the shore of Lake Ontario, and breakout data were mostly in the Lake Erie area (Fig. 3).

Therefore, although the results are interesting, no firm conclusions can be drawn until further research is done. It would not be possible to check effects of, say, depth or geographical location in a statistically reliable way without

more measurements of different types throughout the region. Closer attention should also be paid to any evidence of neotectonic movements and their directionality. This would assist in determining whether the high horizontal stresses in southwestern Ontario have a single regional signature, or whether they can only be predicted locally.

The authors would like to thank Ontario Hydro for funding this project. Thanks also to Dr. Sebastian Bell for his support and advice in this research and to Jonathan Matthews for his assistance in diagram preparation.

References

AADNOY, B. S. 1987. *Modelling of the Stability of Highly Inclined Boreholes in Anisotropic Rock Formations.* Society of Petroleum Engineers paper 16526/1.

ADAMS, J. 1981. *Post-glacial faulting: a literature survey of occurrences in eastern Canada and comparable glaciated areas.* Atomic Energy of Canada Technical Report 142.

—— 1987. *Canadian crustal stress data—a compilation to 1987.* Geological Society of Canada Open File No. 1622.

ADDIS, M. A., BARTON, N. R., BANDIS, S. C. & HENRY, J. P. 1990. *Laboratory studies on the stability of vertical and deviated boreholes.* Society of Petroleum Engineers paper 20406, 19–30.

ALLISON, M. L. & NIELSON, D. L. 1987. Survey of borehole breakouts from active geothermal systems. *EOS,* **68**, 1460–1461.

BABCOCK, E. A. 1978. Measurement of subsurface fractures from dipmeter logs. *AAPG Bulletin,* **62**, 1111–1126.

BASS, J. D., SCHMITT, D. R. & AHRENS, T. J. 1986. Holographic in situ stress measurements. *Geophysical Journal of the Royal Astronomical Society,* **85**, 13–41.

BELL, J. S. 1990. Investigating stress regimes in sedimentary basins using information from oil industry wireline logs and drilling records. *In:* HURST, A., LOVELL, M. A. & MORTON, A. C. (eds) *Geological Applications of Wireline Logs.* Geological Society, London, Special Publication, **48**, 305–325.

BELL, J. S. 1989. Vertical migration of hydrocarbons at Alma, offshore Eastern Canada. *Bulletin of Canadian Petroleum Geology,* **37**, 358–364.

—— & ADAMS, J. 1990. Do the rocks remember? How contemporary are regional stresses in Canada? Submitted to proceedings volume of CANMET conference on Stress Determination Predictions and Monitoring of Stress Redistribution for Underground Structures, Ottawa.

—— & GOUGH, D. I. 1979. Northeast–southwest compressive stress in Alberta: evidence from oil wells. *Earth and Planetary Science Letters,* **45**, 475–482.

—— & MAYERS, I. R. 1990. Stress measurements in sedimentary basins (and applications). Course Notes RPI/I-90-11.

BRADLEY, W. B. 1979. Failure of inclined boreholes. *Journal of Energy Resource Technology, Transaction AIME,* **11**, 232–239.

BRERETON, N. R. & EVANS, C. J. 1987. *Rock Stress Orientations in the United Kingdom from Borehole Breakouts.* Report RG 87/14, British Geological Survey.

COX, J. W. 1970. The high resolution dipmeter reveals dip-related borehole and formation characteristics. *SPWLA 11th Annual Logging Symposium.*

—— 1983. Long axis orientation in elongated boreholes and its correlation with rock stress data. *Transactions of SPWLA 24th Annual Logging Symposium,* 1–17.

ENGELDER, T. & GEISER, P. 1980. On the use of regional joint sets as trajectories of paleostress fields during the development of the Appalachian Plateau, New York. *Journal of Geophysical Research,* **85**, 6319–6339.

GOUGH, D. I. & BELL, J. S. 1982. Stress orientations from borehole wall fractures with examples from Colorado, East Texas and Northern Canada. *Canadian Journal of Earth Science,* **19**, 1358–1370.

—— & —— 1981. Stress orientations from oil well fractures in Alberta and Texas. *Canadian Journal of Earth Science,* **18**, 638–645.

GUENOT, A. 1987. Stress and rupture conditions around wellbores (in French). Proceedings of the 6th Congress of the International Society of Rock Mechanics, Montreal, 1, 109–118.

—— 1989. Borehole breakouts and stress fields. *International Journal of Rock Mechanics Mineral Science and Geomechanics Abstracts,* **26**, 185–195.

—— 1990. General report: Instability problems at great depth drilling boreholes in wells. Proceedings of the ISRM-SPE International Symposium: Rock at Great Depth, Pau, Maury and Fourmaintraux (eds), 3, 1199–128.

HAIMSON, B. C. 1978. Crustal stress in the Michigan Basin. *Journal of Geophysical Research,* **83-B12**, 5857–5863.

—— & HERRICK, C. G. 1986. Borehole breakouts—a new tool for estimating in-situ stress? *Proceedings of the International Symposium on Rock Stress and Rock Stress Measurements, Stockholm,* 271–328.

—— & LEE, C. F. 1980. Hydrofracturing stress determinations at Darlington, Ontario. *13th Canadian Rock Mechanics Symposium,* CIM Vol. **22**, 42–55.

HASEGAWA, H. S., ADAMS, J. & YAMAZAKI, K. 1985. Upper crustal stresses and vertical stress migration in eastern Canada. *Journal of Geophysical Research,* **90-B5**, 3637–3648.

HERGET, G. & ARJANG, B. 1990. Update on ground stresses in the Canadian Shield. Proceedings of Speciality Conference: Stresses in Underground Structures, 33–47.

HOLST, T. B. 1982. Regional jointing in the northern Michigan Basin. *Geology,* **10**, 273–277.

JANOT, P., GAUER, P. & GROSS, E. 1988. Orientation et la contrainte tectonique dans l'Europe de l'ouest a partir des ovalisations de trous de forages. *Revue de l'Institut Français de Pétrole,* **43** (4), 517–522.

LAUBACH, S. E., BAUMGARDNER, R. W. & MEADOR, K. J. 1987. Analysis of natural fractures and borehole ellipticity, Travis Peak formation, East Texas. Report for Gas Research Institute.

LEE, C. F. & WHITE, O. L. 1986. On some geomechanical aspects of high horizontal stresses. Presented at the International Symposium on Geomechanics, Beijing.

LINDER, E. N. & HALPERN, J. A. 1978. In-situ stress in North America: A compilation. *International Journal of Rock Mechanics Mineral Science and Geomechanics Abstracts,* **15**, 183–203.

LO, K. Y. 1978. Regional distribution of in situ horizontal stresses in rocks of southern Ontario. *Canadian Geotechnical Journal,* **15**, 371–381.

MCLEAN, M. R. 1988. *Wellbore stability analysis.* PhD thesis, University of London.

MALONEY, S. & KAISER, P. K. 1989. Results of borehole breakout simulation tests. *In:* PAU, MAURY & FOURMAINTREAUX (eds) *Proceedings of the ISRM-SPE International Symposium: Rock at Great Depth,* **2**.

MASTIN, L. 1988. Effect of borehole deviation on

breakout orientations. *Journal of Geophysical Research*, **93**, 9187–9195.

MEREU, R. F., BRUNET, J., MORRISSEY, K., PRICE, B. & YAPP, A. 1986. A study of the microearthquakes in the Gobles oilfield area of SW Ontario. *Bulletin of the Seismological Society of America*, **76**, 1215–1223.

PAILLET, F. L. & KIM, K. 1987. Character and distribution of borehole breakouts and their relationship to in situ stress in deep Columbia River basalts. *Journal of Geophysical Research*, **92**, 6223–6234.

PLUMB, R. A. & COX, J. W. 1987. Stress distributions in eastern North America determined to 4.5 km from borehole elongation measurements. *Journal of Geophysical Research*, **92**, 485–4816.

—— & HICKMAN, S. H. 1985. Stress-induced borehole elongation: a comparison between the four-arm dipmeter and the borehole televiewer in the Auburn geothermal well. *Journal of Geophysical Research*, **9**, 5513–5522.

SANFORD, B. V., THOMPSON, F. J. & McFALL, G. H. 1985. Plate tectonics—a possible controlling mechanism in the development of hydrocarbon traps in southwestern Ontario. *Bulletin of Canadian Petroleum Geology*, **33**, 52–71.

SANTARELLI, F. J. & BROWN, E. T. 1989. Failure of three sedimentary rocks in triaxial and hollow cylinder compression tests. *International Journal of Rocks Mechanics, Mining Science and Geomechanics Abstracts*, **26**, 401–413.

SBAR, M. L. & SYKES, L. R. 1973. Contemporary compressive stress and seismicity in eastern North

America: An example of intra-plate tectonics. *Geological Society of America Bulletin*, **84**, 1861–1882.

SPRINGER, J. E., THORPE, R. K. & McKAGUE, H. L. 1984. *Borehole Elongation and its relation to Tectonic Stress at the Nevada Test Site*. USGS report UCRL-53528.

SUTER, M. 1987. Orientational data on the state of stress in Northeastern Mexico as inferred from stress-induced borehole elongations. *Journal of Geophysical Research*, **92**, 2617–2885.

WHITE, O. L., KARROW, P. F. & MACDONALD, J. R. 1973. Residual stress relief phenomena in southern Ontario. *Proceedings of the 9th Canadian Rock Mechanics Symposium, Montreal*, 323–348.

—— & RUSSELL, D. J. 1982. High horizontal stresses in Southern Ontario—their orientation and origin. *Proceedings of the 4th Congress International Association of Engineering Geologists, New Delhi*, **2**, 39–51.

WHITHAM, K. & HASEGAWA, H. S. 1975. The estimation of seismic risk in Canada—a review. *Publications of the Earth and Physics Branch, Energy, Mines and Resources Canada*, **45** (2), 136–161.

ZOBACK, M. D., MASTIN, L. & BARTON, C. 1986. In situ stress measurements in deep boreholes using hydraulic fracturing, wellbore breakouts and Stoneley wave polarisation. *Proceedings of the International Symposium on Rock Stress and Rock Stress Measurements, Stockholm*, 289–293.

——, MOOS, D., MASTIN, L. & ANDERSON, R. N. 1985. Wellbore breakouts and in situ stress. *Journal of Geophysical Research*, **9**, 5523–5553.

In situ stress orientations in the Witch Ground Graben, North Sea, revealed by borehole breakouts: preliminary results

S. M. COWGILL,[1] P. G. MEREDITH,[1] S. A. F. MURRELL[1] & N. R. BRERETON[2]

[1] *Department of Geological Sciences, University College London, Gower Street, London WC1E 6BT, UK*

[2] *British Geological Survey, Keyworth, Nottingham NG12 5GG, UK*

Abstract. Directions of minimum horizontal stress (σ_h) derived from borehole breakouts are presented for seven boreholes in the Witch Ground Graben, North Sea. We have found a clear tendency for a dominant N–S (more accurately a NNW–SSE) direction of σ_h, approximately perpendicular to the graben axis. However, significant variations from this general trend were encountered even within this relatively small area, especially within the boreholes from the Tartan and Highlander fields. Contrasts in local structure and lithology within the area may explain these differences although such contrasts have yet to be quantified.

Broad regional trends of σ_h are apparent throughout Northwest Europe, although previously published data show a certain amount of variability. It has been shown that the minimum compressive stress direction changes significantly across many of the graben structures within the North Sea basin, and this also appears to be the case within the Witch Ground Graben.

This report describes a preliminary study of dipmeter data from released wells in the British sector of the North Sea. The data were obtained from the Department of Energy and subsequently analysed using the WELLOG program, devised and designed by the British Geological Survey (BGS).

The North Sea basin is an intra-continental sedimentary basin which developed during the Devonian as one of a number of continental extensional rifts in the region of the Caledonian orogenic belt. It represents a failed rift arm of the Triassic Arctic–Atlantic rift system. The area has therefore had a long and complex geological history (Glennie 1986). The two major rift zones in the North Sea, the Viking and Central Grabens, meet in a triple junction; the third arm extends into the Moray Firth (Fig. 1). It is in this area that the Witch Ground Graben (WGG) developed (Fig. 2).

Previous work on in situ stress in the North Sea and surrounding areas has been carried out by Ranalli & Chandler (1975), Klein & Barr (1986), Clauss et al. (1989) and most recently by Spann et al. (1991). This paper is a continuation of the previous work; the overall aim being to relate local stress directions to both regional tectonic stress and local modifications and perturbations due to both structural and lithological influences.

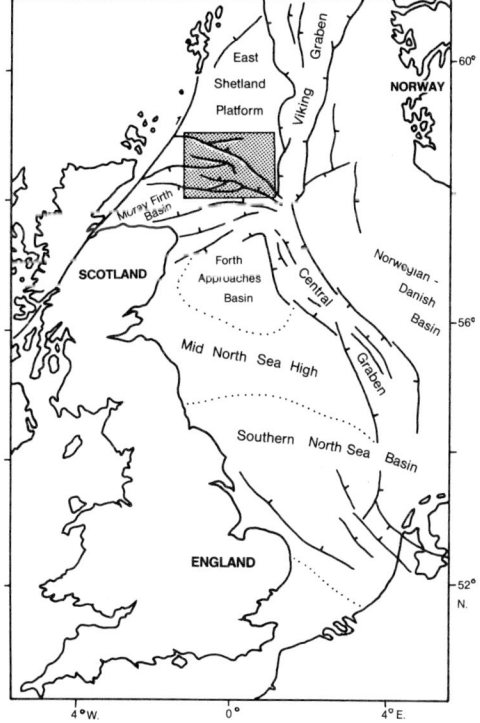

Fig. 1. Location map of the area under study.

From HURST, A., GRIFFITHS, C. M. & WORTHINGTON, P. F. (eds), 1992, *Geological Applications of Wireline Logs II.* Geological Society Special Publication No. 65, pp. 179–184.

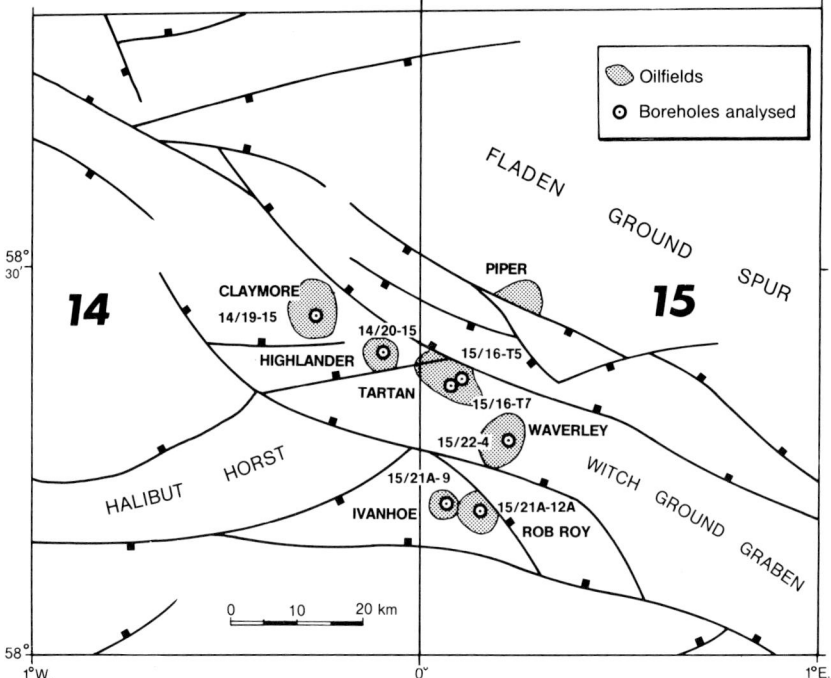

Fig. 2. The Witch Ground Graben, structure, oilfields and boreholes analysed.

Borehole breakouts: theory, measurement and identification

A variety of data indicates that the three principal stresses operating in the Upper Crust lie in approximately horizontal and vertical planes (e.g. Zoback & Zoback 1980). Unequal horizontal stresses imposed on near-vertical boreholes will often cause localized spalling of the wall-rock perpendicular to the direction of maximum compressive stress. The resultant elongation of borehole cross sections has been termed 'breakouts' (Babcock 1978).

The boreholes analysed in this study were drilled as exploration, appraisal or development wells in and around the Witch Ground Graben, and all the data were obtained from conventional dipmeter tools. This tool comprises two pairs of orthogonal caliper pads and a magnetometer; and the tool also measures formation resistivity. As the tool is pulled up the hole it rotates due to cable torque and the four caliper arms are hydraulically pressed against the borehole wall facilitating hole size measurement.

Breakout formation is not an instantaneous process, but takes some time from the initiation of cracking to eventual breakout stabilization

(Zoback *et al.* 1985). Thus, if a borehole is logged soon after completion of drilling, any breakouts towards the top of the section may be well developed whereas any towards the base may still be in the initial stages of development. Resistivity data is therefore potentially useful in the identification of 'zones of incipient spalling' (Brereton *et al.* 1990). Such zones will have been permeated by the drilling fluid and will thus exhibit a lower resistivity but will not yet have 'broken out'. This type of incipient breakout would not normally be detected by immediate post-completion caliper measurements.

The Witch Ground Graben

The Witch Ground Graben (WGG) is a NW–SE trending extensional basin in the central part of the Northern North Sea, lying between the Moray Firth the Viking Graben (Figs 1 & 2).

The structural and stratigraphic character of the WGG occurs as a result of complex tectonic activity and reactivation of Early Palaeozoic and Late Mesozoic fault systems (Harker *et al.* 1987). Its NW–SE trend may possibly have developed as a reactivation of Hercynian features. The primary NE–SW Caledonian structural grain was activated during the Ordovician to Early Devonian during which time the granitic basement of the WGG was emplaced. Following the Caledonian event, continental clastics were deposited until the onset of Hercynian tectonism in the Late Carboniferous. This was followed by terrestrial sedimentation of the Early Permian Rotliegende succession. Late Permian subsidence and deposition of Zechstein evaporites was followed by the basin being ultimately infilled by Triassic red beds.

NW–SE extension occurred from Triassic to Cretaceous times along listric normal faults along the SW basin margin of the developing graben, with downthrow to the NE giving rise to the asymmetry of the basin (McQuillan *et al.*

1982; Beach 1984). By the start of the Upper Cretaceous major rifting seems to have died out, giving way to broad general subsidence which allowed the accumulation of chalk, sands from the NW, and finally a thick sequence of Tertiary clays, shales and siltstones (Christie & Sclater 1980). The WGG essentially achieved its current structural configuration by the close of the Barremian (Harker *et al.* 1987).

Estimates of extension within the WGG range from 40% at the base of the Permian to 15% at the base of the Cretaceous (Beach 1984). It has been shown (Christie & Sclater 1980) that subsidence beneath the graben may be accounted for by a mechanism involving 50–100% stretching of the lithosphere from the Jurassic through to

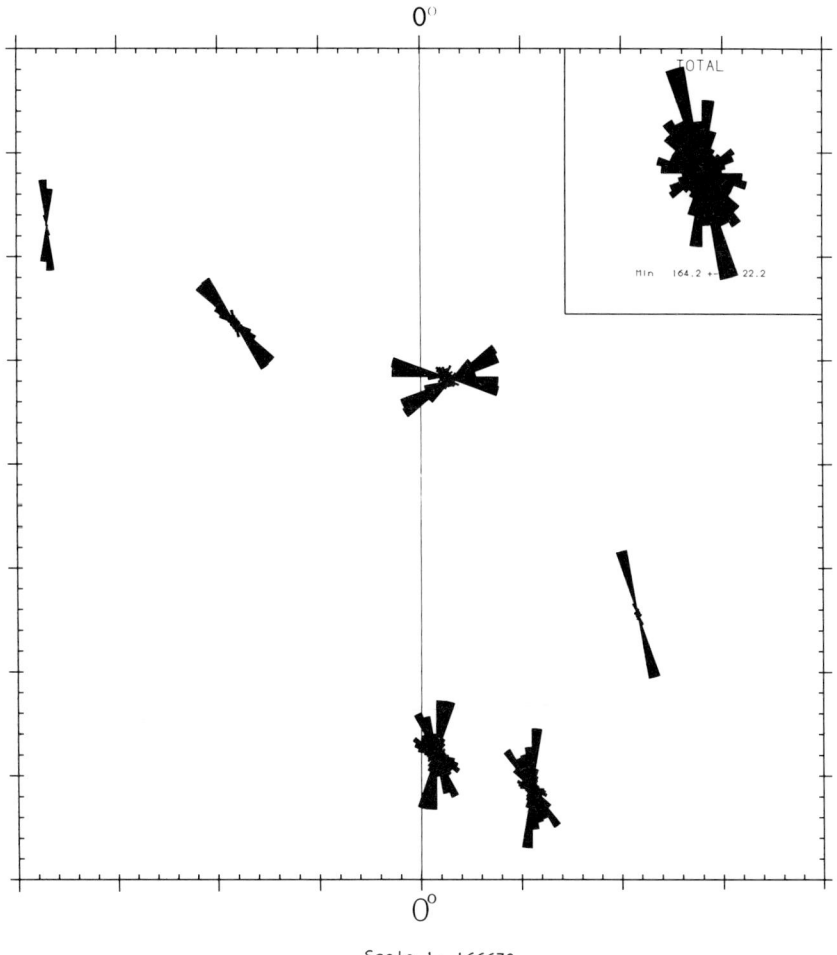

Scale 1: 166670

Fig. 3. Breakout directions in the Witch Ground Graben (caliper data, hole deviation 0–10°).

Table 1. *Breakout azimuths from the Witch Ground Graben*

Hole angle →	Breakout azimuth—caliper data (weighted statistics)			Breakout azimuth—resistivity data (weighted statistics)			Azimuth of deviation (if known)
	0–5°	5–10°	0–10°	0–5°	5–10°	0–10°	
Well number (Latitude/longitude)							
14/19-15	178 ± 4	003 ± 10	177 ± 8	—	—	—	—
58°26'00"N 0°18'44"W							
14/20-15	140 ± 18	—	—	145 ± 25	—	—	—
58°24'10"N 0°7'21"W							
15/16-T5	139 ± 22	100 ± 12	104 ± 18	166 ± 25	105 ± 9	105 ± 21	108°
58°22'11"N 0°4'24"E							
15/16-T7	044 ± 21	060 ± 9	057 ± 15	130 ± 22	158 ± 15	150 ± 21	059°
58°22'11"N 0°4'24"E							
15/21A-9	004 ± 22	—	—	010 ± 22	—	—	—
58°11'37"N 0°3'34"E							
15/21A-12A	007 ± 21	—	—	115 ± 25	—	—	—
58°11'27"N 0°8'30"E							
15/22-4	164 ± 9	163 ± 19	164 ± 20	164 ± 8	161 ± 18	163 ± 14	—
58°15'52"N 0°13'47"E							
Weighted mean	163 ± 24	—	164 ± 22	143 ± 25	—	147 ± 26	

the mid-Cretaceous. This would translate into 25–50 km of stretching across the WGG. It is difficult to reconcile this amount of stretching with the tentative estimate of approximately 5–6 km extension across the Inner Moray Firth basin to the east proposed by McQuillan et al. (1982).

Results

To minimize the likelihood of asymmetric borehole elongations due to wear by drill pipes ('key seats') being confused with breakouts, an upper limit for borehole deviation of 10° from vertical was set for all the boreholes analysed. Of the 12 boreholes used, five were found to be deviated more than 10° (up to 50° in some cases). The remaining seven had deviations of less than 10° for at least part of the section. Only data from sections deviating less than 10°, measured in these seven boreholes have been used in this study.

It has also been shown (e.g. Brereton et al. 1990) that rose diagrams and compressed borehole sections are a very effective means of condensing and displaying large data sets relating to breakout orientations and magnitudes. It is possible to weight the rose diagrams according to the length of individual broken out zones and in this way dominant trends are enhanced. Individual rose diagrams may also be combined into a rose map (Fig. 3) which displays the summed or total rose and indicates the major breakout

direction for a number of holes. This particular option may also be weighted, thereby further enhancing any dominant trends.

The results from this study, as analysed using the WELLOG program, are presented in Table 1 and Fig. 3. Table 1 shows the breakout azimuths (minimum horizontal stress directions), obtained from both caliper and resistivity data, for each borehole along with the azimuth of deviation where applicable. Figure 3 shows the spatial distribution of the boreholes, the azimuths of breakouts and the general trend over the whole area.

Preliminary results are as follows.

1. Higher hole deviation (5–10°) appears to affect the apparent breakout azimuth in some boreholes (see, for example, 15/16-T5 in Table 1).
2. The resistivity data appear to have higher variability than the caliper data. However, the differences between mean breakout azimuths determined by the two methods are significantly less than the variability. Occasionally the azimuth values are very different; see, for example, 15/21A-12A and 15/16-T7 where, even allowing for the maximum variability, the breakout azimuths are at least 60° different.
3. As can be seen from Table 1 and Figure 3, the majority of breakouts analysed are orientated approximately N–S.
4. The weighted average of the data from the caliper logs (where hole deviation is between 0

and 5°) is 163 ± 24°. This is significantly different from the regional average determined for the UK landmass which gives a minimum stress orientation of 054 ± 11°, (Brereton & Evans 1987; Evans & Brereton 1990).

5. Minimum stress orientations reported by Brereton & Evans (1987), Clauss *et al.* (1989), Evans & Brereton (1990) and Spann *et al.* (1991) show a certain amount of variability, but broad regional trends are apparent. Nevertheless, their results show that across many of the graben structures within the North Sea basin the minimum compressive stress direction changes significantly. This may also be the case within the WGG.

6. Breakout directions from northern England and Scotland (Brereton & Evans 1987), from a limited number of wells exhibit similar orientations to those in the WGG.

Conclusions

Breakout orientations so far measured in seven wells in the WGG exhibit an approximate north–south trend over the area, with the average minimum horizontal stress orientation being 163°. However, significantly different orientations occur even within this relatively small area, especially within the boreholes from the Tartan and Highlander fields. This may be due to differences in the local structure and lithology within this central area, and further investigation into the geology, structure and well records is required.

The breakout orientations in the WGG may be influenced by structures at depth. It has been suggested, for example, that the Great Glen Fault and other major faults to the west and south may exert an influence on stress orientation, as may the Viking and Central grabens to the east and north (Spann *et al.* 1991; M. A. Addis, pers. comm. 1991). The structural grain and consequent anisotropy developed during the Caledonian and Hercynian orogenies may also exert a strong influence on the present day stress regime in this part of the North Sea. This also requires further investigation.

Finally, further detailed analysis of the geology and structure of this area is required and it should be stated clearly that the conclusions presented here are tentative. The analysis of additional data from other boreholes in this area is required before we can confidently define the stress regime currently operating in the Witch Ground Graben.

Support for this work has been provided by a NERC Research Studentship for S.M.C. with CASE support from BGS.

References

BABCOCK, E. A. 1978. Measurement of subsurface fractures from dipmeter logs. *AAPG Bulletin*, **62**, 1111–1126.

BEACH, A. 1984. Structural evolution of the Witch Ground Graben. *Journal of the Geological Society, London*, **141**, 621–628.

BRERETON, N. R. & EVANS, C. J. 1987. *Rock Stress Orientations in the UK from Borehole Breakouts. Report of the Regional Geophysics Research Group.* B.G.S. Report RG87/14.

——, ——, PEART, R. J. & HYETT, A. J. 1990. *Mechanisms and theoretical basis governing the creation and development of borehole breakouts. Report of the Regional Geophysics Group.* B.G.S. Technical report WK/90/18.

CHRISTIE, P. A. F. & SCLATER, J. G. 1980. An extensional origin for the Buchan and Witch Ground Graben in the North Sea. *Nature*, **283**, 729–732.

CLAUSS, B., MARQUART, G. & FUCHS, K. 1989. Stress orientations in the North Sea and Fennoscandia, a comparison to the Central European stress field. *In*: GREGERSEN, S. & BASHAM, P. W. (eds) *Earthquakes at North Atlantic passive margins: Neotectonics and Post-glacial rebound.* Kluwer, Dordrecht, 277–287.

EVANS, C. J. & BRERETON, N. R. 1990. *In situ* crustal stress in the United Kingdom from borehole breakouts. *In*: HURST, A., LOVELL, M. A. & MORTON, A. C. (eds) *Geological Applications of Wireline Logs.* Geological Society, London, Special Publication, **48**, 327–338.

GLENNIE, K. W. 1986. The structural framework and Pre-Permian history of the North Sea area. *In*: GLENNIE, K. W. (ed.) *Introduction to the Petroleum Geology of the North Sea.* 2nd edition. Blackwell, Oxford, 25–62.

HARKER, S. D., GUSTAV, S. H. & RILEY, L. A. 1987. Triassic to Cenomanian stratigraphy of the Witch Ground Graben. *In*: BROOKS, J. & GLENNIE, K. W. (eds) *Petroleum Geology of Northwest Europe.* Graham & Trotman, London, 809–818.

KLEIN, R. J. & BARR, M. V. 1986. Regional state of stress in Western Europe. *In*: STEPHANSSON, O. (ed.) *Rock Stress and Rock Stress Measurement. Proceedings of the International Symposium on Rock Stress and Rock Stress Measurements, Stockholm.* Centek, Luleå. 33–44.

McQUILLAN, R., DONATO, J. A. & TULSTRUP, J. 1982. Development of basins in the Inner Moray Firth and the North Sea by crustal extension and dextral displacement of the Great Glen Fault. *Earth and Planetary Science Letters*, **60**, 127–139.

RANALLI, G. & CHANDLER, T. E. 1975. The stress field in the upper crust as determined from *in situ* measurements. *Geologische Rundschau*. **64**, 653–674.

SPANN, H., BRADY, M. & FUCHS, K. 1991. Stress evaluation in offshore regions of Norway. *Terra Nova*, **3**, 148–152.

ZOBACK, M. D., MOOS, D., MASTIN, L. & ANDERSON, R. N. 1985. Wellbore breakouts and *in situ* stress. *Journal of Geophysical Research*, **90**, 5523–5530.

ZOBACK, M. L. & ZOBACK, M. D. 1980. State of stress in the Conterminous United States. *Journal of Geophysical Research*, **85**, 6113–6156.

An integrated interpretation of fracture apertures computed from electrical borehole scans and reflected Stoneley waves

BRIAN E. HORNBY & STEFAN M. LUTHI

Schlumberger Cambridge Research, Madingley Road, Cambridge CB3 0EL, UK

and

Schlumberger-Doll Research, Old Quarry Road, Ridgefield, CT 06877, USA

Abstract. Fractures crossing the borehole are probed using electrical currents generated by Formation MicroScanner* pads and Stoneley waves generated by an acoustic source. New techniques have been developed to invert these measurements for estimates of equivalent fracture aperture, a crucial parameter controlling the fluid flow in fractured reservoirs. The objective of this paper is twofold: first, to compare aperture estimates in a variety of rock types; and second to exploit the different physical measurements in order to improve our understanding of the fracture system. Single fractures in controlled borehole experiments are used to cross-validate the two techniques, and good comparisons were found in the aperture range from $10\,\mu$m to several millimetres. Core data, acoustic borehole televiewer images, and in one case results using the Dual Laterolog* measurement serve as supporting evidence. Three case studies are presented: one from crystalline rocks, another from a chalk and shale sequence, and a third one from a fractured dolomite reservoir. In these case studies we found that a combined analysis of the two methods yields valuable additional information concerning fracture extent, fracture connectivity and borehole effects such as fracture enlargement and borehole rugosity.

Recently, two techniques have been proposed as providing in situ estimations of fracture aperture. Hornby *et al.* (1989) estimate fracture aperture and location using reflected Stoneley wave arrivals. Luthi & Souhaité (1990) estimate fracture aperture from electrical currents, generated by Formation MicroScanner pads. Each technique is limited when used on its own; however, additional information about the fractures can be derived when the techniques are used in combination (Hornby *et al.* 1992). This paper is a further elaboration on this approach with additional new data.

Fracture evaluation using reflected Stoneley wave arrivals has the advantage that the fractures are probed some distance away from the borehole. For fractures with an effective extent of several wavelengths or more away from the borehole, no attenuation is seen, resulting in the same response as of a fracture with infinite extent. For a low-frequency Stoneley wave of 500 Hz, the wavelength is 10 feet and the signal may probe 30 feet or more away from the borehole. The actual depth of penetration of the Stoneley wave may be less and will vary depending upon fracture size, fluid viscosity, frequency of the measurement as well as fracture roughness

(Hornby *et al.* 1989). The low-frequency Stoneley wave arrival is useful to prove fractures away from the borehole, but the result of the measurement—the Stoneley wave reflectivity—has a low vertical resolution along the borehole. For example, a 500 Hz Stoneley wave signal has a maximum vertical resolution of approximately 2.5 feet or 1/4 of the wavelength. Consequently we commonly find that a number of fractures contribute to each Stoneley reflection response. The technique provides no information on the dip angle and azimuth of the fracture, and hence on the geometry of the fracture network. Additionally, an estimation of the fracture dip angle is required as input to compute fracture widths accurately from Stoneley wave reflectivity.

Fracture evaluation using electrical borehole scans has the advantage of high vertical resolution and quantitative information regarding the fracture orientation and morphology. In addition, fracture aperture may be computed as a function of azimuth, and fracture apertures can be displayed in azimuthal plots like normal borehole images (Luthi & Souhaité 1990). The major limitation of fracture evaluation using electrical borehole scans is the depth of penetration. For typical formation parameters, most of the fracture signal is from a depth of perhaps one inch, compared with ten feet or more for the reflected Stoneley wave measurement.

---

* Mark of Schlumberger

From HURST, A., GRIFFITHS, C. M. & WORTHINGTON, P. F. (eds), 1992, *Geological Applications of Wireline Logs II.* Geological Society Special Publication No. 65, pp. 185–198.

The limitations of each technique correspond precisely to the areas of application for the other technique, making the two methods complementary in nature. The objective of this paper is, therefore, to show comparisons of fracture aperture estimations from electrical borehole scans and reflected Stoneley waves in an attempt to understand the similarities and differences between both techniques. We suggest ways in which an integrated interpretation leads to an enhanced understanding of fractured formations. Where comparisons are appropriate, we quantitatively predict location, orientation and apertures of fracture sets and give a qualitative estimate of their connectivity away from the borehole.

Method

Fracture evaluation from electrical borehole scans

The Formation MicroScanner* tool is a wireline device producing electrical scans of the borehole wall (Ekstrom *et al.* 1987) using arrays of small electrodes mounted on pads and held at a known potential with respect to a return electrode in the upper part of the tool. Currents emitted from these electrodes are recorded every 2.5 mm and are used to produce azimuthally oriented conductance images of those parts of the borehole wall scanned by the pads while logging. Open fractures are always the most prominent features seen on electrical images because of the large conductivity contrasts between the fluid in the fracture, typically assumed to be the drilling mud, and the surrounding rock (Plumb & Luthi 1989).

Finite element modelling

The finite element method is used to model the current emitted by a single Formation Micro-Scanner button in front of a fracture. In the approach, which is closely related to the technique Chang & Anderson (1984) used for simulating induction logs, the fracture is modelled as a very thin, planar and parallel sheet element with a uniform resistivity equal to the mud resistivity R_m (Fig. 1a). Current densities are computed on a three-dimensional, adaptive grid consisting of 70 000 nodes within the borehole and the formation surrounding the tool pad. Button currents are obtained by multiplying current densities with their corresponding area. Modelled formation resistivities (R_{xo}) are 10, 100 and 1000 Ωm. Fracture apertures (d) are varied

from 50 μm to 200 μm, fracture dips (θ) from 0° to 40°, and tool stand-offs from 0 to 2.5 mm. The mud resistivity (R_m) is held constant at 0.1 Ωm. It was found that, as the tool approaches a fracture, there is a sharp increase in button current in the immediate vicinity of the fracture. Increasing tool stand-off causes the button current to increase earlier, but the current peak diminishes. A simple expression of the fracture response is the integrated excess conductance A caused by the presence of the fracture, or

$$A = \frac{1}{V_e} \int_{z_0}^{z_n} \{I_b(z) - I_{bm}\}dz. \qquad (1)$$

Here V_e is the potential difference (in volts) between the imaging arrays and the current return at the top of the tool, I_b the button current (in μA) as a function of the vertical position z as the tool moves across the fracture (from location 0 to n), and I_{bm} the button current in the undisturbed zone or matrix. Figure 1b shows how A varies as a function of the resistivity regime and the fracture aperture d (for a tool which, in fact, differs from the commercially available design). Within the studied range, A was found to be related to the principal variables as

$$A = \frac{d}{R_m \cdot c} \left(\frac{R_{xo}}{R_m}\right)^{b-1} \qquad (2)$$

with c and b obtained numerically from the simulations. Fracture dip and tool stand-off were found to have an insigificant effect on the above relation. Equation (2) allows computation of the fracture aperture d if A and the mud and formation resistivities are known.

Image inversion

A three-step procedure is used for inverting Formation MicroScanner images to fracture parameters. Potential fractures are first detected, then traced and finally inverted for fracture aperture using equation (2). Fractures are detected using a statistical method in which there is no *a priori* knowledge of the fracture shape. The method involves computation of a running median of the rock conductance I_b/V_e over all buttons in a number of consecutive rows of the electrical image. Locations are retained for which the measurement exceeds the median passage by a certain number of standard deviations. This procedure, in essence, filters out a high-frequency 'fracture image' from a low-frequency background or 'matrix' image. A is

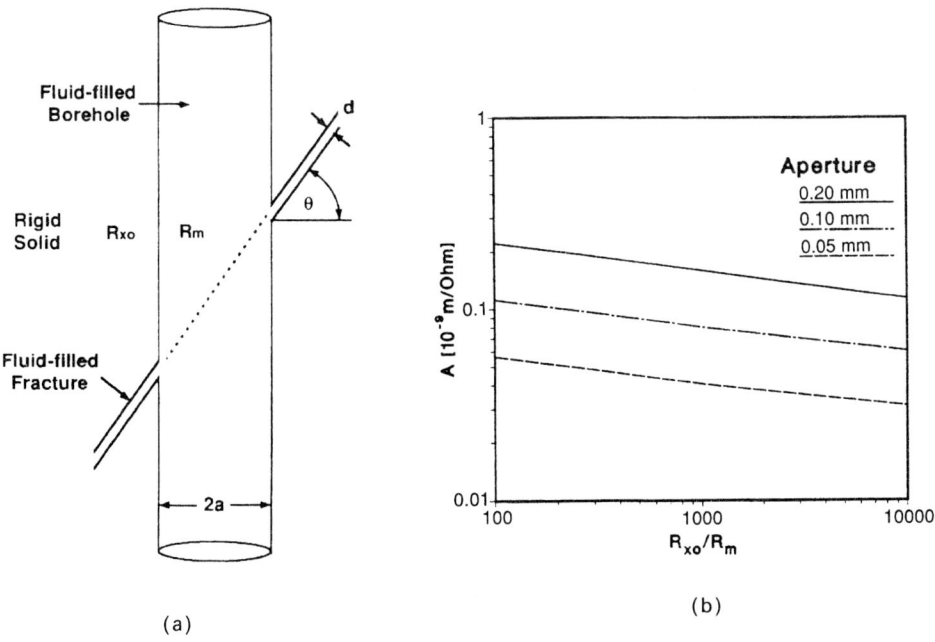

(a)

(b)

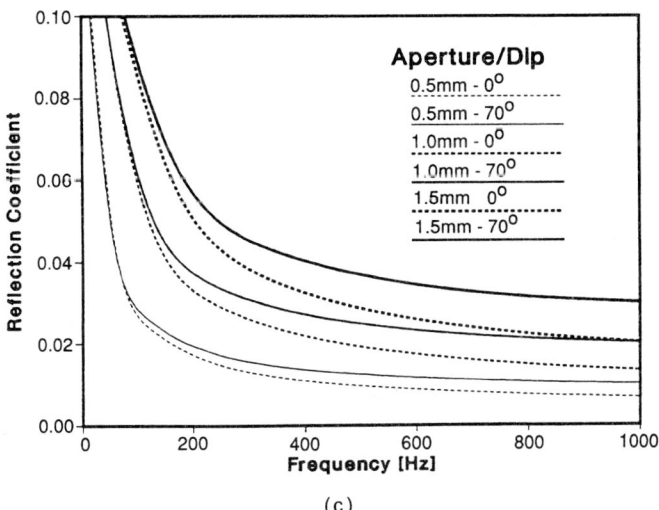

(c)

Fig. 1. Model and theoretical responses for a fluid filled fracture crossing a borehole. (a) Basic model for both techniques. The fracture is modelled as a very thin, planar and parallel plate element filled with the borehole fluid. The surrounding solid is perfectly rigid and the fracture is dipping with angle θ with respect to the horizontal plane. (b) Fracture aperture from electrical borehole scans: plot of the integrated additional conductance A as a function of formation resistivity and fracture aperture d. (c) Fracture aperture from reflected Stoneley waves: theoretical values for Stoneley wave reflectivity as a function of dip angle θ and fracture width d.

computed for each fracture location (x_o, z_o) as an integral of the form

$$A(x_o, z_o) = \frac{1}{r\sqrt{\pi}} \cdot \frac{1}{V_e} \int_{x_o - r}^{x_o + r} dx$$

$$\times \int_{z_o - [r^2 - (x - x_o)^2]^{1/2}}^{z_o + [r^2 - (x - x_o)^2]^{1/2}} dz (I_b(x,z) - I_{bm}). \qquad (3)$$

With this approach the process of estimating local slopes can be avoided, but some local averaging is obtained. Radius r has to be sufficiently large to gather all excessive current, but small enough not to receive interfering signals from any adjoining fractures. The tracing procedure tests all fracture locations for connectivity by eliminating those points which are connected to less than a certain number (typically three) of other fracture locations. This procedure deletes scattered data points. For the computation of the fracture aperture d a standard shallow-resistivity log such as the microspherically focused log (MSFL*) or the spherically focused log (SFL*) is needed to obtain R_{xo}. The mud resistivity (at borehole temperature) is normally known and fairly constant even over long depth intervals. Equation (2) is then applied to each fracture location using a value of A computed through equation (3). This results in a local fracture aperture for each picture element belonging to a fracture. Since the azimuth of each data point is known, the resulting fracture apertures are displayed in an azimuthal plot like normal borehole images. In some of the figures we show a maximum aperture value per unit inverval (e.g. per inch), which is of significance because producibility is more influenced by the larger fracture apertures. In fracture networks, the highest apertures are often found at fracture intersections, in which case the maximum values have to be interpreted with caution. Additionally, we compute a fracture porosity per unit area (Luthi & Souhaité 1990). This value is highly dependent on the size of the window used (the smaller the window, the larger the values in front of the fractures become). We typically use a window of one inch in the vertical, and the width of the images in the horizontal. The resulting fracture porosity values should therefore only be used for comparative purposes.

Fracture evaluation using reflected Stoneley wave arrivals

At low frequencies, the borehole Stoneley (or tube) wave may be considered as a simple pressure pulse propagating in a cylindrical borehole. When the borehole Stoneley wave intersects a permeable fracture crossing the borehole, pressure is released into the fracture. This pressure drop gives rise to both an attenuation of the direct arrival and a secondary (or reflected) Stoneley wave. The reflected Stoneley wave may be considered to be generated by a secondary source located where the fracture plane crosses the borehole. The strength of this secondary source is relative to the amount of pressure released into the fracture and, hence, the fluid conductivity of the fracture. Using this basic principle, an analytical theory has been developed to compute the reflected Stoneley wave response as a function of fracture width and dip angle (Hornby et al. 1989). Again Fig. 1a details the model used for the theory. The fracture is idealized as a planar fluid slab of thickness d and infinite extent, crossing the borehole of radius a at dip angle θ. The formation is assumed to be perfectly rigid and the pressure is spatially uniform throughout the region of the fracture/borehole intersection. We assume that the frequency is well below the cut-off frequency for any mode other than the fundamental mode so that the pressure may be considered uniform across the borehole and across the fracture. Water is assumed to saturate the borehole as well as the fractures, and the effect of the drilling fluid viscosity is ignored. The effects of drilling fluid viscosity were studied in Hornby et al. (1989), who concluded that the effects of viscosity may be ignored as long as the viscous skin depth, $\delta = [(2\eta/\rho_f \omega)]^{1/2}$ is small compared to d. The viscous skin depth at 100 Hz is only 60 μm and, for the analysis of full sonic waveforms, we concentrate on fractures larger than this by using frequencies higher than 100 Hz.

For the special case of a horizontal fracture ($\theta = 0$), the reflection coefficient of the Stoneley wave $\tilde{r}(\omega)$ is

$$\tilde{r}(\omega) = -\frac{idH_1^{(1)}(qa)/aH_0^{(1)}(qa)}{1 + idH_1^{(1)}(qa)/aH_0^{(1)}(qa)} \qquad (4)$$

where $H_n^{(1)}$ is the outgoing Hankel function of order n, a is the borehole radius, and the axial wavenumber q equals ω/c where $c = (K_f/\rho_f)^{1/2}$ is the speed of sound in terms of the bulk modulus of the fluid, K_f, and its density ρ_f.

For $\theta > 0$, a planar fracture crossing the cylindrical borehole creates an elliptical intersect as shown in Fig. 1a. The solution for this problem is more complex, and can be obtained by means of a transformation to elliptic coordinates and separation of the wave equation in terms of Mathieu functions

$$\tilde{r}(\omega) = -\left(\frac{2\pi d}{ia}\sum_{m\,=\,\text{even}}\frac{[D^m_{\,0}]^2\,V_m(q\rho,\xi_0)}{N^e_m Re^3_{\,m}(q\rho,\xi_0)}\right)$$

$$\times\left(1 + \frac{2\pi d}{ia}\sum_{m\,=\,\text{even}}\frac{[D^m_{\,0}]^2\,V_m(q\rho,\xi_0)}{N^e_m Re^3_{\,m}(q\rho,\xi_0)}\right)^{-1} \quad (5)$$

where $\xi_0 = \csc\theta$, θ is the dip angle of the fracture, Re^3_m is the outgoing radial Mathieu function, and D^m_0, $V_m(q\rho,\xi)$ and N^e_m are defined in terms of Mathieu functions and explained in Hornby *et al.* (1989). Figure 1c shows the theoretical values for the magnitude for the Stoneley wave reflection coefficient for a fracture crossing the borehole at two dip angles θ, 0° and 70°, and for three fracture widths of 0.5 mm, 1.0 mm and 1.5 mm.

Fracture width inversion

Inversion for fracture width involves several steps.

1. Full waveform sonic data are initially processed to separate direct and reflected Stoneley wave components.
2. Deconvolution is used to compute the Stoneley wave reflectivity response:

$$\tilde{r}(\omega) = \frac{R(\omega)D^*(\omega)}{\max[D(\omega)D^*(\omega),c\,K]} \quad (6)$$

Here, R is the reflected Stoneley wave signal, D is the direct Stoneley wave signal, r is the Stoneley wave reflectivity, K is the peak value of the spectrum $D(\omega)D^*(\omega)$, and c serves to fill in 'holes' in the spectrum with a fraction (typically 1%) of the maximum signal (Dey-Sarker & Wiggins 1976).
3. Locations of Stoneley wave 'secondary sources' are determined to be at depth locations where the reflected and directed arrivals times are coincident. This is accomplished by a least-squares fit of reflection arrival times. Secondary source locations are used as an initial estimate of fracture locations.
4. Fracture width is estimated at discrete depth locations, determined in step 3, by fitting the measured Stoneley wave reflectivity with a series of theoretical curves, computed as a function of fracture width using borehole diameter, borehole fluid slowness, and the fracture dip angle as inputs. Fracture dip angle is estimated or measured from borehole images. The result is a continuous output of the magnitude of the Stoneley wave reflectivity along with estimations of fracture widths at discrete locations along the borehole.

Cross-validation methodology

Cross-validation of the techniques is performed on single fractures or isolated fracture zones (Hornby *et al.* 1992). We discuss one case from the crystalline rocks of the Moodus well (see below), where a major fracture zone was observed on both the ultrasonic borehole televiewer and the Formation MicroScanner images (Fig. 2), and water influx into the well was reported by the driller. Fig. 2a is a plot of the Stoneley wave results. Stoneley wave reflectivity (RC) and equivalent fracture width were computed using the methods described above. Results are presented along with the borehole diameter (HD), a gamma ray curve, and a single receiver, constant-offset variable-density plot of the raw waveform data. A large Stoneley wave reflection with a coefficient of almost 0.1 is seen at 613 feet. This response is equivalent to the response from an idealized parallel plate fracture with an aperture of 5.6 mm. Figure 2b is an expanded display of the ultrasonic borehole scan (BHTV), the electrical borehole scan, fracture aperture and fracture porosity computation from the electrical borehole scan. Both imaging methods show a zone about five feet thick with complex fracturing. The zone as a whole dips eastward at a relatively modest angle of about 20°. Maximum fracture apertures from the Formation MicroScanner technique range around 1 mm at three depths located at the top, the middle, and the bottom of the zone. At intersections the computed fracture apertures are locally significantly higher than 1 mm. Approximately 25 other fractures seem to have typical apertures of around 200 μm. Depending on the scan line taken through the data, the cumulative fracture aperture from the electrical borehole scan is between 6 and 8 mm. This value compares remarkably well with the equivalent single fracture aperture obtained from the reflected Stoneley wave analysis. Using the same approach, Hornby *et al.* (1992) demonstrate in several other zones the validity of the two techniques to obtain fracture apertures.

Experimental field data

Field data from three different geological settings are examined, one in naturally fractured crystalline rocks, one in a chalk reservoir overlain by a shale cap rock, and one in a fractured dolomite reservoir. For all wells, electrical borehole scans were acquired using a commercial two-pad Formation MicroScanner tool, and full waveform sonic data were obtained using a research prototype full-waveform sonic tool.

The Formation MicroScanner tool used is identical to the one described in Ekstrom *et al.* (1987). It is important to notice that for the fracture computation technique described herein we use the raw button currents (normally measured in μA), compensated by the applied voltage (which itself may vary with depth). A shallow Laterolog has been used in the first and third wells to measure the formation resistivity. In the second well, the shallow resistivity measurement of an experimental Adaptive Induction Tool (AIT*), having a depth of investigation of 10 inches (254 mm), was used. The AIT provides a measurement every 3 inches (76 mm).

The research prototype full-waveform sonic tool used incorporates a single transmitter and 11 receivers spaced at 6 inch (152 mm) intervals. The distance from the transmitter to the first receiver is 10 feet (3 m). Full waveforms are acquired from the entire receiver array at each source firing, and the source is fired at 6 inch (152 mm) intervals. For fracture characterization the source is electronically driven with a low-frequency source drive and data are recorded at each receiver for 25 ms or more. The low-frequency source drive excites a Stoneley wave with significant energy below 1000 Hz.

Moodus No 1 well

The Moodus No 1 well was drilled in east-central Connecticut by the Empire State Electrical Energy Research Corporation (ESEERCO) to understand better the state of stress in southern New England and, in particular, to understand the nature of the local seismicity known as Moodus noises (Naumoff 1988). The well was air-drilled to a depth of approximately 4800 ft into crystalline rocks of Palaeozoic to Precambrian age belonging to the Merrimack and Avalon terrane. During the drilling process, two water producing zones were identified at depths of about 610 ft and 3490 ft. Water influx at the deeper zone was great enough that the well had to be grouted from 3500 ft to 2965 ft and redrilled (P. G. Naumoff, pers. comm.).

Electrical borehole scans and full waveform sonic data were acquired through all intervals. In addition, ultrasonic images of the borehole wall were acquired using a research ultrasonic imaging tool. This tool is based on the borehole televiewer (BHTV) concept developed by Mobil Oil Company (Zemanek *et al.* 1969). The upper water-producing zone was found to consist of a single fractured interval; it is discussed in the preceeding section.

Lower water-producing zone

This zone is located at a depth of around 3490 ft where the driller reported water influx. Reflected Stoneley wave results (Fig. 3) are more complex than in the zone at 610 ft. In addition to a dominating reflection at 3510 ft, a number of smaller reflections can be identified throughout the interval. Computed apertures are 3 mm for the large event, and 0.5 to 1.0 mm for the smaller ones. The borehole images exhibit a variety of features, such as wellbore breakouts (at 3507 ft, oriented approximately N–S), fractures with high dip angles (at 3490), as well as low-angle features which are interpreted as foliation planes. The Formation MicroScanner results suggest a strong correlation of the fracturing events with the foliation planes, perhaps because of shearing along the foliation planes. Typical apertures are around 400 μm to 1 mm. Overall, the Formation MicroScanner results have an excellent correlation with the reflected Stoneley wave fracture aperture estimates—in terms of both location as well as fracture aperture.

One notable exception is at 3510 ft, where the Formation MicroScanner results suggest an aperture of 0.5 mm, while the Stoneley wave results indicate an aperture of 3 mm. Closer analysis of this zone sheds some light on this difference. Using the acoustic velocity of the borehole fluid, a high-resolution, azimuthally averaged borehole radius is computed from the BHTV transit time masurement and displayed in Fig. 3. This curve shows a borehole enlargement of about 0.5 inches at 3510 ft. While this will somewhat increase the Stoneley wave response, the difference cannot be discarded as merely a borehole effect. Firstly, the Formation MicroScanner results show about four features, of which only the largest one is 0.5 mm, while the other ones are about 0.3 mm, bringing the total to about 1.4 mm. Thus, this zone resembles in some ways the zone at 610 ft, where the cumulative fracture apertures are close to the Stoneley results. Secondly, the zone shows intersecting fractures, some of which are steeply dipping. It is possible that these fractures are connected to other fractures above and below this zone. This would increase the response of the Stoneley wave, but not of the electrical borehole scans, which investigate a much shallower volume. Thirdly, the mere presence of a borehole enlargement of this size in a very competent rock indicates failure prior to drilling, perhaps associated with enhanced leaching by circulating groundwater. In such a complex zone perfect agreement between the two methods is unlikely.

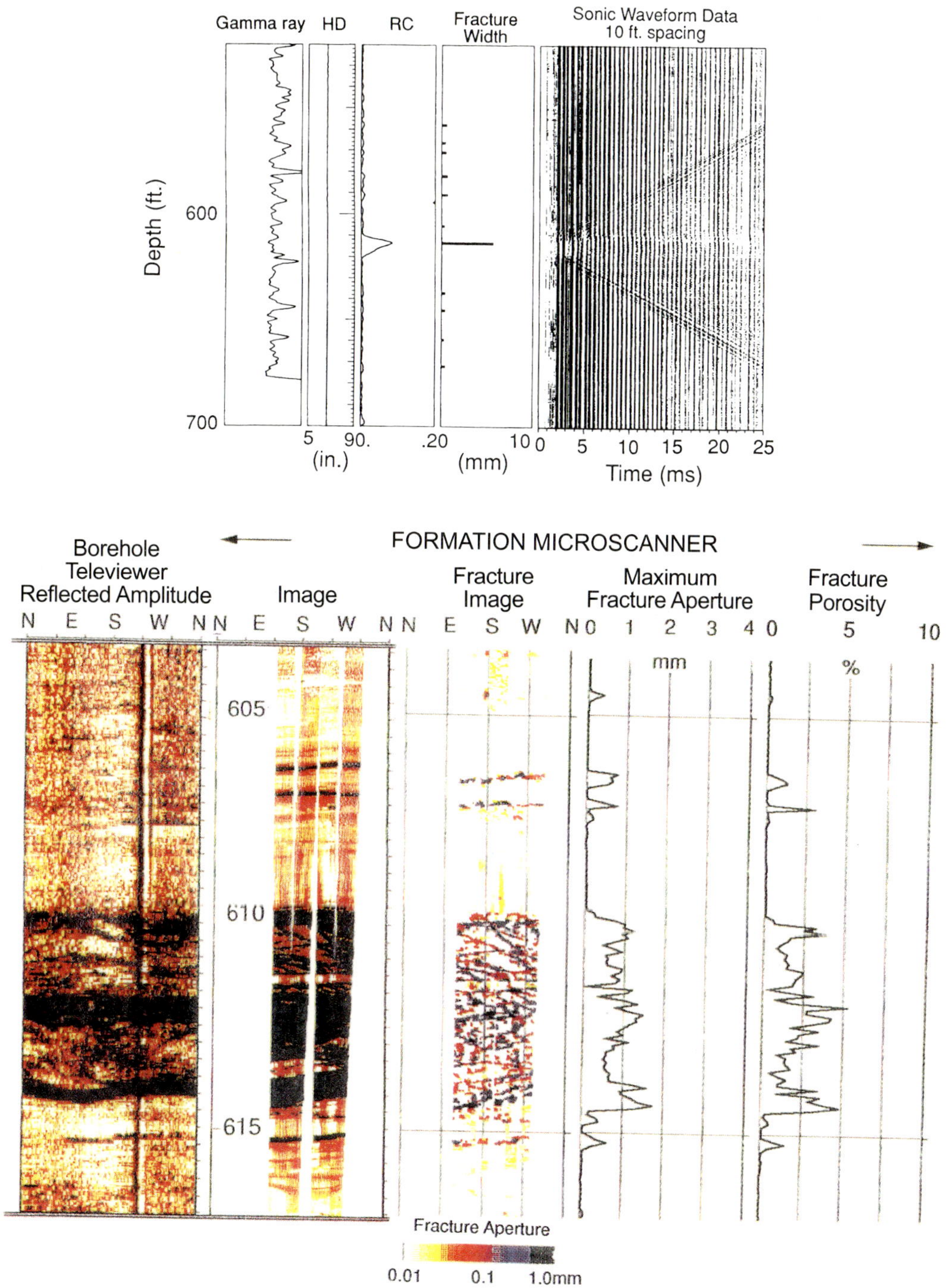

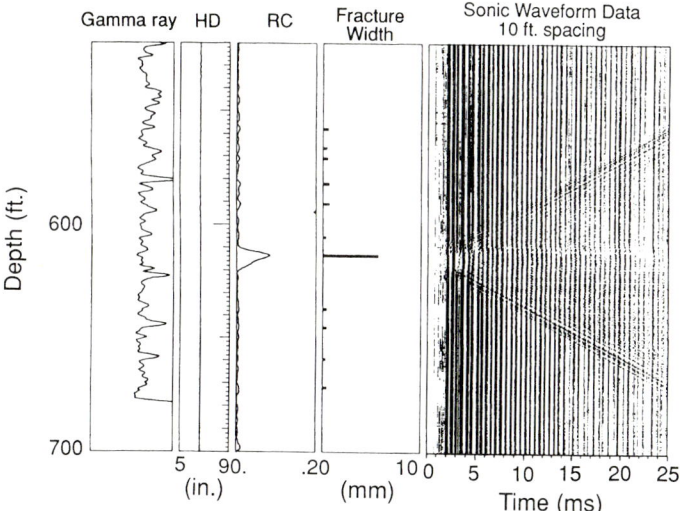

Fig. 2. Comparison of the two techniques using field data acquired in a well penetrating crystalline rock (Moodus No 1 well). Water influx into the well was reported by the driller at this interval. (a) Reflected Stoneley wave results: Stoneley wave reflectivity (RC), fracture width, and waveform plot and (b) an expanded display of the ultrasonic borehole scan (BHTV), the electrical borehole scan, fracture aperture and fracture porosity computation from the electrical borehole scan. The source of water influx is clearly the fracture interval at 610 feet.

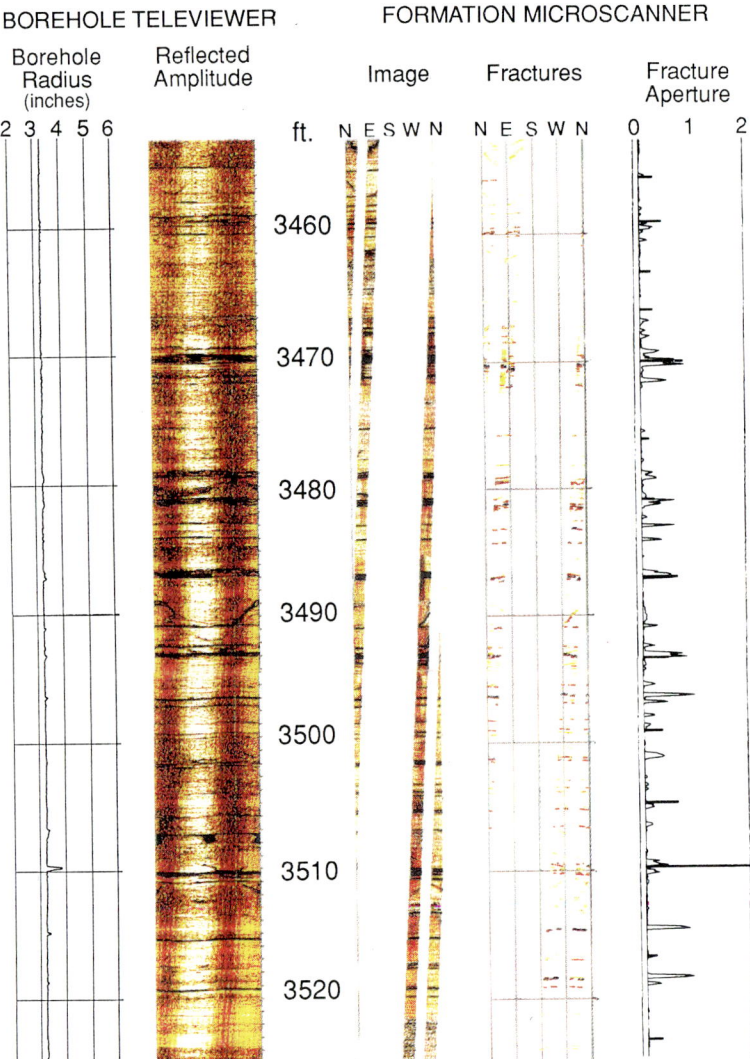

Fig. 3. Moodus No 1 well—lower-water producing zone. Comparison of fracture apertures from the Formation MicroScanner (solid curve) with fracture apertures from reflected Stoneley waves (bars). Water influx, reported at a depth of approximately 3500 feet, correlates with estimates of large (400 μm–2.0 mm) fractures from both techniques. For both techniques, an excellent comparison is seen for both fracture aperture estimates and fracture locations.

Discussion of producibility

The zone at 610 ft offers an interesting opportunity to assess its fluid producibility, assuming either a single fracture or a set of fractures as obtained in the current analysis (Fig. 2). Assume that the producibility, or fluid flow conductance, is simply $d^3/12\eta$, an expression which is valid for smooth, parallel-plate fractures. Brown (1987) has shown that deviations from this cubic law due to the roughness of the fracture walls are less than 10% as long as the mean aperture is greater than five standard deviations of the wall roughness. The roughness of the two fracture zones in situ are not known at present. If we assume the producibility of the zone at 610 ft to be due to a single fracture with 6 mm aperture, its fluid flow turns out to be 36 times larger than for six individual fractures of one millimetre aperture each. If, however, we assume from our previous analysis that this zone consists of three fractures of one millimetre aperture each, and 15 fractures of 200 μm aperture (again for a total of six millimetres), the flow rate would be almost seventy times smaller than for the case of a single fracture. We conclude that knowledge of the individual fracture apertures can substantially refine fluid flow estimates. Calculation of the actual flow rate has to take into account the pressure drop from the formation to the wellbore.

Austin chalk well

Our second example is from a research well drilled by Mobil Oil Company in 1988 for the purpose of investigating the effects of permeability on sonic log responses. The well was first drilled with fresh mud and cored with a $6\frac{1}{8}$ inch core barrel. The well was then reamed out to $8\frac{1}{2}$ inches using salt water. This exercise was only partially successful, since shale sloughing occurred and polymer had to be added to stabilize the well. Next, the salt water was replaced with mud in order to build up the mudcake again, and lastly the well was reamed out to $9\frac{7}{8}$ inches diameter. Logging runs, including full waveform sonic and Mobil's acoustic borehole televiewer (BHTV), were taken after each borehole diameter or mud change. During the last logging run, electrical borehole scans were acquired using the Formation MicroScanner tool. The well was not expected to penetrate permeable fractures, but analysis of whole core and borehole scans revealed several intervals of apparently open fractures, leading to the following combined fracture analysis of the recorded Formation MicroScanner and full waveform

sonic data. Efforts were concentrated on the data acquired in the $9\frac{7}{8}$ inch hole, where both full waveform sonic data and Formation Micro-Scanner data are available. However, analysis of the stand-alone full waveform sonic data in the different borehole diameters and mud systems provided valuable insights into the sensitivity of the inversion technique to borehole size and mud type. Furthermore, there are clear indications of drilling-enhanced fractures as the drilling of this well proceeded. Several fractured zones were studied in detail by Hornby et al. (1992) and good agreement of the two techniques was found.

Overview of results in the Austin chalk well

Figure 4 shows an ELAN* volumetric analysis from open hole logs (Quirein et al. 1986), showing the shale section of the Taylor marl in the uppermost part, underlain by the Austin chalk. The borehole diameter (HD), the Stoneley wave reflection coefficient (RC), the fracture porosity computed using the Formation MicroScanner tool, as well as the fracture apertures from the two methods are also shown. There is reasonable to good agreement between the Stoneley wave reflection curve and the fracture porosity calculated from the Formation MicroScanner in the upper half of the section. In the lower half, a washout in a sandstone layer at 2386 ft and a brecciated, partly vuggy zone around 2350 ft account for some differences. There is also a generally good agreement of the two fracture aperture results, most notably for a known, steeply dipping fracture at 2145 ft. Intersecting fractures at a shale interbed at around 2240 ft sees a higher Stoneley wave derived fracture aperture than the aperture computed from the Formation Micro-Scanner, as does the washout zone mentioned above. Due to a rifling pattern on the borehole wall, caused by the repeated reaming of the borehole, the noise level in the Formation MicroScanner is higher than usual and accounts for the elevated background of the aperture curve in Fig. 4. There is, however, still an inherently higher resolution of the Formation Micro-Scanner results as demonstrated by the activity of the curves. Some of the fractures are clearly associated with lithological features, such as the calcareous bed in the Taylor marl, the base of the marl and the contact of the Austin chalk with the shale layer around 2242 ft. These are all interpreted to be locations where stresses concentrated and fracturing occurred. Such failure was initially perhaps only in the form of hairline cracks, but was subsequently widened in the drilling process.

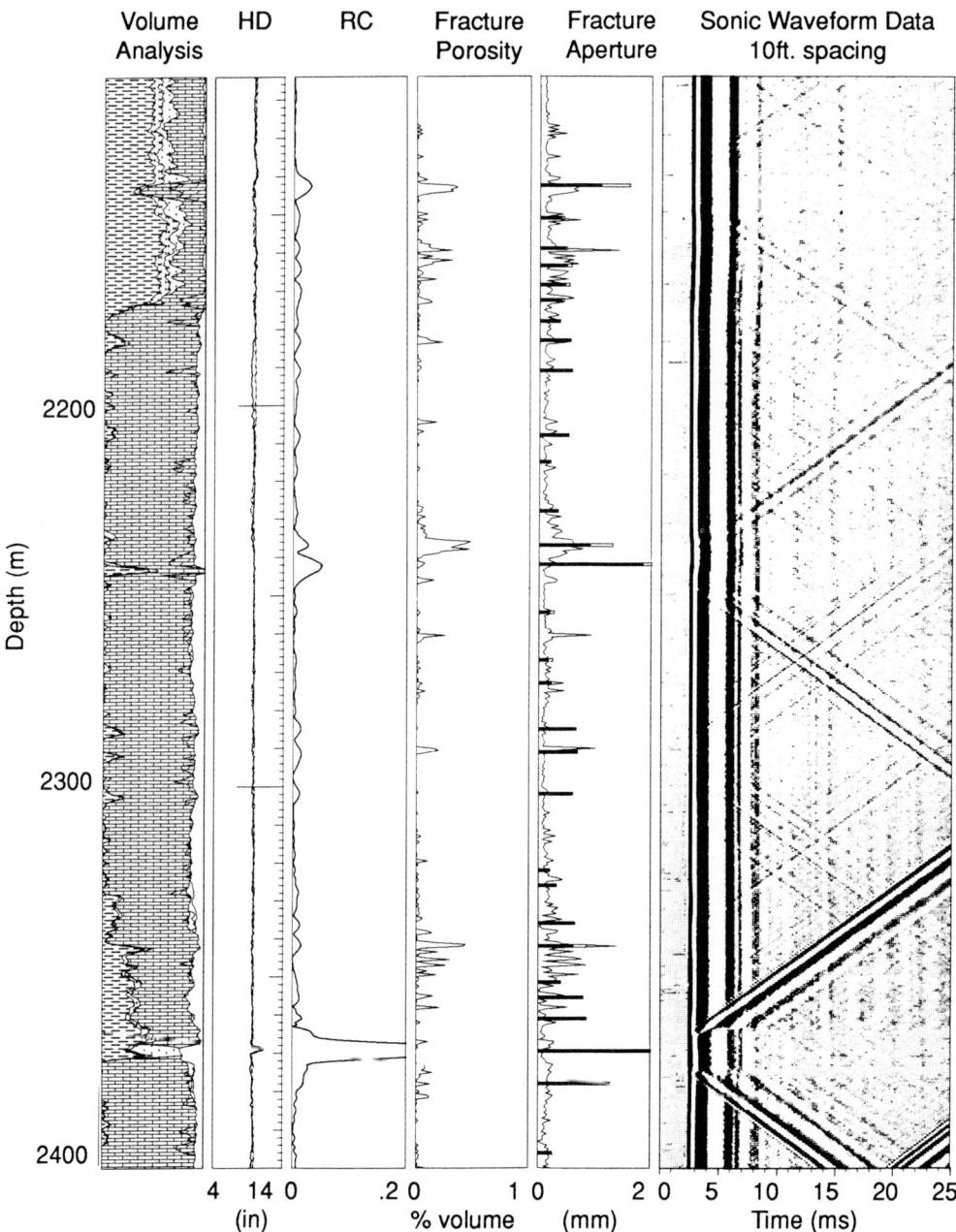

Fig. 4. Austin chalk well. ELAN volumetric analysis, borehole diameter (HD), Stoneley wave reflection coefficient (RC), fracture porosity computed from the Formation MicroScanner, fracture apertures from the two methods, and a variable density plot of the sonic waveform for a constant source-receiver offset (lithology key: dots = sand, dashes = clay, dot-dash = siltstone, bricks = limestone, blank = water).

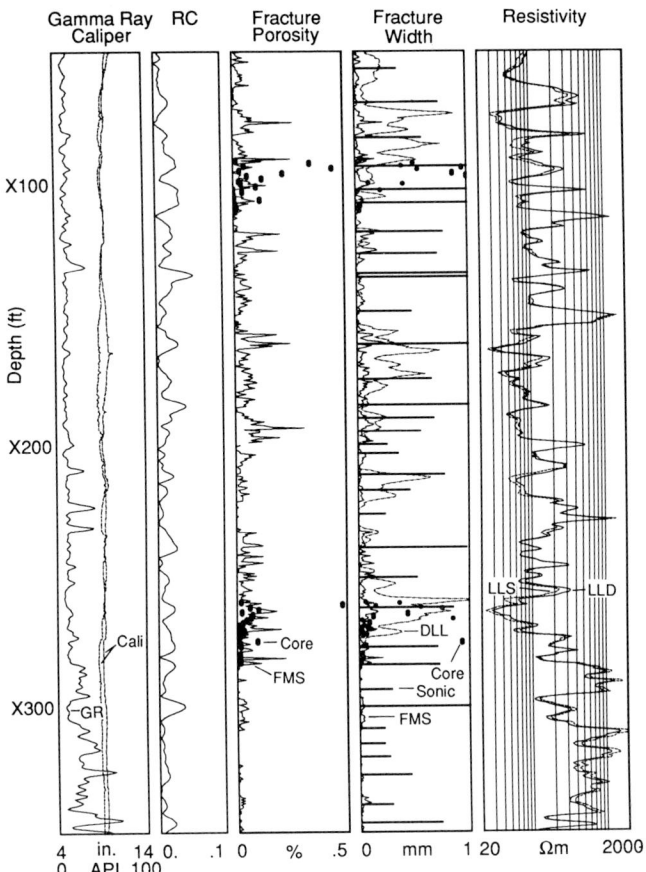

Fig. 5. Ellenburger well. Integrated results with, from the left, gamma ray (solid curve) and two calipers (dashed), reflection coefficient RC, fracture porosity calculated from Formation MicroScanner results (solid curve) and core analysis (solid circles), fracture widths from reflected Stoneley waves (bars), electrical borehole scans (solid curve), Dual Laterolog (dashed curve) and core analysis (solid circles), and finally Laterolog deep (LLD, solid curve) and Laterolog shallow (LLS, dashed curve) logs.

Ellenburger Group well

The Lower Ordovician Ellenburger Group is a sequence of mostly dolomitic and fractured shallow-water lithofacies assemblages which form prolific oil reservoirs in West Texas. Fracturing is both tectonic and karst-controlled, with the latter having developed through caves, sinkholes and collapse-breccias during a prolonged period of subaerial exposure in the Middle Ordovician (Kerans 1988). We applied our techniques in a partially cored well through the Ellenburger, drilled by Phillips Petroleum Company in 1988. All cores exhibit strong intersecting fracturing, mostly at relatively high dips. Phillips Petroleum Company provided us with digital fracture data measured on the core such as dip angle, dip

azimuth, aperture and length. Formation Micro-Scanner images also indicate intricate fracturing patterns throughout most of the sequence. In addition to these techniques we applied a method proposed by Sibbit & Faivre (1985), which evaluates the effective fracture apertures in highly fractured reservoirs such as the Ellenburger using responses of the Dual Laterolog. Thus, we had four different indicators of fracturing in this well.

Fracture evaluation using the Dual Laterolog

The technique developed by Sibbit & Faivre (1985) uses finite-element modelling results to establish a relationship between fracture aper-

tures and conductivities of the mud and the
various parts of the formation. Two cases are
distinguished: horizontal and vertical fractures.
Since our example features mostly steeply dipping fractures we applied the equation given by
Sibbit & Faivre (1985) for vertical fractures,
which in our notation reads

$$d = aR_m \left(\frac{1}{R_{LLS}} - \frac{1}{R_{LLD}} \right) \qquad (7)$$

where R_{LLS} is the shallow resistivity and R_{LLD} the
deep resistivity measured by the Dual Laterolog.
Coefficient a is obtained numerically from
modelling. The results are shown in Fig. 5 alongside our previously obtained fracture apertures.

Discussion of Ellenburger results

Figure 5 shows an interval of about 300 feet (of a
total thickness of 700 feet of the Ellenburger
Group) which we evaluated. Fracture apertures
from Stoneley reflections are found to be significantly higher—often by an order of magnitude—than those obtained from Formation
MicroScanner images, which are generally
around 100 μm or less.

The Dual Laterolog technique, which in some
senses is a hybrid of the other two methods, often
gives intermediate results. Fracture aperture estimations using the Dual Laterolog have to be
carefully interpreted since they may result from
features other than fracturing such as fluid
invasion effects. Core data are available in this
section over two intervals from X090 to X120 ft
and from X260 to X280 ft. Both fracture aperture and porosity were estimated. These data
corroborate the in situ analysis available from
the other three techniques; however, the core
analysis should be interpreted with great care
because of possible variations in the results due
to surface stress relaxation and additional
coring-induced fracturing. Using the core-derived
fracture apertures and fracture lengths, fracture
porosity was estimated over six-inch intervals
throughout the cored sections. The core-derived
fracture porosity gives results which are surprisingly close to the fracture porosity calculated
from the Formation MicroScanner data. The
fracture porosity estimations lie generally
around 0.2%, with outliers reaching as high as
0.5%.

The Ellenburger well is a typical example of
a hydrocarbon reservoir whose production is
entirely dominated by fractures. The intricate
and complex fracturing network observed both
on cores and borehole images is probed at different scales by our methods. The Formation Micro-
Scanner, the Dual Laterolog and the Stoneley
reflection probe increasingly deeper into the
formation and, therefore, average more of the
fracture network. In view of the good hole
conditions no effects other than fractures and
related features such as large-scale solution-
enhanced porosity are believed to account for
the responses. Together with fracture estimation
using the Dual Laterolog measurement, these
techniques provide strong and convincing evidence of the existence of fracture networks in the
Ellenburger Group. The number of Stoneley
reflections, the magnitude of the fracture apertures as well as the fracture porosity calculated
from the electrical borehole scans provides a
clear picture of which zones have higher fracture
densities and larger fracture apertures (Fig. 5).
Together with the analysis of the fracture orientations (Plumb & Luthi 1989), this provides a
comprehensive characterization of the fracturing
in reservoirs.

Integrated interpretation: a discussion

Our observations on these and other field experiments lead us to propose an integrated interpretation of the combined results of fracture
aperture obtained from electrical borehole scans
and Stoneley wave reflections. Three interpretation scenarios are proposed (Fig. 6, left).

1. *Fracture aperture estimates agree.* Where
the fracture aperture estimates for both techniques are in reasonable agreement, fractures are
likely to be isolated planar features and of
large extent, i.e. greater than the Stoneley wave
penetration away from the borehole. For the
cases studied here, a number of fractures and
fractured intervals show excellent agreement in
the fracture aperture estimates. Examples include
the fractures at around 3500 ft in the Moodus
borehole in crystalline rock, a fracture at 2145 ft
in the Austin chalk well, and other examples
cited in earlier work (Hornby et al. 1992).

2. *Fracture aperture from electrical borehole
scans greater than fracture aperture from reflected Stoneley waves.* This occurs when the feature is considerably shorter than the wavelength
of the Stoneley wave (10 ft at 500 Hz)
and includes features such as vugs (Fig. 6, right),
which are not fractures *sensu stricto*, as well as
drilling-induced fractures of shallow extent. In
the Austin chalk well, a local feature at 2159 ft
and the zone at 2350 ft are examples of this
effect. We believe, in fact, that such a combined
interpretation of shallow and deep fracture aperture estimates may be an effective way of identi-

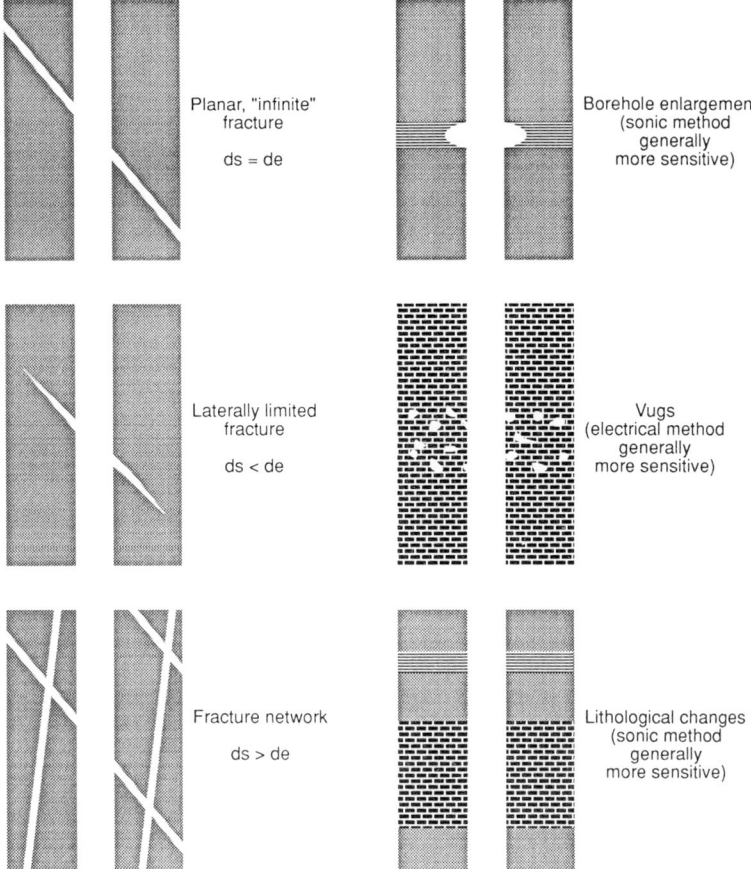

Fig. 6. Comparison of sonic (ds) and electrical (de) fracture apertures as a function of fracture types and other effects.

fying drilling-induced fractures which are not connected to other fractures away from the wellbore.

3. *Fracture aperture from electrical borehole scans less than fracture aperture from reflected Stoneley waves.* This may occur where the fractures are connected to a network of fractures sufficiently far away from the borehole that they are not probed by the Formation MicroScanner measurement (Fig. 6, left). The Ellenburger well demonstrates this effect over about 700 feet of data. Another example is the fracture zone at 3510 ft in the Moodus well. This scenario, however, has to be evaluated carefully, since a local borehole enlargement may give rise to a similar effect (Fig. 6, right). High-resolution caliper measurements are necessary for a proper interpretation, as is illustrated by the only washed-

out zone in the Austin chalk borehole at 2368 ft.

Lithology effects. Although not seen in these data sets, note that strong changes in lithology that result in a strong change in the formation shear modulus (i.e. from calcite to soft shale. Fig. 6, right) can give measurable reflected Stoneley wave responses but will have little effect on the Formation MicroScanner results.

It is important to recognize that one should not rely on any single technique for fracture evaluation; borehole images, high-resolution calipers, lithology indicators, and Stoneley wave data should all be used together in order to discriminate against environmental effects and to arrive at an intelligent interpretation of fracture properties.

Estimates of producibility have to be done

with great care from these results. The higher resolution of the Formation MicroScanner generally allows one to identify most fractures individually, but in the case of intersecting fractures, the flow performance may differ significantly from the one for parallel fractures. Additionally, one has to take into account the effects of surface roughness (Brown & Scholz 1985) as well as the fact that fracture aperture estimates differ when determined by electrical and hydraulic methods (Brown 1989). Nevertheless, in some cases, such as the zone discussed in the Moodus well, simple predictions of relative producibility can be done.

Conclusions

Estimations of fracture apertures were computed using electrical borehole scans and reflected Stoneley waves. In three separate field examples both techniques successfully identified features evident on core and/or borehole televiewer analysis. Comparison of fracture aperture estimates from the two techniques were in good agreement for a number of fractures and fractured intervals. For intervals where fracture aperture estimation differed, a combined interpretation analysis attempts to explain these differences in terms of fracture extent, fracture connectivity, and borehole effects such as fracture enlargement and borehole rugosity.

We acknowledge permission to publish these data by Mobil Oil, the Empire State Electrical Energy Research Corporation and Phillips Petroleum Company. We thank R. Anderson, J. Hensley, D. Johnson, N. Kuich, C. Mennig, P. Pezard, R. Plumb and J. Zemanek for providing and discussing the data presented in this paper. The authors, however, take full responsibility for all interpretations offered herein.

References

BROWN, S. R. 1987. Fluid flow through rock joints: the effects of surface roughness: *Journal of Geophysical Research*, **92**, 1337–1347.

—— 1989. Transport of fluid and electrical current through a single fracture. *Journal of Geophysical Research*, **94**, 9429–9438.

—— & SCHOLZ, C. H. 1985. Broad bandwidth study of the topography of natural rock surfaces: *Journal of Geophysical Research*, **90**, 12575–12582.

CHANG, S. K. & ANDERSON, B., 1984. Simulation of induction logs by the finite-element method. *Geophysics*, **49**, 1943–1958.

DEY-SARKER, S. K. & WIGGINS, R. A. 1976. Source deconvolution of teleseismic P wave arrivals between 14° and 40°. *Journal of Geophysical Research*, **81**, 3633.

EKSTROM, M. P., DAHAN, C., CHEN, M. Y., LLOYD, P. & ROSSI, D. J. 1987. Formation imaging with microelectrical scanning arrays. *The Log Analyst*, **28**, 294–306.

HORNBY, B. E., JOHNSON, D. L., WINKLER, K. W. & PLUMB, R. A. 1989. Fracture evaluation using reflected Stoneley wave arrivals. *Geophysics*, **54**, 1274–1288.

—— LUTHI, S. M. & PLUMB, R. A. 1992. Comparison of fracture apertures computed from electrical borehole scans and reflected Stoneley waves: An integrated interpretation. *The Log Analyst*, Jan–Feb, 50–66.

KERANS, C. 1988. Karst-controlled reservoir heterogeneity in Ellenburger Group carbonates, West Texas. *AAPG Bulletin*, **72**, 1160–1183.

LUTHI, S. M. & SOUHAITÉ, P. 1990. Fracture apertures from electrical hole scans: *Geophysics*, **55**, 821–833.

NAUMOFF, P. 1988. Lithology and structure identified in a 1.5 km borehole near Moodus, Connecticut (Abstract), *EOS*, **69**, 491.

PLUMB, R. A. & LUTHI, S. M. 1989. Analysis of borehole images and their application to geological modelling of an eolian reservoir. *SPE Formation Evaluation*, **4**, 505–514.

QUIREIN, J., KIMMINAU, S., LAVIGNE, J., SINGER, J. & WENDEL, F. 1986. A coherent framework for developing and applying multiple formation models: *Transactions of SPWLA 27th Annual Logging Symposium*.

SIBBIT, A. M. & FAIVRE, O. 1985. The Dual Laterolog response in fractured rocks. *Transactions of SPWLA 26th Annual Logging Symposium*, Paper T.

ZEMANEK, J., CALDWELL, R. L., GLENN, E. E., HOLCOMB, S. V., NORTON, L. J. & STRAUS, A. J. D. 1969. The Borehole Televiewer—a new concept for fracture location and other types of borehole inspection. *Journal of Petroleum Technology*, **25**, 762–774.

Fracture permeability and alteration in gabbro from the Atlantis II Fracture Zone

D. GOLDBERG,[1] C. BROGLIA,[1] & K. BECKER[2]

[1] Borehole Research Group, Lamont-Doherty Geological Observatory, Palisades, NY 10964, USA

[2] Rosenstiel School of Marine and Atmospheric Science, University of Miami, Miami, FL 33149, USA

Abstract. Geophysical logs recorded over 500 m in Ocean Drilling Program Hole 735B have been analysed to identify the in situ properties and the fracture permeability of basement rocks in the Atlantis II Fracture Zone. Hydrous alteration minerals were most abundant in the upper section of the hole. Packer tests also indicate the highest permeabilities in this zone. This interval is characterized by the presence of several open fractures imaged by the acoustic televiewer and by temperature log indications of fluid flow through transmissive fractures. Negative temperature gradient and large semblance anomalies were found to correlate well with the largest of the observed fracture zones. Most of the 70 identified fractures strike subparallel to the north–south fracture zone and dip steeply WSW. These fractures probably formed as a result of normal faulting on the transform side of the uplifted ridge in the Atlantis II Fracture Zone, where 735B was sited, rather than as a direct result of horizontal plate tectonic motion.

During Leg 118 of the Ocean Drilling Program, the *Joides Resolution* drilled into crystalline rocks in the vicinity of the Atlantis II Fracture Zone near the South West Indian Ridge. This fracture zone is one of a series of north–south trending transform faults that offset the slow-spreading South West Indian Ridge (Fig. 1). Site 735 is located to the east of the Atlantis II Fracture Zone on a platform lying in 731 m of water. The position of the platform in the magnetic anomaly pattern on the transform wall suggests a crustal age of about 12 Ma (Robinson, Von Herzen *et al.* 1989). As the gabbro was emplaced high on the flank of the fracture zone, it was probably subjected to stress, which resulted in a variety of deformational textures and fractures.

Hole 735B penetrated primarily gabbro to a depth of 500 mbsf and a complete suite of in-hole electrical, acoustic and nuclear logs and hydraulic packer tests were recorded (Robinson, Von Herzen *et al.* 1989). The overall recovery of core was 87%. Below 100 mbsf, recovery was 95%. Thus, Hole 735B gave the first nearly continuous recovery of core in fresh and altered oceanic gabbro. This study focuses on the processing and interpretation of the downhole data, in particular, the Schlumberger neutron porosity log, the acoustic borehole televiewer log, the multichannel sonic log, the borehole temperature profile and four hydraulic packer tests.

Direct comparison of the core with the extensive suite of downhole logging measurements provides for the first time, characterization of the in situ physical properties of oceanic gabbro and its relationship to tectonics and hydrogeology at a fracture zone site.

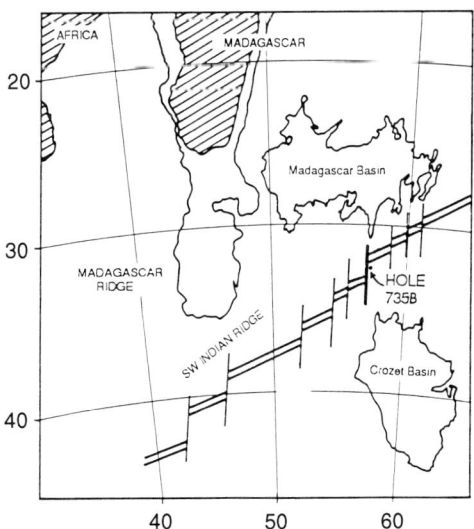

Fig. 1. Location map of Hole 735B in the Atlantis II Fracture Zone.

From HURST, A., GRIFFITHS, C. M. & WORTHINGTON, P. F. (eds), 1992, *Geological Applications of Wireline Logs II.* Geological Society Special Publication No. 65, pp. 199–210.

Downhole measurements

Alteration and porosity profile

The Schlumberger neutron porosity tool used in Hole 735B uses a chemical source of americium–berillium to bombard the formation with fast neutrons, and four detectors to measure their concentration in epithermal (>0.2 eV) and thermal (<0.025 eV) energy ranges (Ellis *et al.* 1988). During the scattering process, the neutrons interact elastically with nuclei in the formation, losing part of their kinetic energy with each collision; eventually, they are reduced to the thermal energy level at which they can be absorbed by the surrounding nuclei. Because its mass is equal to that of a neutron, the hydrogen atom is most efficient in slowing down neutrons. The slowing-down length, L_s, is the average distance a neutron must travel to reach the thermal energy level, usually between 10 and 30 cm (McKeon & Scott 1988). This parameter depends largely on the hydrogen concentration in the rock and also on the relative concentration of its other constituents; thus it is different for basalts and for most sedimentary rocks.

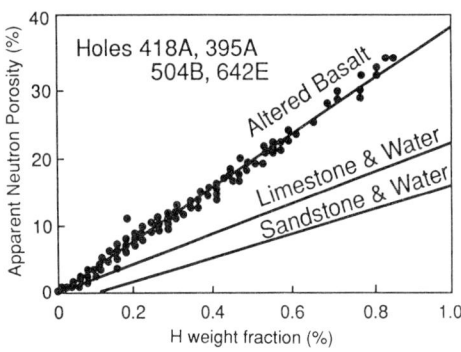

Fig. 2. The linear relationship between computed apparent porosity and hydrogen content at four DSDP-ODP Sites (after Broglia & Ellis 1990). The apparent neutron porosity of altered basalts exceeds the values of the water-filled porosity by an amount related to the bound-water content. This relationship has been used in Hole 735B to estimate the neutron response to hydrous minerals from the hydrogen weight fraction of selected samples (also see Fig. 4, track 2).

Most rocks, however, also contain hydrogen in the form of bound water associated with clays and other alteration minerals. Broglia & Ellis (1990) have presented in detail the technique necessary to correct the neutron porosity

measured in basaltic and gabbroic rocks for the presence of hydrous minerals. In Fig. 2, the apparent porosities of basaltic samples, computed from L_s, are plotted as a function of their hydrogen weight fraction measured by X-ray analysis (after Broglia & Ellis 1990). The apparent porosity of basalts that contain no water-filled porosity and only bound hydrogen ranges from 1% in fresh massive units to about 35% in highly altered units. The positive linear relationship implies that apparent neutron log porosity may exceed the value of the water-filled porosity if hydrous minerals are among the rock constituents. Hence, basaltic rocks that have undergone extensive alteration in the form of hydrous clay mineralization (e.g. celadonite, palagonite, amphiboles, talc, zeolites, and iron-hydroxides) will exhibit elevated apparent porosities by an amount linearly related to their content of bound-water. Gabbro, having the same elemental composition as basalt, is assumed to have a similar neutron log response. With this relationship established, shipboard X-ray measurements of the hydrogen weight fraction on core samples from Hole 735B were used to estimate directly the effect of bound hydrogen on the neutron log response (Fig. 3, track 2, dots).

Estimates from core, however, do not provide a continuous profile of bound water such as is traditionally obtained from the gamma-ray log in sedimentary formations (Poupon & Gaymard 1970). Furthermore, the gamma ray log in Hole 735B has a very low signature because amphiboles, the major hydrous mineral present in the gabbros, lack potassium. Therefore, in order to predict continuously the effect of bound hydrogen on the neutron log, alternative logs that are sensitive to the alteration minerals present must be used (Broglia & Ellis 1990). The hydrogen yield log recorded with the induced gamma-ray spectrometry tool, a pulsed-neutron device that measures the relative concentration of some of the rock constituents (Hertzog *et al.* 1987), was chosen as a continuous measure of hydrous mineral alteration in Hole 735B. In the absence of large changes in borehole size (see caliper and density log correction in Fig. 3, track 6), both the hydrogen yield and neutron porosity logs measure the concentration of hydrogen in formation pores and hydroxyls bound in alteration minerals. The correlation between the hydrogen yield and the neutron log indicates that both are sensitive to the same parameters. Differences between them may be attributed to increases in the hydrogen yield response from seawater in the borehole or from the sensitivity of the measurement to certain heavy elements, such as titanium, in the formation.

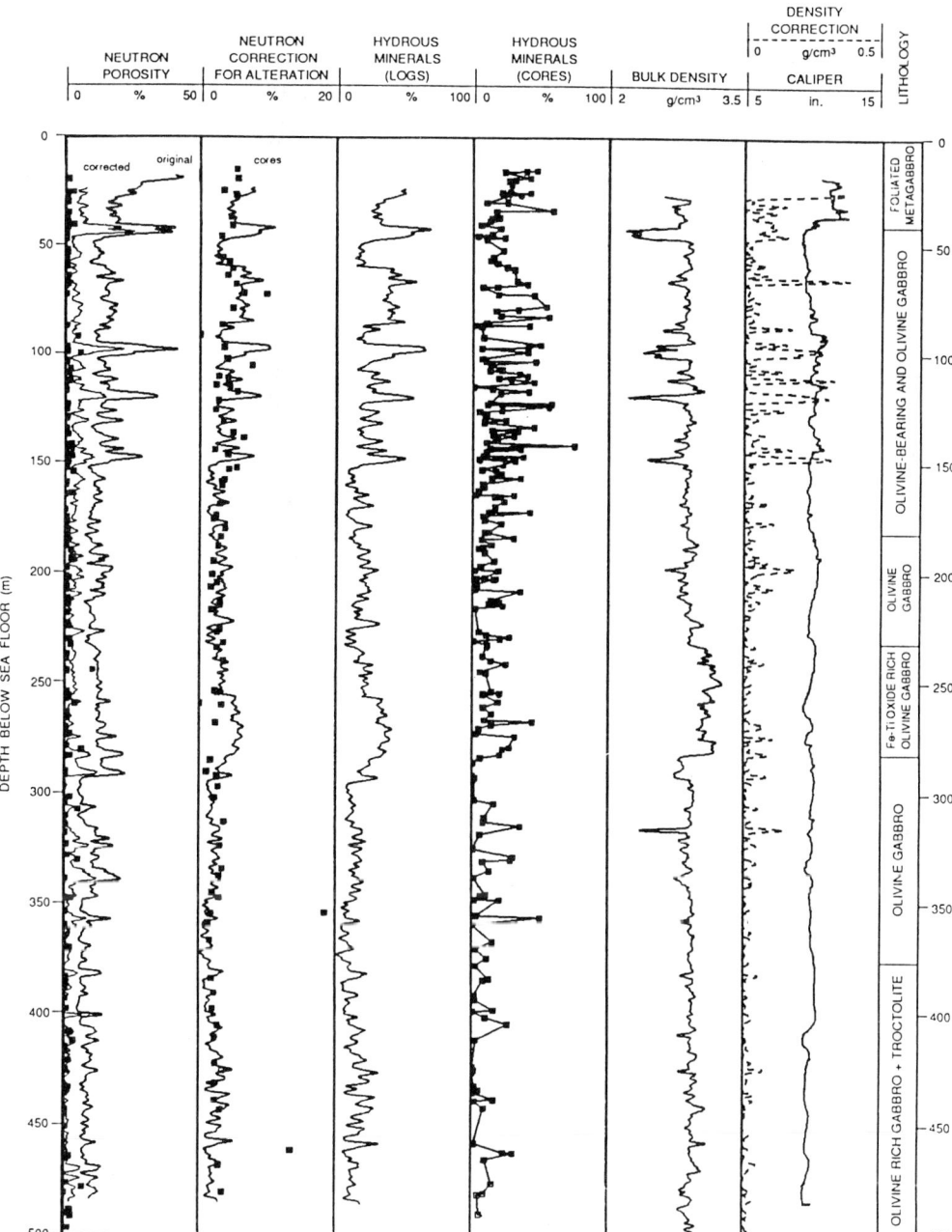

Fig. 3. Porosity and alteration profiles at Hole 735B (tracks 1 through 4). Also displayed are the bulk density (track 5), and caliper and density correction (track 6). Locally, the alteration can be as high as 60%, which corresponds to a neutron porosity correction of about 10%. Differences between log and core porosities are attributed to the presence of open fractures. Alteration and porosity decrease downhole, and the olivine-rich gabbros and troctolite layers in the lower part of the hole exhibit porosities of less than 2%.

A continuous alteration estimate was computed by re-scaling the mean hydrogen yield (Fig. 3, track 3) to match the hydrogen weight fraction measurements of core samples (Fig. 3, track 2, dots). The re-scaled curve (Fig. 3, track 2, line) roughly represents the response of the neutron log to hydrous alteration minerals. This continuous curve then can be used to correct the neutron log for the porosity effect of the bound hydrogen. The corrected porosity, ϕ_c, is computed simply by

$$\phi_c = \phi_1 - \phi_a$$

where ϕ_1 and ϕ_a are the original neutron log and the re-scaled hydrogen yield log, respectively. ϕ_c, ϕ_1 and core-derived porosities are plotted in Fig. 3 (track 1). Clearly, the minimum ϕ_c values correspond reasonably well with the core-derived values; peaks may be attributed to fracture porosity unsampled by coring.

An estimate of the volume of alteration V_{alt} can be computed from ϕ_a by

$$V_{alt} = \phi_a / \phi_m$$

where $\phi_m = 16\%$ is the neutron response to actinolite-tremolite amphiboles (Ellis *et al.* 1988). In Fig. 3, the volume of alteration estimated from the logs (track 3) compares on average with the results from shipboard core-sample modal analyses (track 4). The correlation of ϕ_c and V_{alt} with other data is discussed below.

Acoustic borehole televiewer

The borehole televiewer is an ultrasonic device that is used to image the wall of a borehole, producing oriented information about the distribution and orientation of natural and induced features intersecting the hole. The mechanical operation of the analog borehole televiewer is fully described by Zemanek *et al.* (1970). The present authors use digital image processing of the borehole televiewer data to determine the location and orientation of fractures intersecting the wellbore with superior resolution to other logging tools (Barton 1988). Ideally, televiewer images can resolve open fractures as fine as 5 mm in a water-filled borehole. The televiewer can also provide an accurate measurement of the diameter, surface roughness and ellipticity of the borehole. Depending on the logging speed, a minimum horizontal dip of about $20°$ can be resolved from an unwrapped image. Features that do not affect either the roughness, reflecti-

vity or radius of the borehole, cannot be resolved at all.

Over the 500 m interval logged in Hole 735B, geometrical analysis of televiewer amplitude images identified 70 fractures having measurable strikes, dips and apparent apertures (Table 1). Figure 4 shows the dip vector of each fracture plotted on an equal-area, lower-hemisphere stereonet. The contours illustrate the number of fractures in overlapping circles of radii equal to one fifth of the radius of the stereonet (after Seeberger & Zoback 1982). The centroid of the entire fracture distribution indicates a NNW strike and a WSW $75°$ dip. Analysis of these data over smaller depth intervals shows no distinct trends in orientation with depth.

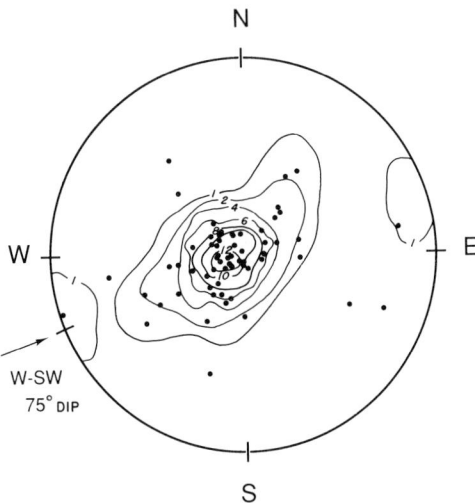

Fig. 4. Lower hemisphere plot of fracture dips computed from televiewer image analysis. Contours indicate the frequency of occurrence (number) of fractures in overlapping circles having radii of about 1/5 the radius of the stereonet. The centroid of the distribution dips approximately $75°$ WSW. Fracturing is likely to have resulted from uplift of the walls of the Atlantis II Fracture Zone, not horizontal plate-tectonic motion.

Tectonic models of the Atlantis II Fracture Zone indicate that significant uplift of the eastern transform wall, where Hole 735B was sited, occurred after a readjustment of north–south tectonic plate motion and induced extension across the transform valley (Dick *et al.* 1991). Wellbore breakouts, which could corroborate the orientation of present-day horizontal stresses in the vicinity of the wellbore, were not observed in travel time images from the televiewer data. But because the orientation of macroscopic

Table 1. *Tabulated depth, azimuth, and plunge of fracture dip vectors measured from borehole televiewer images in Hole 735B*

Depth (mbsf)	Azimuth (degrees)	Dip (degrees)	Depth (mbsf)	Azimuth (degrees)	Dip (degrees)
28.60	235.7	38.2	195.30	330.0	80.8
30.15	111.4	21.5	203.06	244.3	75.9
31.90	115.7	38.2	203.78	227.1	82.9
35.25	351.4	81.5	204.22	38.6	77.6
35.65	162.9	83.7	263.51	240.0	72.0
39.81	287.1	78.3	280.32	342.9	88.5
39.57	94.3	65.8	239.78	180.0	63.9
40.38	34.3	48.1	277.46	312.9	76.1
40.66	34.3	48.1	280.55	312.8	74.2
41.60	240.0	48.1	281.40	287.1	75.7
41.88	240.0	48.1	294.05	300.0	74.5
41.55	252.9	0.0	303.34	252.9	67.3
42.59	252.9	0.0	313.11	257.1	83.1
42.88	252.9	0.0	315.26	180.0	85.2
43.77	261.4	61.7	317.21	308.6	76.4
43.63	265.7	79.1	317.74	231.4	80.2
45.23	77.1	65.8	327.32	278.6	68.1
45.82	261.4	29.1	347.92	304.3	84.6
47.21	197.1	36.6	351.95	300.0	78.3
55.50	214.3	86.8	380.62	72.9	75.1
59.16	218.6	68.8	404.41	278.6	68.2
60.56	330.0	81.2	408.32	278.6	68.2
60.72	107.1	82.0	418.69	252.9	79.7
70.57	261.4	57.8	442.27	90.0	81.3
86.20	201.4	68.3	439.13	175.7	86.7
87.92	240.0	57.6	458.04	261.4	84.8
92.99	197.1	70.7			
98.07	248.6	43.6			
108.40	38.6	65.6			
116.16	64.3	79.7			
117.93	235.7	82.8			
122.40	184.3	81.0			
123.93	201.4	72.4			
128.98	42.9	66.8			
129.24	312.9	52.3			
136.41	210.0	70.7			
141.70	321.4	38.2			
145.77	222.9	73.7			
166.33	30.0	52.7			
174.62	42.9	70.7			
183.98	210.0	57.9			
186.16	81.4	17.7			
189.67	227.1	70.3			
192.15	261.4	78.6			

fractures provides information about the stress orientation at the time of their formation, it seems likely that the transform side of this uplifted ridge is in extension, sub-parallel to the transform, causing west-dipping normal faults, rather than in north–south compression that would be expected as a direct result of horizontal plate motion.

The mechanical and hydrological characteristics of a fracture are controlled primarily by the degree of contact of its walls and the size of the resulting permeable pathways (Brown & Scholz 1985). The width of fractures measured by the borehole televiewer as they intersect the wellbore, their 'apparent aperture', may be used as an estimate of these characteristics (Barton 1988; Barton & Moos 1988). The apparent aperture of a fracture in a televiewer image differs from the true fracture width as a result of fracture dip and borehole geometry effects, wall destruction and

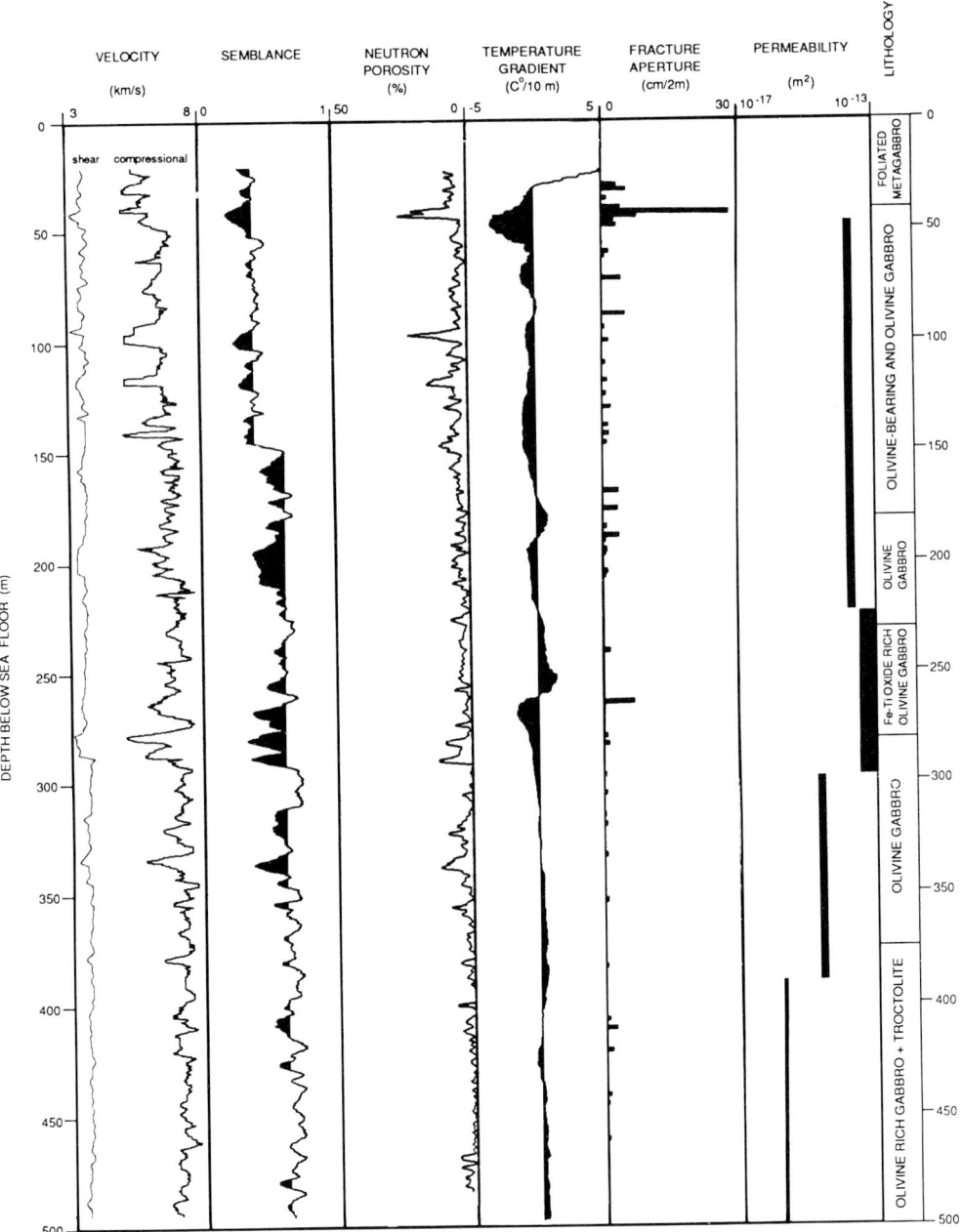

Fig. 5. Geophysical logs in Hole 735B: V_p and V_s, (track 1), mean semblance (track 2), corrected neutron porosity (track 3), temperature gradient (track 4), 2 m cumulative fracture aperture (track 5), and bulk permeability (track 6). Mean semblance (shaded below the mean) decreases systematically at 150 mbsf due to a malfunction of the pre-set receiver gains. Uncertainty in permeability estimates is represented by the line width. Supported by coincident low-velocity and low semblance, high porosity, and negative temperature-gradient perturbations, large fractures identified in the BHTV log at 40 and 264 mbsf probably contribute most to the measured permeability.

fluid injection during drilling, and dispersion of the acoustic televiewer beam. Even after correction for geometric effects, the apparent aperture is only a rough estimate of true fracture width and probably does not indicate transmissivity through rock away from the borehole walls. The apparent aperture of fractures in Hole 735B was measured from unwrapped amplitude images which could resolve a fracture within a 2 cm wide error band. A log of the cumulative fracture aperture, the running sum of individual apertures over 2 m intervals, is displayed in Fig. 5 (track 5). Although the mean cumulative aperture in the hole is less thn 2 cm, its greatest value (28 cm) is reached near 40 mbsf due to a large concentration of small fractures. Elsewhere, large-aperture (2–8 cm) intervals generally result from single fractures, and as a first approximation, the fracture density follows a trend similar to the cumulative aperture log.

As seen in other studies of fracturing in crystalline rock wells, there is an overall trend in decreasing fracture aperture and fracture density with depth (Seeberger & Zoback 1982; Haimson & Doe 1983). This is expected for sub-horizontal fractures closing gradually with increasing overburden stress. In this shallow hole, however, many fractures are high-angle and remain open with depth, an expected result from tectonic block uplift. In addition, hydrothermal alteration through open conduits may fill fractures with mineral deposits as fluid circulation continues. The correlation of the fracture distribution with the other logs in this hole is discussed below.

Multichannel sonic log

The multichannel sonic (MCS) logging tool used in this study has a monopole source separated from an array of hydrophone receivers suspended vertically in the borehole. Acoustic energy from the source travels as a compressional pulse in the borehole fluid and converts at the borehole wall into a series of body and guided wave modes. The velocities and amplitudes of compressional and shear body waves are primarily a function of the formation properties. The guided modes are controlled by the properties of both the formation and the fluid-filled borehole (e.g. Chen & Toksoz 1981). Shear waves are most influenced by high-amplitude guided modes at nearby frequenies and direct V_s measurements are usually contaminated (e.g. Paillet & White 1982). These V_p and V_s estimates, however, can be used to characterize many formations by their elastic properties (Goldberg & Gant 1988).

The MCS tool used in Hole 735B has a single source located 2 m above a 12-receiver array. The array spans 1.65 m, and both the source and the array are centred in the borehole by means of bowspring centralizers. Full waveforms are transmitted from each of the receivers through a standard logging cable and are digitally recorded uphole. In this hole, the source was fired at 30 cm intervals producing nearly 20 000 waveforms in less than 3 hours over the 500 m depth interval.

The full-waveform acoustic data were analysed to yield compressional and shear velocities across the receiver spread using a modified semblance calculation (Kimball & Marzetta 1984). Semblance was applied here as a measure of the coherence between channels to calculate the compressional and shear group velocities across the receiver array and as a normalized measure of the amplitudes of these waves. At Site 735, the MCS waveforms required preprocessing to remove unwanted noise and the guided wave modes from the signal. Time domain filters were used to remove noise outside the principal 2.5 to 17.5 kHz frequency band of the signal. An electronic malfunction precluded computer control for data acquisition in the upper 140 m, and waveforms were not recorded in their usual increasing-offset sequence with pre-set gains. A pre-semblance process to reorder the sequence of recordings by threshold picking of the high-amplitude fluid arrival was designed to return the data to sequential order. The data were then processed normally for velocities by semblance correlation of waveforms having increasing moveout.

The compressional (V_p) and shear (V_s) velocities and their mean semblance in Hole 735B are displayed in Fig. 5 (tracks 1 and 2, respectively). V_p shows considerable variation throughout the hole and averages 6.7 $\pm$ 0.3 km/s. Overall, V_p from the MCS log agrees well with V_p from discrete laboratory measurements made at ultrasonic frequencies and from vertical seismic profiling results in this hole (Kirby et al. 1991). V_s is generally less variable than V_p, due to the strong influence of guided modes in the MCS log, but is representative of the average V_s in the hole (3.5 $\pm$ 0.15 km/s). V_p/V_s and Poisson's ratio are thus reliable only for large interval averages and are not shown.

Several low V_p zones occur throughout the hole (e.g., 40, 90, 120, 140 mbsf, etc.) and nearly all correspond with semblance anomalies (shaded below the mean) and elevated porosities. The mean semblance is broadly lower in the upper 140 mbsf, where the pre-set gain of waveform was disabled. The correlation of the V_p and semblance logs with porosity, alteration and fracturing will be discussed below.

Borehole temperature profile

High-precision temperature logs (within 0.01°C) were run before and after the completion of the other logging experiments in Hole 735B (Robinson, Von Herzen, *et al.* 1989). A temperature gradient profile was then computed by spatial differentiation of the temperature log over 10 m depth intervals. In Fig. 5 (track 4), the temperature-gradient profile shows negative gradient anomalies from 30–50, 65–70, 90–160, 190–210, 260–270, and 430 mbsf. The repeat temperature profile indicates that the borehole cooled about 1.5°C overall during the 38-hour logging period, but temperature-gradient anomalies were observed at the same depths. An extrapolation of the temperature log to equilibrium was computed by Von Herzen & Scott (1991) and indicates that the temperature gradient anomalies are preserved at equilibrium.

In dense crystalline rocks, fluid flow is likely to occur through fractures, which shifts the temperature profile and yields an anomalously low or negative temperature gradient. Here, negative gradient anomalies occur over intervals containing fractures, although not all of the observed fractures correspond to negative gradient anomalies. Negative gradient anomalies cannot be caused by variations in thermal conductivity with depth, since all earth materials have conductivities greater than zero. The observed negative-gradient anomalies in this hole are probably due to the presence of transmissive fractures contributing cool fluid back into the borehole which was displaced by warm bottom water during drilling (Von Herzen & Scott 1991). It is likely that the same transmissive fractures influencing the temperature profile are hydraulically active during packer tests.

Hydraulic testing

An inflatable single-element drill string packer was used to measure the average in situ transmissivities and to calculate the average permeabilities of four different intervals isolated in Hole 735B: 49–500, 223–500, 299–500, and 389–500 mbsf. These intervals correspond to the zones between the bottom of the hole and the depths at which the packer was inflated. The packer inflation depths were chosen to isolate the major lithologic units, which also showed different character in the downhole logs: the inflation at 389 mbsf isolated lithologic Unit VI, the inflation at 299 mbsf isolated Units V and VI, the inflation at 223 mbsf isolated Units III to VI, and the inflation at 49 mbsf isolated Units II to VI.

As described in detail by Becker (1991), these measurements were performed either by monitoring the decay of nearly instantaneous pressure pulses applied to the isolated zone (slug tests) or by monitoring the gradual pressure increases as seawater was pumped at a constant rate into the isolated zone for up to 30 min (injection tests). These are standard techniques which yield the bulk transmissivities and permeabilities of the respective isolated zones with accuracies of about ±30% (for slug tests see Bredehoeft & Papalopulos 1980; for injection tests see Matthews & Russell 1967; for processing details see Becker 1991).

Results show that permeability decreases by over two orders of magnitude from its maximum value in the deepest 200 m of Hole 735B. The permeabilities of the smaller overlapping intervals (e.g., 49–223, 223–299, and 299–389 mbsf) could be estimated by taking the differences in transmissivities measured in the isolated zones. These interval permeabilities are plotted in Fig. 5 (track 6), and both the measured and calculated interval permeabilities are summarized in Table 2. The relationship of these measurements with other logging data is interpreted below to determine the influence of alteration and fracturing on the in situ permeability.

Table 2. *Measured and calculated hydraulic transmissivity and permeability over various depth intervals in Hole 735B (after Becker 1991)*

	Interval (mbsf)	Transmissivity $(10^{-6}\ m^2/s)$	Permeability $(10^{-15}\ m^2)$
Measured:	49–500	54	24
	223–500	32	24
	299–500	2.0	2.1
	389–500	0.1	0.2
Calculated:	49–223	22	24
	223–299	30	81
	299–389	1.9	4.4

Discussion

Fracturing, alteration, and permeability

The variation of in situ permeability with depth in Hole 735B suggests that certain intervals are hydraulically transmissive and others are tight. To differentiate these intervals, subdivisions of Hole 735B based on the average porosity,

Table 3. *Mean and* (standard deviation) *of log and core measurements over various depth intervals and lithologic units in Hole 735B*

Unit	Interval (mbsf)	Porosity (%) log	Porosity (%) core	Alteration volume (%) log	Alteration volume (%) core
I/II	23–150	5.6 *(4.7)*	1.7 *(1.2)*	32.2 *(13.3)*	23.4 *(15.4)*
II/III	150–230	3.4 *(2.0)*	1.5 *(0.7)*	16.6 *(6.5)*	13.5 *(9.7)*
IV	230–280	2.6 *(1.7)*	2.4 *(1.8)*	26.9 *(9.0)*	15.2 *(10.0)*
V	280–375	3.4 *(3.2)*	1.5 *(1.3)*	13.6 *(6.9)*	11.4 *(12.3)*
VI	375–500	1.7 *(1.8)*	1.5 *(1.0)*	14.3 *(6.6)*	8.2 *(8.2)*

volume of alteration, and fracture aperture logs are compared with the measured interval permeability (Table 3 and Fig. 5). We find that subdivisions based on the log responses correspond nearly, but not exactly, with the lithologic units (Units I through VI) described in Robinson, Von Herzen, *et al.* (1989). Differences can be largely attributed to fracturing which is not accurately represented in the recovered core. We also find that the weak correlation of in situ permeability with porosity and its strong correlation with temperature and televiewer logs gives evidence that fracturing controls the measured permeability in Hole 735B.

The upper 150 m of Hole 735B, most of Units I and II, are characterized by 32.2% average alteration volume and an average porosity of 5.6%. Core values in this interval average 23.4% alteration by volume and 1.7% porosity. In this most highly altered interval, the presence of open fractures, that are not measured in laboratory core samples, can account for the core vs. log discrepancy. In addition, the caliper and the density correction logs (Fig. 3, track 6) tend to show spikes in intervals where hole diameter varies sharply, likely due to fracturing. In the interval from 150 to 230 mbsf, made up of Units II and III, the volume of hydrous alteration minerals decreases dramatically to averages of about 16.6% and 13.5% from logs and cores, respectively. While the core-porosity is quite similar to that determined for the interval above, the average log-porosity is 3.4%, suggesting that one or more open fractures affect the log response. From the seafloor to 230 mbsf, the corrected porosity is anomalously high, as expected, where isolated, open fractures are observed in the televiewer log (Fig. 5, tracks 4 and 5). The permeability is high in this interval, $24 \times 10^{-15} \, m^2$, indicating that one or more fractures are conductive even though alteration minerals are abundant (Fig. 5, track 6). However, if the hydrological structure of the site does not change dramatically, it is likely that

increasing hydrothermal mineralization through open fractures will continue to reduce the high permeability.

Borehole Azimuth (Deg)

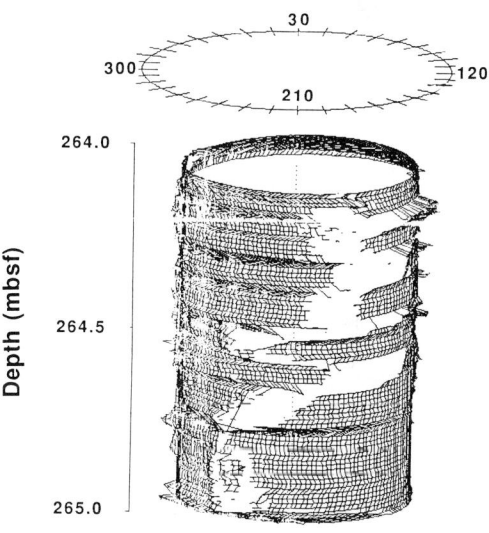

Fig. 6. Cyclindrical wireframe projection of a high-angle (72° west-southwest), wide-aperture (8 cm) fracture near 264 mbsf. Horizontal offsets in the image fracture of this fracture result from misorientation of the north pulse by strong remanent magnetization in this Fe–Ti oxide-rich zone.

The interval from 230 to 280 mbsf, spanning the Fe–Ti oxide gabbro identified as Unit IV, is differentiated from the rocks above by a increase in the average alteration volume measured on cores (26.9%) and from the logs (15.2%), but displays the highest interval permeability, $81 \times 10^{-15} \, m^2$, determined from the packer tests (Fig. 5, tracks 3 and 6). The televiewer log in this zone indicates that it contains a few small fractures near 290 mbsf and a wide-aperture, high-angle fracture at 264 mbsf shown by a cylindrical wireframe projection in Fig. 6. The dip of this fracture is 72° WSW and the apparent aperture is about 8 cm. Due to the associated thermal anomaly at this depth, it is likely that this fracture is mostly open and contributes significantly to the measured hydraulic permeability in the interval. Note that the abundance of dense, magnetic Fe–Ti oxides below 260 mbsf in Unit IV affects the north orientation pulse of the televiewer and results in azimuthal offsets of up to 30° in the image of this fracture.

Below 280 mbsf, the olivine gabbro in Unit V is distinguished by a small increase in average log-porosity to 3.4% and a large decrease in permeability to 4.4×10^{-15} m^2 (see Tables 2 and 3). At 375 mbsf, the Unit VI olivine gabbros and troctolites have lower average log-porosity (1.7%) and significantly lower permeability (0.2×10^{-15} m^2). The volume of alteration in these two units remains nearly constant and is the lowest in the hole. Core and log porosities also agree well in both Units V and VI, indicating a minimal effect of unsampled fractures, and only a few high-porosity anomalies occur at depths where small-aperture fractures were observed in the televiewer log (Fig. 5, tracks 3 and 5).

Downhole logs as permeability indicators

Indicators of fracturing from downhole experiments (e.g. acoustic, porosity, temperature and televiewer logs) may be used to distinguish fractures that are transmissive from those that are not. For example, the causal relationship between fluid flow through fractures and negative temperature-gradient anomalies probably indicates which transmissive fractures contribute cool fluid back into the borehole. As shown in Fig. 5 (tracks 4, 5, and 6), negative gradient anomalies occur in Units I through IV at depths where the porosity is high, fractures are observed, and the measured permeability is high. The magnitude of the temperature-gradient anomalies also shows a general correlation to fracture aperture. Because the temperature gradient in a borehole is assumed to be affected by fluid flow through even a small open fracture, the gradient log effectively indicates depths at which fractures are transmissive, but smooths their response over a wider interval. It is likely that depths where the negative gradient anomalies are greatest, specifically near 40 and 264 mbsf, are near to multiple or large open fractures, which contribute most to fluid flow in this hole.

In a similar manner, acoustic properties, particularly guided-wave amplitudes, have been related to hydraulic transmissivity in previous studies of fractured rocks (e.g. Paillet 1980; Hardin et al. 1987; Barton & Moos 1988). Such correlations result from measurable decreases in signal amplitude resulting from elastic and inelastic dissipation across impedance boundaries and porous fractures. Reasonable mechanisms controlling this effect may result from a combination of changes in Poisson's ratio, scattering and interference, and dissipation by formation and borehole fluid movements. Observations suggest that velocity and semblance also decrease as a result of these mechanisms (e.g., Moos & Zoback 1983; Goldberg & Gant 1988). Ideally, with constant lithology and borehole size, decreases in velocity and semblance could be attributed largely to the effects of fractures and correlated to their transmissivity.

Because acoustic logs can achieve greater depth resolution than hydraulic or temperature measurements, they are useful for detailed correlation of fracture transmissivity to the measured permeability. In Hole 735B, fracture occurrences generally correspond to zones of high porosity, low V_p and low semblance (Fig. 5; tracks 1, 2, 3 and 5). In most of these zones, V_p anomalies correspond with the fracture location, but without a clear relationship to fracture aperture. Semblance anomalies, however, occur at depths where large fracture apertures and negative temperature gradient anomalies are spatially coincident, notably at 40, 100–150, 200 and 264 mbsf, as well as over several other depth intervals. These coincident anomalies may be the best indicators as to which fractures intersecting the borehole are transmissive, because they both reflect the result of fluid movement through fractures by independent physical measurements.

Conclusions

From our interpretation of the geophysical logs recorded in Hole 735B, we were able to conclude the following about the in situ properties in oceanic layer 3 gabbros.

(1) The greatest abundance of hydroxyl-bearing alteration minerals are observed in the upper section of the hole. The highest permeabilities also occur in this interval.

(2) The 70 fractures identified from the televiewer images, which strike predominantly NNW and dip steeply WSW, are not likely the result of north–south, horizontal stresses from plate tectonic motion on the uplifted ridge of the Atlantis II Fracture Zone. No wellbore break-

outs were observed in the televiewer data to confirm the horizontal stress orientation in the vicinity of the Hole 735B.

(3) Anomalies in the temperature gradient and mean semblance, and to a lesser extent velocity and porosity, correspond to the cumulative fracture aperture log computed from televiewer images recorded in the hole.

(4) Fracturing near 40, 100–150, 200 and 264 mbsf corresponds to large and coincident negative temperature gradient and semblance

anomalies that indicate active fluid flow and is consistent with high packer-test transmissivities measured in these intervals.

D. Goldberg and C. Broglia were supported during this research by the National Science Foundation and Joint Oceanographic Institutions, Inc. under contract JOI 66-87D. K. Becker was supported by the National Science Foundation under contracts OCE 85-13537 and OCE 88-00077. The authors thank C. Wilkinson and C. Barton for their assistance with the televiewer image processing.

References

BARTON, C. A. 1988. *Development of* in-situ *stress measurement techniques for deep drillholes.* PhD Thesis, Stanford University.

—— & Moos, D. 1988. Analysis of macroscopic fractures in the Cajon Pass scientific drillhole: over the interval 1829–2115 metres. *Geophysical Research Letters*, **15**, 1013–1016.

BECKER, K. 1991. In-situ bulk permeability of gabbros in Hole 735B, ODP Leg 118. In: *Proceedings of the Ocean Drilling Program, Scientific Results.* Ocean Drilling Program, College Station, TX, **118**.

BREDEHOEFT, J. D. & PAPADOPULOS, I. S. 1980. A method for determining the hydraulic properties of tight formations. *Water Resources Research*, 16: 233–238.

BROGLIA, C. & ELLIS, D. 1990. The effect of alteration, formation absorption, and standoff on the response of the thermal neutron porosity log in gabbros and basalts: examples from DSDP-ODP Sites. *Journal of Geophysical Research*, **95**(B6): 9171–9188.

BROWN, S. R. & SCHOLZ, C. H. 1985. Broad bandwidth study of the topography of natural rock surfaces. *Journal of Geophysical Research*, 90: 12,575–12,582.

CHENG, C. H. & TOKSÖZ, M. N. 1981. Elastic wave propagation in a fluid-filled borehole and synthetic acoustic logs. *Geophysics*, **46**, 1042–1053.

DICK, H. J. B., SCHOUTEN, H., MEYER, P. S., GALLO, D. G., BERGH, H., TYCE, R., PATRIAT, P., JOHN-SON, K. T. M., SNOW, J. & FISHER, A. 1991. Tectonic evolution of the Atlantis II Fracture Zone. In: *Proceedings of the Ocean Drilling Program, Scientific Results.* Ocean Drilling Program, College Station, TX, **118**.

ELLIS, D. V., FLAUM, C., McKEON, D., SCOTT, H., SERRA, O. & SIMMONS, G. 1988. Mineral logging parameters: nuclear and acoustic. *The Technical Review*, Schlumberger Educational Serevices, Houston, TX, 38–52.

GOLDBERG, D. & GANT, W. T. 1988. Shear-wave processing of sonic log waveforms in a limestone reservoir. *Geophysics*, **53**, 668–676.

HARDIN, E. L., CHENG, C. H., PAILLET, F. L. & MENDELSON, J. D. 1987. Fracture characterization by means of attenuation and generation of tube waves in fractured crystalline rock at Mirror Lake, New Hampshire. *Journal of Geophysical Research*, **92**, 7989–8006.

HAIMSON, B. C. & DOE, T. W. 1983. State of stress, permeability and fractures in Precambrian granite of northern Illinois. *Journal of Geophysical Research*, **88**, 7355–7371.

HERTZOG, R., COLSON, L., SEEMAN, B., O'BRIEN, M., SCOTT, H., McKEON, D., WRAIGHT, P., GRAU, J., ELLIS, D., SCHWEITZER, J. & HERRON, M., 1987. Geochemical logging with spectrometry tools. SPE 16792.

KIMBALL, C. V. & MARZETTA, T. L. 1984. Semblance processing of borehole acoustic array data. *Geophysics*, **39**, 587–606.

KIRBY, S. H., ITURRINO, G., GOLDBERG, D. & SWIFT, S. 1991. Intercomparison between compressive-wave velocities measured by ultrasonic methods on layer 3 gabbroic drillcore with multichannel sonic and VSP logs at ODP Site 735: Evidence for unsaturated *in-situ* porosity below 60 m depth. *EOS*, **44**, 178.

MATTHEWS, C. S. & RUSSELL, D. G. 1967. Pressure build-up and flow tests in wells. *American Institute Min. Met. Pet. Eng.*, SPE monograph, 1.

McKEON, D. C. & SCOTT, H. D. 1988. SNUPAR-A nuclear parameter code for nuclear geophysics applications. *Nuclear Geophysics.*, **2**(4), 215–230.

Moos, D. & ZOBACK, M. D. 1983. *In-situ* studies of velocity in fractured crystalline rocks. *Journal of Geophysical Research*, **88**, 2345–2358.

PAILLET, F. L. 1980. Acoustic propagation in the vicinity of fractures which intersect a fluid-filled borehole. *Transactions of SPWLA 21st Annual Logging Symposium*, paper D.

—— & WHITE, J. E. 1982. Acoustic modes of propagation in the borehole and their relationship to rock properties. *Geophysics*, **47**, 1215–1228.

POUPON, A. & GAYMARD, R. 1970. The evaluation of clay content from logs. *Transactions of SPWLA 11th Annual Logging Symposium*, Paper G.

ROBINSON, P. T., VON HERZEN, R. P. *et al.* 1989. Site 735. In: ROBINSON, P. T., VON HERZEN, R. P., *Proceedings of the Ocean Drilling Program Initial Reports.* Ocean Drilling Program, College Station, TX, **118**, 89–816.

SEEBERGER, D. A. & ZOBACK, M. D. 1982. The distribution of natural fractures and joints at depth in crystalline rock. *Journal of Geophysical Research*, **87**, 5517–5534.

VON HERZEN, R. P. & SCOTT, J. 1991. Thermal modeling for Hole 735B. *In: Proceedings of the Ocean Drilling Program Scientific Results*, Ocean Drilling Program, College Station, TX, **118**, in press.

ZEMANEK, J., GLENN, E. E., NORTON, L. J. & CALDWELL, R. L. 1970. Formation evaluation by inspection with the borehole televiewer. *Geophysics*, **35**, 254–269.

Attempts to detect open fractures and non-sealing faults with dipmeter logs

J. S. BELL,[1] G. CAILLET,[2] & J. ADAMS[3]

[1] Geological Survey of Canada, 3303 33rd St. N.W., Calgary, Alberta T2L 2A7, Canada
[2] Elf-Aquitaine, CSTJF Avenue Larribau, 64018 Pau Cedex, France
[3] Geological Survey of Canada, 1 Observatory Crescent, Ottawa, Ontario K1A 0Y3, Canada

Abstract. If enough stress-generated breakouts have been identified on four-arm dipmeter logs to map regional horizontal stress trajectories, it may be possible to use anomalous orientations to identify open fractures and non-sealing faults. These act as free surfaces and, if they were oriented obliquely to regional stresses, deflect them locally. This type of analysis is simple to apply and appears to offer a novel way to obtain key information required for optimizing the production of oil and gas. In offshore eastern Canada, the regional stress regime is well established and local deflections appear to be caused by near-by open fractures and non-sealing faults. In the Aquitaine Basin in France the regional stress signature is not homogeneous. Though some anomalous orientations appear to be caused by open fractures and faults, it is clear that many are not. Hence the method may not be applicable everywhere. Possible reasons include differences in the $\sigma_H : \sigma_h$ ratios of the two areas and the geotectonic settings.

Today it is recognized that a large number of productive hydrocarbon reservoirs owe a significant portion of their permeability to open fractures. To maximize recovery from such reservoirs we need to know where the open fractures are most abundant and how they are oriented. Much useful data can be obtained from oriented cores and from borehole wall scanning tools. Nevertheless, there is still a need for techniques that can sense directly the presence of open fractures in the subsurface. Here we describe a simple approach that, in favourable situations, should provide good indications of open fractures at low cost, using dipmeter logs.

All buried rocks are in a state of stress and are subjected to compression. In most sedimentary basins, the compressive stress is not isotropic and the principal stresses differ in magnitude. One principal stress is usually approximately vertical (σ_V) and the other two (σ_H and σ_h) are more or less horizontal (McGarr & Gay 1978; Hoek & Brown 1980).

The technique described herein for recognizing open fractures and non-sealing faults depends first on knowing the regional stress orientations, and then on identifying areas where the stress trajectories are anomalous. The rationale for this approach is illustrated in Fig. 1. It is assumed that open fractures in a rock body constitute, in a geomechanical sense, tabular-shaped voids enclosed by a free surface. These free surfaces will deflect stress trajectories in the immediately

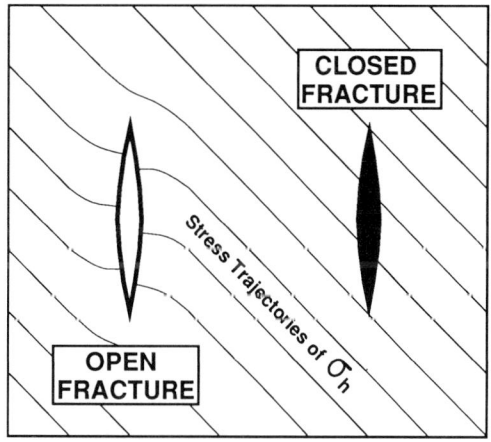

Fig. 1. Deflection of stress trajectories in the neighbourhood of open fractures that are not oriented perpendicular to the smallest principal stress. Vertical fractures viewed in the horizontal plane are illustrated, one open (white interior) and one closed (solid black). The closed fracture does not deflect stress trajectories.

surrounding area so that the smallest principal stress approaches the free surface at right angles. In the case of vertical and steeply dipping open fractures, σ_h will be deflected unless it is oriented exactly normal to them (Fig. 1). On the other hand, if the fractures are sealed, stress will be

From HURST, A., GRIFFITHS, C. M. & WORTHINGTON, P. F. (eds), 1992,
Geological Applications of Wireline Logs II. Geological Society Special Publication No. 65, pp. 211–220.

transmitted across them with no deflection of trajectories, unless the sealing minerals exhibit strong elastic contrasts with the country rock. What is required therefore is to map the regional stress trajectories and then to obtain enough local stress orientation measurements in the area of interest to indicate if directional anomalies are present.

The most widespread indicators of stress orientations in buried sediments are borehole breakouts (Bell & Gough 1983). Breakouts are intervals where the wall of a well has caved so as to produce arcuate extensions on opposite sides (Fig. 2). In vertical and near-vertical wells, the long axis of these extensions has been shown to correspond to σ_h (Bell & Gough 1979). Breakouts are relatively common phenomena in oil wells and their long axes are recorded by such tools as four-arm dipmeters, Formation Micro-Scanners and borehole televiewers. With appropriately located wells, breakouts can be used to map regional stress directions as well as local stress trajectories.

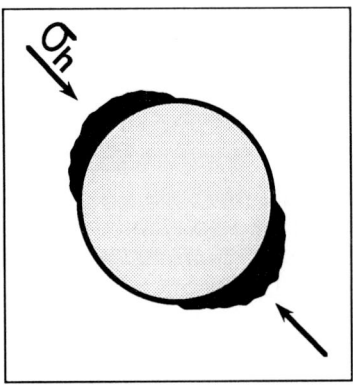

Fig. 2. Vertical view of an elliptically caved breakout zone in a well. The smaller horizontal principal stress, σ_h is parallel to the long axis of the breakout.

Breakouts respond to local changes in stress orientation, an effect likely to be produced by the presence of open fractures, and they will therefore diagnose a condition rather than identify a specific feature. However, as Fig. 1 indicates, if the regional stress directions are well known, anomalous breakout orientations may give an idea of the orientation of open fractures. However, breakout data alone will not indicate fracture length, height or width, or give any quantitative measure of permeability. This technique differs from the direct approaches, such as borehole wall scanning, which use imagery to locate individual open fractures.

Methodology

At present, this method of locating open fractures is very much on trial. For it to be successful, we need to know the regional stress trajectories in an area and to be able to recognize local anomalies.

To map σ_H and σ_h trajectories across a basin, or within a significant part of one, extensive measurements of breakout orientations from a large number of well spaced holes are needed. Depending on the size of the area, at least 25 to 50 wells should be studied, they should be near vertical holes, and each should have breakout data distributed over several thousand metres. Furthermore, the breakout orientations should give a coherent and homogeneous picture on a regional scale. In other words, there should be an inter- and intra-well consistency in the σ_H and σ_h orientations across the basin. These conditions have been met in several North American basins (e.g. Bell & Babcock 1986; Plumb & Cox 1986; Zoback et al. 1988).

Identifying local anomalies that might be diagnostic of open fractures requires densely distributed breakout data. Analyses from closely spaced wells are desirable and it is essential to have enough reflection seismic coverage to identify and map the main faults. Ideally 3-D seismic coverage should be used. Anomalous breakout orientations should be related to established fault planes to assess whether they, or similarly oriented fractures, are likely to be acting as free surfaces and causing stress deflections. Special attention should be paid to faults that show evidence of recent activity. Such faults may well act as free surfaces themselves or, strictly speaking, as a series of free surfaces. They are unlikely to be single free surfaces, unless movement is imminent! More probably, recently active faults consist of a semi-connected network of similarly oriented open fractures, planally aligned within a tabulate zone, that form a disconnected array of free surfaces (Fig. 3). Wells particularly worth studying are those that cross suspected fractured intervals, or that cut faults with indications of recent movement.

While breakouts are ubiquitous in some sequences they are not found everywhere. Rocks most prone to breaking out in response to far field stresses seem to be those that are sufficiently brittle to transmit anisotropic stress and that are also weak enough to fail around borehole walls (Gough & Bell 1982; Zoback et al. 1985). More breakouts are found in shales and carbonates than in sandstones. The contrast between the magnitudes of σ_H and σ_h must also be large enough so that local stress amplification close to the borehole wall exceeds the cohesive strength of the surrounding rocks (Zoback *et al.* 1985). If this is not the case, stress-induced breakouts will rarely occur, no matter how geomechanically suitable the rocks are. Moreover, where horizontal-stress-magnitude stress contrasts are low, processes other than amplification of far field stresses are likely to initiate asymmetric caving and produce oval well bores (e.g. Santarelli 1987). This last consideration is of particular concern because there is currently very little information on horizontal principal stress contrasts in sedimentary basins. Measurements of σ_h are relatively common because the information is needed to assess how induced fractures will propagate, but there are few direct measurements of σ_H in sediments, and none in deep wells. With the increasing industrial use of borehole breakouts, it is becoming important also to obtain measurements of the magnitude of σ_H. Tools need to be developed.

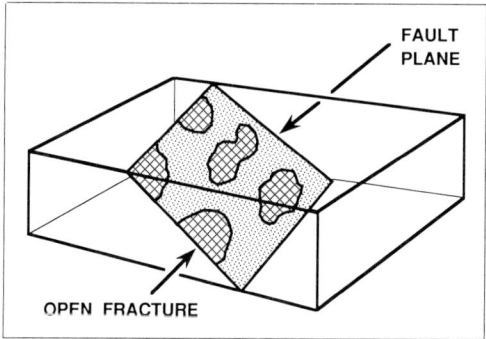

Fig. 3. Schematic representation of the geomechanical character of the plane of a non-sealing fault. It contains a series of aligned discontinuous open fractures.

Stress deflections around large planar openings and fracture networks have not been widely investigated. Their general form is known from finite element simulations such as those of Wu & Chang (1978) and Hoek & Brown (1980), but we do not yet have enough simulations of deflections near fractures to appreciate in detail how stress trajectories are affected by fracture orientation and stress contrast. A better understanding is needed to help characterize signals given by anomalous breakout orientations.

It is also essential that the structural style of the area under investigation be well known and that maps show where there are likely to be juxtapositions of rock bodies with markedly different elastic properties, such as shale abutting halite, since they can also give rise to stress deflections (Bell & Lloyd 1989).

Open fractures and non-sealing faults in eastern Canada

Thirty vertical and near-vertical wells drilled in the Sable Sub-basin on the Scotian Shelf, offshore eastern Canada, exhibit consistently oriented breakouts that demonstrate that σ_h trends NW–SE across the basin (Bell 1990). Standard deviations of measurements in wells with more than ten breakout intervals range from 5.5° to 20.0°. This regional stress regime (Fig. 4) is documented by breakouts that range in depth from 351 to 5134 m, but are largely concentrated between 1000 and 4000 m. Anomalously oriented breakouts occur locally in several wells.

At Alma F-67 (Bell 1989), anomalously oriented breakouts occur between 3127 and 3618 m (Fig. 5) and document clockwise and anticlockwise swings of σ_h within this interval. ENE–WSW trending down-to-south listric faults cut the well between approximately 3200

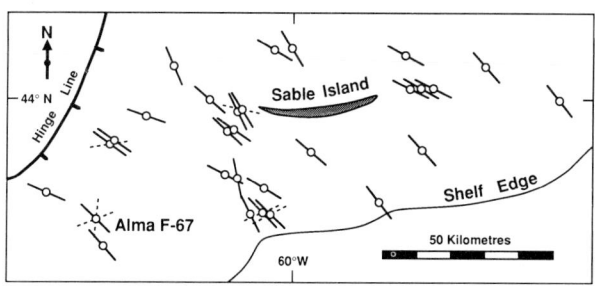

Fig. 4. Mean breakout azimuths from 29 wells in the Sable Sub-basin of the Scotian Shelf, offshore eastern Canada. Solid lines denote major populations, dashed lines show mean directions of minor populations. Note the well developed NW–SE directional signature of σ_h.

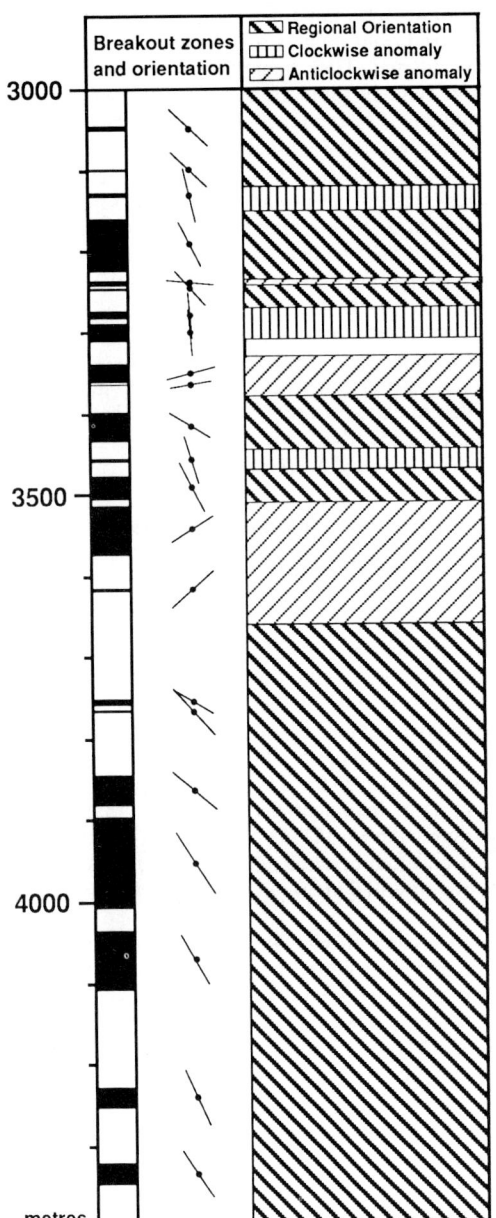

Fig. 5. Breakouts encountered between depths of 3000 and 4400 metres in the Alma F-67 well. The individual breakout orientations are depicted as compass azimuths. In the column on the right their directions are classified as regional, as clockwise anomalies, and as anticlockwise anomalies.

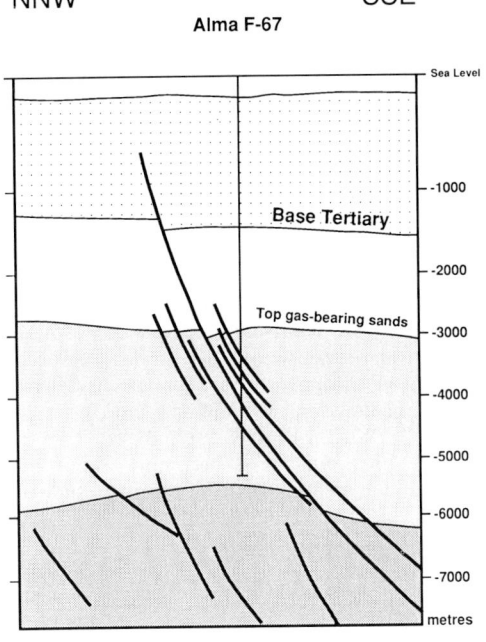

Fig. 6. NNW–SSE cross section across the Alma F-67 well, interpreted from a seismic line. Note the listric faults cutting the well between depths of 3300 and 4600 metres.

depth to within 600 m of the seafloor, and may extend higher, suggesting that it is a young, possibly still active, structure.

If portions of some of the above ENE–WSW trending faults contained open fractures filled with fluids, could the resulting free surfaces produce stress deflections that correspond to the observed anomalous breakout directions? Assuming the faults consist of a planar array of disconnected fluid-filled fractures, the regional stress trajectories documented in the Alma area are likely to be distorted locally around them as shown in Fig. 7. If these open fractures dipped to the SSE, as the mapped faults do, a well crossing between them might exhibit a clockwise σ_h stress orientation anomaly. The change in orientation down the hole in such a case can be appreciated in two dimensions by considering the changes in stress trajectories betwen point B and A in Fig. 7. On the other hand, a well that passed close to an open fracture could show an anticlockwise σ_h stress orientation anomaly. This change in orientation can be depicted in two dimensions by the changes in stress trajectories between point D and C in Fig. 7. Thus, in this local environment, if only ESE–WSW open frac-

and 4350 m according to seismic reflection data (Fig. 6). Indications of faults and fractures are found on logs over the same interval. The main listric fault climbs from deeper than 8000 m

tures are present, we should expect to encounter two types of anomalies, clockwise anomalies when the well crosses an open fracture and anticlockwise anomalies when it passes beside or betweeen fractures (Fig. 8). Between 3000 m and 4000 m in Alma F-67, there are three clockwise and three anticlockwise anomalies (Fig. 5), suggesting that the well may be cutting through three near-vertical, ENE–WSW aligned, free surfaces, and that it may pass close to three others. In the local structural environment it is reasonable to suspect that the inferred free surfaces correspond to fluid-filled open fractures that may be spatially associated with non-sealing faults.

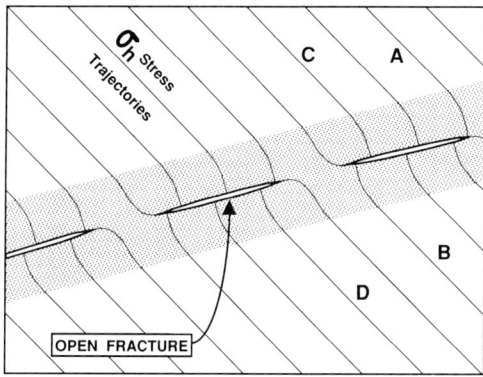

Fig. 7. Schematic representation of σ_h stress trajectories intersecting open fractures at Alma. The diagram is drawn in the horizontal plane. Open fractures are shown as discontinuous and aligned parallel to the major listric normal faults at Alma. Regional NW–SE trajectories are deflected as they approach the zone of open fractures and there is a distinct zone (speckled) where this phenomenon occurs. No sense of scale is intended.

The major listric normal fault offsetting the Alma structure appears to cross the Alma F-67 well between 3976 and 4011 metres depth, because this zone is badly caved (Bell 1989). According to the seismic interpretation, it is this fault that has experienced the greatest displacement, and that can be traced closest to the seafloor. Yet, there is no significant reorientation of breakouts associated with it (Fig. 5). We can only speculate why this is so. It is possible that, because of its displacement, which presumably occurred during repetitive movements, a 'smoothed out' fault zone has been developed locally that is now lined with 'sheared shale'. Sealed faults of this type occur in the Gulf coast, offshore of Louisiana (Smith 1978). If this sealing mechanism has occurred, parts of the fault

could now be geomechanically 'intact' and need not be acting as stress-deflecting free surfaces. The Alma F-67 well could have passed through such a zone. Nevertheless, despite a lack of breakout evidence favouring openness of this particular structure as was previously suggested (Bell 1989), there are good indications of free surfaces that are intersected higher in the well. As has been documented elsewhere (Bell 1989), these inferred free surfaces are believed to correspond to a system of open fractures through which gas has migrated to charge an overlying sandstone unit.

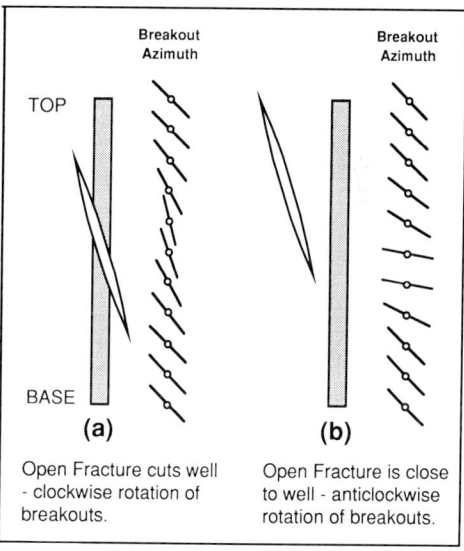

Fig. 8. Schematic representation of how breakout long axis orientations at Alma could rotate over an interval where (a) the well cuts through an open fracture, and (b) passes close to one.

It should be emphasized that the above explanation is not unique in terms of diagnosing fracture geometry; differently oriented open fractures could also be causing the anomalous σ_h directions in Alma F-67. Furthermore, clockwise anomalies do not necessarily suggest that a well has passed through an open fracture, or vice versa; it depends on the horizontal principal stress geometry of the setting and the orientation of the fractures. Nor are fluid-filled open fractures the only geomechanical interfaces that can deflect stresses. However, at Alma, they are the best candidates.

Open faults and fractures in the Aquitaine Basin of France

Following encouragement at Alma, we

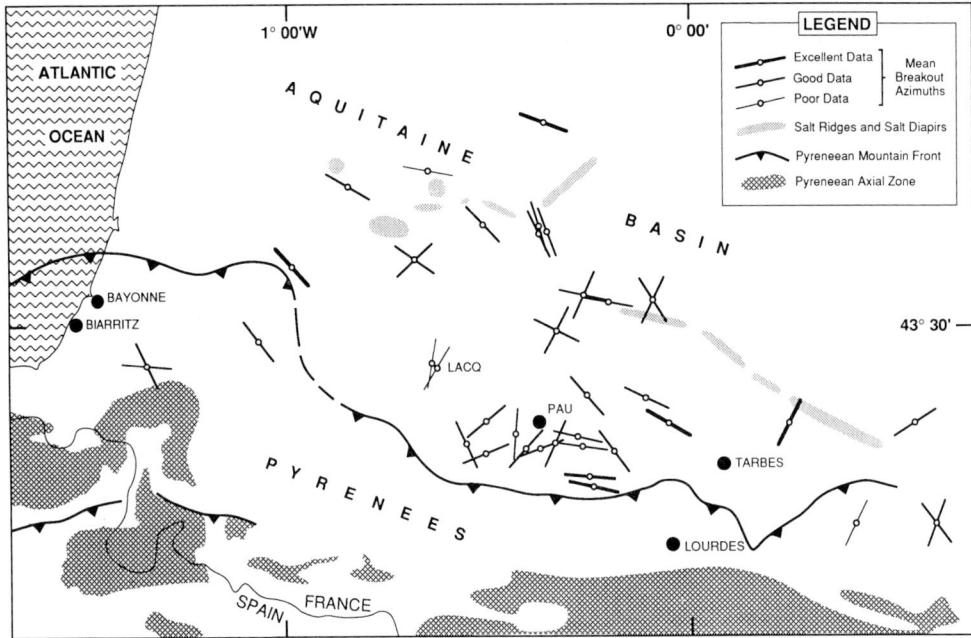

Fig. 9. Mean breakout azimuths from 36 wells in the southern part of the Aquitaine Basin, France. The stress regime does not appear to have a homogeneous directional signature, as is discussed in the text.

attempted to use breakouts to identify open fractures in a small oil field in the southwestern part of the Aquitaine Basin in France (Fig. 9). Unfortunately, in this area, the regional stress trajectories are less homogeneous than on the Canadian Scotian Shelf. Breakout long axes were measured in 40 vertical and near vertical wells, chosen for their depths and locations. These wells record breakouts between depths of 23 metres and 6175 metres and, in many, the net thicknesses of the breakout intervals exceed 300 metres. Special attention was paid to well deviation, and breakouts were ignored if their long axes corresponded to directions of deviation, particularly if the well was inclined at greater than 5° from the vertical.

As can be seen from Figure 9, the stress regime does not appear to be homogeneous in the southwestern part of the Aquitaine Basin, north of the Pyrenees. Breakouts indicate that σ_h is oriented both NW–SE and NNE–SSW (see also Maury & Sauzay 1987). In the western part of the area, breakouts give a clear indication of NW–SE to WNW–ESE σ_h trajectories with an indication of a small clockwise swing in orientation from north to south. To the east, σ_h is oriented NNE–SSW. In the central area, both orientations, as well as some intermediate direc-

tions, are present. Clearly, the stress regime is not homogeneous, or unimodal, everywhere and this makes it difficult to recognize anomalous orientations that might be diagnostic of open fractures.

The oil field that we studied is located in the central part of the area, where both σ_h orientations are documented by breakouts. Oil is prduced from Lower Cretaceous and Jurassic reservoirs that have been exploited by several semi-vertical wells and a much larger number of deviated completions. Differential caliper readings were ubiquitous in most of the wells. However, it was necessary to ignore most of these 'breakout intervals' because their long axes lay within 20–30° of the direction of well deviation. So consistently did the elongated zones and the well deviation directions parallel each other that it seemed clear that most of the ovalization was due to drill pipe wear on the borehole walls, and had nothing to do with local stress directions. Only locally, and in a few semi-vertical sections, were breakouts encountered which could confidently be attributed to horizontal stress anisotropy. This significantly diluted the data set.

Despite these difficulties, we encountered some indications of open fractures and non-sealing faults. Well A is a semi-vertical hole which crosses a high-angle fault at 2184 metres. The fault is mappable on reflection seismic lines and indicated by dipmeter inclinations. Above and below this depth, breakout orientations are anomalous (Fig. 10), suggesting that this fault is deflecting horizontal stresses locally. The implication is that the fault zone is acting as a free surface and is partially, at least, open. Higher in the well, between depths of 300 and 500 metres, are other anomalous breakout orientations which may also be due to stress deflections by non-sealing faults and/or open fractures, but no offsets are shown seismically.

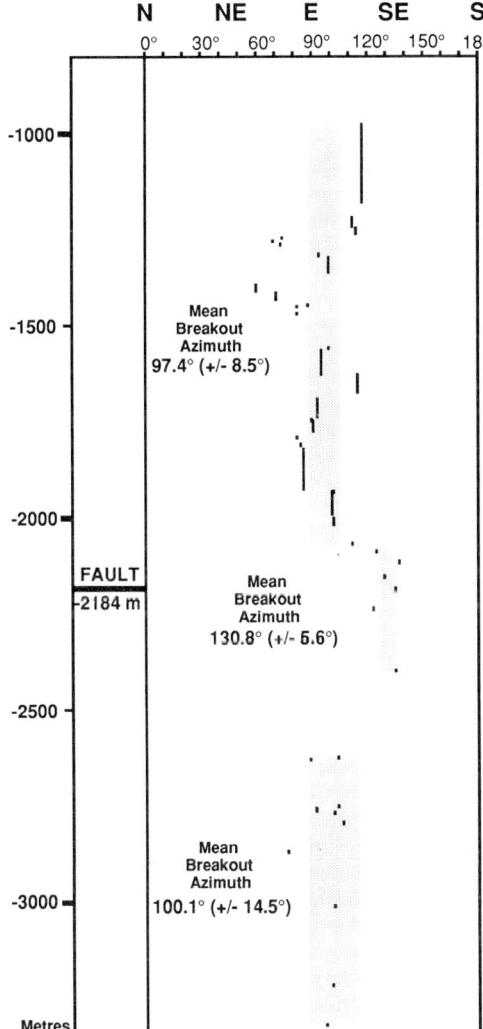

Fig. 10. Breakout orientation of Well A, drilled in an oil field in the central part of the southern Aquitaine Basin. Note how the orientation changes as the well passes through a fault.

In the reservoir rocks themselves, where evidence for open fractures was sought, unequivocal stress-induced breakouts were not abundant. Figure 11 shows mean breakout orientations from wells which had more than 20 metres of clearcut stress-induced breakouts, and traces of faults cutting the adjacent rocks. The faults are well established from 3-D seismic coverage. WNW–ESE orientations of σ_h are believed to coincide with regional stresses in the area and hence are not interpreted as anomalous deflections. There are also several wells with NNE–SSW oriented breakouts including Well B, which is located immediately adjacent to a WNW–ESE oriented fault. The breakout indications that σ_h is oriented approximately perpendicular to the fault plane are interpreted in terms of stress deflection caused by this fault. If true, this suggests that the fault is non-sealing and open in part, or else that similarly oriented open fractures are present near-by. This interpretation is supported by information from oriented cores taken in Well B. They contain E–W trending open fractures. The NNE–SSW breakouts in other wells may also be diagnostic of open fractures, as may the NW–SE oriented breakouts in Well C, although seismic data do not suggest that any major faults pass close to them.

Thus, in the southern part of the Aquitaine Basin, breakouts do not appear to be as powerful indicators of open fractures as they are in offshore eastern Canada. It is worth considering briefly why this may be so.

Discussion

The breakout-bearing sedimentary sequences in offshore eastern Canada and the Aquitaine Basin are similar in many ways. Burial depths and geologic ages are comparable. The stratigraphic sections are lithologically similar, though the rocks in the Aquitaine Basin are slightly more indurated. On the other hand, the structural settings are very different. The Scotian Shelf is an extensional passive margin, flanking the Atlantic Ocean, that has experienced a relatively simple geological history in Mesozoic and Tertiary time (Wade 1981). The southern part of the Aquitaine Basin, on the other hand, has experienced four episodes of Alpine compression involving multiple phases of faulting and salt movement (Durand Delga 1980). Structurally, it is far less homogeneous than the Scotian Shelf.

In the Aquitaine Basin, there is a strong tendency for long axes of borehole wall caved zones to be aligned with the well deviation direction, even when actual deviation in a vertical sense is minimal. This tendency is far less apparent on

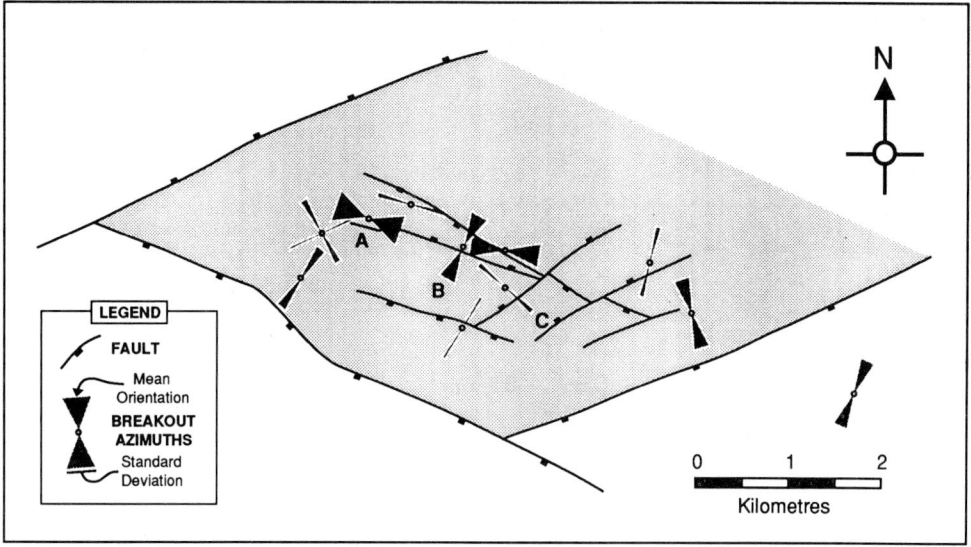

Fig. 11. Mean azimuths of breakouts measured in a fractured reservoir in the southern Aquitaine Basin. Only 10 wells exhibited more than 20 metres net thickness of breakouts in several hundred metres of reservoir. The NNE–SSW oriented breakouts are believed to record deflected stress trajectories caused by open fractures oriented approximately E–W. The WNW–ESE breakouts are interpreted as recording undeflected regional stress trajectories.

the Scotian Shelf (Table 1). It is unlikely that this is due to different drilling practices in the two areas, because the rocks in the Aquitaine Basin are more indurated than those on the Scotian Shelf and are thus likely to have relatively greater cohesive strengths. There is also less homogeneity between wells in terms of principal horizontal stress directions (Table 1). Standard deviations of breakout populations exhibit a wider range in the Aquitaine Basin than on the Scotian Shelf.

Table 1. *Comparison of stress regime indicators for the Scotian Shelf and the Aquitaine Basin*

Indicators of STRESS REGIME	SCOTIAN SHELF	AQUITAINE BASIN
Breakout long axis orientation is similar to the direction of well deviation	RARE	FREQUENT
Breakout standard deviations in specific wells (measure of variation in breakout orientation)	5.5° - 20.0° Mean: 12.3°	8.1° - 32.0° Mean: 15.7°
σ_h / σ_v at 3000 m	$\dfrac{46\ \text{MPA}}{64\ \text{MPA}}$ 0.7	$\dfrac{?61\ \text{MPA}}{72\ \text{MPA}}$?0.8
σ_H / σ_h at 3000 m	$\dfrac{c.63\ \text{MPA}}{46\ \text{MPA}}$ 1.4	?
Earthquake Incidence	RARE	REASONABLY FREQUENT

What about principal stress magnitudes? On the Scotian Shelf, density logs give consistent figures for overburden loads, which are assumed to correspond to σ_V. Leak-off tests provide a reasonable measure of σ_h and the present day extensional tectonics suggest that at depths of less than 4000 metres, σ_H is slightly less than σ_V in magnitude (Bell 1990). With this reasoning, we can infer a horizontal principal stress contrast, $\sigma_H : \sigma_h$, at 3000 metres of approximately 1.4 (Table 1). In the southern part of the Aquitaine Basin, there is little information on stress magnitudes. There have been no measurements made from hydraulic fracturing. Few leak-off tests have been run. Those which are available suggest relatively higher values for σ_h of the order of, perhaps, 61 MPA at 3000 metres depth, compared to 46 MPA on the Scotian Shelf (Table 1). A compressive stress regime is indicated. It is not clear what order of horizontal stress contrast, or magnitude of anisotropy, prevails. It is unfortunate that this information is not available. Horizontal stress contrast is the single factor which most determines how easily stress-induced breakouts form in wells. If the contrast is low, the tendency for borehole walls to cave elliptically *in response to stress amplification* is significantly reduced (Zoback *et al.* 1985). They can still cave asymmetrically, so as to give rise to oval cross-sections, but the initiating mechanisms will involve factors other than stress anisotropy (Santarelli 1987). Breakout formation is obviously not inhibited in the Aquitaine Basin, but it would appear that it is frequently promoted by mechanisms other than the contrast in magnitudes of the horizontal principal stresses acting on the far field rocks.

This discussion raises the question of what factors determine horizontal stress anisotropy in sedimentary basins. Is stress anisotropy related to absolute stress magnitudes? Is it a consequence of the local neotectonics in the upper lithosphere? What is the role of fluids, particularly overpressured fluids? We do not plan to assess this issue in depth here, but would like to raise one possibility. M-L Zoback and others (1989) have shown that, for the faster moving lithospheric plates, there is a parallelism in the orientations of σ_H and their absolute plate motions (Minster & Jordan 1978). This is particularly evident in the case of the North American and South American plates, which are two of the faster moving plates with respect to the underlying Mantle (2.6 cm/year beneath the Scotian Shelf area). The Eurasian plate is moving much more slowly (0.3 cm/year beneath the Aquitaine Basin), and the relationship between stress orientations and absolute motions is not so clearcut, nor is there the same homogeneity in orientation. If the stress signature of the upper crust reflects drag on the base of the lithosphere imposed as a result of plate motions, is it possible that horizontal stress anisotropy in sedimentary basins depends on how fast the plate is moving? In this scenario, stress anisotropy would be relatively small in Europe (compared to North America) because of the Eurasian Plate's low absolute velocity. If this is the case, we would expect breakout formation to be promoted frequently by factors other than stress anisotropy, for example drillpipe wear, mechanical anisotropy and pre-existing fractures in the rocks. It would also be likely that breakouts would not give a very homogeneous signal and that there would be significant directional variation. What we have observed could fall into a pattern of this type, so it is critical to obtain good measurements of σ_H in sedimentary basins in Europe. This is essential, no matter what parameters control stress anisotropy. Until such information is available, we shall not have a clear understanding of the dependability of breakouts as indicators of horizontal stress directions in Europe.

Conclusions

Recognition of open fractures using breakouts identified on four-arm dipmeter logs has been shown to be a viable technique. However, it requires abundant and excellent data, and will be most effective in areas where the majority of the breakouts are stress-induced. In areas where this is not the case, for whatever reason, the approach is likely to be less definitive.

Many of the mean breakout azimuths for wells on the Scotian Shelf were initially measured by A. Podrouzek. Mean breakout azimuths from several wells south of Pau and in the central part of the Aquitaine Basin were derived from unpublished in-house studies by D. Fontanet and C. Tourneret. The authors are very grateful to the Mission France Group of Elf-Aquitaine for their help in obtaining original well data, and to the Management and Directors of Elf-Aquitaine for permission to publish this paper. The manuscript has been improved by the helpful criticism of V. M. Maury, J-M. Pierron and J. P. Richert. JSB and JA thank the Geological Survey of Canada for research support provided through the Frontier Geoscience Program.

References

BELL, J. S. 1989. Vertical migration of hydrocarbons at Alma, offshore eastern Canada. *Bulletin of Canadian Petroleum Geology*, **37**, 358–364.

—— 1990. The stress regime of the Scotian Shelf, offshore eastern Canada, to 6 kilometers depth and implications for rock mechanics and hydrocarbon migration. *In*: MAURY, V. & FOURMAINTRAUX, D. (eds) *Rock at Great Depth*, vol. 3. Rotterdam, 1243–1265.

—— & BABCOCK, E. A. 1986. The stress regime of the Western Canadian Basin and implications for hydrocarbon production. *Bulletin of Canadian Petroleum Geology*, **34**, 364–378.

—— & GOUGH, D. I. 1979. Northeast–southwest compressive stress in Alberta: Evidence from oil wells. *Earth and Planetary Science Letters*, **45**, 475–482.

—— & —— 1983. The use of borehole breakouts in the study of crustal stress. *In*: ZOBACK, M. D. & HAIMSON, B. C. (eds) *Hydraulic Fracturing Stress Measurements*, National Academy Press, Washington, D.C., 201–209.

—— & LLOYD, P. F. 1989. Modelling of stress refraction in sediments around the Peace River Arch, western Canada. *Current Research, Geological Survey of Canada* Paper 89-1D, 49–54.

DURAND DELGA, M. 1980. Itinéraire géologiques: Aquitaine, Languedoc, Pyrénées. *Bulletin du Centre de Recherche Exploration-Production, Elf-Aquitaine*, Memoir 3, Pau, 439.

HOEK, E. & BROWN, E. T. 1980. *Underground Investigations in Rock*. Institute of Mining and Metallurgy, London.

MAURY, V. M. & SAUZAY, J-M. 1987. Borehole Instability: Case Histories, Rock Mechanics Approach,
and Results. Paper SPE/IADC 16051, SPE/IADC
Drilling Conference, New Orleans, U.S.A., March
15–18, 1987.

McGARR, A. & GAY, N. C. 1978. State of stress in the
Earth's crust. *Annual Review of Earth and Planetary Sciences*, **6**, 405–436.

MINSTER, J. B. & JORDAN, T. H. 1978. Present-day
plate motions. *Journal of Geophysical Research*,
83, 5331–5354.

PLUMB, R. A. & COX, J. W. 1987. Stress distributions
in eastern North America to 4.5 km from borehole elongation measurements. *Journal of Geophysical Research*, **92**, 4805–4816.

SANTARELLI, F. J. 1987. *Theoretical and Experimental
Investigation of the Stability of the Axisymmetric
Wellbore*. PhD Thesis, University of London.

SMITH, D. A. 1978. Sealing and Nonsealing Faults
in Louisiana Gulf Coast Salt Basin. *AAPG Bulletin*,
64, 2, 145–172.

WADE, J. A. 1981. Geology of the Canadian Atlantic
margin from Georges Bank to the Grand Banks.
In: KERR, J. W. and FERGUSON, J. (eds) *Geology of
the North Atlantic Borderlands*, Memoir 7, Canadian Society of Petroleum Geologists, 447–460.

WU, H-C. & CHANG, K-J. 1978. Angled elliptical notch
problem in compression and tension. *Journal of
Applied Mechanics*, **8**, 393–401.

ZOBACK, M. D., MOOS, D., MASTIN, L. & ANDERSON,
R. N. 1985. Wellbore breakouts and in-situ stress.
Journal of Geophysical Research, **90**, 5523–5530.

ZOBACK, M. L. *et al.* 1989. Global patterns of tectonic
stress. *Nature*, **341**, 291–298.

Lithological and fracture response of common logs in crystalline rocks

MARTIN H. BREMER, JOHANNES KULENKAMPFF &
JÜRGEN R. SCHOPPER

*Technological University of Clausthal, Institute of Geophysics, Arnold-Sommerfeld-Str. 1,
D-3392 Clausthal-Zellerfeld, Germany*

Abstract. Log responses in crystalline rocks are a function of the lithology (rock matrix), macro- and microfractures. Basically, the porespace-geometry dependent petrophysical parameters (porosity, internal surface area, permeability, formation factor) are similarly interrelated as in sediments, but mineralogy and fractures exert greater control than in sediments.
A qualitative log interpretation method was developed and tested. Fractured and otherwise deteriorated zones with anomalous properties are identified, using a suitable combination of five standard logs. Before a quantitative well-log analysis of petrophysical parameters is possible, the logs are corrected for lithology effects, using a quick-look lithology identification program. For lithology identification seven logs are considered sufficient. Their mean lithology-dependent response in the German KTB borehole is reported.

As the largest field of application for wireline logging is in oil exploration and production, petrophysical modelling and interpretation of logs in sediments is highly developed, although the basic formulae and models are mostly empirical.

Crystalline rocks have been generally less comprehensively studied, although this interest has now increased due to challenging geotechnical questions (waste disposal, geothermal heat production) and scientific projects (German Deep Drilling Project (KTB) and International Ocean Drilling Program (ODP)).

Crystalline rocks are considered as tight rocks with very low porosity and permeability, and a complex mineralogy. Voluminous fluid transport over larger distances may be possible only within some macrofractures. This transport does not affect the bulk rock and only locally influences rock alteration. However, macrofractures are normally surrounded by a zone of microfractures (Kranz 1983) and our measurements of BET-N_2-adsorption-desorption isotherms (Gregg & Singh 1979) show clearly that more than 50% of the pore volume can commonly be attributed to capillaries with characteristic radii less than 100 nm. Figure 1 is a typical example from our results. The properties of the bulk rock, which is not necessarily macroscopically fractured, are strongly influenced by the properties of these small capillaries and their fine structure. Even water properties have to be considered, as they appear to be altered in these capillaries (Clifford 1975). The fluid transport is more akin to a diffusion process in a concentration gradient, than to Darcy flow under a differential pressure. Even diffusion processes such as electrical conductivity, normally taking place predominantly in the pore volume, are modified and are more strongly influenced by surface effects than in more porous rocks. The formation factor is of order 10^3, so that resistivity is almost independent of salinity, being predominantly a function of surface conductivity. Surface effects, which are often negligible in sediments when the pore sizes are larger, are of vital importance in crystalline rocks. In order to understand transport processes, alteration of rocks and elastic or anelastic properties (Bremer *et al.* 1992), these effects have to be taken into account.

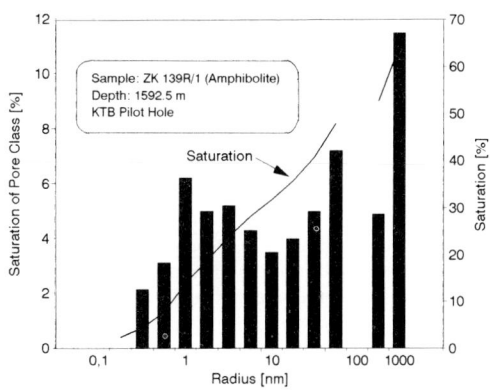

Fig. 1. Pore-size-distribution from N_2-adsorption–desorption measurements and saturation curve.

From HURST, A., GRIFFITHS, C. M. & WORTHINGTON, P. F. (eds), 1992,
Geological Applications of Wireline Logs II. Geological Society Special Publication No. 65, pp. 221–234.

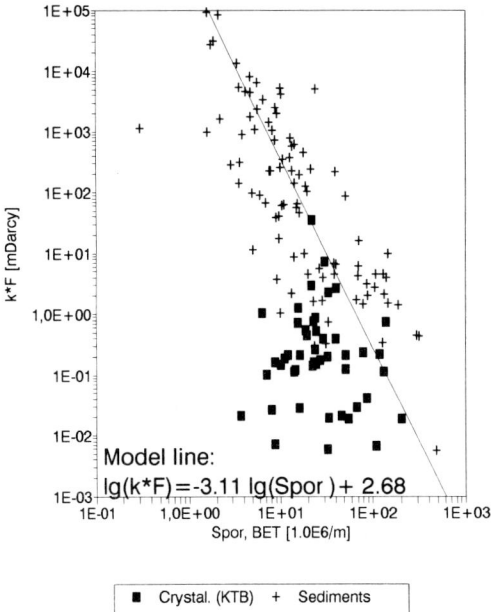

Model line:

$$\lg(k*F) = -3.11 \lg(S_{por}) + 2.68$$

k*F [mDarcy] (vertical axis)

Spor, BET [1.0E6/m] (horizontal axis)

■ Crystal. (KTB)   + Sediments

Fig. 2. The first PaRiS equation. Crosses: diverse sedimentary rocks; squares: crystalline samples from the KTB pilot hole. The line is predicted by the pigeon-hole model. F, formation factor; S_{por}, pore volume-specific internal surface area $(1/\mu m)$; k, permeability (Darcy).

The study of physical rock properties and processes within rocks with the help of theoretical models has always been one of our principal objectives (see Pape *et al.* 1982, 1984, 1987). Pape, Riepe & Schopper (1982) developed the self-similar pigeon-hole-model for sediments, which allows the calculation of permeability from measurements of the internal surface area with the help of the 'First PaRiS' equation. This model was extended (Pape *et al.* 1987), for both sedimentary and igneous rocks, to take into consideration secondary mineralization with chlorites and iron oxide crusts. Figure 2 shows that the 'First PaRiS' equation is a good approximation to the behaviour of sediments. In crystalline rocks the increased scatter is due to the lithological factor g_0, which describes deviations from the self-similarity of the model, either by smoothing of some generations of structure, or by some lamellae-structures, or even by anisotropic effects. The factor can be measured independently. A larger shift to the left of the model line can also be due to hydraulically non-effective pore space, which can be calculated from permeability, N_2-desorption and mercury porosi-

metry measurements (Schopper *et al.* 1990). This is the probable reason for the scattering of the data in crystalline rocks, but we could only test this possibility in a few cases, as the measurements are very expensive.

Knowledge of the internal surface area or, better, of the whole pore size distribution, is the key to describing the transport process(es), but it is difficult to derive these parameters from wireline logs with sufficient accuracy, especially in crystalline rocks.

As a necessary first step, therefore, we are investigating qualitatively the log response of the crystalline rocks in the pilot hole of the German deep drilling project (KTB) in Bavaria, with the aim of identifying lithologies and locating fracture zones. One advantage of this project is the great diversity of geoscientific studies and data from the same rock samples, which can be compared with the wireline logs.

At a later stage, these studies are to be used for deriving petrophysical data such as porosity, permeability and internal surface area from wireline logs.

Fracture detection

As a first stage, a quick and simple algorithm for reliably detecting fracture zones was developed. A parameter describing the intensity of fractures is needed as a cut-off in the lithology recognition program, and it can be used for finding the petrophysically more interesting zones of higher porosity, conductivity and permeability.

Qualitatively, bulk density and resistivity decrease, and acoustic transit times and porosity increase, in fractured rock. Therefore, the bulk density, neutron porosity, compressional and shear transit times, and the logarithm of the deep laterolog, are normalized to an approximately equal standard deviation in order to have equal influence from each log. In equation (1) the product of the parameters increasing with fracture intensity (POS_i) is divided by the product of the parameters decreasing with it (NEG_i), so that their response to fractures is amplified.

$$FRAC = \frac{\Pi \, POS_i}{\Pi \, NEG_i} \qquad (1)$$

During the first stage, no regard is paid to lithology. At the second stage, it is intended to eliminate the lithological influence on the logs before they are combined. This will be done with an iterative approach, first eliminating the influence of fractures on identifying lithology, then of lithology on detecting fractures.

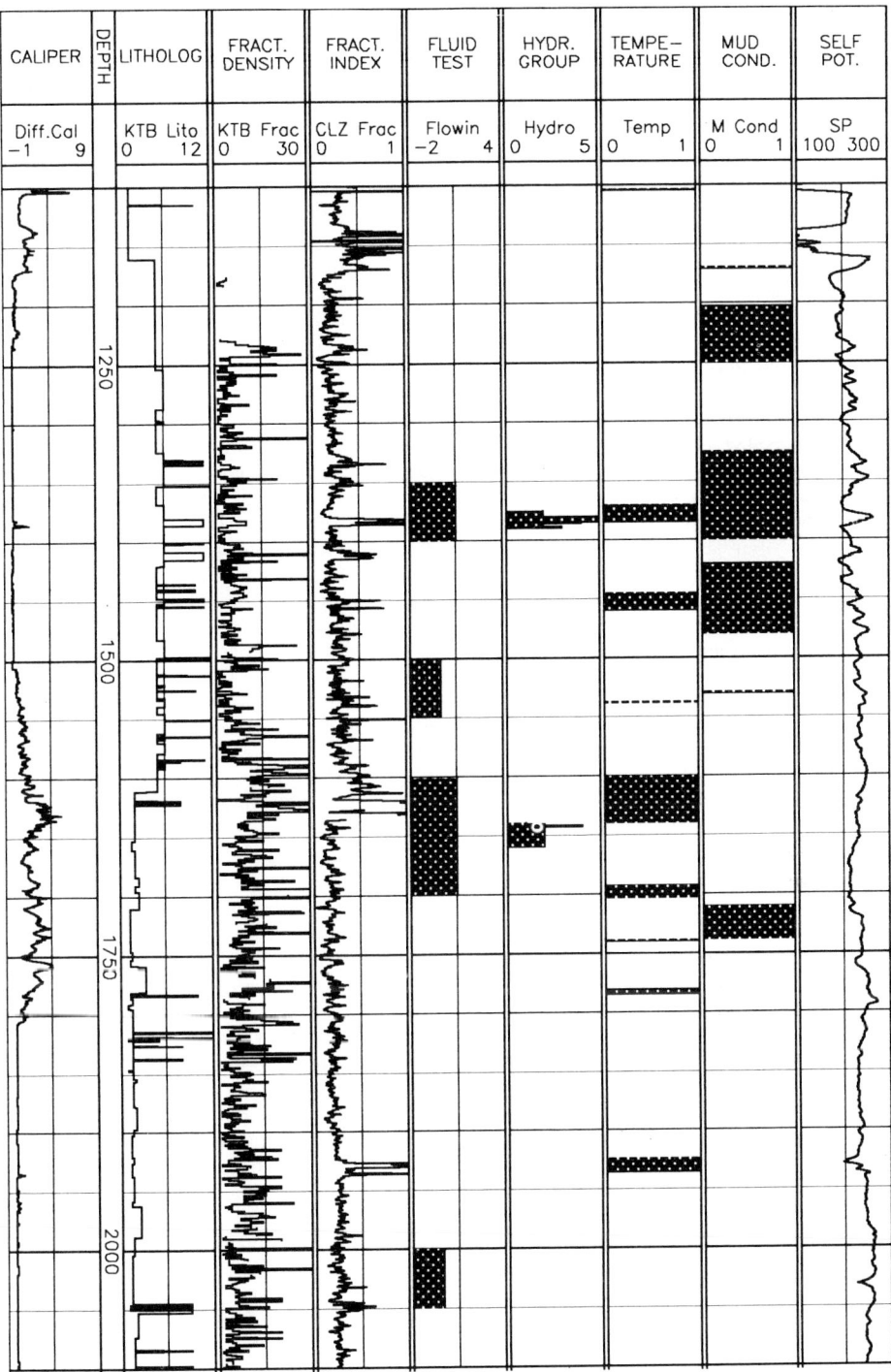

Fig. 3. Comparison of the fracture index 'CLZ Fracs' with the fracture density from core analysis 'KTB Fracs' (Fracs/m), and with results from fluid testing (0, Very low; 4, very high inflow) and predicted inflow (5 : high) (Kessels *et al.* 1990), temperature and mud conductivity anomalies (Jobmann *et al.* 1990).

A qualitative FRAC curve is created in this way. In a later step it will be calibrated to a quantitative measure of fracture intensity with the help of N_2-pore-size-distributions (Bremer *et al.* 1992), but some intensive laboratory work is needed first.

The derived fracture index (FRAC) curve does not always correlate with the fracture density (Fig. 3) from laboratory observations on the cores. The core-derived fracture density data are given as the number of core breaks and latent fractures per metre of core (Massalsky *et al.* 1988; Röhr *et al.* 1989), so that one large open fracture per metre looks like almost unfractured rock. Some of the laboratory-identified fractures were induced after coring by pressure release. Microfractures too small to be seen by the naked eye and not recorded, can strongly influence log responses (e.g. acoustic transit time and resistivity). One large open fracture can have a strong effect on the logs, either by itself or by associated microfractures, but as only one fracture is counted in the laboratory, the fracture density is very low. For independent detection of such macroscopic fractures, imaging tools such as microscanning resistivity tools or acoustic televiewer tools are to be preferred (Draxler *et al.* 1990).

Our FRAC log is intended to be sensitive to microfractures as well as macrofractures. It can be derived from standard logging suites, it does not depend on the more complicated imaging tools and it is intended to be a better measure of the 'fracturedness' of the rock than the fracture density of cores. The vertical resolution is not as good as with imaging tools, and the orientation of the fractures cannot be determined, but almost all fracture zones may be detected without regard to their inclination and position in the borehole wall. Regular and pad tool measurements are combined, the pad tools with a higher resolution but measuring only a small sector of the wall, the free-running tools with less resolution but measuring across the whole circumference of the borehole. If macrofractured zones are determined independently, the FRAC log can be calibrated by volume of microfractures, otherwise it can serve as a qualitative measure of fracturedness.

Figure 3 shows the FRAC log together with other results of investigations relating to fractured zones, e.g. hydraulic tests and flowmeter measurements (Kessels and Pusch 1990; Jobmann and Reifenstahl 1990). Although there is hardly any correlation with the fracture density, there is quite a good agreement with fracture-dependent anomalies of temperature and mud conductivity.

Lithological classification and responses

Before petrophysical parameters such as internal surface area, porosity and permeability can be derived from the logs, and the fracture response can be enhanced, it appears necessary to remove the influence of the different lithologies from the log values. Lithology recognition was carried out with the help of at least one representative depth section of the core lithology column. Histograms were created for the different lithologies and logs, excluding bad hole sections. For most well logs, the influences of the pore liquids are negligible in the first approximation, as mentioned in the introduction. Most logs mainly respond to the matrix and interface properties, and for lithology detection the tools responding most sensitively to the matrix were chosen. Such measurements include gamma-spectrometry, density measurements and also the neutron porosity measurements (responding to both water of crystallization and pore fluids).

In similar lithologies the influence of fractures on individual logs appears to be small in the KTB pilot hole (Fig. 4). Otherwise the FRAC curve could be used as a cutoff for log quality in the same manner as the caliper log.

The characteristic response of the selected unfractured section serves as a database (Table 1). This is then used by our program CLIP (Clausthal Lithology Identification Program) for lithology identification in other regions (Figs 5 & 6).

CLIP sums the absolute values of the relative deviations (relative errors) between all selected logs and the lithology database values, identifying the most probable lithology as being that with the smallest sum. It is possible to enter the probability of occurrence as weight for each lithology. Then, if the sum of deviations of two or more lithologies is equal or very close to each other, the more probable one will be selected (Figs 5 & 6).

The algorithm does not compensate for the influence of bad hole conditions, and it is sensitive to the confidence of the lithology column used to create the database. But CLIP was not designed to serve as such a thorough lithology prediction algorithm as that published by Haverkamp *et al.* (1990), but serves as a quick-look algorithm, and for approximate compensation for lithological effects prior to fracture and petrophysical evaluation.

Precise depth matching of the wireline logs and cores is critical and is still a problem for the moment, at least for thin layers. Another problem and error source at this stage is the presence of minerals (e.g. graphite and pyrite) that have

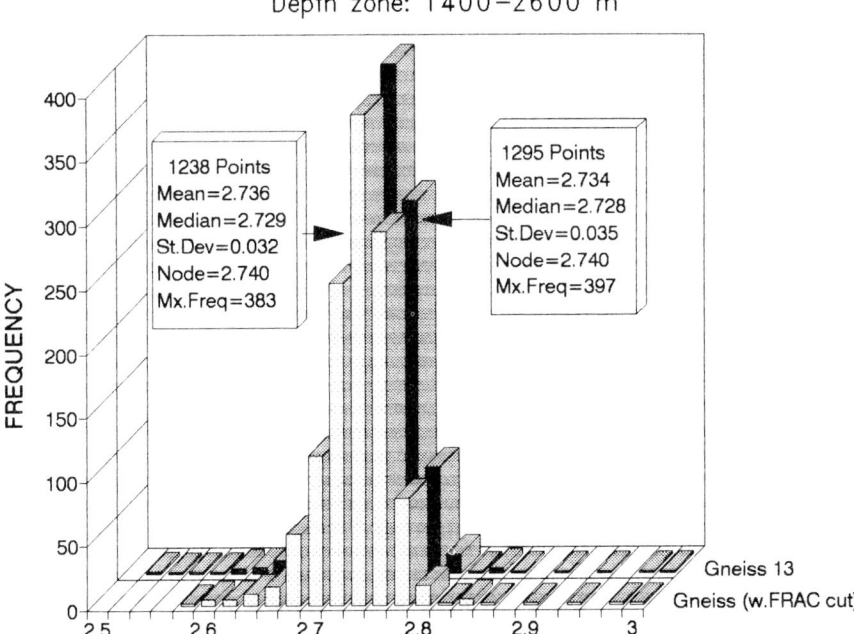

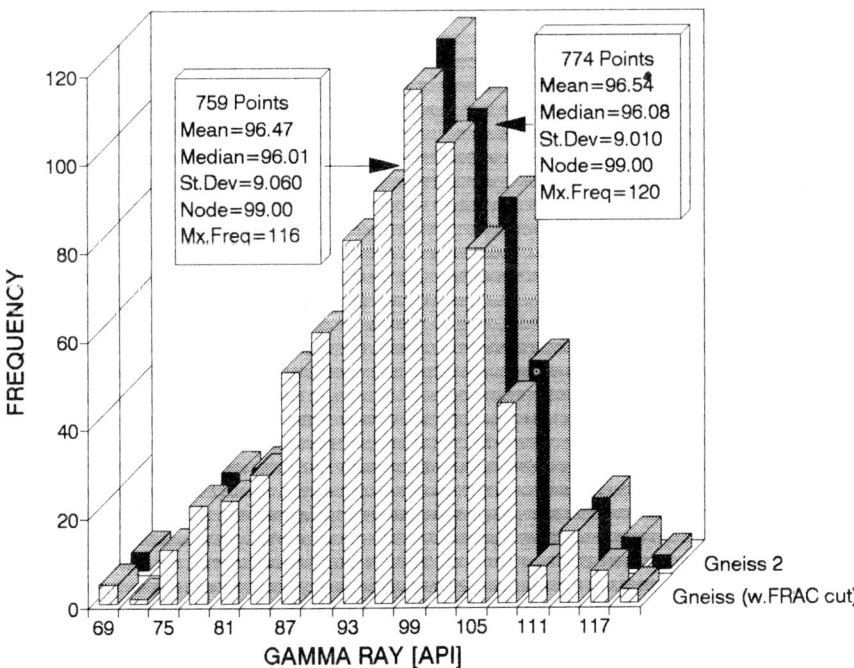

Fig. 4. Histograms with statistic parameters for two different lithologies. Comparison between equal log, lithology and depth both with FRAC cut-off ($\leqslant 0.5$) and without.

Table 1. *Lithological responses. Mean and standard deviation (second line). For lithology code and log description, see Tables 2 and 3. The number of samples is printed in parentheses under the lithology code. The depth interval is 1400–2600 m, and a cut-off from the differential caliper $\leqslant 1$ inch was used.*

Lithology	Gns. 10 (150)	Gns. 13 (1295	Gns. 15 (3808)	Gns.20 (774)	Gns. 25 (412)	Gns. 30 (307)	Gns. 33 (638)	Amphi. (1064)	Metag. (550)	A. Granit (25)	Catacl. (37)	Lampro. (126)	Qz. gan (42)	U. Mafit (46)
RHOB	2.72	2.73	2.73	2.73	2.76	2.82	2.88	2.90	2.93	2.70	2.80	2.73	2.75	2.99
	0.05	0.04	0.05	0.04	0.05	0.06	0.05	0.07	0.08	0.03	0.10	0.04	0.06	0.06
PEF	3.28	3.37	3.41	3.40	3.60	4.23	4.70	4.39	4.43	3.01	3.93	3.60	3.03	4.47
	0.45	0.27	0.27	0.30	0.19	0.48	0.42	0.53	0.50	0.35	0.69	0.41	0.74	0.37
U	8.90	9.19	9.32	9.26	9.91	11.92	13.47	12.70	12.94	8.11	11.07	9.80	8.35	13.24
	1.36	0.81	0.81	0.84	0.57	1.54	1.34	1.72	1.64	1.04	2.35	1.16	2.15	1.19
GR	89.6	95.6	96.9	96.5	103.0	71.1	64.4	26.8	19.6	107.0	102.1	94.3	47.3	19.6
	10.5	9.8	10.1	9.0	6.6	9.2	9.1	11.4	4.9	29.6	19.4	9.8	30.0	10.7
THOR	8.90	9.90	9.78	9.69	10.67	7.05	5.83	2.66	1.52	9.15	8.91	11.36	5.07	1.65
	1.97	1.04	1.13	1.04	1.04	1.16	1.00	1.25	0.64	2.44	2.39	1.29	3.11	1.24
URAN	2.01	1.87	2.02	1.95	1.98	1.56	1.50	0.63	0.32	2.52	1.63	1.94	1.07	0.52
	0.29	0.45	0.47	0.40	0.40	0.24	0.40	0.38	0.22	0.31	0.23	0.53	0.60	0.36
POTA	2.34	2.59	2.63	2.62	2.88	2.02	1.81	0.86	0.75	2.21	2.25	2.11	1.32	0.65
	0.34	0.32	0.30	0.31	0.22	0.26	0.20	0.27	0.18	0.49	0.47	0.32	0.74	0.30
NPHI	10.4	11.4	11.4	11.0	13.1	11.1	11.5	11.8	13.1	7.4	10.6	14.9	7.9	19.5
	3.2	2.0	2.8	2.4	1.5	1.7	1.6	2.6	3.2	1.6	0.9	3.1	2.1	6.1
SIGMA	19.9	20.8	21.2	20.6	22.3	22.3	24.5	22.1	22.3	18.2	22.8	20.4	19.1	22.0
	1.6	1.4	1.7	1.3	0.8	1.8	1.4	1.8	2.6	1.1	3.6	1.1	1.2	0.8
IP	38.0	26.8	27.9	20.8	27.4	58.8	72.1	43.2	37.9	10.4	55.0	17.1	39.9	56.2
	38.1	25.2	22.9	7.8	6.4	13.5	19.5	13.3	10.7	6.4	44.2	13.8	15.1	11.1
KAPPA	1.461	1.173	0.995	1.021	1.203	1.075	0.741	0.890	0.859	1.235	1.261	1.003	0.771	0.891
	0.959	0.538	0.521	0.393	0.885	0.504	0.289	0.265	0.197	0.451	0.266	0.428	0.069	0.195
SUS	0.649	0.564	0.691	0.607	0.633	0.795	1.028	0.859	0.834	0.222	0.924	0.496	0.809	0.767
	0.250	0.196	0.238	0.246	0.079	0.139	0.256	0.188	0.145	0.068	0.285	0.100	0.145	0.345
DT	173	169	169	167	167	167	164	166	163	168	168	168	168	162
	5	5	5	4	3	5	4	5	4	2	7	3	4	3
DTPM	169	166	167	165	164	164	162	164	162	165	165	166	165	160
	4	4	5	4	2	4	4	5	4	1	8	3	4	3
DTSM	286	278	275	272	274	290	288	298	294	273	278	297	284	293
	13	12	12	9	9	7	9	11	11	5	8	12	12	7
DTST	661	678	683	666	706	633	625	659	661	640	694	665	654	667
	19	49	50	30	71	5	2	16	13	4	54	11	16	9
VPVSR	1.70	1.67	1.65	1.65	1.67	1.77	1.78	1.82	1.82	1.66	1.68	1.79	1.72	1.83
	0.09	0.05	0.06	0.05	0.05	0.03	0.04	0.04	0.04	0.02	0.08	0.06	0.07	0.02
PR	0.229	0.219	0.209	0.207	0.220	0.263	0.268	0.284	0.283	0.214	0.223	0.270	0.242	0.287
	0.032	0.026	0.029	0.023	0.014	0.013	0.014	0.013	0.012	0.008	0.036	0.023	0.034	0.007
CECO	29.6	30.2	30.1	29.4	32.6	29.8	29.8	36.2	35.9	32.2	29.8	30.9	33.2	40.0
	3.8	3.2	3.5	3.1	3.2	1.7	1.6	7.1	6.5	1.5	2.8	3.4	7.2	5.7
CESH	47.1	48.2	48.2	48.6	45.2	52.0	50.2	53.0	53.1	48.6	47.9	47.8	52.6	56.6
	5.8	3.8	3.4	2.8	3.3	3.6	4.5	5.4	6.1	2.0	2.3	6.0	5.4	3.5
CEST	51.5	52.3	52.6	51.6	51.9	54.1	53.9	56.4	56.2	52.5	52.7	50.9	56.0	59,8
	?.4	1.9	1.6	2.1	1.1	0.6	0.5	5.4	5.2	0.5	1.4	2.7	4.6	2.5
MSFL	2.47	3.04	3.04	3.17	2.98	3.23	3.64	3.34	3.34	3.35	2.83	2.81	3.55	3.02
(Log)	0.96	0.63	0.76	0.65	0.67	0.69	0.54	0.47	0.46	0.54	0.61	0.74	0.55	0.47
LLD	2.70	3.34	3.52	3.55	3.42	3.48	3.59	3.42	3.43	3.59	3.21	3.13	3.87	3.17
(Log)	0.87	0.65	0.73	0.44	0.46	0.57	0.46	0.43	0.38	0.22	0.45	0.65	0.41	0.38
LLS	2.67	3.40	3.55	3.58	3.41	3.46	3.62	3.44	3.47	3.42	3.26	3.16	3.90	3.07
(Log)	1.06	0.60	0.71	0.45	0.47	0.56	0.45	0.44	0.38	0.36	0.46	0.65	0.43	0.42

responses similar to those of fractures. Other minerals, traces or impurities such as gadolinium affect the response of the density logs, and have a strong influence on the absorption of thermal neutrons (Serra 1990). In these cases the lithology determination may be incorrect. However, the uranium curve of the gamma-spectrometer log, the spontaneous potential, the induced polar-

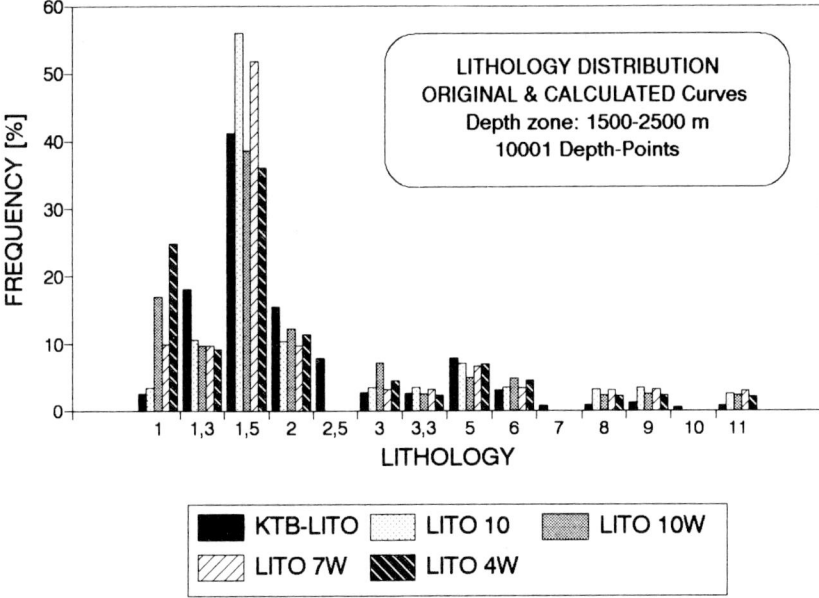

Fig. 5. Lithology distribution from core analysis 'KTB LITO' and the lithologies determined by CLIP: 'LITO 10W' with ten weighted input curves (Table 4a); 'LITO 10' similar, but not weighted; 'LITO 7W' with seven weighted input curves (Table 4b); 'LITO 4W' with four weighted input curves (Table 4c).

ization (IP), the photoelectric effect and magnetic susceptibility, might improve differentiation in such cases.

Some of the log responses have very large standard deviations or are bimodal. Occasionally this may be due to measurement characteristics (depth mismatch, misfunction of the tool etc., see for example IP, DTST; Fig. 7). Sometimes it is due to unpredicted changes in the mineralogical composition (e.g. lamprophyr, cataclasite; Fig. 8). Thin layers of some lithologies together with a slight depth-mismatch cause errors in the database. On the other hand, there are some lithology groups with very similar characteristic responses and it is therefore difficult to separate them.

Results

Wireline log response in a crystalline environment appears to be mainly a function of lithology and fractures. Only large fractures show log responses which are a function of the material filling their volume. The more important classes of microfractures show log responses that are a function of fracture surface effects.

The log measurements and tools used in this study are listed in Table 2 and the lithology classification in Table 3. The code shown in this table is used for all tables and figures.

Figure 6 shows four synthesized lithology curves, with the number of input logs given as an extension of the name (e.g. LITO 10 was calculated from ten input logs). The letter "W" additionally indicates that the lithologies were weighted with the probability of occurrence. For the curve without "W" all lithologies were supposed to occur with the same probability and have the same weight. The input logs and weights are listed in Table 4.

The histogram in Fig. 5 shows the frequency distribution of the calculated curves compared to the original lithology curve. The effect caused by the probability weighting is also shown in Fig. 5 (LITO 10 and LITO 10W). Comparison with the original curve shows a good agreement even when the database consists of only seven logs. Figures 5 and 6 together show that the lithology recognition algorithm works satisfactorily, since both the depth matching and the frequency of the recognized lithologies is mostly in agreement with core data.

A comparison of curves calculated from ten input logs weighted with their probability of occurrence and the core lithology distribution is shown in Fig. 9, where 'a' is valid for the whole borehole, and 'b' only for the interval which was used to create the database.

Not all the lithologies which were found in the laboratory were used for the calculations. Very

Table 2. *Measurements whose response was calculated for the lithologies listed in Table 3.*

Tool	Measurement	Description
LDT*	RHOB	Bulk density (g/cm^3)
	PEF	Photoelectric absorption index
	U	Photoelectric cross section
GR*	GR	Gamma ray (API)
GST*	THOR	Thorium (ppm)
	URAN	Uranium (ppm)
	POTA	Potassium (ppm)
CNT*	NPHI	Neutron porosity (PU)
TDT-P*	SIGMA	Neutron capture cross section
SDT*	DT	Delta transit time (μs/m)
STC*	DTP	Delta TT P-wave (μs/m)
Proc.	DTS	Delta TT S-wave (μs/m)
	DTST	Delta TT Stoneley-wave (μs/m)
	VPVS	P/S velocity ratio
	PR	Poisson's ratio
	CEC	Coherence energy compressional
	CES	Coherence energy shear
	CEST	Coherence energy Stoneley
DLL*	LLD	Laterolog deep [ohm m]
	LLS	Laterolog shallow [ohm m]
MSFL*	MSFL	Microspherical log [ohm m]
IP[+]	IP	Induced polarization [mV/V]
IP[‡]	KAPPA	Induced polarization [%]
SUS[§]	SUS	Susceptibility × 1000

* Schlumberger
[+] NLfB, Germany
[‡] ELGI, Hungary
[§] Uni. Munich

rare lithologies were omitted (e.g. aplite granite) while others were linked together with other lithologies of similar response (e.g. gneiss 20 and gneiss 25), because some of them produced large estimation errors.

The FRAC curve (Fig. 3) correlates better with fractured zones detected by geotechnical indicators such as the geothermal flowmeter, hydraulic testing, etc. (Kessels *et al.* 1990; Jobmann *et al.*, 1990) than with the fracture density measured on the cores (reported by Massalsky *et al.* 1988; Röhr *et al.* 1989). The reason for this has been discussed.

Conclusions

In crystalline rocks, wireline logs mainly respond to lithology, petrophysical properties of the microfractured rock matrix and macrofractures. Either one of these responses can provide useful information. However, separation of the individual log influences is necessary.

Our research has resulted in a log evaluation procedure for common logging tools for:

(1) roughly recognizing the main lithologies in a metamorphic crystalline environment;

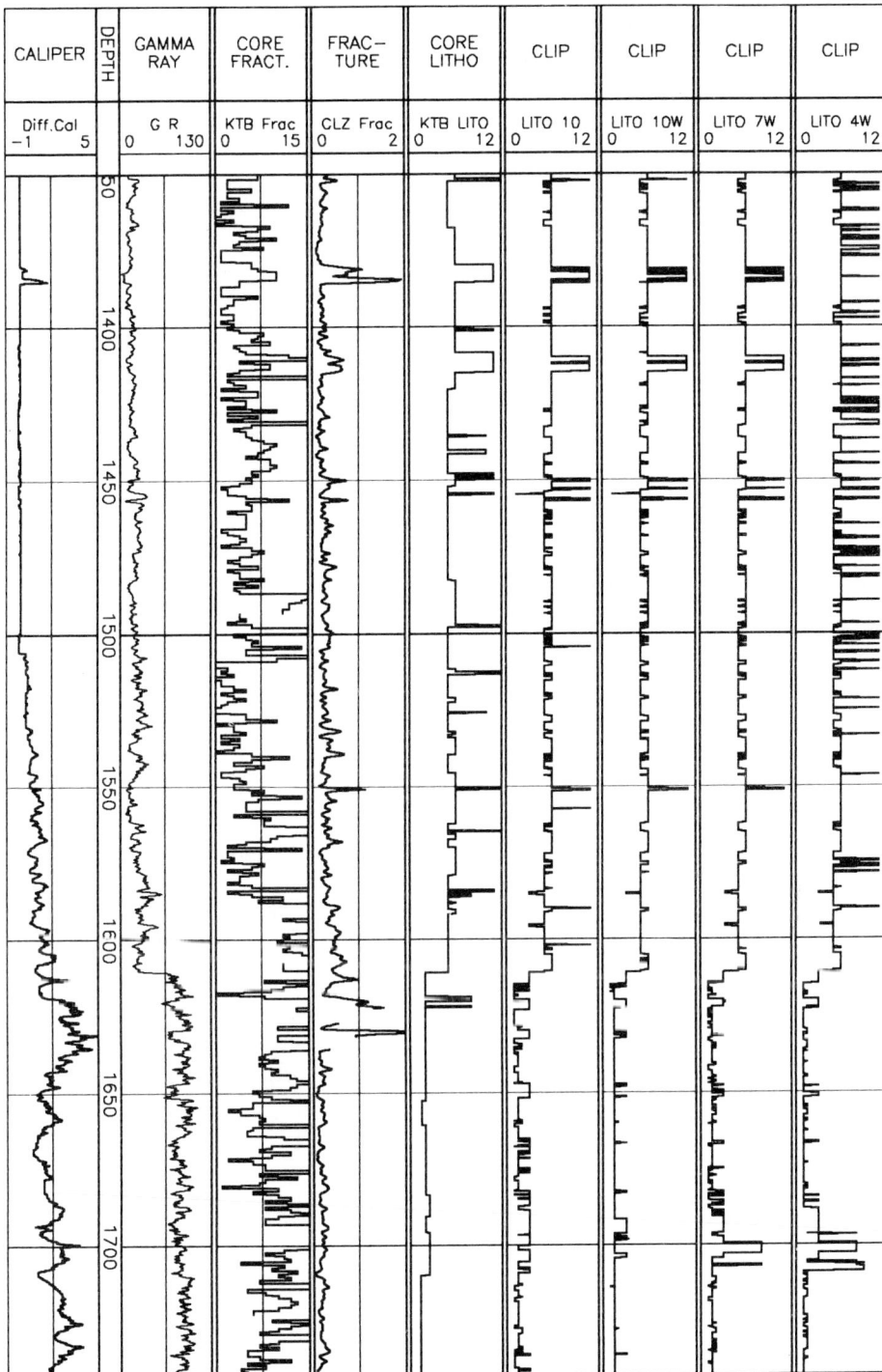

Fig. 6. Comparison between the lithology from KTB pilot hole core analysis 'KTB LITO' and determined lithologies (see Fig. 5 for description). Diff. Cal.: differential caliper in inches and gamma ray in API.

(2) detecting macrofractures and microfractured zones; and a petrophysical model for relating physical, structural and textural properties of microfractured rocks.

Eleven different lithologies can be recognized and a set of seven standard logs has been shown to be sufficient for identifying these lithologies. The lithological identification is used to control for lithology during fracture evaluation and is acquired with minimum computational effort.

Five standard logs suffice for the location of fracture zones. The resulting FRACLOG is more sensitive to microfractured zones than to single macrofractures. In practice, however, it has been found that macrofractures are usually surrounded by a zone of microfractures. In such zones, the individual logs can be used as a source of quantitative information on petrophysical properties. The lithology recognition is disturbed by the presence of fractures, and fracture evaluation is disturbed by changes of lithology. This conflict can be corrected iteratively if necessary.

Table 3. *Lithology classification groups and the code numbers for all tables and figures. The lithology groups and their depth positions where taken from Massalsky et al. (1988) and Röhr et al. (1989).*

Code No	Lithology
1	Biotit Gneiss (10)
1,3	Biotit Garnet or Sillimanit Gneiss (13)
1,5	Biotit Garnet and Sillimanit Gneiss (15)
2	Biotit Garnet Sillimanite Muskovite Plagioclas Gneiss (20)
2,5	Biotit Garnet Sillimanite Muskovite Disthen Gneiss (25)
3	Biotit Garnet Hornblende Gneiss (30)
3,3	Biotit Hornblende Gneiss (33)
3,5	Biotit Garnet Hornblende Plagioclas Gneiss (35)
5	Amphibolite
6	Metagabbro
7	Aplite Granite
8	Cataclasite
9	Lamprophyre
10	Quarzgang
11	Ultramafite
12	Other lithologies

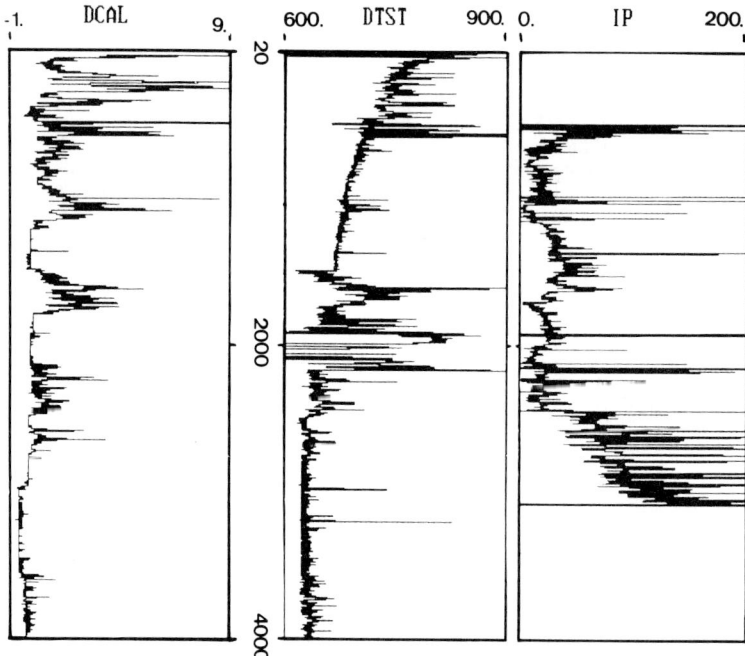

Fig. 7. Differential caliper (DCAL) [in], transit time of Stoneley waves (DTST) [μs/m], and induced polarization (IP) [mV/V]. Depth interval 20 to 4000 [m]. At this scale calibration errors, regional effects or malfunction of the tool are clearly visible.

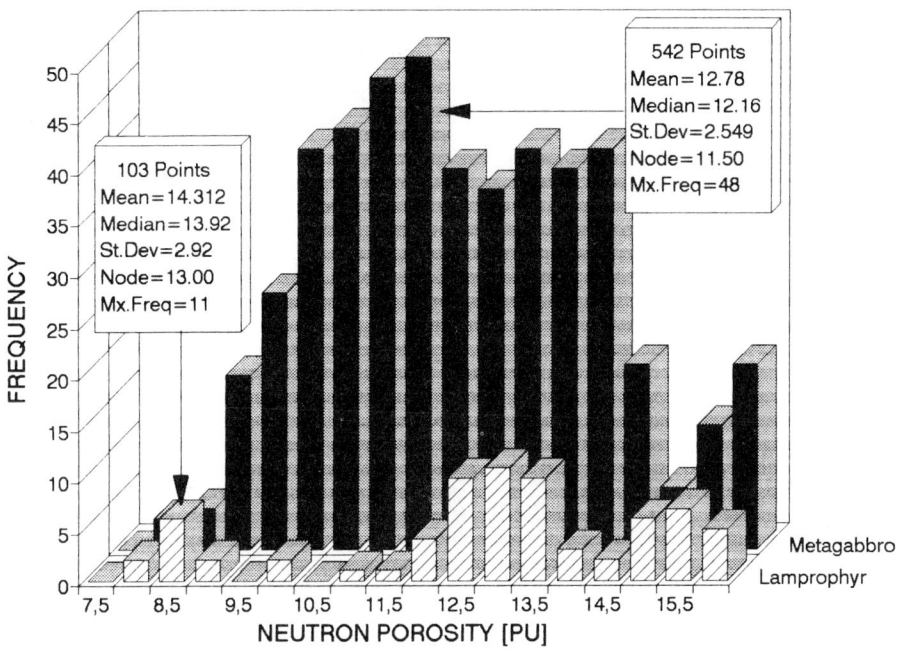

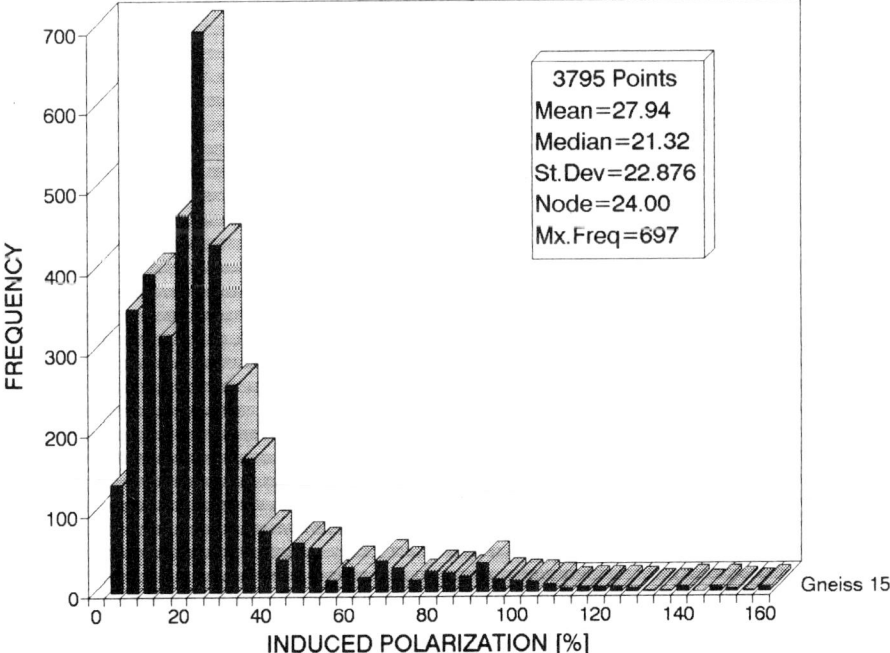

Fig. 8. Histograms with multimodal distribution and high peaks in alteration zones.

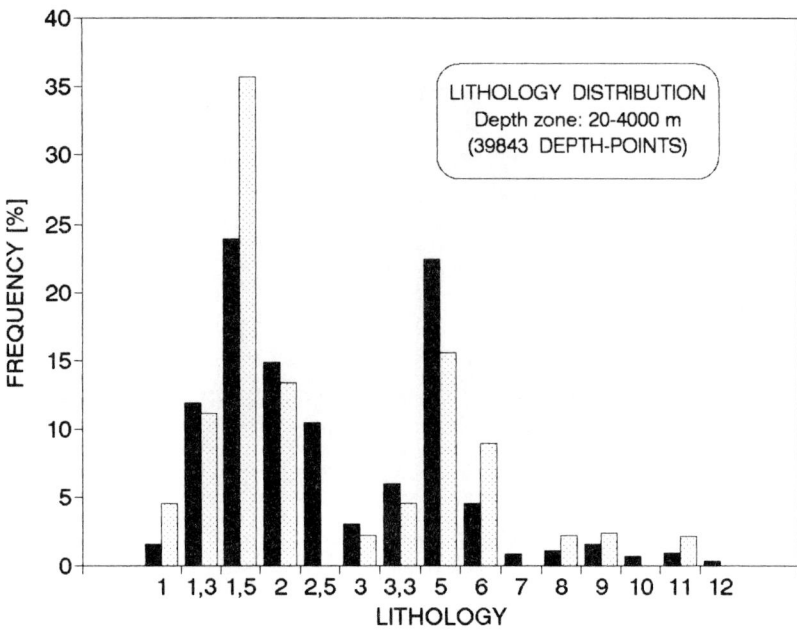

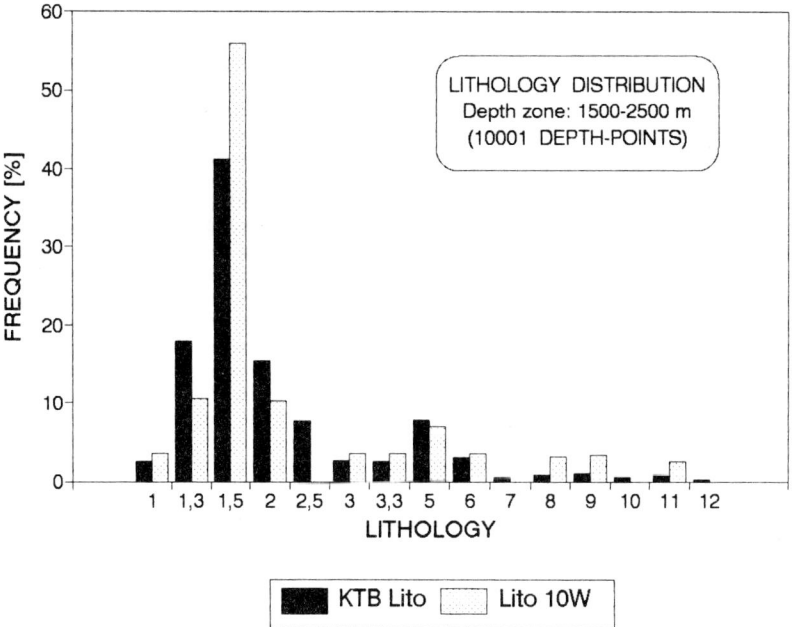

Fig. 9. Lithology distribution histograms for: the KTB core analysis lithology 'KTB LITO' and the determined curve 'LITO 10W' (input curves in Table 4a). (a) Whole depth interval (20–4000 m). (b) Depth interval 1500–2500 m, which was used to construct the data base.

Table 4. *Input logs, lithologies and their matrix response used for the curves shown in Fig. 6 In parentheses: weights = probability of occurrence. (a) LITHO 10 and LITHO 10W; (b) LITHO 7W; (c) LITHO 4W.*

Litho	Gns. 10 (1.5)	Gns. 13 (12.0)	Gns. 15 (24.0)	Gns. 20 (15.0)	Gns. 30 (2.6)	Gns. 33 (4.9)	Metag. (4.1)	Amphi. (22.0)	Catacl. (0.7)	Lampro. (0.9)	U.Mafit (0.6)
(a)											
RHOB	2.72	2.74	2.74	2.74	2.80	2.86	2.94	2.90	2.90	2.73	3.02
PEF	3.22	3.35	3.40	3.40	4.08	4.70	4.43	4.29	4.80	3.80	4.62
GR	89.6	94.3	93.9	98.1	69.5	64.4	18.4	23.8	90.0	84.0	15.0
THOR	8.89	10.08	9.81	9.69	6.90	5.98	1.30	2.66	10.00	12.00	1.96
URAN	1.75	1.83	1.90	1.99	1.56	1.48	0.29	0.55	1.70	1.50	0.35
POTA	2.34	2.65	2.70	2.77	1.95	1.78	0.72	0.81	2.50	2.00	0.54
NPHI	12.0	12.4	11.5	11.0	10.9	12.5	11.6	11.0	10.6	15.6	24.0
SIGMA	20.2	20.1	21.8	21.1	22.2	24.2	21.4	22.1	27.0	19.4	22.2
KAPPA	0.720	0.780	0.540	0.670	0.700	0.480	0.660	0.720	0.960	0.600	0.750
SUS	0.480	0.400	0.680	0.501	0.800	0.940	0.836	0.845	0.780	0.480	0.767
(b)											
RHOB	2.72	2.74	2.74	2.74	2.80	2.86	2.94	2.90	2.90	2.73	3.02
PEF	3.22	3.35	3.40	3.40	4.08	4.70	4.43	4.29	4.80	3.80	4.62
GR	89.6	94.3	93.9	98.1	69.5	64.4	18.4	23.8	90.0	84.0	15.0
THOR	8.89	10.08	9.81	9.69	6.90	5.98	1.30	2.66	10.00	12.00	1.96
URAN	1.75	1.83	1.90	1.99	1.56	1.48	0.29	0.55	1.70	1.50	0.35
POTA	2.34	2.65	2.70	2.77	1.95	1.78	0.72	0.81	2.50	2.00	0.54
NPHI	12.0	12.4	11.5	11.0	10.9	12.5	11.6	11.0	10.6	15.6	24.0
(c)											
PEF	3.22	3.35	3.40	3.40	4.08	4.70	4.43	4.29	4.80	3.80	4.62
THOR	8.89	10.08	9.81	9.69	6.90	5.98	1.30	2.66	10.00	12.00	1.96
URAN	1.75	1.83	1.90	1.99	1.56	1.48	0.29	0.55	1.70	1.50	0.35
POTA	2.34	2.65	2.70	2.77	1.95	1.78	0.72	0.81	2.50	2.00	0.54

A system for common-log evaluation in crystalline rocks has been demonstrated that permits recognition of lithology, detection of macro- and microfractures and determination of petrophysical properties of porous, microfractured or otherwise deteriorated sections.

The authors are greatly obliged to Western Atlas Wireline Services for their help and software support with the Well Data System, and to the institutions of CONACyT in Mexico and Alfried Krupp von Bohlen und Halbach Stiftung in Germany for their financial support in the form of M. Bremer's fellowship. Our thanks are also due to the DFG (German Research Foundation) for supporting the research projects and to the KTB Project Management for providing the data.

References

BREMER, M. H., KULENKAMPFF, J. & SCHOPPER, J. R. 1992. An attempt at deterministic interpretation of KTB-pilot hole standard logs. *Scientific Drilling*, **3**, 6–15.

CLIFFORD, J. 1975. Properties of water in capillaries and thin films. *In*: FRANKS, F. (ed.) *Water—a comprehensive treatise.* Plenum, New York.

DRAXLER, J., HEINSCHILD, H. J., HIRSCHMANN, G., KESSELS, W., KOHL, J. & WÖHRL, T. 1990. *Klufterkennung durch Bohrlochmessungen, Gasanalyse und Kernaufnahme.* KTB-Report 90-4, Geological Survey of Lower Saxony (NLfB) Hannover.

GREGG, J. S. & SINGH, K. S. W. 1979. *Adsorption, Surface Area and Porosity.* Academic, London.

HAVERKAMP, S., WOHLENBERG, J. & WALTER, R. 1990. *FACIOLOG-Korrelation Bohrlochgeophysikalischer Messungen mit Kristallinem Gestein aus der KTB-Vb.* KTB-Report 90-4, Geological Survey of Lower Saxony (NLfB) Hannover.

JOBMAN, M. & REIFENSTAHL, F. 1990. *Vergleich der Ergebnisse von Absenk- und Injektionstest im Hinblick auf Klufterkennung.* KTB-Report 90-5, Geological Survey of Lower Saxony (NLfB) Hannover.

234 M. H. BREMER *ET AL*

KESSELS, W. & PUSCH, G. 1990. *Auswahl hydraulischer Testzonen in der KTB-Oberpfalz VB anhand von Bohrlochmessungen.* KTB-Report 90-5, Geological Survey of Lower Saxony (NLfB) Hannover.

KRANZ, R. L. 1983. Microcracks in rocks: a review. *Tectonophysics,* **100**, 449–480.

MASSALSKY, T., MÜLLER, H., RÖHR, C., GRAUP, G., HACKER, W., KEYSSNER, S. & KOHL, J. 1988. *Ergebnisse der geowissenschaftlichen Bohrungsbearbeitung im KTB-Feldlabor (Windischeschenbach), Teufenbereich von 1530 bis 1998 m.* Paper B, KTB Report 88-9, Geological Survey of Lower Saxony (NLfB) Hannover.

PAPE, H. G., RIEPE, L. & SCHOPPER, J. R. 1982. A pigeon-hole model for relating permeability to specific surface. *Log Analyst,* **23**, 5–13.

——, —— & —— 1984. The role of fractal quantities, as specific surface and tortuosities, for physical properties of porous media. *Particle Characterization* **1**, 66–73.

——, —— & —— 1987. Theory of self-similar structures in sedimentary and igneous rocks. *Journal of Microscopy,* **148**, 121–147.

RÖHR, C., HACKER, W., KEYSSNER, S., KOHL, J. & MÜLLER, H. 1989. *Ergebnisse der geowissenschaftlichen Bohrungsbearbeitung im KTB-Feldlabor (Windischeschenbach), Teufenbereich von 1709 bis 2500 m.* Paper B, KTB Report 89-2, Geological Survey of Lower Saxony (NLfB) Hannover.

SCHOPPER, J. R., DEBSCHÜTZ, W. & KULENKAMPFF, J. 1990. Lithological interpretation of ASW-mechanisms by petrophysical measurements and logging data. *In: Application of the Absorption of seismic waves in hydrocarbon exploration,* DGMK-Report 386, Deutsche wissenschaftliche Gesellschaft für Erdöl, Erdgas und Kohle, Hamburg.

SERRA, O. 1990. *Element Mineral Rock Catalog.* Schlumberger.

Identification of tectonic rotations in boreholes by the integration of core information with Formation MicroScanner and Borehole Televiewer images

C. J. MacLEOD,[1] L. M. PARSON,[1] W. W. SAGER[2] & the ODP Leg 135 Scientific Party

[1] *Institute of Oceanographic Sciences, Deacon Laboratory, Wormley, Surrey, GU8 5UB, UK*
[2] *Department of Oceanography, Texas A&M University, College Station, Texas 77843-6331, USA*

Abstract. Palaeomagnetic studies on land have shown that block rotations on a variety of scales are commonplace in many tectonic regimes. In the oceans, however, analogous rotations have rarely been documented. This is due principally to the difficulty of their measurement rather than to their likely absence. Palaeomagnetic measurements can be, and are, readily made on borehole cores. However, the difficulty in reliably reorienting these cores, and hence their magnetization vectors, back to geographical coordinates severely hampers their use for the purposes of addressing tectonic rotations. Several techniques for orienting cores are commonly in use, none of which is without its limitations for palaeomagnetic purposes. A technique is outlined here that potentially allows the reliable orientation of sections of core by matching distinctive inclined planar features measured on the core with their images on Formation MicroScanner and Borehole Televiewer wireline logs. Its methodology is described and an example, from Ocean Drilling Program Leg 135, presented to illustrate its application.

Structural and tectonic information may be acquired from boreholes in two principal ways: either from structures measurable in the core itself or from features detected on 'images' of the borehole wall generated with speciality tools such as the Formation MicroScanner (or its predecessor, the Stratigraphic High Resolution Dipmeter) and Borehole Televiewer. The former method has the advantage of direct observation, but requires the core to be oriented, whereas the latter method has the advantage of continuous coverage and, because the logging tools are equipped with three-axis magnetometers, the reliable orientation of the observed features relative to geographical coordinates (e.g. Pezard & Luthi 1988; Serra 1989; Lehne 1990). Although some mesoscopic features, such as bedding and fractures, can be readily measured from downhole logs (see below), many more subtle features cannot be detected. These include, for example, the measurement of slip lineations on fracture planes, microfabrics and preferred crystallographic orientations (e.g. Moore 1986; Lundberg & Moore 1986), anelastic strain measurements (e.g. Teufel 1982), study of the relative ages of different generations of veins and their diagenetic histories (e.g. Knipe 1986; Agar 1991), palaeomagnetic measurements and other spatially anisotropic physical properties of the cores

(e.g. permeability etc.). In order to consider these features in their spatial context, the borehole cores must be oriented back to geographical coordinates. Several techniques for core orientation can be used, each of which has its advantages and disadvantages. They are discussed briefly below.

Multishot orientation tool

Oriented core may be obtained using a multishot orientation tool. In its simplest form, this device consists of a camera that is installed in a pressure housing together with a compass and pendulum, and lowered down the hole with the inner core barrel assembly. It takes photograhs, at preset intervals, of the compass together with a reference line, whose orientation is determined with respect to the core. The pendulum device determines the magnitude and direction of deviation of the borehole from the vertical. Non-magnetic drill collars must be included in the bottom-hole assembly when the multishot tool is employed. Various types of reference markers are in use: these may either be grooves carved into the core with special knives set into the coring assembly (Eastman Whipstock 1982, described in Nelson *et al.* 1987); or simply lines previously drawn

From HURST, A., GRIFFITHS, C. M. & WORTHINGTON, P. F. (eds), 1992,
Geological Applications of Wireline Logs II. Geological Society Special Publication No. 65, pp. 235–246.

onto core liners. The former type of marker is commonly used in the oil industry; the latter by the Ocean Drilling Program.

Multishot tools used by the oil industry are designed with a non-rotating inner core barrel so that they can be used in conjunction with rotary drilling (e.g. Nelson *et al.* 1987); however, older multishot tools, such as the one used by ODP, are simpler devices that are restricted to use with hydraulic piston coring systems (Ocean Drilling Program 1990). These lattermost therefore have the significant disadvantage of being usable only in soft, unlithified sediments from the uppermost parts of the borehole.

The multishot device has many advantages in its simplicity. Its small number of moving parts, relative ease of fitment, adaptability to most gauges of core barrel, and almost real-time acquisition of results all mean that it can be widely used. However, in practice it has several, potentially serious, shortcomings that restrict its usefulness. Because of the need to use non-magnetic drill collars in the bottom-hole assembly, the decision to use the multishot tool must be made before drilling commences, as it is impractical to trip the pipe in order to change the bottom-hole assembly at a later stage. The simple mechanical operation of the multishot tool is not infallible, and in the past human error during fitment of the system has resulted in misalignment of the tool by 180°. The reliability of the tool is critically dependent on the physical alignment between the groove or line on the core or core liner itself and the reference line photographed by the camera. This is dependent both upon the initial alignment of the tool during assembly and on operating procedures: torquing of spacer rods within the coring assembly and excessive vibration whilst coring, for example, can both have significant effects on the quality of the data. The frequent observation of the reference grooves spiralling up the core (e.g. Nelson *et al.* 1987) suggests that torque may be transmitted to the inner core barrel under certain circumstances. If the spiralling is too severe, the orientation data for those affected sections of core are unlikely to be reliable unless the multishot photographs are taken very frequently. In order to avoid the problem of vibration corrupting the photographic record, it is normally necessary to cease coring at the time the multishot photographs are taken. This not only wastes rig time, and is therefore expensive, but increases the likelihood of the core becoming jammed or broken. It also means that it is unlikely that oriented data will be obtained from the shallowest cores at any one site, as this is a time when the drillers are anxious to proceed with as few halts as possible in order to ensure a successful spudding of the hole.

Under ideal circumstances, orientation errors with the conventional multishot tool are claimed to be in the order of ±3° (Ocean Drilling Program 1990) to ±5° (Nelson *et al.* 1987) but are probably greater under normal operating conditions. Because of problems such as those outlined above, the success rate of the tool when used by commercial operators in the oil industry is generally only in the region of 50% (N. Brown, Sperry-Sun Drilling Services, pers. comm. 1991).

In order to circumvent the limitations of the conventional camera-based multishot orientation tools, electronic core orientation devices have recently been developed for use in the oil industry (Sperry-Sun 1991). These employ solid state magnetometers and accelerometers in the bottom-hole assembly that make continual, direct records of the orientation of the reference knives as drilling is proceeding; they therefore obviate the need to cease coring and, because they can make far more measurements than the camera-based devices, can be used with much longer core barrels and yield usable data even from the most severely spiralled cores.

Palaeomagnetism

Ideally, the horizontal components of the stable remanent magnetization vectors of samples taken from borehole cores can be assumed to indicate north, if the magnetization was acquired during a period of normal geomagnetic field polarity (south if during a reversed polarity period), and structures may therefore be oriented relative to this direction. Several problems are, however, inherent in using palaeomagnetics in this way, not least that a strong magnetic field emanating from the drill string frequently has the effect of partially remagnetizing the samples, giving rise to a steeply inclined magnetic component in the core. Stepwise demagnetization of the samples by alternating field or thermal techniques is therefore often necessary to remove this component and isolate a stable primary magnetization vector.

In addition to the problem of remagnetization, this method of reorienting structures cannot account for secular variations of the earth's magnetic field. Such variations may be of considerable significance, of the order of ±15° in declination and inclination, particularly in those lithologies that have acquired their magnetic characteristics rapidly, such as lava flows. Typically, a period of about 10^4 to 10^5 years is considered sufficient to average out secular variations; thus slowly deposited sediment samples are considered to give more reliable indications of north than rapidly deposited sediments or lava flows.

Most important of all for our present purposes, the assumption that the primary magnetization vector of the sample will point due north or south can hold only if no tectonic rotation of the sample is suspected (see below). Nevertheless, under favourable circumstances, reorientation of core using palaeomagnetic data can be a simple and inexpensive technique in, for example, relatively undisturbed sedimentary sections from which palaeomagnetic measurements are already being made as a matter of routine (as with the Ocean Drilling Program).

Comparison with wireline logging data

A third approach, and one which mates the directly and remotely sensed methods of acquiring structural information, is to reorient cores by comparing them with downhole images obtained from modern speciality logging tools, such as the Formation MicroScanner (FMS) and Borehole Televiewer (BHTV). Data from these tools can be processed to generate 'images' of the borehole wall, from small-scale variations in, respectively, resistivity and acoustic backscattering (Ekstrom et al. 1986; Zemanek et al. 1970). The resolution of the BHTV is estimated to be in the order of 3 mm in a 40 cm (15.7 inches) diameter hole (Zemanek et al. 1970), and that for the FMS, 2.5–5 mm, although considerably finer features can be sensed by the latter if they have a sufficiently high resistivity contrast (Ekstrom et al. 1986; Serra 1989).

Both tools carry triaxial fluxgate magnetometers, so that the borehole images may be oriented to geographical co-ordinates, provided the formation is not too highly magnetic. Reorientation of core to the FMS and BHTV images requires the recognition of a sufficient number of orientable features that can be correlated between both core and downhole log. Regularly inclined bedding planes or consistently oriented fracture sets are ideal for this purpose. A 1:1 correlation between features in the cores and on the images is desirable, but in practice may be possible only for certain intervals of the borehole. Note that the substantially larger external diameter of the borehole compared to that of the core (typically 25 cm as opposed to 8 cm) means that the oblique features will have much greater vertical extents on the FMS and BHTV images than on the cores or core photos; only the strike azimuth of the feature will appear at the same depth (e.g.

Adams et al. 1990). In addition, it is important to realize that in areas of less than 100% recovery, most operators assume that all lost core comes from the bottom of the cored zone, when in reality it could have come from anywhere within the cored interval.

The technique by which reorientation of core using FMS and BHTV images is carried out is described in more detail below.

Measurement of planar features in cores

Several methods of measuring planar features in cores are in use. The most accurate and sophisticated way is to use a goniometer, either mechanical (e.g. Nelson et al. 1987) or electronic (Bergosh et al. 1985). Nelson et al. (1987) claim a reproducibility error of $\pm 2.75°$ in azimuth and $\pm 0.75°$ in dip for the mechanical instrument, and about half of that with the electronic device.

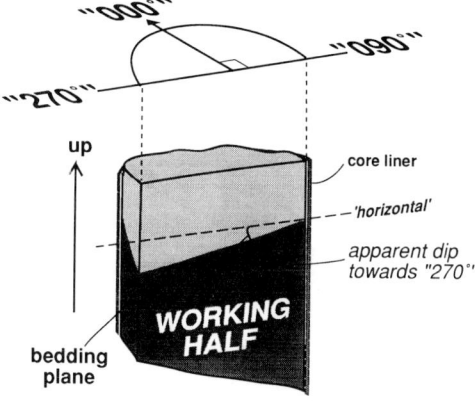

Fig. 1. Definition of artificial coordinates in core, as used, for example, with Ocean Drilling Program cores. ODP cores are routinely sectioned along their axes, with one half reserved for archiving and the other used for sampling. The artificial coordinates are usually defined as perpendicular to the cut face of the core in the working half. Spatial information (such as the apparent dip shown here) is initially recorded relative to these coordinates prior to attempts to reorient the core to real geographical coordinates.

An alternative technique, less precise but usable when no specialist equipment is available, is to measure the features by hand by combining apparent dip measurements. This technique has been used successfully with, for example, Ocean Drilling Program (ODP) cores (e.g. Leg 135: Parson, Hawkins, Allan et al. 1992; and see below), but is equally applicable to any core.

ODP cores, which are recovered in plastic liners, are routinely sectioned along the axis of the core. One half of the sectioned core is reserved for archiving and the other used for sampling purposes. The plastic core liners have orientation marks, and a nominal 'core liner north', which has no geographical significance, is chosen relative to these marks. Usually this 'north' is taken to lie in the working half of the core, perpendicular to the plane along which the core was split (Fig. 1). Standard procedure is initially to make all structural and palaeomagnetic measurements relative to these synthetic coordinates, prior to any attempt to reorient them to geographical co-ordinates by any of the techniques described here. Note that, although most sections of soft unlithified sediment recovered by hydraulic piston coring techniques should be contiguous within any one core barrel, in those more lithified formations recovered by rotary coring (particularly igneous units) in which recovery is usually poor, the core is broken up by the drilling process and each individual piece is rotated independently of its neighbours.

In unlithified sediments dips and strikes of bedding are made by the measurement of two apparent dips: one on the cut ('east–west') surface of the core, and another on a surface perpendicular to this, made by removing a quarter round of core using a scoop or some such apparatus (Fig. 2). With a moderate amount of practice these measurements can easily be made to the nearest degree. The two apparent dips are combined into a single dip and strike (relative to the synthetic coordinates) using simple stereographic techniques. Significant errors in the strike measurement in near-horizontal bedding are to be expected by this method, but these decrease rapidly as the dip increases (Fig. 3). Dip estimates are much more well constrained, and should be no worse than ± 1–$2°$.

In lithified but coherent cores synthetic orientations can be measured according to a similar principle, but instead by sawing the core to obtain the second apparent dip surface. Often a horizontal surface may be cut and a strike measurement obtained instead or as well, and this constrains much better the orientation of the feature, the estimated uncertainty in strike azimuth being reduced to no more than ± 1–$2°$.

Measurement of planar features on FMS and BHTV images

The FMS and BHTV carry triaxial fluxgate magnetometers that allow orientation of their resistivity and acoustic backscatter measurements respectively. The accuracy of the magnet-

ometers is claimed to be in the order of $\pm 2°$ (Schlumberger 1986), i.e. good relative to the orientation of features measured by hand from the core (Fig. 3), although the possibility of systematic errors in or adjacent to highly magnetic formations (e.g. basaltic lava flows) should be noted.

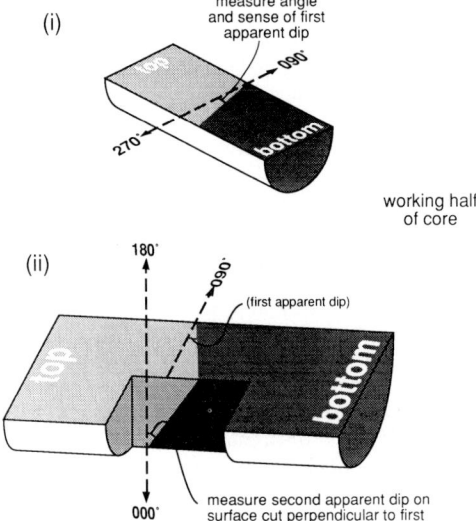

Fig. 2. Method of determining bedding direction in cores (relative to the core-liner coordinates) by the combination of two apparent dip measurements. (i) The apparent dip and direction of apparent dip are measured on the cut ('E–W') surface of the core. (ii) A quarter-round of core is removed perpendicular to the previous face to expose a 'N–S' section. A second apparent dip is measured on this surface. The two apparent dips are combined into a composite dip and strike using stereographic techniques.

Processed FMS and BHTV images are displayed on a rectangular plot that represents an unwrapped image of the inside of the borehole. The x-axis, equivalent in width to the circumference of the borehole, gives the azimuth of the image relative to true north, and the y-axis the borehole depth. Inclined planar features intersecting the borehole will appear with sinusoidal traces on plots (Fig. 4). Calculation of the orientations of planes is made simply by measuring the azimuth of the lowest point of the sinusoidal features to obtain the dip direction, and working out the arctangent of the difference in height of the top and bottom of the trace (h) divided by the borehole diameter (d) (Fig. 4). Acetate overlays with sinusoids of known dip may be constructed for use with hardcopy images, but are specific to given vertical and horizontal scales

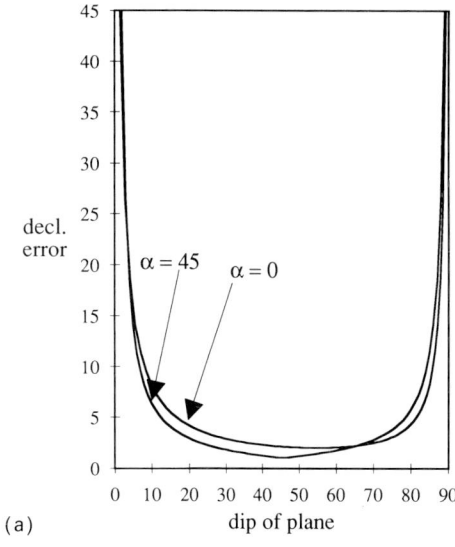

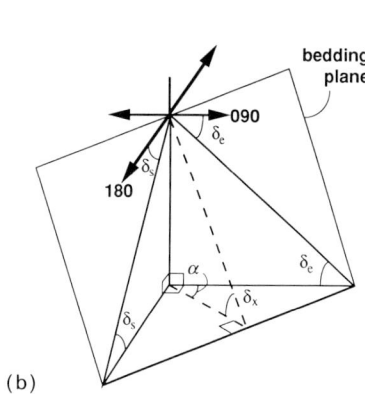

(a)   dip of plane   (b)

Fig. 3. Estimation of the errors inherent in the calculation of the orientation of a plane from the combination of two orthogonal apparent dip measurements, assuming that each apparent dip can be measured to the nearest degree. (a) Declination error as a function of dip of the plane. Curves are calculated for a plane dipping towards one of the apparent dip axes (labelled $a = 0°$), and for a plane dipping at 45° to these axes (labelled $a = 45°$). Note the extremely high potential declination errors for near-horizontal strata, but much lower errors as the dip of the plane increases. (b) Calculation of curves given in (a). δ_e and δ_s are the two orthogonal apparent dips, here shown relative to core liner 'east' and 'south' respectively, and δ_x the true dip of the plane; a is the dip direction or declination of the plane, for convenience here measured from 'east'. The maximum declination error is calculated as either a-a' or a-a'', where $\tan a' = \tan (\delta_s + 1)/\tan (\delta_e - 1)$ and $\tan a'' = \tan (\delta_s - 1)/\tan (\delta_e + 1)$. The two curves given in (a) are for $a = 0°$ and $a = 45°$; intermediate values of a will yield comparable results.

and borehole diameter. Software is now commonly in use that allows the matching of sinusoids interactively with computer-displayed images (e.g. the Schlumberger 'FMS Image Examiner': Serra 1989; Bourke *et al.* 1989, or the 'BHTV Image' interactive workstation: Barton *et al.* 1990). Errors introduced at this stage of the process, particularly with the computer-picked sinusoids, are considered to be minimal.

The BHTV presents coverage of 100% of the borehole wall; in contrast, an FMS image from a single pass of the tool covers approximately 20% (for the older two-pad tool) or 40% (for the more modern four-pad tool) of the surface of the borehole only. Whereas sinusoidal features can be observed unambiguously on BHTV images, on FMS plots they are discontinuous and must be interpolated between the individual pad traces. In order to minimize errors resulting from false correlations, multiple passes of the FMS tool are usually made and their resistivity images merged into a single plot to increase the effective coverage of the borehole.

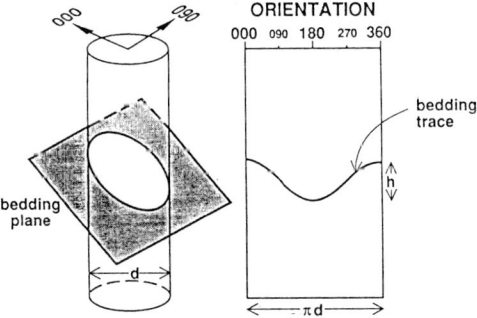

Fig. 4. Sketch of a hypothetical bedding plane intersecting a borehole; and the appearance of such a plane on an FMS- or BHTV-style plot. FMS and BHTV data are usually presented in the form of an unwrapped cylinder; planes intersecting the borehole will therefore appear as sinusoids on plots. The dip of the plane is calculated as the arctangent of the height of the trace on the borehole image (h) divided by the diameter of the hole (d).

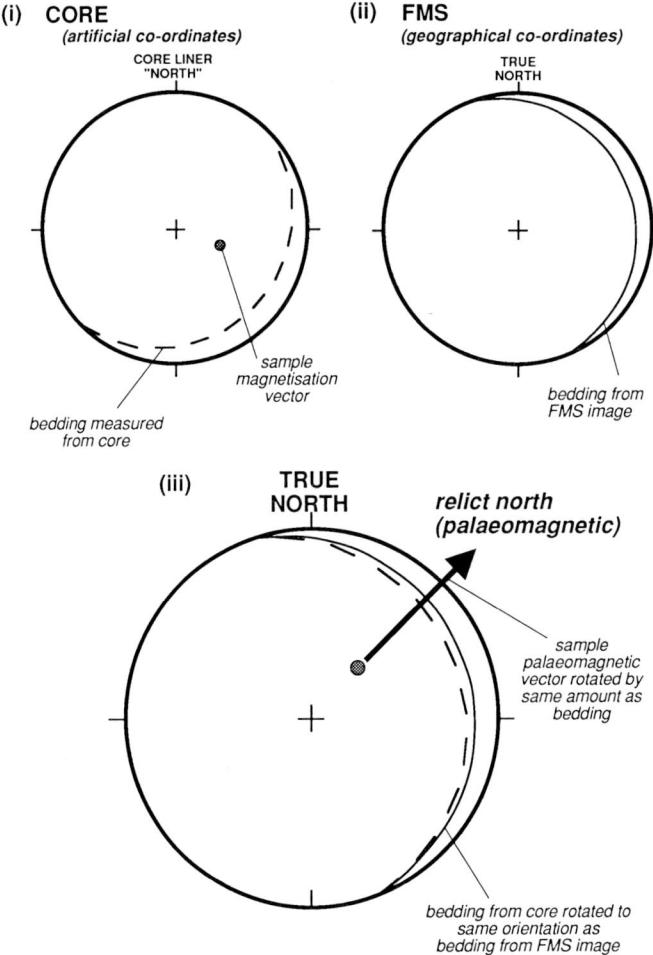

Fig. 5. (a) Sedimentary bedding measured from the core (Figs 1 & 2), together with the cleaned stable remanent magnetization vector from a sample from that interval, are plotted relative to core liner coordinates (lower hemisphere, equal angle projection). (b) The true sedimentary bedding dip and dip direction are known from FMS or BHTV images from the same interval. (c) The core bedding is therefore rotated to the true bedding orientation and the sample magnetization vector rotated by the same amount. The deviation of the restored magnetization vector from north (or from south if the sample is of reversed polarity) may be due to a tectonic rotation.

In addition to the manual picking of dips from FMS images, whether by hand or with a workstation, FMS data from the four-pad tool can be processed to yield plots of bedding dip and strike analogous to those produced by the Dual Dipmeter (or Stratigraphic High Resolution Dipmeter) Tool. In essence, this processing involves the comparison and correlation of resistivity responses between the four pads over a given interval of the borehole. The resistivity response of a dipping feature with characteristic signature will be recorded by the pads at different depths, and from this its dip and strike can be calculated. A number of techniques for the computation of dips from the resistivity data are in use, based upon different methods of correlation between pads. The techniques, and interpretation of their results, are discussed in detail in Schlumberger (1986). However, it is pertinent to note that most of these methods (including the standard 'Mean Square Dip' or MSD computation) involve calculation of a single 'best fit' average dip measurement over a given search interval, typically 4 feet (1.2 m); therefore smaller scale features that may be visible in the core and on the FMS images may not be discriminated.

Integration of log-reoriented core with palaeomagnetic measurements

If an inclined planar feature visible on FMS or BHTV images of an interval of borehole can also be recognized in core from the same interval, then that section of core can be reoriented to geographical coordinates. Not only are the inclined features reoriented (be they bedding, joint planes or whatever), but also all other more cryptic features within the same contiguous core piece that are too small or indistinct to be imaged by the logging tools; e.g. fault slip lineations, microfabrics etc. The data thus oriented may be of enormous value for a wide variety of applications. In addition, and for the present purposes most significant of all, palaeomagnetic data can also be reoriented to geographical co-ordinates.

The reorientation technique is illustrated in Fig. 5, and demonstrated more fully in the example given below. Systematic deviations of restored, cleaned palaeomagnetic declinations away from north may be indicative of tectonic rotations about vertical or inclined axes. Analysis of changes in restored declination directions with respect to the chronostratigraphy of the borehole should show when such tectonic rotations took place.

The errors inherent in the reorientation technique are estimated to be typically in the order of ± 8–$12°$ (at two standard deviations) for hand-measured dips (depending whether a dip and a strike or two apparent dips are used to define the core measurement; the stated figure is for a plane dipping at $15°$), or ± 7–$8°$ using an electronic goniometer. These figures assume an uncertainty in the FMS/BHTV magnetometer measurement of $\pm 2°$, and $\pm 3°$ in the palaeomagnetic measurement.

In practice, definite correlations between inclined features in core and on log images may only be possible at certain intervals within the borehole. This does not necessarily restrict the usefulness of the technique, for in determining the magnitude and timing of tectonic rotations it is better to obtain accurately restored declinations from fewer discrete, well constrained intervals in the borehole than to attempt to make correlations of dubious reliability from the entire depth range. The principal uncertainty in the technique is likely to be in making inappropriate

matches between features in the core and on FMS or BHTV logs; thus as large a number of reorientations as possible should be made from the better constrained intervals (typically regular, parallel sedimentary bedding over a depth range of several metres to tens of metres) to minimize these spurious matches.

This method of core orientation can potentially be applied to any core for which logging data are available, and at any stage, for example if the need for obtaining oriented material was not appreciated at the time drilling took place and the multishot tool therefore not employed. It is also an extremely cheap means of acquiring oriented core when compared to the operating costs of either the camera-based or electronic multishot tools, and does not suffer the disadvantage of the older (ODP-type) multishot orientation tool of being restricted in use to soft sediment.

Example

The Lau Basin is a triangular-shaped active backarc marginal basin in the southwestern Pacific, forming at the leading edge of the Indo-Australian plate as it overrides the subducting Pacific plate (Fig. 6a). The basin lies between a north-trending remnant arc (the Lau Ridge) to the west and a north-northeast-trending active island arc/forearc complex (the Tonga Ridge) in the east (Fig. 6b). It is generally assumed (following Karig 1970 and Parson, Hawkins, Allan et al. 1992) that the opening history of the basin, which spans a period of 6 Ma, involved a protracted period of attenuation and rifting of the original Lau/Tonga protoarc prior to initiation of the backarc seafloor spreading known to be active at the present day (e.g. Collier & Sinha 1990). Sidescan sonar studies have shown that the spreading axis is segmented, and a large, southward-propagating rift identified (Parson et al. 1990; Fig. 6b). The extent of opening of the basin has clearly been greater in the north-central part than in the south, but the timing and mechanism of this apparent fan-shaped opening is not well understood. Various models have been proposed that involve rifting and translation of the active and remnant arcs during basin formation: either of the Lau Ridge away from the (essentially static) Tonga Ridge (Packham 1978; Herzer & Exon 1985) or *vice versa* (Karig 1970; Hawkins et al. 1984). All of these hypotheses require the bulk rotation of either one or both of the Lau and Tonga Ridges as the Lau Basin opened. Although the end result is similar from the point of view of the morphology of the basin, the implications for the

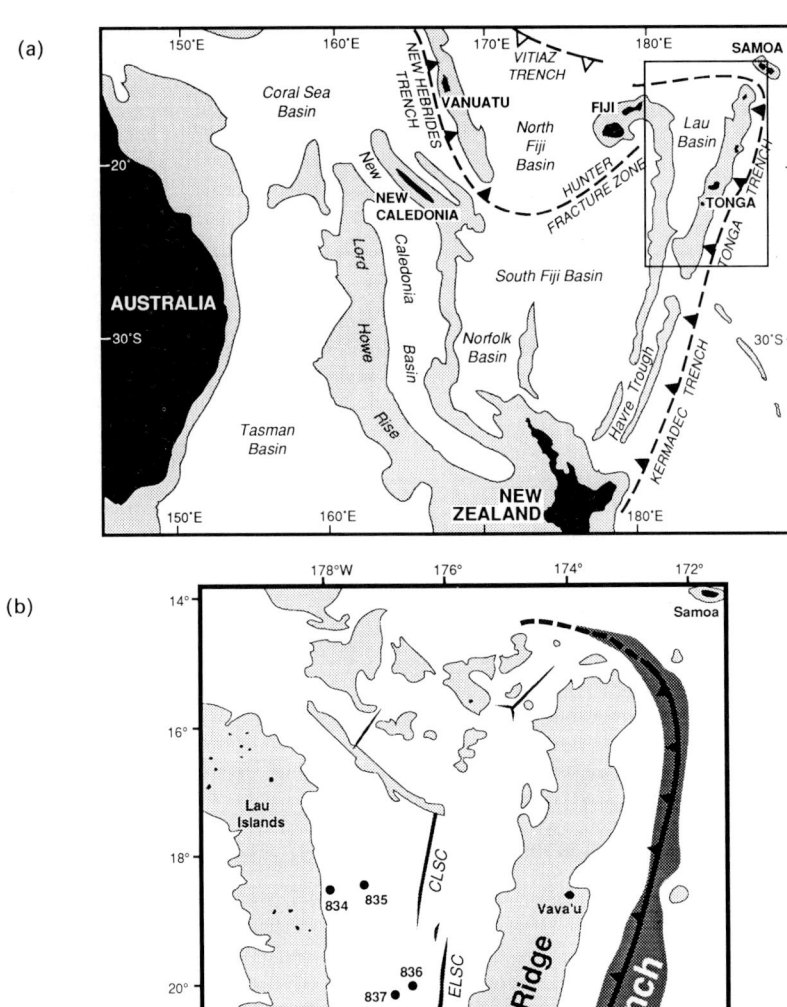

Fig. 6. (a) Location of the Lau back-arc basin. The shaded areas represent those with bathymetries of less than 2000 m. Active subduction zones are marked with black barbs; inactive subduction zones with white barbs. The area of map (b) is shown by the box at upper right. (b) Detail of the Lau Basin area. Locations of sites drilled on ODP Leg 135 are indicated, together with DSDP Site 203. The data considered in this paper come from Site 841, on the upper trench slope of the Tonga forearc. CLSC, Central Lau Spreading Centre; ELSC, Eastern Lau Spreading Centre.

Table 1. *Sediment dips from the interval 490–550 mbsf, Hole 841B, ODP Leg 135, are calculated using the method of two apparent dips, or an apparent dip and a strike measurement, as per Figs 1 & 2, and subsequently combined into an auxiliary dip and dip direction. 'B', 'R', 'T' and 'L' suffixes refer to apparent dip directions towards the temporary core-liner coordinates '000', '090', '180' and '270' respectively. Palaeomagnetic data (from the same core pieces) are stable, assumed primary, magnetization directions isolated after alternating field demagnetization. All these palaeomagnetic and auxiliary bedding measurements are made relative to core-liner coordinates. The true dip directions of the sediments are obtained from Formation MicroScanner data from the same interval of adjacent Hole 841C (Parson, Hawkins, Allan et al. 1992). From these data corrected magnetic azimuths can be calculated, and these are plotted in Fig. 7.*

leg-hole-core-sect-locn	depth mbsf	app dip 1	app dip 2	app str	aux dip	dip dirn	cleaned dec; inc	FMS dip dirn	corr azi (true N)
135-841B-35R-2-113	489.73	22L		013	23	283	173; −29	113	016
135-841B-35R-3-123	491.33	18R	28T		32	149	047; −37	113	024
135-841B-35R-3-125	491.35	25R	37T		42	148	047; −37	113	025
135-841B-36R-1-56	497.36	25R	2T		25	94	000; −15	095	014
135-841B-36R-1-58	497.38	21R	0		21	90	000; −15	095	018
135-841B-37R-1-76	507.16	27L	2T		27	266	131; −48	135	013
135-841B-37R-3-114	510.54	30L	1B		30	272	159; −48	135	022
135-841B-37R-3-119	510.59	24L		013	26	283	159; −48	135	024
135-841B-37R-4-90	511.80	25L	11T		27	247	170; −54	135	071
135-841B-38R-1-45	516.55	21R	11T		23	115	012; −22	135	027
135-841B-38R-1-57	516.67	31R	6T		31	100	012; −22	135	055
135-841B-38R-2-135	518.95	12R		000	12	90	358; −45	135	051
135-841B-38R-3-8	519.18	23R	4T		23	99	337; −39	135	021
135-841B-38R-4-21	520.81	18R	5T		19	105	002; −54	135	040
135-841B-40R-1-50	535.80	8L	8T		11	225	171; −21	135	094
135-841B-40R-1-51	535.81	8L	5T		9	238	171; −21	135	081
135-841B-40R-1-114	536.44	21L	2B		22	283	191; −66	135	043
135-841B-40R-1-120	536.50	13L	8B		15	301	191; −66	135	038
135-841B-40R-2-49	537.29	29L	3B		29	275	185; −38	135	058
135-841B-40R-3-4	538.34	25L	3B		25	276	191; −35	135	063
135-841B-40R-3-7	538.37	33L	1T		33	268	191; −35	135	058
135-841B-40R-3-11	538.41	32L	0		32	270	191; −35	135	069
135-841B-41R-1-40	545.40	21L	27B		33	323	222; −66	165	077
135-841B-41R-2-29	546.79	32R	24T		37	125	335; −63	165	028
135-841B-41R-2-31	546.81	33R	16T		35	114	335; −63	165	026
135-841B-41R-2-34	546.84	30R		026	33	116	335; −63	165	024
135-841B-41R-3-23	548.23	29L	27B		37	313	197; −59	165	062

mechanism of back-arc basin formation are important.

ODP Leg 135 drilled six sites in the western part of the Lau Basin, between the Lau Ridge and active Lau spreading centres, and two on the Tonga platform/forearc (Fig. 6b). The latter two sites were of particular interest with regard to the above debate; specifically, to attempt to determine whether or not the Tonga Ridge had rotated clockwise during opening of the Lau Basin. Site 841 was drilled in approximately 4800 m of water on the upper trench slope of the Tonga forearc, through late Eocene to Recent turbidites and oozes into a dacitic arc complex (Parson, Hawkins, Allan et al. 1992). Active normal faulting has been documented at the site

and has given rise to tilting of the succession, such that sediment dips in excess of 30° are not uncommon, even close to seafloor. This tectonic tilting makes the site ideal for application of the reorientation technique described herein, as errors due to the manual measurement of core dips are minimized (cf. Fig. 3).

In Table 1 cleaned remanent magnetization vectors of discrete samples from planar-bedded Lower-Middle Miocene sediments in the interval 490–500 metres below seafloor (mbsf) in Hole 841B are restored to geographical coordinates. This particular depth range was chosen because a very regular direction of dip to the sediments is indicated on shipboard FMS-derived dipmeter plots from this interval. Systematic 1:1 correlation between individual strata in the recovered core and on the FMS images is yet to be made (MacLeod *et al.*, work in progress); instead, because of the regularity of the FMS dipmeter dip direction over a scale of metres to tens of metres in this section of the borehole, it is considered justified to restore individual sediment dips (and their accompanying palaeomagnetic vectors) to an average dip direction for each core.

Restored azimuths of cleaned magnetization directions from samples in the 490–550 mbsf interval are plotted in Fig. 7, using data taken from Table 1. They show a consistent deviation of the horizontal component of the magnetization vector to the east of north, with an average declination of 031.8°, inclination of −44.9°, and a_{95} cone of confidence of 8.5°. Application of conventional bedding corrections to these data reduces both the declination and inclination of the mean magnetization vector by a few degrees; however, the uncorrected data are preferred here as we cannot easily constrain the relative contributions of tectonic tilting versus primary depositional dip to the overall dip of the sediments. This is discussed further in a forthcoming publication (MacLeod *et al.* in preparation, ODP Leg 135 Scientific Results).

We interpret the northeasterly mean declination in Fig. 7 to result from the clockwise rotation by tectonic means of at least this portion of the borehole. Preliminary palaeomagnetic results from unlithified sediments at both Leg 135 sites on the Tonga Ridge (i.e. Sites 840 and 841; Fig. 6b) that were oriented using the ODP multishot tool also show statistically significant clockwise deviations of magnetic declinations away from true north (mean declination 036° ± 13°: Sager *et al.* 1991). These data confirm the FMS-oriented results given here, supporting the suggestion that clockwise rotation of the entire Tonga Ridge has occurred. They contrast markedly, however, with palaeomagnetic results from the western part of the Lau Basin (Sites 834–839: Fig. 6b), which exhibit consistent northerly declinations (Sager *et al.* 1991) and therefore do not appear to have suffered comparable tectonic rotations.

Discussion

The use of palaeomagnetic techniques in land-based field study has become commonplace, and has shown that tectonic rotations on a variety of scales are an integral part of the deformation process in continental tectonic belts (e.g Kissel & Laj 1989, and papers therein). It is clear that such rotations do not occur only on land, nor are they restricted to continental crust: structural and palaeomagnetic studies of ophiolites have shown that rotations respectively about sub-horizontal and sub-vertical axes are significant at least at some ocean ridges and transform faults (Allerton & Vine 1987; Bonhommet *et al.* 1988, Allerton 1989, MacLeod *et al.* 1990). Anomalous palaeomagnetic inclinations in basalts from some Deep Sea Drilling Program cores have been attributed to the effects of listric normal faulting (Verosub & Moores 1981, 1985); nevertheless, although vertical axis rotations have been postulated from analysis of magnetic lineations (e.g. Tamaki & Larson 1988; Searle *et al.* 1989) and seamount magnetic anomalies (Sager & Pringle 1987), they have largely been inferred by indirect means only. This is mostly for practical reasons, due principally to the difficulty of the recovery and orientation of samples from the seafloor.

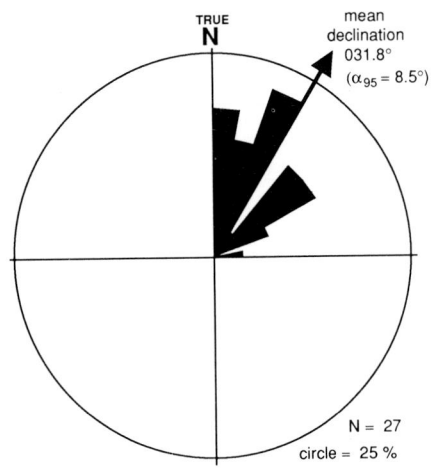

Fig. 7. Restored magnetic declinations for data from the interval 490–550 mbsf from Site 841, ODP Leg 135. The restored data show a clear deviation away from north, and it is suggested that this is a consequence of a clockwise rotation of the Tonga Ridge upon opening of the Lau Basin. See text for discussion.

The technique outlined here offers one of the first direct approaches to the analysis of tectonic rotations in the oceanic domain. The Ocean Drilling Program is perhaps uniquely suited to such study; hence much of the description of the method given above has been geared towards ODP-type cores. Application of the technique is not, however, restricted to study of the oceanic crust. Analysis of restored declinations in sedimentary basins or at passive margins, and particularly of changes in restored declination across faults in boreholes, can help constrain the kinematics of rifting and potentially identify oblique components of extension within rift zones. Rotation of terrains or microplates should also be recognizable, enabling better understanding of, for example, the plate tectonic evolution of oceanic plateaux, or (as illustrated above) of the arcs and marginal basins of the West Pacific.

We would like to thank the following: the organizers of the symposium and the editors of this volume, for their patience; R. B. Whitmarsh, B. J. Murton and two anonymous reviewers, whose comments significantly improved the manuscript; M. Rees and N. Brown of Sperry-Sun Drilling Services in Aberdeen, for supplying unpublished information on the modern electronic multishot devices; and special thanks are offered to N. Abrahamsen and R. Reynolds of the Leg 135 Scientific Party for performing some of the sample demagnetizations and processing the FMS/dipmeter data respectively on board ship.

References

ADAMS, J., BOURKE, L. & BUCK, S. 1990. Integrating formation images and cores. *Oilfield Review* 2, 52–65.

AGAR, S. M. 1990. Fracture evolution in the upper ocean crust: evidence from DSDP Hole 504B. *In*: KNIPE, R. J. & RUTTER, E. H. (eds) *Deformation Mechanisms, Rheology and Tectonics*. Geological Society, London, Special Publication, **54**, 41–50.

ALLERTON, S. 1989. Fault block rotations in ophiolites: results of palaeomagnetic studies in the Troodos Complex, Cyprus. *In*: KISSEL, C. & LAJ, C. (eds) *Paleomagnetic Rotations and Continental Deformation*. NATO ASI Series **C254**, Kluwer, Dordrecht, 393–410.

—— & VINE, F. J. 1987. Spreading structure of the Troodos ophiolite, Cyprus: some palaeomagnetic constraints. *Geology* **15**, 593–597.

BARTON, C. A., TESLER, L. G. & ZOBACK, M. D. 1990. Interactive image analysis of Borehole Televiewer data. *In*: PALAZ, I. & SENGUPTA, S. K. (eds) *Automated Pattern Recognition in Exploration Geophysics*. Geophysical Press.

BERGOSH, J. L., MARKS, T. R. & MITKUS, A. F. 1985. New core analysis techniques for naturally fractured reservoirs. SPE 13653.

BONHOMMET, N., ROPERCH, P. & CALZA, F. 1988. Palaeomagnetic arguments for block rotations along the Arakapas fault (Cyprus). *Geology*, **16**, 422–425.

BOURKE, L., DELFINER, P., TROUILLER, J.-C., FETT, T., GRACE, M., LUTHI, S., SERRA, O. & STANDEN, E. 1989. Using Formation MicroScanner images. *Technical Review*, **37**, 16–40.

COLLIER, J. & SINHA, M. 1990. Seismic images of a magma chamber beneath the Lau Basin back-arc spreading centre. *Nature*, **346**, 646–648.

EASTMAN WHIPSTOCK, 1982. Eastman Whipstock Inc. general catalogue.

HAWKINS, J. W., BLOOMER, S. H., EVANS, C. A. & MELCHIOR, J. T. 1984. Evolution of intra-oceanic arc-trench systems. *Tectonophysics* **102**, 175–205.

HERZER, R. H. & EXON, N. F. 1985. Structure and basin analysis of the southern Tonga forearc. *In*: SCHOLL, D. W. & VALLIER, T. L. (eds) *Geology*

and Offshore Resources of the Pacific island arcs–Tonga region. Circum-Pacific Council for Energy and Mineral Resources, Earth Science Series **2**, 55–74.

KARIG, D. E. 1970. Ridges and basins of the Tonga–Kermadec island arc system. *Journal of Geophysical Research* **75**, 239–254.

KISSEL, C. & LAJ, C., 1989 (eds) *Paleomagnetic Rotations and Continental Deformation* NATO ASI Series **C254**, Kluwer, Dordrecht.

KNIPE, R. J., 1986. Microstructural evolution of vein arrays preserved in Deep Sea Drilling Project Cores from Japan Trench, Leg 57. *In*: MOORE, J. C. (ed.) *Structural Fabric in Deep Sea Drilling Project Cores from Forearcs*. Geological Society of America, **166**, 75–87.

LEHNE, K. A. 1990. Fracture detection from logs of North Sea chalk. *In*: HURST, A., LOVELL, M. A. & MORTON, A. C. (eds) *Geological Applications of Wireline Logs*. Geological Society, London, Special Publication, **48**, 263–271.

LUNDBERG, N. & MOORE, J. C., 1986. Macroscopic structural features in Deep Sea Drilling Project cores from forearc regions. *In*: MOORE, J. C. (ed.) *Structural Fabric in Deep Sea Drilling Project Cores from Forearcs*. Geological Society of America Memoir 166, 13–44.

MACLEOD, C. J., ALLERTON, S., GASS, I. G. & XENOPHONTOS, C. 1990. Structure of a fossil ridge-transform intersection in the Troodos ophiolite. *Nature*, **348**, 717–720.

MOORE, J. C. (ed.) 1986. *Structural Fabric in Deep Sea Drilling Project Cores from Forearcs*. Geological Society of America Memoir 166.

NELSON, R. A., LENOX, L. C. & WARD, B. J. 1987. Oriented core: its use, error and uncertainty. *AAPG Bulletin* **71**, 357–367.

OCEAN DRILLING PROGRAM, 1990. *Wireline logging manual*. Borehole Research Group, Lamont-Doherty Geological Observatory, Palisades, New York.

PACKHAM, G. H. 1978. Evolution of a simple arc: the Lau–Tonga Ridge. *Bulletin of Australian Society Explor. Geophys.* **9**, 133–140.

PARSON, L. M., HAWKINS, J. W., ALLAN, J. & the ODP Leg 135 Scientific Party, 1992 (in press). *Proceedings of the Ocean Drilling Program, Initial Reports*. Ocean Drilling Program, College Station, TX, **135**.

——, PEARCE, J. A., MURTON, B. J., HODKINSON, R. & the RRS Charles Darwin Scientific Party, 1990. The role of ridge jumps and ridge propagation in the tectonic evolution of the Lau backarc basin. *Geology*, **18**, 470–473.

PEZARD, P. A. & LUTHI, S. M. 1988. Borehole electrical images in the basement of the Cajon Pass scientific drillhole; fracture detection and tectonic implications. *Geophysical Research Letters*. **15**, 1017–1020.

SAGER, W. W., ABRAHAMSEN, N. &. MACLEOD, C. J. 1991. Tectonic rotation of the Tonga arc, Southwest Pacific, from ODP Leg 135 paleomagnetic data. *EOS, Trans. Amer. Geophys. Union*, **72**, 541.

—— & PRINGLE, M. S. 1987. Paleomagnetic constraints on the origin and evolution of the Musicians and South Hawaiian seamounts, central Pacific Ocean. *In*: KEATING, B. H., FRYER, P., BATIZA, R. & BOEHLERT, G. W. (eds) *Seamounts, Islands and Atolls*. Geophysics Monograph Series **43**, A.G.U., Washington D.C., 133–162.

SCHLUMBERGER, 1986. *Dipmeter Interpretation: Fundamentals*. Schlumberger Ltd.

SEARLE, R. C., RUSBY, R. I., ENGELN, J., HEY, R. N., ZUKIN, J., HUNTER, P. M., LeBAS, T. J., HOFF-MAN, H-J. & LIVERMORE, R. 1989. Comprehensive sonar imaging of the Easter microplate. *Nature*, **341**, 701–705.

SERRA, O. 1989. *Formation MicroScanner image interpretation*. Schlumberger Educational Services.

SPERRY-SUN, 1991. *ESS-Electronic Core Orientation*. Sperry-Sun Drilling Services, Aberdeen.

TAMAKI, K. & LARSON, R. L. 1988. The Mesozoic history of the Magellan microplate in the western central Pacific. *Journal of Geophysical Research*. **96**, 2857–2874.

TEUFEL, L. W. 1982. Prediction of hydraulic fracture azimuth from anelastic strain recovery measurements of oriented core. *In*: GOODMAN, R. E. & HEUZE, F. E. (eds) *Issues in Rock Mechanics*. Proceedings of 23rd U.S. Symposium on Rock Mechanics, 238–245.

VEROSUB, K. L. & MOORES, E. M. 1981. Tectonic rotations in extensional regimes and their paleomagnetic consequences for oceanic basalts. *Journal of Geophysical Research* **86**, 6335–6349.

—— & —— 1985. "Tectonic rotations in extensional regimes and their paleomagnetic consequences for oceanic basalts." Reply to Comment by S. C. Cande & D. V. Kent. *Journal of Geophysical Research* **90**, 4652–4654.

ZEMANEK, J., GLENN, E. E., NORTON, L. J. & CALDWELL, R. L. 1970. Formation evaluation by inspection with the Borehole Televiewer. *Geophysics*, **35**, 254–269.

Application of dipmeter data in structural interpretation, Niger Delta

J. T. ADAMS,[1] J. K. AYODELE,[2] J. BEDFORD,[3] C. H. KAARS-SIJPESTEIJN[2]
& N. L. WATTS[2]

[1] *Schlumberger Exploration and Reservoir Services (UK) Ltd., Woodlands Drive,
Kirkhill Industrial Estate, Dyce, Aberdeen, AB2 0ES, UK*
[2] *Shell Petroleum Development Company of Nigeria Ltd., Freeman House, 21/22 Marina,
P.M.B. 2418, Lagos, Nigeria*
[3] *Schlumberger (Nigeria) Ltd., NRJ, c/o AFP, Batiment H, 50 Avenue Jean Jaures,
B.P. 362, 92541 Montrouge, France*

Abstract. Modern 3D seismic techniques provide both high resolution and dense coverage for interpretation of subsurface structures, and are a great improvement on older 2D methods. Features at the lower limit of seismic resolution, however, are often not resolved and interpretation can be ambiguous. Structures at this scale, in particular small-scale faults, are critical for optimum placement of production wells and enhanced field development. Integration of structural analysis techniques from borehole dipmeter data with 3D seismic data allows re-examination of an interpretation and enhancement of such small-scale features.

Dipmeter analysis of a faulted structure in the Niger Delta has enhanced the structural model derived from 3D seismic. The faulted intervals were first identified using vertical dip trend analysis, and the faults were orientated by applying stereographic techniques to bedding dips in zones of fault-related deformation. Near-wellbore structure was then modelled by constructing dipmeter-based cross-sections using an interactive workstation, and minimum fault throw was estimated from these models. Accurate positioning of the faults on the seismic data was achieved by time-indexing the dipmeter log, and the structure was integrated with the present-day stress regime using borehole breakout analysis.

Application of these methods has allowed faults that were below seismic resolution to be identified and oriented, and for these to be explained in terms of the regional structural regime.

Advances in seismic acquisition and processing technology have allowed 3D seismic surveys to become the standard method of evaluating the structure of a field during the appraisal and subsequent development stages; in fact, many companies are now realizing the importance of 3D seismic for exploration drilling, particularly with smaller, complex or subtle traps. 3D techniques have many advantages over older 2D surveys, including clearer delineation and orientation of structural features due to the increased sampling density (commonly 12.5 m × 25 m or 25 m × 25 m; Boreham *et al.* 1991), and the ability to carry out sophisticated seismic attribute processing for reservoir property analysis (Bouvier *et al.* 1989; Ruijtenberg *et al.* 1990; Brown 1988). Features at the lower limit of seismic resolution (i.e. less than about 150 feet (50 metres) thick) or with low acoustic impedance contrast are often missed and interpretation can become ambiguous in areas of complex faulting (Bouvier *et al.* 1989).

Structures, especially faults, at this scale are important to reservoir production (Hardman & Booth 1991), as they may form permeability barriers which can lead to increased reservoir compartmentation, or juxtapose reservoir rocks providing increased communication. Identification of small-scale faults is particularly important for optimal placement of development wells and for subsequent recognition in drilled wells for incorporation into field volumetrics. Highly permeable small-scale faults can give positive influence on well productivity provided such increases in effective vertical permeability do not result in early water or gas breakthrough. By contrast healed or smeared faults can introduce vertical barriers to horizontal cross-flow with a strong effect on well drainage volumes and reservoir sweep. Finally, recognition and quantification of small-scale faulting is becoming crucial for planning of high cost horizontal wells. Whether such wells are drilled for increased productivity (higher well rate) or

From HURST, A., GRIFFITHS, C. M. & WORTHINGTON, P. F. (eds), 1992,
Geological Applications of Wireline Logs II. Geological Society Special Publication No. 65, pp. 247–264.

247

reduced drawdown (constant well rate but longer dry oil period), the presence of such faulting will be crucial to the performance of the well. It is therefore important to gain as much detailed information as possible on such structures to optimize reservoir development.

The aim of this paper is to show how advanced dipmeter interpretation techniques can be applied in complex structural settings, and how faults interpreted at the borehole scale may then be integrated with a 3D seismic interpretation to help improve structural control. The example illustrated here involved detailed examination of a single well in the Akaso Field, Nigeria, but the techniques used are applicable in many tectonic and depositional settings.

Location and structural setting

The Akaso Field is situated in oil mining lease OML-18, some 12 miles (20 km) south of Port Harcourt in the eastern coastal swamp area of the Niger Delta (Fig. 1). Hydrocarbons occur in a stacked series of footwall closures against a major antithetic fault, the Akaso boundary fault, which separates Akaso from the larger Cawthorne Channel Field to the north (Fig. 2). Both fields are typical of the megastructural style of the Niger Delta (Evamy *et al.* 1978) where syndepositional growth faulting in the delta has resulted in a range of closures from simple rollover anticlines, through stacked footwall/ hangingwalls to complex, collapsed crest structures of the Cawthorne Channel type (Weber 1986).

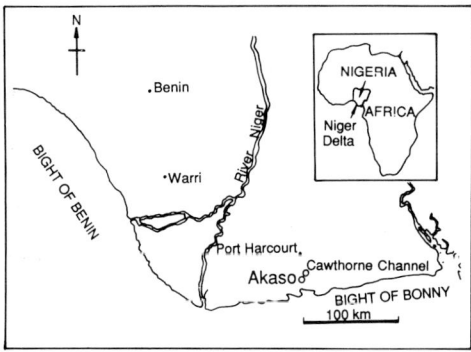

Fig. 1. Location map, Akaso Field.

The Akaso Field is a relatively recent discovery. The first well was drilled vertically in a downdip position in 1979 and found relatively small volumes of oil (and gas). Appraisal drilling in 1988/1989 increased these volumes fifteen-fold

through the placement of deviated fault scooping wells backed up by improved 2D seismic. This encouragement was followed by the acquisition of 3D seismic in 1988 as a southern rim to a large survey over Cawthorne Channel. Full coverage over Akaso was completed in 1990.

The 3D seismic has resulted in even larger oil and gas volumes through the identification of elongate closures and by adventurous appraisal drilling backed up by sophisticated amplitude and fault seal studies (Jev *et al.* 1991). The field is now recognized as being divided into a number of appraised and unappraised blocks. Backsplits of the main antithetic fault and a series of NE–SW oriented faults have been recognized. Seismic resolution is hampered, however, by the shadow effects of the main antithetic fault (Fig. 3) and interference from a second large fault (Bonny Fault) which swings northeastward and results in Akaso being a back-to-back structure (Fig. 2b; Weber 1986).

Development wells drilled in 1990 generally confirmed the structural interpretation but revealed considerably greater complexity than hitherto realized. All six development wells were faulted over the main reservoir interval and led to re-evaluation of the earlier four wells. The resulting structural picture is highly complex with a number of cut-outs, not immediately recognizable on seismic, being identified. Additional faults could only be resolved using advanced dipmeter analysis.

The latter is the focus of this paper, based upon a suite of logs taken in well Akaso-10, drilled in early 1990.

Dipmeter data

The well examined in this study was drilled to a total depth of 11 700 feet (3566 m), and was logged using an 8-curve Schlumberger Stratigraphic High Resolution Dipmeter Tool (SHDT*), as described in Schlumberger (1986) and Höcker *et al.* (1990). The well had a maximum deviation of 8°, and was drilled with a 12.25 inch (312 mm) bit using water-based mud.

Interpretation of faults and other major structures requires that large-scale processing parameters be used, and in this case a Mean Square Dip (MSD) interval correlation processing using a 4 feet (1.2 m) correlation interval, 2 feet (0.6 m) step distance, and 40° × 2 search angle was run. Variation of the processing parameters did not produce any significant improvement in result quality for the purpose of structural interpretation.

---

* Mark of Schlumberger

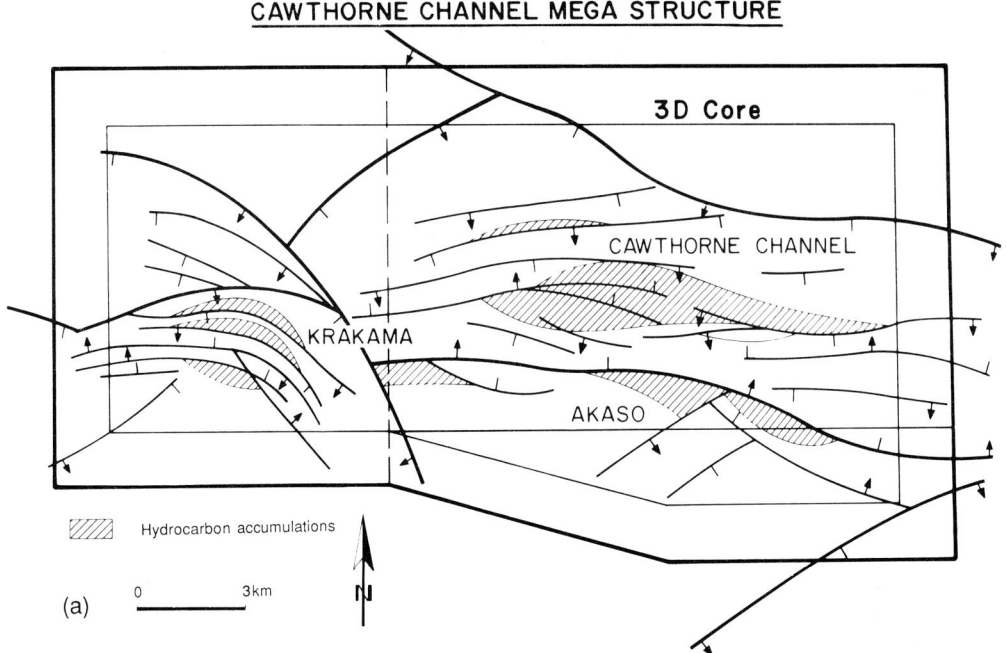

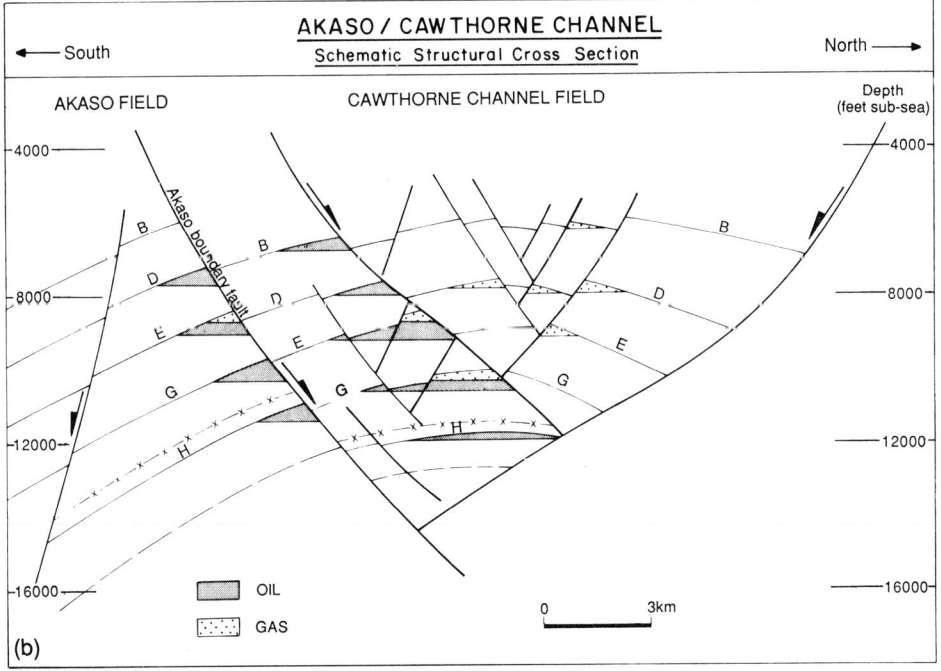

Fig. 2. Structural setting of the Akaso and Cawthorne Channel Fields. (a) Map of Cawthorne Channel megastructure. (b) Schematic structural cross section, Akaso and Cawthorne Channel Fields.

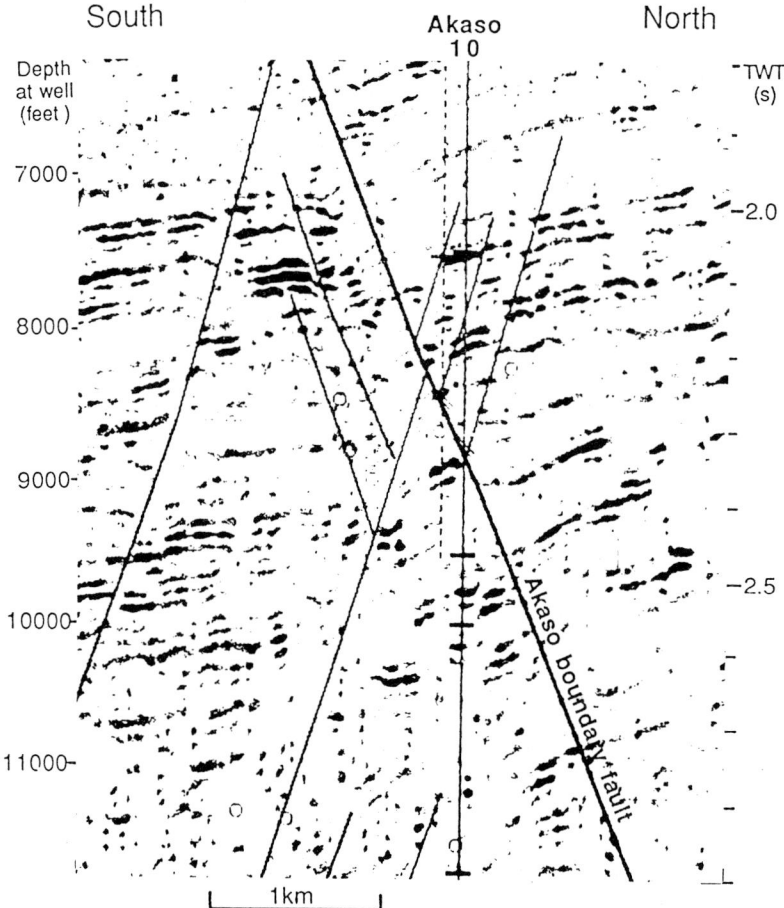

Fig. 3. Seismic line over the Akaso Field and southern portion of the Cawthorne Channel Field, showing the original (pre-dipmeter) structural interpretation.

Data quality in this well is affected mainly by hole conditions. Over most of the interval examined, hole conditions were excellent, with the hole remaining cylindrical and in-gauge, however several distinct washed out zones were observed, related to zones of major dip change. In these zones the hole was enlarged over the maximum extension of the dipmeter caliper arms (21 inches (54 cm)). The significance of these zones will be discussed later; however, it should be noted that within these intervals the dipmeter curves show no visible cross-borehole correlation, and consequently no reliable dip results are generated.

Dipmeter interpretation approach

The following interpretation sequence was followed to derive maximum structural information from the wellbore data. Dipmeter interpretation involves examination and re-examination of the data using different techniques and therefore the order specified below should be regarded as fully iterative.

1. Identification of structural dip zones and faulted intervals from vertical dip trends.
2. Detailed examination of faulted intervals on dipmeter curves.
3. Stereoplot analysis to determine fault orientation.
4. Zonation of major structural blocks using borehole breakout orientations.
5. Cross section modelling.
6. Estimation of minimum fault throw.

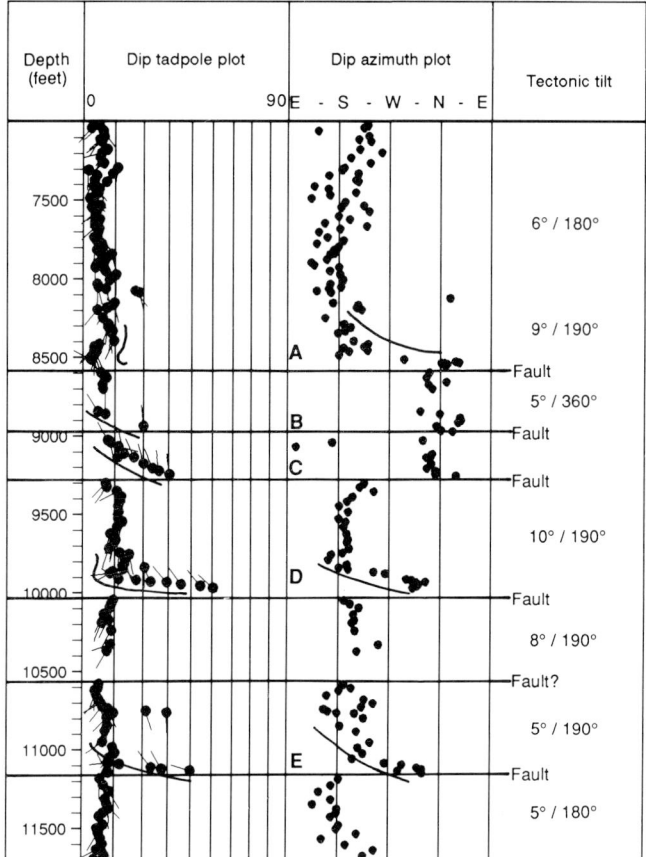

Fig. 4. Summary dipmeter log, well Akaso 10, showing interpreted fault cuts and zones of bedding rotation (A–E).

Vertical dip trend analysis

Analysis of vertical dip trends is an essential first stage in structural interpretation of dipmeter data, and allows the accurate depth-positioning of fault cut-outs in the wellbore. The objective of this stage is to identify zones where dip remains reasonably constant, thus defining tectonic tilt of the structural block. Zones where dip rotation occurs provide evidence of the position and nature of structural block boundaries.

Fault block tilt is identified from zones of uniform dip (equivalent to traditional "green" patterns: Schlumberger 1986) in beds that were originally deposited horizontally, such as mudstones, well laminated heterolithics or parallel-laminated sandstones. In the well described in this paper tectonic tilt ranges from 4–10° South (Fig. 4). In this case, where dip magnitude is relatively constant at less than 10°, but dip

direction shows considerable variation, bedding rotation is better observed using dip azimuth plots (Devilliers & Werner 1990) in conjunction with the more usual tadpole or arrow plots. Five zones of bedding rotation are observed (Fig. 4), which characteristically show a rotation from a southerly tectonic dip to a northerly/northwesterly dip. These may represent fault-drag intervals. Note that examination of the traditional dip-arrow 'tadpole' plot alone would not have provided such resolution.

Figure 5 shows one such interval (interval "D" on Fig. 4). Here bedding dip rotates from a tectonic tilt of 10°/200° (true dip magnitude/dip direction measured from true north) above 9810 feet (2990 m) to a maximum dip of 50°/310° at 9960 feet (3036 m). The interval immediately below the zone of dip rotation is badly washed out, with only scattered and low quality math-

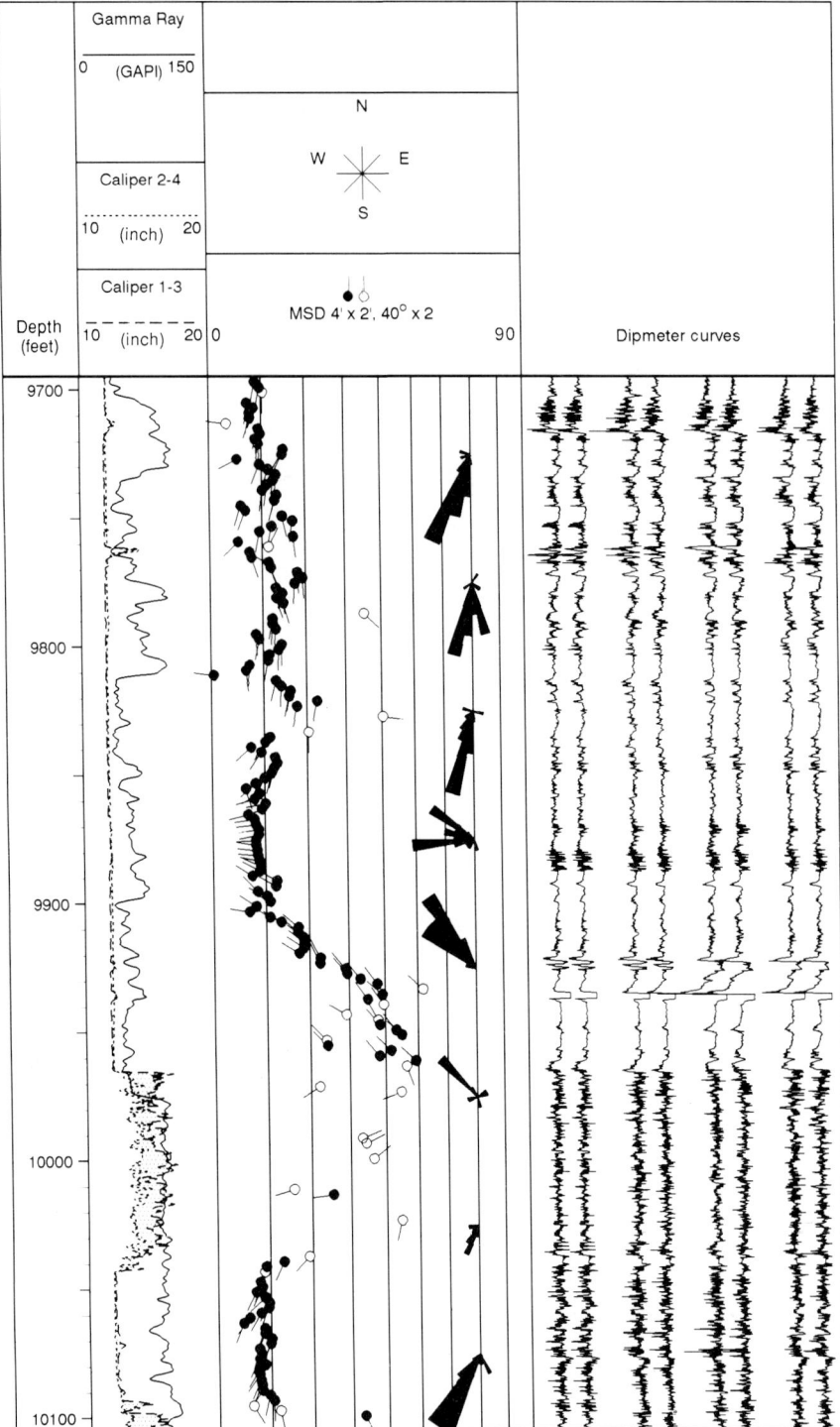

Fig. 5. Caliper, gamma ray log, dipmeter computation results and speed corrected dipmeter curves, 9700–10 100 feet. Black dip arrows are high quality, white arrows are low quality.

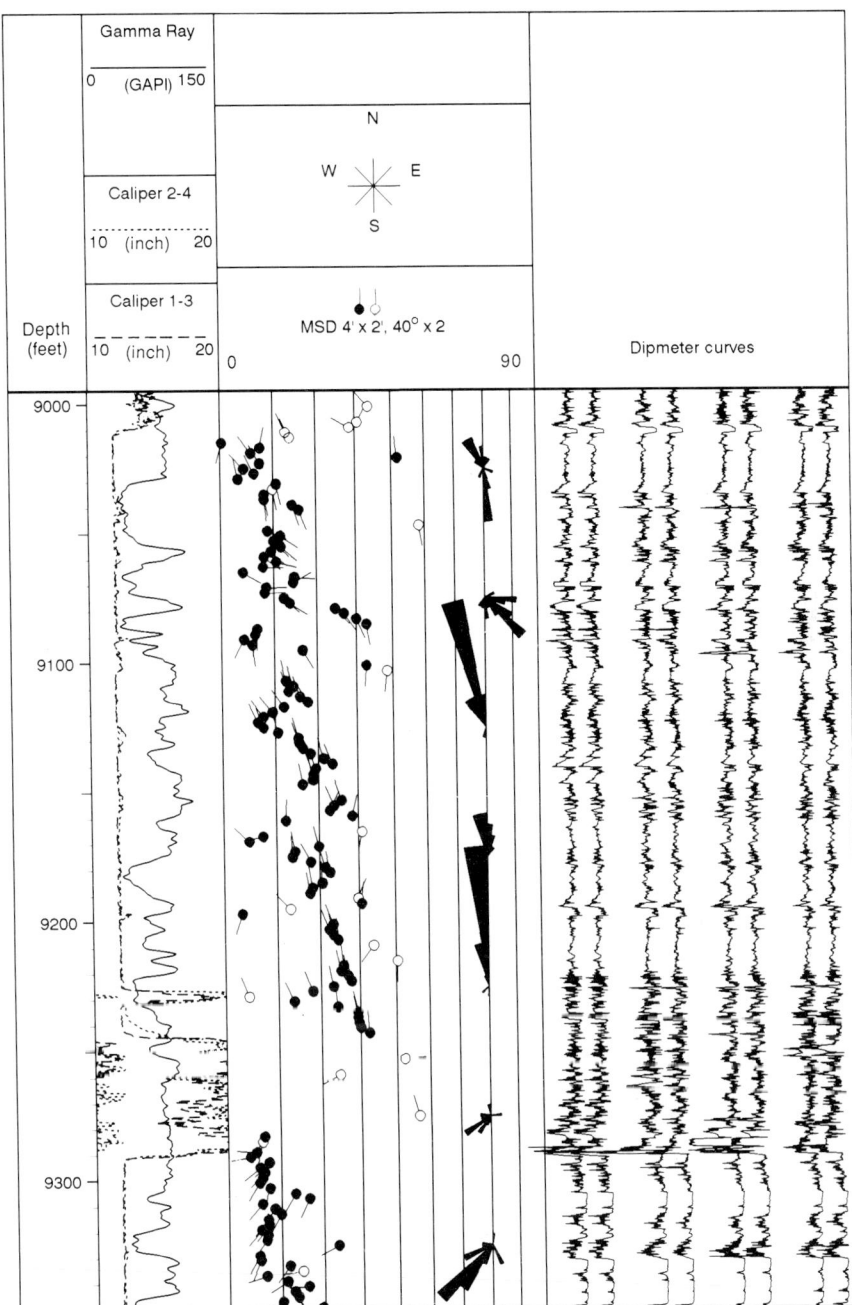

Fig. 6. Caliper, gamma ray log, dipmeter computation results and speed corrected dipmeter curves, 9000–9350 feet. Black dip arrows are high quality, white arrows are low quality.

ematical dips computed, and below 10 040 feet (3060 m) a consistent (tectonic) dip of 8°/220° is noted. The fault plane is not picked up on the dipmeter plot; it is interpreted from this data at the top of a consistent rotation in the footwall at approximately 10 020 feet (3054 m). The washed-out interval probably represents failure near the fault zone, which is up to 40 feet (12 m) wide.

A similar zone of rotation is illustrated in Fig. 6 (interval "C" on Fig. 4). In this case bedding dip shows a gradual rotation from 6° north-northwest to 30°/350° in the interval 9010–9245 feet (2746–2818 m), followed by a 45 feet (14 m) thick washed out interval. Below 9290 feet (2831 m), a tectonic tilt of 7°/200° is observed. The fault plane lies in the interval 9245–9290 feet (2818–2831 m), probably around 9280 feet (2828 m). Some southerly and southwesterly dips occur in the interval 9030–9105 feet (2752–2775 m). However, openhole logs indicate this to be a sandy interval. The dips are interpreted as sedimentological in origin and do not represent major structural features in the wellbore.

Examination of dipmeter curves

Once faults (fault zones) have been identified using dip trend analysis, it is useful to examine the raw dipmeter curves over the faulted interval, as these curves can provide additional information on fault orientation. In many cases steeply dipping events such as faults and fractures are not picked up by standard correlation procedures due to the different response of the tool on the opposite sides of the borehole (Frisinger & Gyllensten 1986), but may be identified by visual inspection of the curves. In this well, each fault zone was examined, but due to the effects of borehole washout in these intervals, no clear curve features were observed (Figs 5 & 6).

A logical extension of this technique is the examination of borehole images, such as Formation MicroScanner* or Borehole Televiewer logs, where it may be possible to image the fault plane directly (Koepsell *et al.* 1989; Harker *et al.* 1990). Due to the adverse borehole conditions encountered in the zones of interest in this particular well, however, these techniques would not have assisted fault identification.

Stereographic analysis

Stereoplots of dipmeter results are commonly used to help define structural dip within a well-bedded sequence, but can also be used to orient

faults by examination of zones of fault-generated deformation.

By plotting poles to bedding within the drag zones on a stereoplot, it is possible to define the strike of the associated fault plane using the π-circle technique (Ramsay & Huber 1987). As an example, the poles to bedding in the drag zone illustrated in Fig. 5 were plotted on a stereonet (Fig. 7), and a best-fit great circle was drawn through these points. The great-circle fit indicates that these bedding planes form part of a cylindrical structure, and the pole to the great circle represents the orientation of the structural axis or rotation axis (Etchecopar & Bonnetain 1989), which, it is assumed, is parallel to the strike of the fault that generated the drag fold.

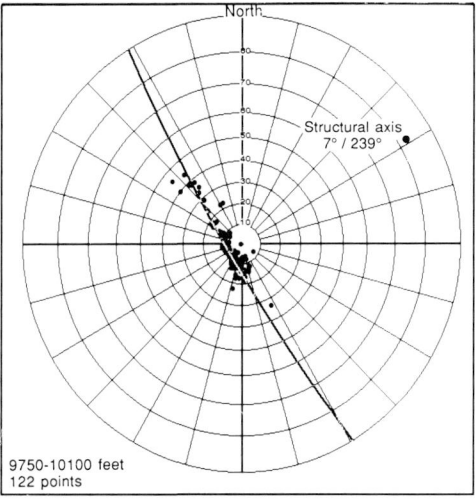

Fig. 7. Stereoplot (Wulff upper hemisphere, poles to planes) and best fit great circle for the interval 9750–10 100 feet.

Unless the fault plane can be recognized from examination of the dipmeter curves, it is not possible to determine the dip or dip direction of the fault. Therefore local geological knowledge must be applied. In this case the maximum dip in the drag zone is 50°, which is the lowest possible fault dip. Seismic data throughout the Delta reveal listric growth faults generally oriented east–west consistent with an extensional stress regime with maximum principal stress vertical and minimum principal stress north–south (see below). In the main paralic sequence faults generally dip at 60–65° but flatten out with depth. In the Akaso Field, using data from the heavily faulted Cawthorne Channel Field, fault dips of 60–65° would be expected.

Measured fault orientations, however, change

---

* Mark of Schlumberger

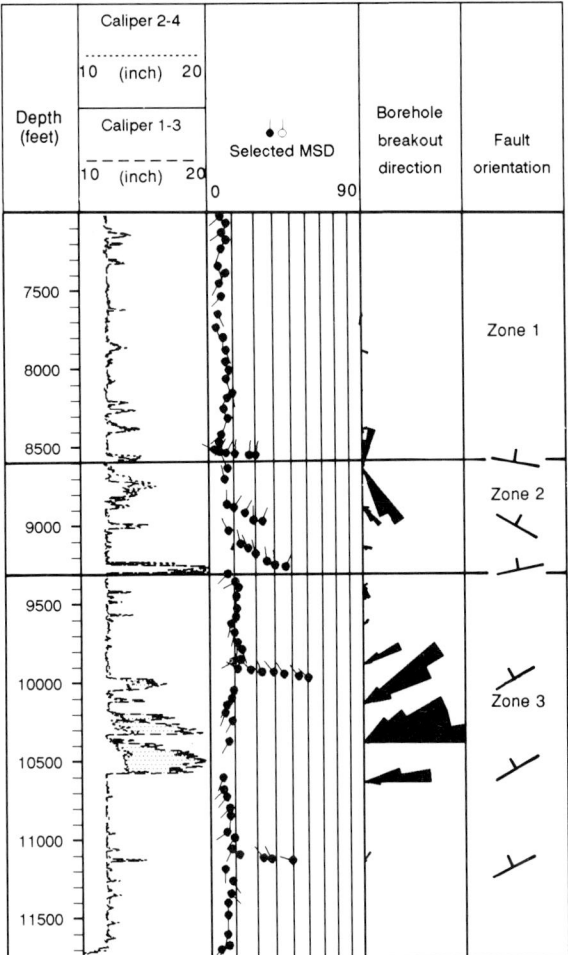

Fig. 8. Summary log showing calipers, computed dip results, borehole breakout data and interpreted fault orientations.

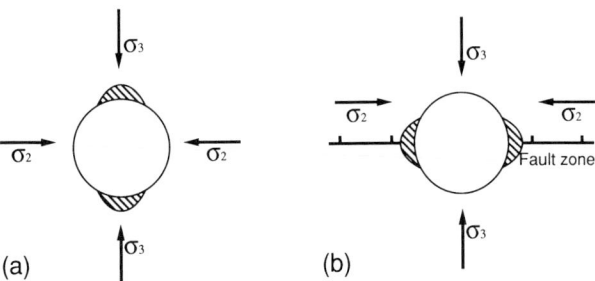

Fig. 9. Two possible scenarios for borehole elongation in an extensional regime with maximum stress axis ($\sigma 1$) vertical: (a) breakout along the minimum horizontal stress direction ($\sigma 3$); (b) elongation along planes of weakness parallel to fault strike.

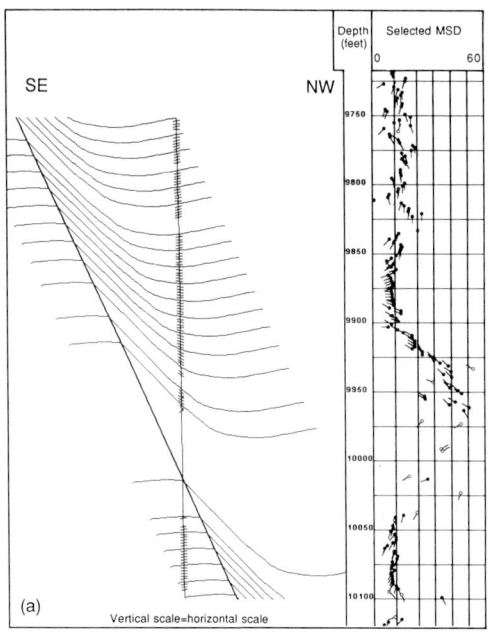

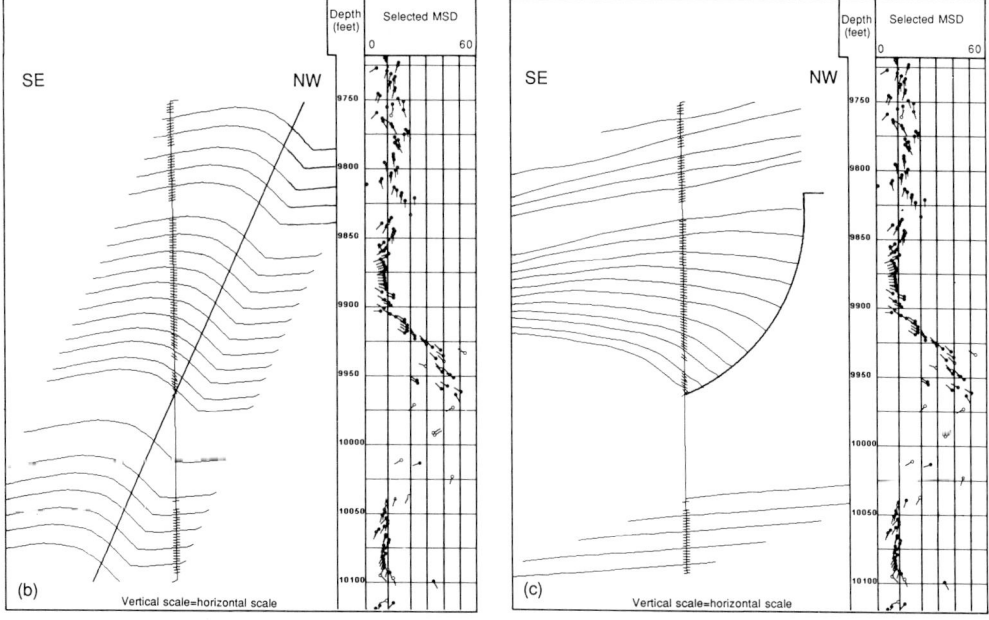

Fig. 10. Southeast–northwest cross section models for the interval 9750–10 100 feet. In all cases, vertical and horizontal scales are identical: (a) normal drag model; (b) reverse drag model; (c) listric rollover model.

with depth in this well. The uppermost observed faults show a northwest–southeast strike (Fig. 8) rotating through depth to a northeast–southwest strike below 9290 feet (2831 m). This is interpreted as being due to penetration of a different fault block below 9290 feet (2831 m) and is confirmed by breakout orientation data. This is examined later by integration with the seismic data.

In situ stress from borehole breakouts

The washed-out sections of borehole do not, in this case, provide any useable dip data. However, analysis of the direction of breakout may be useful in determining the present-day stress regime. Use of borehole breakouts is a recognized method for determining in situ stress within boreholes (Babcock 1978; Plumb & Cox 1987; Bell 1990; Evans & Brereton 1990). During the drilling process, shear failure of the borehole wall will tend to occur parallel to the local minimum horizontal stress direction, causing elongation of the borehole in that direction (Fig. 9a). This borehole elongation is easily measured using the orientated calipers of the dipmeter log (Babcock 1978).

The criteria used for selecting breakouts in this well were similar to those used by Evans and Brereton (1990) and Plumb & Hickman (1985), i.e. tool rotation must slow down or stop in the zone of elongation; the difference between the two calipers should be greater than 0.25 inches (7.5 mm); one caliper must remain smaller than and exhibit less variation than the other in the zone of elongation; and the length of the zone of elongation must be greater than 1 foot (30 cm). Key-seating is not expected in this well, as deviation was less than 8° throughout.

From the breakout data (Fig. 8) three distinct intervals were identified. This allows the subdivision of the well into three major structural blocks. In the uppermost part of the hole (Zone 1: 7000–8600 feet (2133–2621 m)) a few breakouts are observed. These show a north-northeast to south-southwest elongation direction. A larger number are observed in Zone 2 (8600–9290 feet (2621–2831 m)), elongated northwest to southeast. Breakout directions in these zones, which are in the hangingwall of the Akaso boundary fault (9290 feet (2831 m)) are perpendicular to the interpreted fault strike and correspond to the expected north–south extension (Fig. 9a).

In the footwall of the Akaso boundary fault (Zone 3: 9290–11 700 feet (2831–3566 m)) the maximum breakout elongation occurs and is oriented northeast–southwest (i.e. parallel to the

fault strike), rotating deeper in the well to an east–west direction. A 90° shift in principal stress directions over such a short interval is unlikely and would indicate that the system would be actively shearing along the major fault at 9290 feet (2831 m). The change in breakout orientation is caused by fracturing and hole washout along the plane of least resistance which is parallel to the fault zone (i.e. perpendicular to the minimum horizontal stress direction; Fig. 9b). Caution must therefore be exercised in the interpretation of such measurements in close proximity to fault zones.

Near-wellbore cross-section modelling

Using the fault orientation information derived from stereoplot analysis of the drag zones, it is possible to construct near-wellbore cross sections. This technique is described fully in Etchecopar & Bonnetain (1989), and allows the near-wellbore structure to be visualized and different interpretation models generated.

Cross-section construction is best carried out on a dipmeter analysis workstation. The technique assumes that in any cross section all deformation structures can be described as similar folds. The position of a layer boundary can therefore be derived from others by translation (along dip isogons) in a constant direction which is parallel to the fold axis.

As no information has yet been made available regarding the dip direction of the faults (only the strike has been identified so far), several alternative models may be constructed for each interval where dip rotation has been observed. For example, the rotation observed in the zone 9800–10 050 feet (2987–3063 m) may be explained using a normal drag model with the fault dipping towards the north (Fig. 10a). Alternatively, a reverse drag model (Fig. 10b) or a listric rollover model (Fig. 10c) may be generated, both with the fault dipping towards the south. It is therefore important to incorporate data from other sources, such as local geological knowledge, seismic data and interwell correlation, to decide which model is most appropriate. In this structural context, the seismic section shows at least one major fault dipping towards the north, though local southerly dipping faults are possible. Both reverse drag and listric fault models are discounted in this sequence. The high magnitude of dip change that is observed precludes a reverse drag model (Barnett et al. 1987). Similarly, the listric rollover model, although apparently appropriate in the overall large-scale structural regime of the Delta, is unsuitable for the small-scale deformation adjacent to the

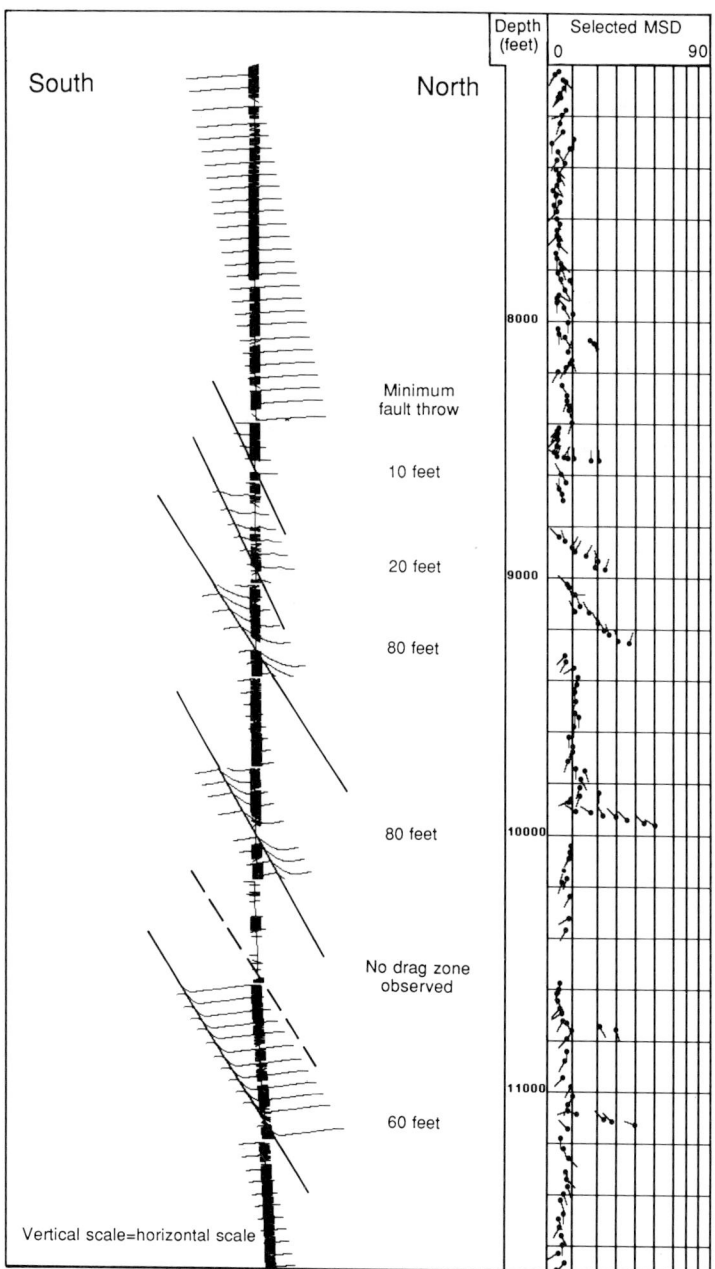

Fig. 11. Composite south–north cross section, illustrating fault related deformation near the wellbore. A normal fault model was used in all cases as this is the most likely interpretation. Minimum throw values have been estimated for all faults.

faults where drag and clay smear are demonstrated phenomena (see Bouvier *et al.* 1989).

The normal drag model is therefore the most likely model for the data recorded in this well. Moreover, for those faults clearly visible on seismic, use of this model agrees well with observed fault dips. The proposed structural interpretation for this well is illustrated by the composite cross section in Fig. 11.

Estimation of minimum fault throw

Displacement of beds by faulting is accommodated by two mechanisms, slip along the fault surface (discontinuous displacement), and ductile shearing of the beds in the 'drag zone' adjacent to the fault (continuous displacement; Fig. 12a). Using a cross-section model constructed from single-well dipmeter data, the amount of throw that is accommodated by ductile shearing may be measured (Fig. 12b) and this element of continuous displacement or 'minimum throw' may be estimated. The fault illustrated in Fig. 10a thus has a minimum throw of 80 feet (24 m). Values for other modelled faults are indicated on Fig. 11.

Except where the well is deviated sub-parallel to the fault plane, the same bed cannot be encountered in both the footwall and hangingwall of a normal fault. Consequently total displacement (i.e. continuous plus discontinuous) cannot be measured directly, except from seismic data. From seismic interpretation, the Akaso boundary fault has a total throw of 1200 feet (365 m) and where it crosses the well at 9290 feet (2831 m) the minimum throw is measured as 80 feet (24 m; Fig. 11). The ratio between the two measurements is dependent on the ductility of the rocks in the drag zone, which in turn is dependent on a number of factors such as lithology and the timing of fault movement.

Dipmeter-seismic integration

Direct integration

It is difficult to integrate directly dipmeter and surface seismic data due to two problems: that of scale, and the fact that log data are usually recorded in terms of depth, whilst seismic data is in two-way-time. The problem of scale is overcome by selecting those dipmeter results which represent the major structural changes (either by averaging mathematically or by direct selection by the interpreter) and displaying them at a much reduced scale (e.g. 1:12 500). The depth–time index problem can be overcome either by carrying out depth-conversion on the seismic

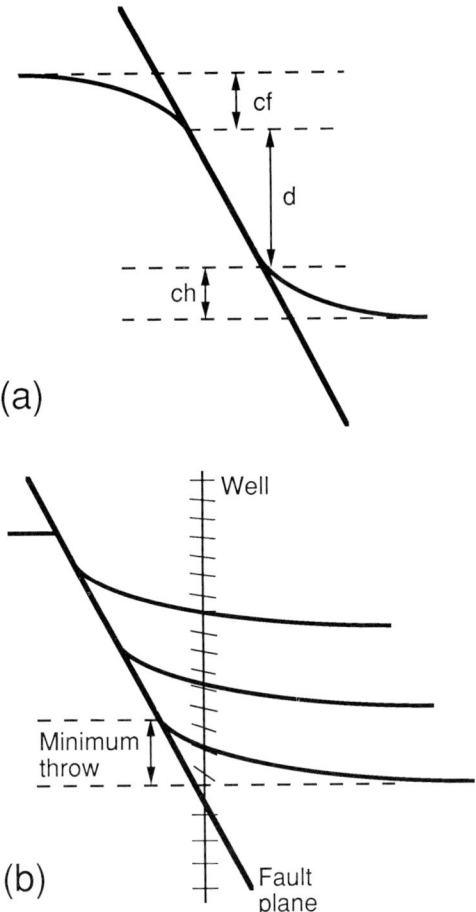

Fig. 12. Schematic diagrams related to normal faulting. (a) Continuous and discontinuous throw related to normal faulting: cf, continuous throw accommodated in the footwall; ch, continuous throw accommodated in the hangingwall; d, discontinuous throw accommodated by slip along the fault plane. (b) Calculation of minimum fault throw (continuous throw) from a cross section model with a passive footwall.

data using wellbore velocity information, or by converting the wellbore data into a time index (Werner *et al.* 1987). Both conversion techniques can be carried out quickly using workstations. In this case the interpreted fault orientations were converted to time and are displayed as a stick plot (apparent dip in the line of section), directly overlain on the seismic data (Fig. 13a). By comparing the two datasets in this manner, it is possible to identify faulted intervals on the seismic data that were not previously resolvable, by direct comparison of the seismic section and the interpreted wellbore data.

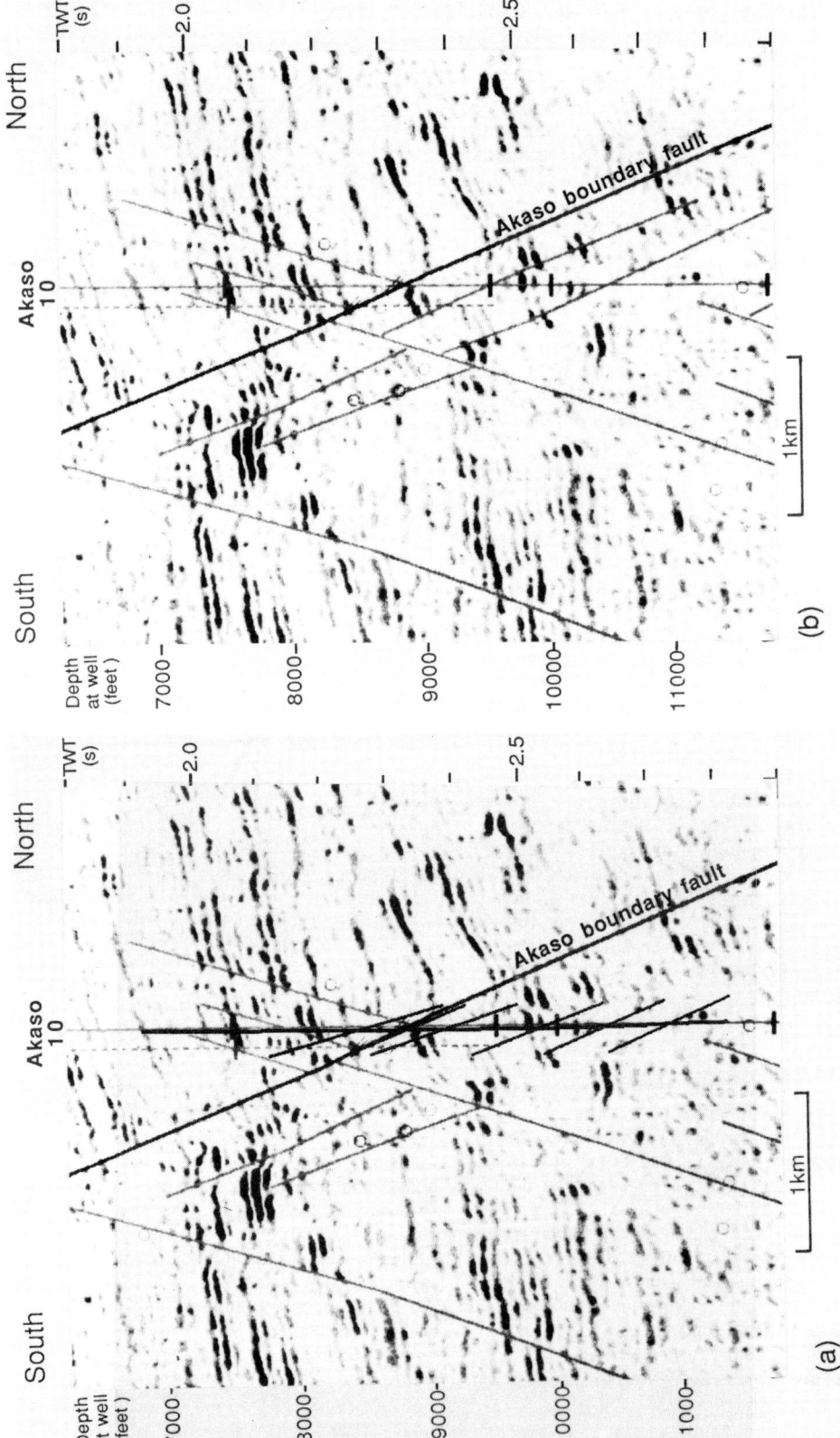

Fig. 13. Dipmeter-seismic integration and final interpretation. (a) Time-indexed dipmeter stick plot overlain on the original interpreted south–north seismic section (see Fig. 3). (b) Re-interpreted seismic section, following integration of dipmeter data. Note the number of faults appearing in the footwall of the Akaso boundary fault.

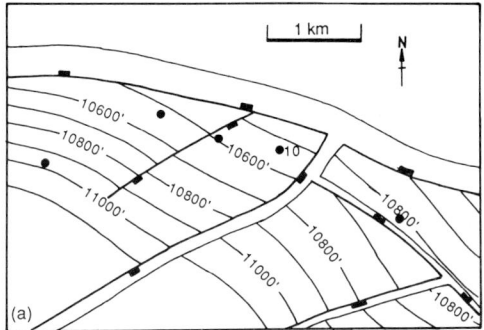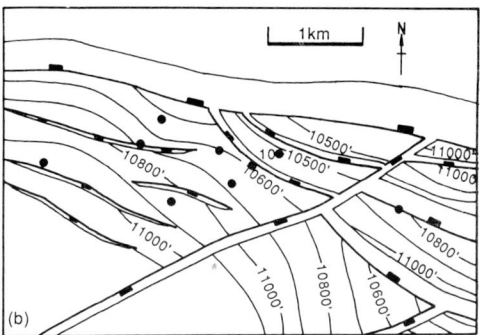

Fig. 14. Comparison of interpretations of horizon G1.1 in part of the Akaso Field. (a) Initial interpretation from 2D seismic data, 1989. (b) Final interpretation using 3D seismic enhanced with dipmeter interpretation, 1991.

As can be seen from the above, dipmeter data can be processed and analysed through a variety of steps and important input can be given to problematic or complex seismic interpretation. It should be noted, however, that feedback from the seismic during the dipmeter modelling is essential and an iterative approach will prove to be most successful. This is particularly important in the selection of modelling parameters and assumptions of likely fault dip.

The Akaso results

The dipmeter data clearly reveal a major block-bounding fault at *c.* 9290 feet (2.3 s). This fault corresponds with the main northerly dipping antithetic fault which effectively separates the Akaso and Cawthorne Channel Fields (Akaso boundary fault). Other faults above and below this level are probably backsplits and shear zones associated with this antithetic fault such that the boundary 'fault' probably more closely resembles a wide zone of antithetic faulting. The measured fault strike orientation *c.* NW–SE and the northeasterly fault dips are consistent with this interpretation.

The integration of the dipmeter interpretation with the 3D seismic data allows these faults to be picked from the seismic with increased confidence (Fig. 13b). The final interpretation involving both 3D seismic and wellbore dipmeter input shows a significant enhancement of detail compared to the interpretation made from 2D seismic data alone (Fig. 14). This enhanced interpretation allows wells to be positioned to maximize hydrocarbon recovery.

Limitations

The techniques described above for identification, orientation and modelling of faults from dipmeter data will only work where the dipmeter picks up bedding deformation associated with faulting. Where no deformation is present, or where chaotic dips result from shattering or poor hole conditions, it is not possible to characterize faults fully from dip data alone. For example, on Fig. 13b, seismic interpretation has identified a southerly dipping fault crossing the wellbore at around 2.14 s TWT, which is approximately 7850 feet. No indication of drag or other bedding deformation is noted on the dipmeter plot at this depth (Fig. 13) and therefore no fault was identified. Closer examination of the dipmeter plot reveals a zone of poor quality dips at *c.* 7850 feet, due to washout at the fault, however no further orientation information may be gleaned from this interval and the fault would not have been recognized from dipmeter alone. Other similar features are noted in the interval 7900–8400 feet and may represent a series of small faults with the same characteristics. The fault at 10 580 feet (Fig. 11) also displays no coherent bedding deformation and was only identified due to the large washout occurring above it, similar to that noted above other faults in this well.

The only way to solve this problem of non-identification of faults from dipmeter data is to progress to borehole images, where fault identification and orientation is not dependent on bedding deformation (Koepsell *et al.* 1989; Harker *et al.* 1990).

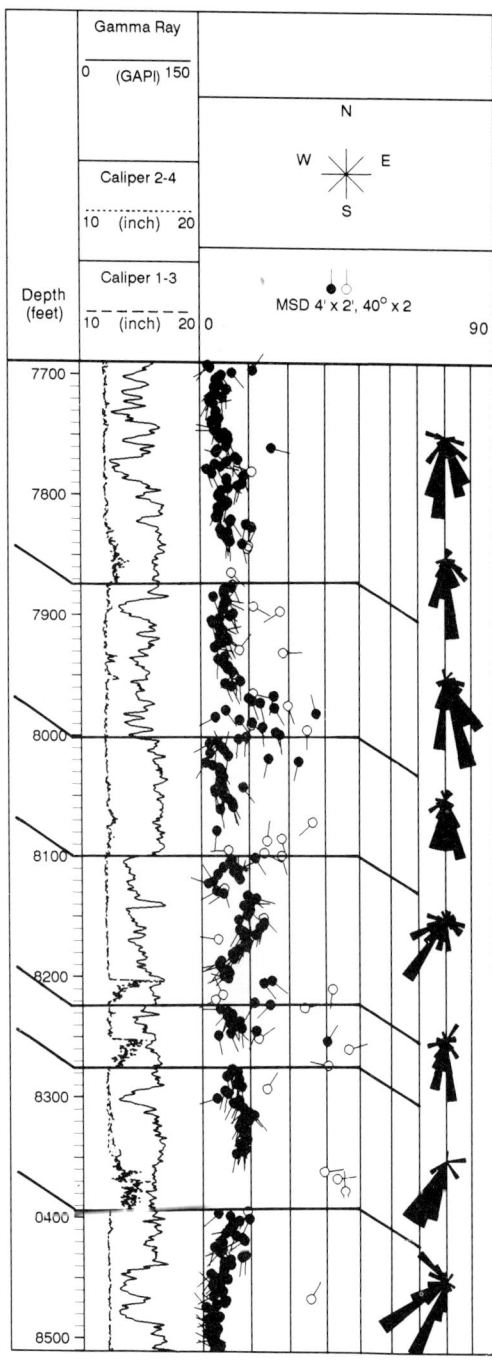

Fig. 15. Caliper, gamma ray log and dipmeter computation results, 7700–8500 feet. Minor faults could be interpreted where marked, but cannot be investigated further due to lack of clear fault-associated deformation.

Conclusions

The well examined penetrated the major fault zone (Akaso boundary fault), separating the Akaso Field to the south from the Cawthorne Channel Field to the north.

Detailed dipmeter interpretation provided increased structural control during interpretation by proving a number of faults below the effective resolution of 3D seismic data. This was important especially in the footwall of the major fault, where shadow effects had prevented confident fault picking from seismic, and the dipmeter interpretation was used to place these faults accurately. Identification of these faults is important for optimum well positioning to maximize hydrocarbon recovery.

Such interpretation requires the iterative use of several techniques, including vertical dip trend analysis, examination of dipmeter curves, breakout analysis, stereographic analysis and modelling, and generation of time-indexed data. Computer workstations are essential to manipulate the large volumes of data required in a commercially viable time span.

Breakout analysis in heavily faulted zones such as this can give misleading results for the minimum horizontal stress direction, and must be interpreted in context.

Minimum throw (i.e. continuous displacement) may be measured from cross-section models constructed from dipmeter data. This value can then be compared with the total displacement measured from seismic sections to provide information on fault movement.

Fault identification from dipmeter data is limited to those faults which display clear bedding deformation, and which will allow orientation. Where such deformation is not present, or where chaotic dips result from shattering or poor hole conditions, it is not possible to characterize faults fully from dip data alone.

The authors thank Tim Chapman and John Walsh for their constructive critique and helpful review.

The authors are indebted to Shell International Petroleum Maatshappij B.V., Shell Petroleum Development Company of Nigeria Limited, Nigerian National Petroleum Corporation, Nigerian Department of Petroleum Resources and Schlumberger for permission to publish this paper.

References

BABCOCK, E. A. 1978. Measurement of subsurface fractures from dipmeter logs. *AAPG Bulletin*, **62**, 1111–1126.

BARNETT, J. A. M., MORTIMER, J., RIPPON, J. H., WALSH, J. J. & WATTERSON, J. 1987. Displacement geometry in the volume containing a single normal fault. *AAPG Bulletin*, **71**, 925–937.

BELL, J. S. 1990. Investigating stress regimes in sedimentary basins using information from oil industry wireline logs and drilling records. *In*: HURST, A., LOVELL, M. A. & MORTON, A. (eds) *Geological Applications of Wireline Logs*. Geological Society, London, Special Publication, **48**, 305–325.

BOREHAM, D., KINGSTON, J., SHAW, P. & VAN ZEELST, J. 1991. 3D marine seismic data processing. *Oilfield Review*, **3**, January, 41–55.

BOUVIER, J. D., KAARS-SIJPESTEIJN, C. H., KLUESNER, D. F., ONYEJEKWE, C. C. & VAN DER PAAL, R. C. 1989. Three-dimensional seismic interpretation and fault sealing investigations, Nun River Field, Nigeria, *AAPG Bulletin*, **73**, 1397–1414.

BROWN, A. R. 1988. *Interpretation of three-dimensional seismic data*. AAPG Memoir 42, Second Edition.

DEVILLIERS, M. C. & WERNER, P. 1990. Example of fault identification using dipmeter data. *In*: HURST, A., LOVELL, M. A. & MORTON, A. (eds) *Geological Applications of Wireline Logs*, Geological Society, London, Special Publication, **48**, 287–295.

ETCHECOPAR, A. & BONNETAIN, J-L. 1989. Cross-section construction from dipmeter data. *SAID/SPWLA Logging Symposium*, Paris, Paper BB.

EVAMY, B. D., HAREMBOURE, J., KAMERLING, P., KNAPP, W. A., MOLLOY, F. A. & ROWLANDS, P. H. 1978. Hydrocarbon habitat of Tertiary Niger Delta, *AAPG Bulletin*, **62**, 1–39.

EVANS, C. J. & BRERETON, N. R. 1990. In situ crustal stress in the United Kingdom from borehole breakouts. *In*: HURST, A., LOVELL, M. A. & MORTON, A. (eds) *Geological Applications of Wireline Logs*, Geological Society, London, Special Publication, **48**, 11–25.

FRISINGER, M. R. & GYLLENSTEN, A. 1986. Fracture detection in North Sea reservoirs. *SPWLA Tenth European Formation Evaluation Symposium*, Aberdeen, April 22–25.

HARDMAN, R. F. P. & BOOTH, J. E. 1991. The significance of normal faults in the exploration and production of North Sea hydrocarbons, *In*: ROBERTS, A. M., YIELDING, G. & FREEMAN, B.

(eds) *The Geometry of Normal Faults*. Geological Society, London, Special Publication, **56**, 1–13.

HARKER, S. D., McGANN, G. J., BOURKE, L. T. & ADAMS, J. T. 1990. Methodology of Formation MicroScanner image interpretation in Claymore and Scapa Fields (North Sea). *In*: HURST, A., LOVELL, M. A. & MORTON, A. (eds) *Geological Applications of Wireline Logs*. Geological Society, London, Special Publication, **48**, 11–25.

HÖCKER, C. F. W., EASTWOOD, K. M., HERWEIJER, J. C. & ADAMS, J. T. 1990. Use of dipmeter data in clastic sedimentological studies. *AAPG Bulletin*, **74**, 105–118.

JEV, B. I., KAARS-SIJPESTEIJN, C. H., PETERS, M. P. M. A. & WILKIE, J. T. 1991. (Abs) The Akaso Field, Nigeria: Use of integrated 3-D seismic/fault slicing/clay shearing on fault trapping and dynamic leakage. *AAPG Bulletin*, **75**, 602–603.

KOEPSELL, R. J., JENSON, F. E. & LANGLEY, R. L. 1989. Gulf Coast fault orientation determined by formation imaging techniques. *SPWLA Transactions of 30th Annual Logging Symposium*, Paper VV.

PLUMB, R. A. & COX, J. W. 1987. Stress directions in eastern North America determined to 4.5 km from borehole elongation measurements. *Journal of Geophysical Research*, **92**, 4805–4816.

—— & HICKMAN, S. H. 1985. Stress induced borehole elongation: A comparison between the four arm dipmeter and the borehole televiewer in the Auburn Geothermal Well, *Journal of Geophysical Research*, **90**, 5513–5521.

RAMSAY, J. G. & HUBER, M. I. 1987. *Techniques of Modern Structural Geology, Volume 2: Folds and Fractures*. Academic, New York.

RUIJTENBERG, P. J., BUCHANAN, R. & MARKE, P. A. B. 1990. Three-dimensional data improve reservoir mapping. *Journal of Petroleum Technology*, **42**, 22–25, 59–61.

SCHLUMBERGER, 1986. *Dipmeter Interpretation—Fundamentals*. Schlumberger, New York.

WEBER, K. J. 1986. Hydrocarbon distribution patterns in Nigerian growth-fault structures controlled by structural style and stratigraphy. *Journal of Petroleum Science and Engineering*, **1**, 91–104.

WERNER, P., PIAZZA, J-L. & RAIGA-CLEMENCEAU, J. 1987. Using dipmeter data for enhanced structural interpretation from the seismic. *Transactions of SPWLA 28th Annual Logging Symposium*, Paper II.

Enhanced resolution resistivity logging for fracture studies

P. D. JACKSON,[1] S. SHEDLOCK,[1] J. WILLIS-RICHARDS[2] & A. S. P. GREEN[2]

[1] British Geological Survey, Keyworth, Nottingham NG12 5GG, UK

[2] CSM Geothermal Energy Project, Rosemanowes Quarry, Herniss, Penryn, Cornwall
TR10 9DU, UK

Abstract. Electrical resistivity techniques have been developed which have a higher resolution than standard logging methods. The paper describes an enhanced focusing technique which is shown to increase the resolution to an extent where individual fractures can be identified within what appears to be a single fracture from conventional focused logs. The measurements were made in the Carnmenellis Granite and the problems of making resistivity measurements in this resistive environment are discussed in terms of electric current flow in the borehole which is controlled by the proximity of connected fractures and the resistivity contrast between the fluid and the formation.

The investigation of thin layers using instrumentation within boreholes has traditionally been limited by decreasing depths of investigation as the resolution has been increased. Small sidewall tools have been used to obtain high-resolution data which now have the capability of imaging the borehole wall at 0.1 inch (2.5 mm) resolution (Ekstrom *et al.* 1986). This is achieved at the expense of penetration which is the 10 mm range. Non-pad resistivity logging tools typically have resolutions in the metre range and are not suitable for the very detailed assessments of the resistivity structure downhole that are now required.

The use of resistivity logging to aid fracture studies in crystalline rock is hampered by the lack of high-resolution measurements that can investigate outside the zone of drilling disturbance. Drilling induced fractures present problems when interpreting electrical well-bore images (Standen 1991), masking the natural fractures but enabling predictions to be made regarding stress directions.

If the plane of a fracture is normal to the borehole it may be considered to be electrically equivalent to a thin layer having a lower resistivity than the formation. If the rock matrix is far more resistive than the formation fluid then such fractures may act as major conduits for the flow of electrical current. The resistivity of the rock matrix in crystalline formations is known to be high (Pezard & Luthi 1988; Zablocki 1964), resulting in a short-circuit path in the borehole even when the fluids are of low salinity. In such cases focused resistivity methods are preferred as they counteract this short circuit in the borehole (Doll 1951; Moran & Chemali 1979; Samworth & Cherrie 1976).

The present work seeks to demonstrate that the resolution of a focused array can be increased by additional focusing within the original focused current beam. The depth of investigation of a focused array is known to be controlled by its overall dimensions (Ajam 1978). The benefits of such a technique are illustrated by logging a shallow borehole containing subhorizontal fractures that are also known to be hydraulically connected to the far field (Parker 1989); better estimates may be obtained of the number of fractures and the extent of zones which are unaffected by fracturing, without relying on measurements controlled by the disturbed near-borehole zone. Examples are presented from shallow boreholes in the Charnwood Forest and the Carnmenellis granite at the Rosemanowes Quarry, which is used for HDR geothermal research by the Camborne School of Mines.

Fractures in a high-resistivity environment

It is necessary to have a reasonable estimate of the resistivity of the rock matrix and the formation fluids in order to assess the proportion of electric current flowing in a given fracture system, as opposed to the matrix. The apertures of the joints/fractures at Rosemanowes have been estimated at 400 μm, with an approximate spacing of 10 m and filled with formation fluid having a resistivity of 20 ohm m (Parker 1989). For a matrix resistivity of greater than 50 kohm m and a formation fluid resistivity of 20 ohm m, simple geometrical calculations indicate that at least 40% of the electric current will flow in such joints in the far field (Jackson *et al.* 1989). However, there is some difficulty in assessing the resistivity of the granite matrix in situ. Down-

From HURST, A., GRIFFITHS, C. M. & WORTHINGTON, P. F. (eds), 1992,
Geological Applications of Wireline Logs II. Geological Society Special Publication No. 65, pp. 265–274.

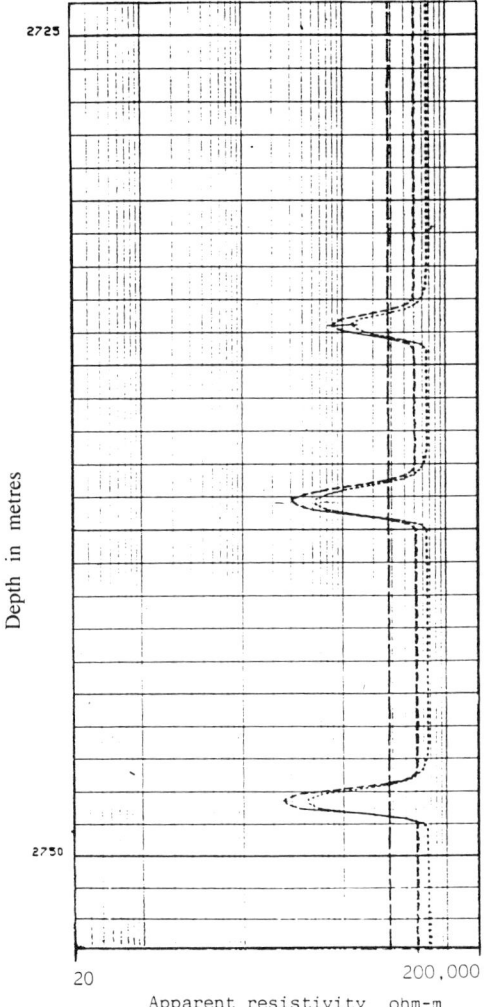

Depth in metres

2725

2730

20   200,000

Apparent resistivity   ohm-m

Fig. 1. An example of a seven-electrode Schlumberger 'Dual Laterolog' response, taken from logs from the Carnmenellis Granite at Rosemanowes Quarry showing the minimum resolution to be approximately 1 m and the 'saturation' of the measurement at 70 kohm m.

hole electrical resistivity logs are normally run in lower resistivity environments. The example shown in Fig. 1, a Schlumberger Dual Laterolog log from the 2 km deep borehole array at Rosemanowes, shows what appears to be saturation at a value of 70 kohm m. The rock matrix resistivity would be expected to be higher than this value.

The following factors control the electrical resistivity of the matrix of crystalline rocks including granites (Brace *et al.* 1965; Parkho-

menko 1967; Skagius & Neretnieks 1986; Evans 1980):

(1) pore fluid resistivity;
(2) surface conduction;
(3) connected or effective porosity;
(4) tortuosity of the pore channels.

Laboratory measurements of electrical resistance of granite core plugs are degraded by the alteration caused by the sampling process, and it is difficult to predict in situ values from laboratory measurements. Also, the porosity is difficult to assess as the effective porosity is likely to be very much smaller than total porosity (Evans 1980). Unfractured resistivities are best assessed in situ, however at present there are difficulties due to:

(1) large volumes of investigation that will contain some fractures;
(2) disturbance of the rock mass by the drill hole resulting in a de-stressed zone of disturbance near the borehole;
(3) most proven commercial equipment is not designed to measure the very high resistivities.

Brace and his co-workers (Brace *et al.* 1965; Brace & Orange 1968) reported a series of experiments that investigated the effects of pressure on the resistivity of crystalline rock cores. This work provides a good account of the effects of porosity changes and surface conduction and suggests how to describe the tortuosity.

Their results using salt solution obeyed the Archie equation:

$$F = \phi^m$$

where F is the formation factor and ϕ is the fractional porosity. F is defined as R_o/R_w where R_o is the resistivity of the formation, and R_w is the resistivity of the pore fluid. The porosity of each sample was changed by the application of pressure which tended to close the microfractures. Brace and his co-workers showed the resulting F/ϕ relationships to obey the Archie equation, and that the value of the exponent m was 2. Their data using tap water, having a resistivity of 20 ohm m, deviated from this relationship by about an order of magnitude, when the fractional porosity was less than 0.05. This decrease in resistivity was attributed to surface conduction.

Using the work of Brace *et al.* (1965) one can arrive at a simplified equation relating the rock matrix resistivity (R_T) to effective porosity ϕ, for $R_w = 20$ ohm m.

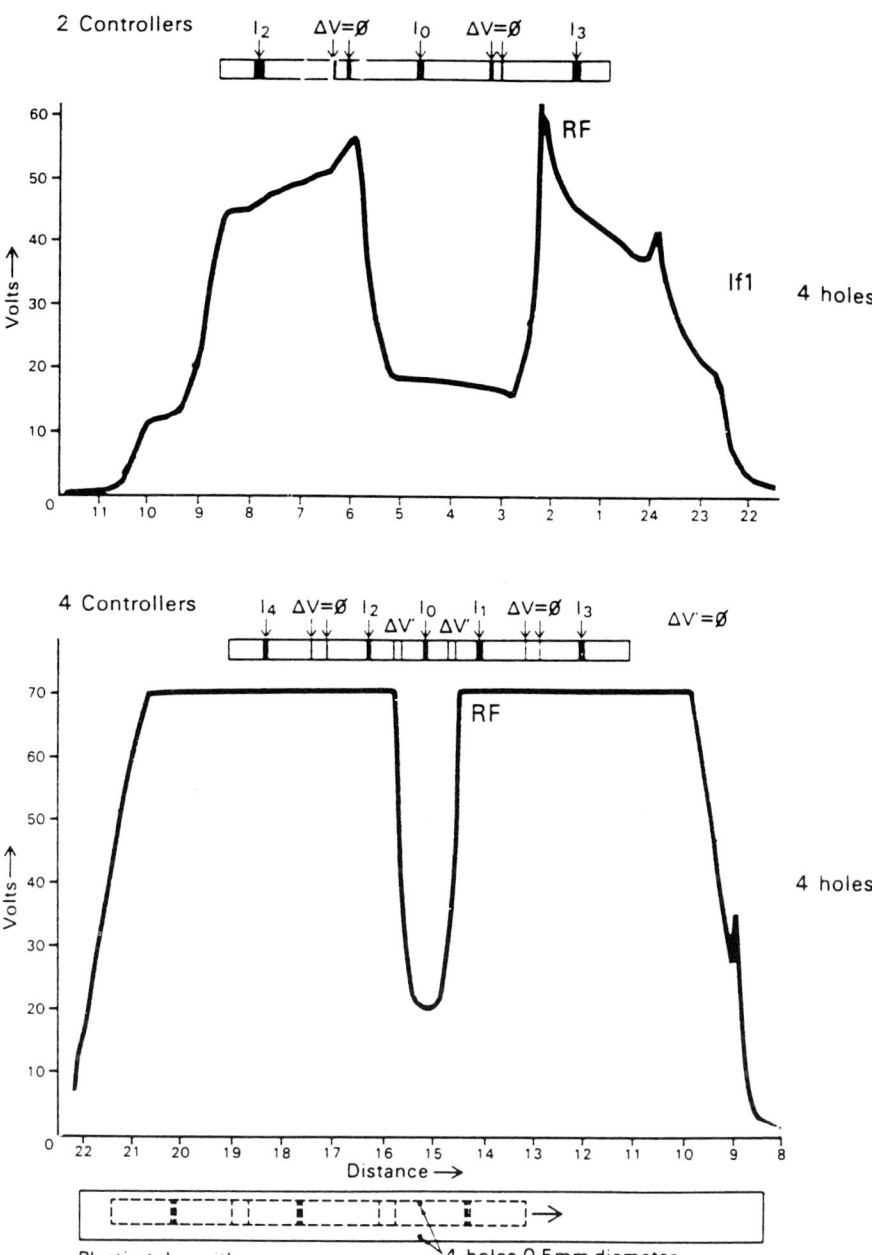

Position of 0.5mm diameter holes in the fixed plastic tube that simulates a fracture at the middle of of the tube. The distances the electrode arrays have moved within the tube are shown in cms at the same scale as the electrode separations.

Fig. 2. Tank experiments with focused and enhanced focused electrode arrays having 7 and 13 electrodes respectively, showing responses to a simulated conductive fracture within a resistive borehole.

$$R_T = R_w/(10 \, \phi^2).$$

Here, ϕ is connected porosity and the factor 10 approximates to the surface conduction effect. Using this formulation one would expect

$$10^6 > R_T > 2 \times 10^4 \text{ ohm m}$$

for fractional porosity in the range 0.01 to 0.001 and a pore fluid resistivity of 20 ohm m. Core samples will necessarily have been de-stressed and disturbed and so there will always be uncertainties in these laboratory measurements due to the difficulty in assessing drilling induced microfracturing. Thus, laboratory measurements will tend to underestimate the in situ value. The matrix resistivity of the granite at Rosemanowes is high enough to sugest that more than 40% of the electrical conduction will take place within the joint system, indicating joints will behave as thin, low-resistivity layers with respect to focused logging devices.

Enhancing the resolution of a focused array

The logs in Fig. 1 indicate the resolution of the Dual Laterolog is in the range 0.6–1.0 m, with a smooth response that is consistent with that of a thin bed of low resistivity (Moran & Chemali 1979; Koithara & Chandra 1977).

The response of the 'laterolog type' of array to thin relatively low resistivity layers, in a high resistivity environment, was investigated using a water-filled model tank. Focused arrays were passed through a plastic tube to simulate a borehole passing through a high-resistivity formation. A ring of four 0.5 mm diameter holes was made at the middle of the tube and arranged evenly around the circumference, to simulate the electrical effect of a low-resistivity fracture. The disposition of these holes was such that they simulated a layer that was very thin compared to the thickness of the focused current beam of the 'laterolog type' of array (Fig. 2).

The results in Fig. 2 show the anomaly produced by a seven electrode focused array, with two independent current controllers, to be a convolution of the width of the current beam with the 'thin layer', if the beam being defined by the zero potential difference at either side of the central electrode from which the current I0 emanates (Doll 1951). All electrodes of the array are independently connected, with two isolated current controllers, which allows effective focusing in the case of unsymmetrical resistivity distributions (Moran & Chemali 1979; Jackson 1976).

An attempt was made to increase the resolution of the 7-electrode array by adding extra focusing electrodes inside the existing current beam, and two more controllers to maintain independent focusing (Jackson 1975, 1981). The relative positions of the current and potential electrodes were maintained in the new array in order to minimize distortions of the current beam, compared to the 7-electrode 'laterolog type' array (Serra 1982). The principle of independent control was maintained with all electrodes isolated from each other with respect to the electrical connections.

The response of the new 13-electrode focused array to the same simulated fracture is shown in Fig. 2, where the width of the anomaly can be seen to be equal to the thickness of the inner current beam. Thus, the resolution of the new focused measurement had been substantially increased by partitioning the current beam of a seven-electrode 'laterolog type' device using two extra independent focusing current controllers and their associated isolated electrodes.

An experimental borehole tool was developed, based on the 13-electrode array discussed above, and is shown in Fig. 3. The constant current I0 flows from the central electrode A0. The vertical extent of the current beam is defined by M1,M1' and M2,M2', with M3,M3' and M4,M4' providing the seven-electrode array focusing that determines the overall depth of investigation. The measurements of resistance were made using one potential electrode on the sonde and a reference electrode placed near the borehole at the ground surface.

Table 1. *The four modes of operation of the sonde*

Description	Currents used	Potential electrode
1. Enhanced focused	I0,I1,I2,I3,I4	M1
2. Inner focused	I0,I1,I2	M1
3. Focused (LL7)	I0,I3,I4	M3
4. Unfocused (normal)	I0	M3

The sonde operated in four modes (see Table 1). Different modes were included in order to compare different resolutions and to allow at least two highest-resolution measurements but different investigation depths, the 'inner focused' being a shorter array of shallower investigation depth.

The prototype array was tested in the upper part of a borehole in the Charnwood Forest, UK, where there were fractured and highly metamorphosed Precambrian sedimentary rocks, and data from the Schlumberger Dual

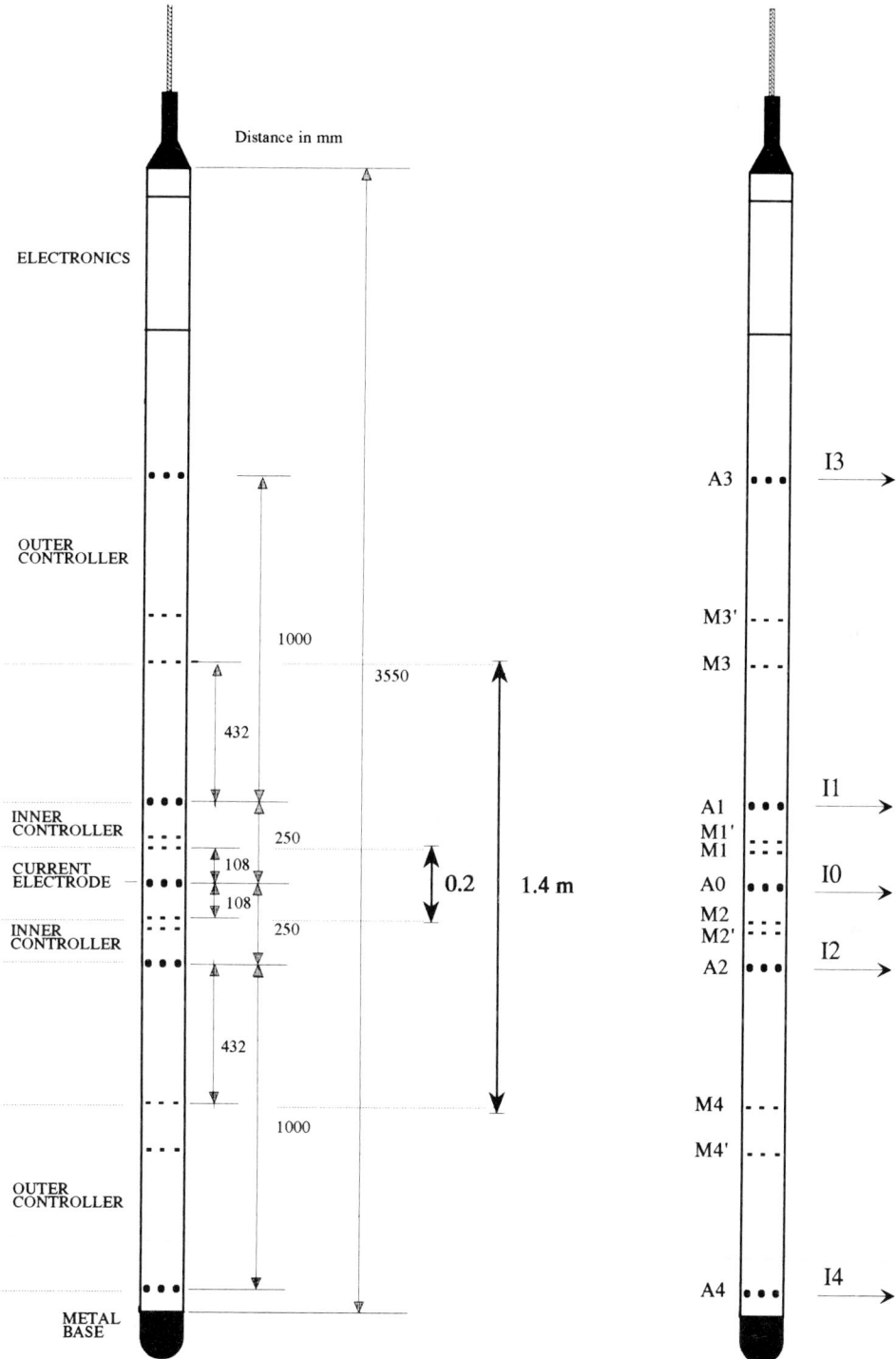

Fig. 3. A schematic diagram showing the enhanced focusing sonde with current beams of 0.2 and 1.4 m widths.

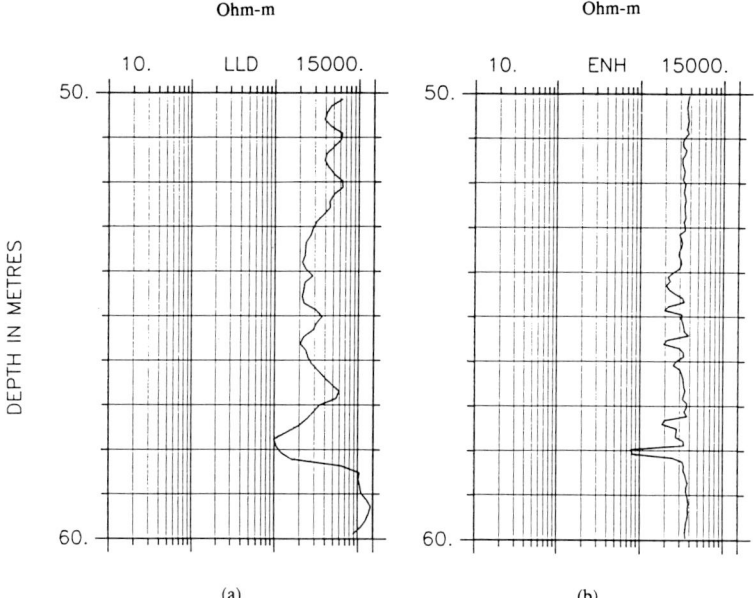

Fig. 4. A comparison of the Schlumberger deep laterolog (Dual Laterolog) (a) with the enhanced focused array (b) using data gathered from a borehole at Morley Quarry, Charnwood Forest, UK.

Laterolog were available. The field test of the enhanced focusing array is compared with the deep trace of the Dual Laterolog in Fig. 4. Both logs have measurements every 0.1 m and show the resolution of the enhanced focusing method to be 3–4 times higher than that of the deep Dual Laterolog, considering the low-resistivity zone at a depth of 58 m. At this depth the deep laterolog exhibits a single layer response over a depth of approximately 1 m, while the enhanced array provides detail that suggests there are at least two thin low resistivity zones present. During this test the enhanced array was found to be limited to an upper apparent resistivity of 3 kohm m. This effect was minimized by reducing the value of the I0 current before later experiments in the Carnmenellis granite.

Enhanced focused logging in the Carnmenellis Granite

Experiments were undertaken in the shallow borehole array at the Camborne School of Mines Geothermal Energy Project, Rosemanowes Quarry. The boreholes are approximately 300 m deep and have undergone a series of stimulation experiments where the cross-borehole permeability has been enhanced by explosive initiation and fluid injection.

Rosemanowes Quarry is situated within the outcrop of the Carnmenellis granite, several kilometres away from the nearest contact with the country rock. Three regular joint/fracture directions dominate the rock permeability; sub-horizontal and two near vertical directions striking approximately 330° and 070° on average, but with wide variation. The sub-horizontal joint set becomes less frequent with depth but is the major conduit for flow within the target zone between 200 and 300 m below the surface. The history of the boreholes suggests substantial fracturing below 200 m with discrete fractures or joints expected to intersect the borehole.

Figure 5 displays the three types of focused log suitable for characterization in borehole RH8B for the depth range 200–280 m. The enhanced focused log has a beam width of 0.2 m and greatest depth of investigation, while the inner focused maintains the resolution at a smaller depth of investigation, and the focused 'LL7-type' had reduced resolution (focused beam width of 1.4 m) but an equivalent depth of investigation to the enhanced focused device (see also Fig. 3).

The results are consistent with the expectations of the different techniques with the focused array having a marked averaging effect compared to the other two higher resolution devices. This averaging is such that the responses

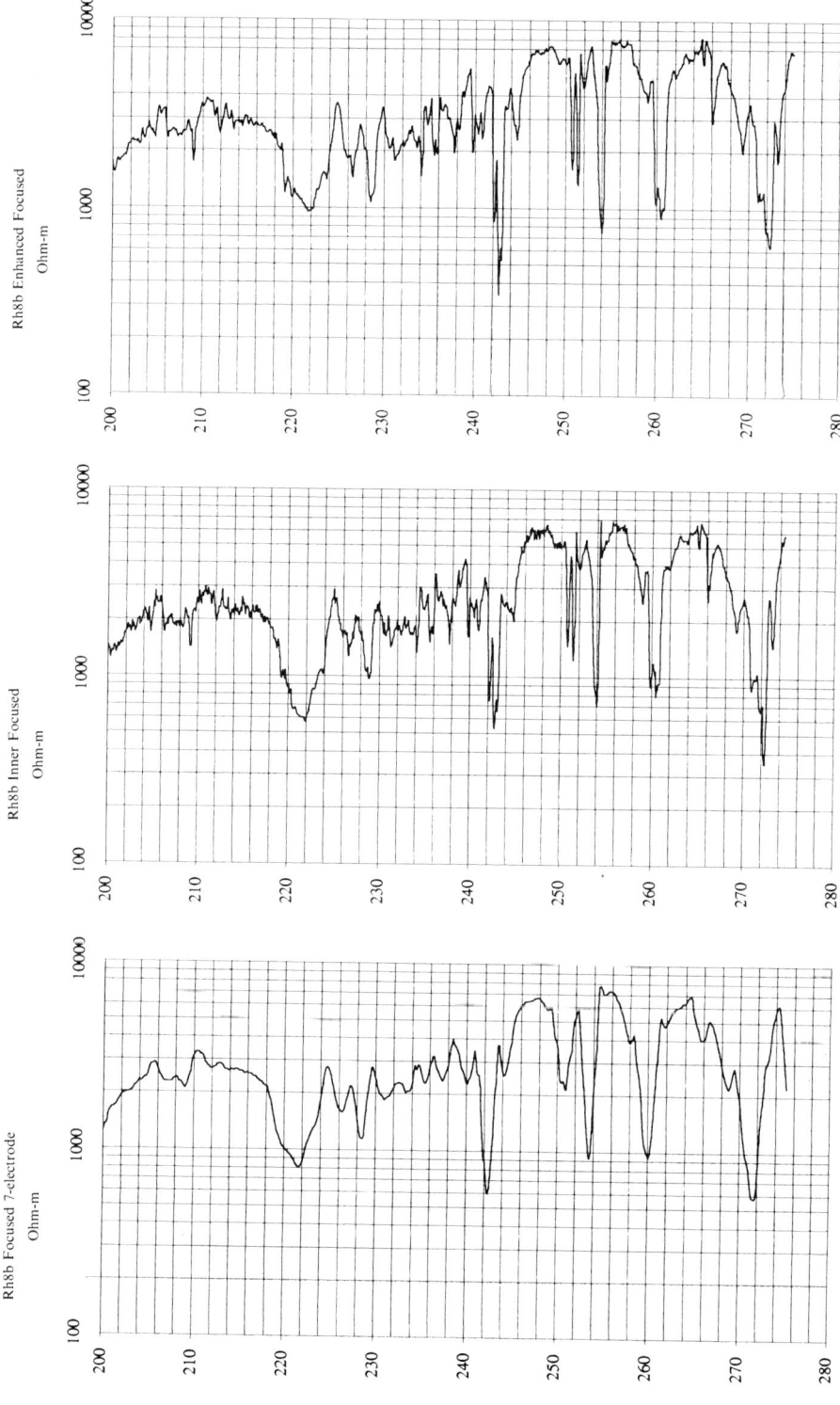

Fig. 5. Resistivity logging using three focused arrays showing increased resolution when using thinner current beams in fractured granite. The enhanced resolution has higher values than the shallower inner focused array where there do not appear to be fractures.

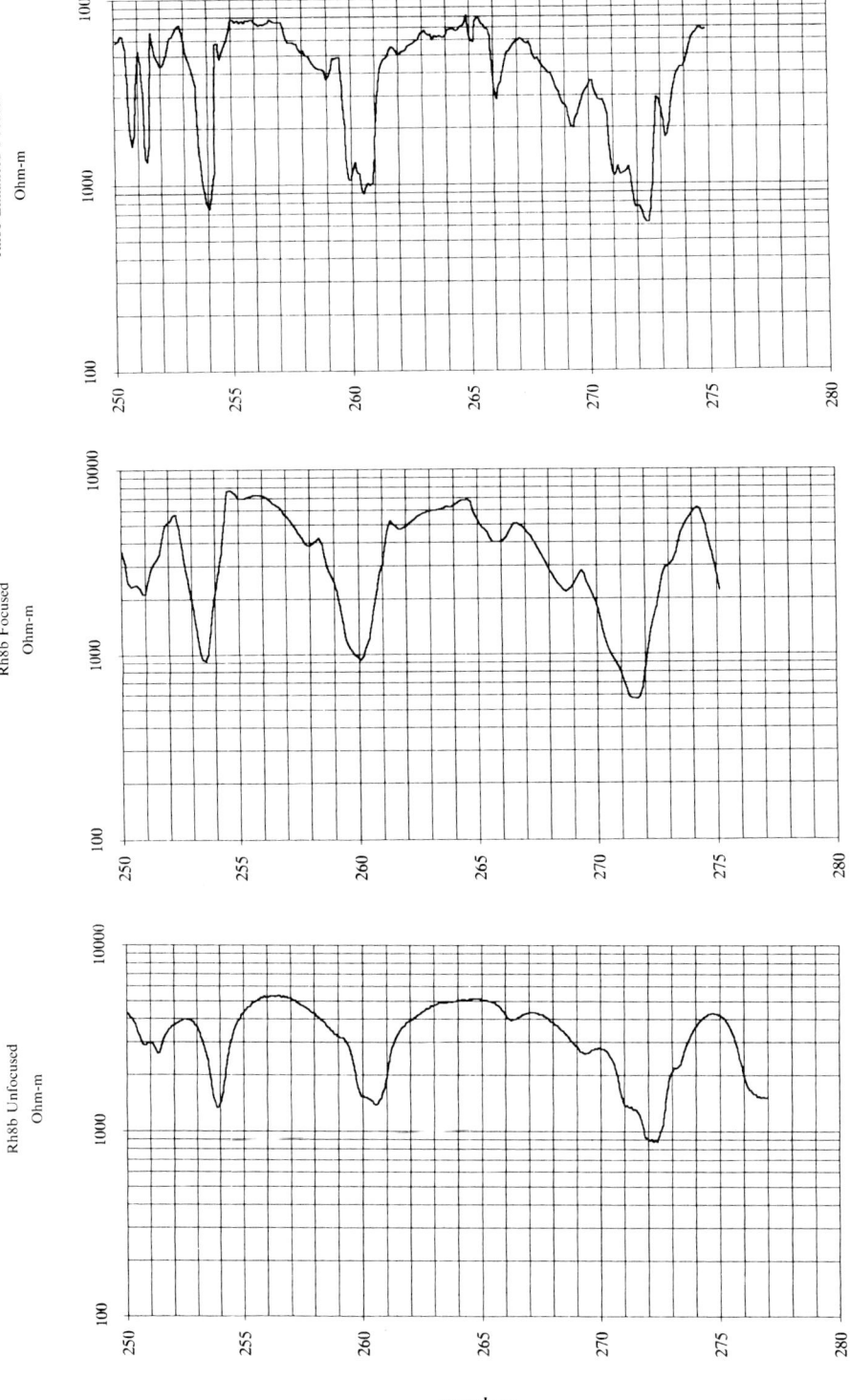

Fig. 6. The increase in resolution obtained using focusing and enhanced focusing compared to unfocused measurements all made with similar penetrations. The enhanced focusing provides better quantitative information on individual fractures.

Rh8b Enhanced Foc.

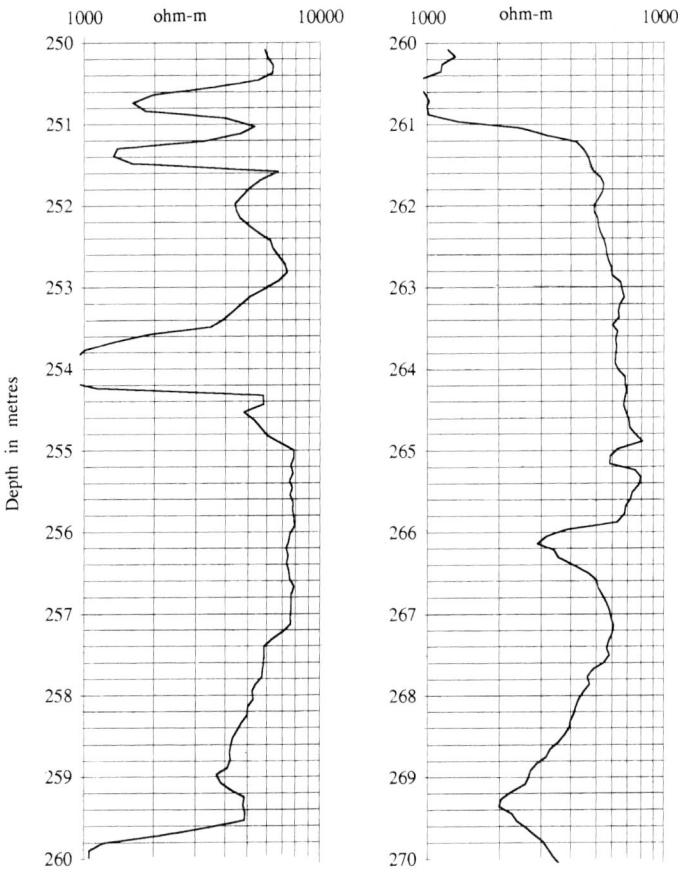

Fig. 7. An expanded portion of Fig. 6 showing the enhanced focused section between 250–260 m downhole in borehole Rh8B. The results show a minimum anomaly width of 0.2 m equal to the width of the current beam.

of many individual fractures have been lost. For example, between 250–252 m in Fig. 5 there are clearly two responses in the enhanced and inner focused logs, while the lower resolution focused array displays only one. Significant structures are seen at 260–272 m in Fig. 5 which are consistent with a number of fractures. Even the highest resolution (200 mm) may not be capable of resolving the fractures in these instances.

A comparison of resolutions is shown in Fig. 6 where the zone 250–280 m is displayed for unfocused, focused, and enhanced-focused measurements. The results show an increase in resolution as focusing is introduced, and then enhanced. There are a number of sections where the enhanced log shows structure within a single

anomaly, indicating the presence of more than one thin, low-resistivity horizon, such as a fracture. As the fluid has been shown to be a major carrier of electric current, due to the high resistivity of the granite matrix, it is essential to investigate single fractures if resistivity methods are to be used to infer fracture densities and relative apertures.

The performance of the enhanced focusing was further investigated by plotting a portion of the data at an expanded scale, over the depth interval 255–265 m (Fig. 7). The results depict a small number of responses 0.2 m wide, or less, which indicates the logging has achieved a resolution equal to the width of the enhanced focused current beam. There is additional very fine scale resistivity structure that could be quantitatively investigated if higher resolution still, was available.

Conclusions

The introduction of an enhanced focusing electrode array under computer control, has enabled the resolution of resistivity logs to be increased by a factor of five without loss of 'penetration' into the formation.

The improvement of resolution without loss of penetration enables the number of fractures to be more accurately determined in the undamaged zone, which is not possible using arrays with very shallow depths of investigation, such as the borehole wall imaging devices.

The feasibility of independently controlling the resolution and depth of investigation of focused arrays has been demonstrated. Thus, there is a means by which constant, high resolution measurements can be made with variable penetrations. This in turn will allow the resistivity structure to be investigated as a function of distance into the formation at a resolution approximately five times higher than that commonly used at present. The improvement in resolution demonstrated here suggests that the resistivity structure of individual fractures may be studied, which will in turn aid the prediction of fluid flow, particularly if sealed and weathered fractures can be distinguished from fresh open ones.

This paper is published with the permission of the Director of the British Geological Survey (NERC). The project was undertaken in collaboration with the Camborne School of Mines HDR Project which is funded by the UK Department of Energy through the Energy Technology Support Unit (ETSU).

References

AJAM, S. O. 1978. The Dual Guard/FoRxo logging device. *Log Analyst.* September–October, 1978.

BRACE, W. F., ORANGE, A. S. & MADDEN, T. R. 1965. The effect of pressure on electrical resistivity of water saturated crystalline rocks. *Journal of Geophysical Research*, **70**, 5669–5678.

—— & —— 1968. Further studies of the effect of pressure on electrical resistivity of water saturated crystalline rocks. *Journal of Geophysical Research*, **73**, 5407.

DOLL, H. G. 1951. The Laterolog: A new resistivity logging method with electrodes using automatic focusing system. *Transactions of the American Institute of Mining & Metallurgical Engineers*, **198**, 17–32.

EKSTROM, M., DAHAN, C. A., CHEN, M., LLOYD, P. M. & ROSSI, D. J. 1986. Formation imaging with microelectrical scanning arrays. *Transactions of SPWLA*, Paper BB.

EVANS, C. J. 1980. *The Seismic and Electrical Properties of Crystalline Rocks Under Simulated Crustal Conditions.* PhD Thesis, University of East Anglia.

JACKSON, P. D. 1975. An electrical resistivity method for evaluating the in-situ porosity of clean marine sands, *Marine Geotechnology*, **1**, 91–115.

—— 1976. Comments on "New results in resistivity well logging": *Geophysical Prospecting*, **24**, 407–408.

—— 1981. Focused electrical resistivity arrays, some theoretical and practical experiments. *Geophysical Prospecting*, **29**, 601–626.

——, BUSBY, J. P., RAINSBURY, M., REECE, G. & MOONEY, P. 1989. The potential of electrical resistivity methods to detect hydraulically connected fractures. *In*: BARIA, R. G. (ed.) *Proceedings of the Camborne School of Mines International Hot Dry Rock Geothermal Energy Conference, 27–30 June 1989.* Robertson Scientific Publications.

KOITHARA, J. & CHANDRA, D. 1977. Field results on the relative performance of laterolog 7 and normal devices. *Geophysics*, **42**, 1478–1483.

MORAN, J. H. & CHEMALI, R. E. 1979. More on the Laterolog Device, *Geophysical Prospecting*, **27**, 902–930.

PEZARD, P. A. & LUTHI, S. M. 1988. Borehole electrical images in the Basement of the Cajon Pass scientific drill holes, California; fracture identification and tectonic implications. *Geophysical Research Letters*, **15**, 9, 1021–1024.

PARKER, R. H. 1989. *Hot Dry Rock Geothermal Energy, Phase 2B Final Report of the Camborne School of Mines Project*, Vol. 1, Pergamon, Oxford.

PARKHOMENKO, E. I. 1967. *Electrical Properties of Rocks*. Plenum, New York. (Translated from Russian and edited by G. V. Keller.)

SAMWORTH, J. R. & CHERRIE, M. A. 1976. A focused resistivity tool for slimline coal logging systems, *SPWLA London Chapter, Transactions of 4th European Formation Evaluation Symposium, London*, paper H.

SERRA, O. 1982. *Fundamentals of Well-Log Interpretation*, vol. 1. Elsevier, Amsterdam.

SKAGIUS, K. & NERETNIEKS, I. 1986. Diffusivity measurements and electrical resistivity measurements in rock samples under mechanical stress. *Water Resources Research*, **22**, 570–580.

STANDEN, E. 1991. Tips for Analysing Fractures on Electrical Well Bore Images. *World Oil*, April 1991, 99–117.

ZABLOCKI, C. J. 1964. Electrical properties of serpentine from Mayaguez, Puerto Rico, in a study of serpentinite. *Natl. Acad. Sci. Rep.* **1118**, 107.

Physical properties

High-frequency pseudo-Rayleigh waves as a new indicator of shear velocity

ARNE MARIUS RAAEN

IKU, N-7034 Trondheim, Norway

Abstract. In slow formations, where the shear velocity is lower than the mud compressional velocity, several indirect methods have been used to estimate shear velocity from full waveform acoustic logs. These methods include Stoneley wave velocity and inversion of refracted P-wave amplitude. In this paper we show that in addition to the methods previously reported, the amplitude of the leaky pseudo-Rayleigh waves may in some cases be strongly dependent on the shear wave velocity of the formation. Numerical analysis has shown that there is also some dependence on the velocity and attenuation of the mud, and the borehole radius, whereas the amplitude of the leaky pseudo-Rayleigh waves is quite insensitive to the compressional velocity of the formation and the intrinsic anelasticity of the formation. The predictions have been confirmed by model experiments in a slow formation. It is felt that the effect may be a useful tool, for example, in the study of stress dependent velocities around a borehole.

The traditional acoustic logging tools are monopole tools, which excite an axisymmetric wavefield. Acoustic compressional and shear velocities are inferred from the arrival times of refracted waves. For a *fast* or *hard* formation, in which the shear wave velocity of the formation higher than the velocity of the borehole fluid, both compressional and shear refractions exist. For this case, one is generally able to infer the formation S-wave velocity by identifying the shear refraction, although overlap with P-wave ringing may cause problems.

In a *slow* or *soft* formation, in which the shear wave velocity is lower than that of the borehole fluid, no shear refraction exists. One must then use indirect methods to infer the shear wave velocity. Several methods have been published for this purpose.

The low-frequency limit of the Stoneley mode is sensitive to the shear velocity via the relation (see e.g. White 1983)

$$v_{St}/v_m = [1 = (\rho_m/\rho)(v_m/v_s)^2]^{-1/2}$$

where v_{St} is the Stoneley wave velocity, v_s is the shear velocity, v_m is the mud velocity and ρ_m and ρ are the densities of the mud and the formation, respectively.

A more advanced version of this method uses a numerical inversion of the Stoneley wave shape in the frequency band of the signal. This method gives better results, since it includes the dispersion of the Stoneley wave.

Both the above methods are normally based on an elastic model of the formation, whereas in practice the Stoneley wave at low frequencies is sensitive to the formation permeability (see e.g. Schmitt *et al.* 1988). This means that shear wave estimates from the Stoneley mode are at best uncertain in high permeability formations.

Another method for estimating the shear velocity in a slow formation is the inversion of the P-refraction. The basis for this method is that the amplitude and duration of the P-wave refraction is mainly sensitive to the shear velocity and the attenuation of the P-wave. By using an anelastic forward model, one may thus estimate the shear velocity and the quality factor for the P-wave by inversion (see Cheng *et al.* 1983; Cheng 1989).

The most direct and precise method for determining shear velocity in slow formations is by direct shear wave logging using a dipole or quadrupole logging tool (see White 1967; Zemanek *et al.* 1984; Harrison *et al.* 1990). These tools excite non-axisymmetric modes (the flexural mode or the screw mode) which for low frequencies propagate at the formation shear wave velocity. When this tool has been run, the indirect methods for estimating shear wave velocity are redundant.

In this paper we describe how the leaky pseudo-Rayleigh wave in certain cases is sensitive to the shear velocity of the formation. (For a discussion of nomenclature for borehole modes, see Cheng & Toksöz 1981). The effect may in given cases be used to estimate roughly the shear wave velocity.

From HURST, A., GRIFFITHS, C. M. & WORTHINGTON, P. F. (eds), 1992, *Geological Applications of Wireline Logs II.* Geological Society Special Publication No. 65, pp. 277–283.

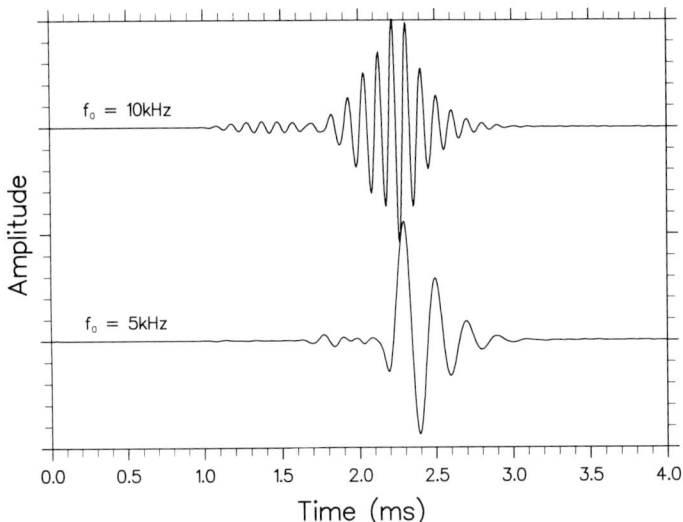

Fig. 1. Examples of acoustic wavetrains in a fast formation. $v_P = 3400$ m/s, $v_S = 2000$ m/s, $v_m = 1500$ m/s, $\rho = 2.2$ g/cm^3, $\rho_m = 1$ g/cm^3. Borehole radius is 10 cm. Transmitter–receiver separation is 3 m.

Theory

The acoustic wavefield in a borehole may be classified into two main groups of contributions (see e.g. White 1983): the refractions and the borehole eigenmodes. The *refractions* may be understood in a ray model. A ray emitted from the transmitter hits the borehole wall at a critical angle, and propagates along the borehole wall at the formation velocity while it continuously emits energy into the borehole, which may be detected by the receiver. The refractions are thus attenuated due to geometrical spreading even for a model with no intrinsic attenuation.

The *borehole modes* are modes that are confined to the borehole, and which in some cases propagate without geometrical attenuation. For an axisymmetric wavefield, the eigenmodes are the Stoneley mode and the family of pseudo-Rayleigh modes. The *Stoneley* mode is slightly dispersive, and exists as a propagating mode down to zero frequency. The *pseudo-Rayleigh* modes exist as undamped modes only in a fast formation (in which the shear velocity is higher than the mud velocity), and above a given cut-off frequency which increases with the mode number. The cut-off frequency may be roughly estimated by assuming that an odd number of half-wavelengths should fit into the borehole. For a 20 cm borehole, the cut-off frequency for the lowest mode is thus of order 10 kHz.

In a fast formation, the wavetrain will thus consist of P and S refractions, and contributions

from the pseudo-Rayleigh modes and the Stoneley mode. Figure 1 shows typical examples of wavetrains in a fast formation. (All synthetic wavetrains shown in this paper have been computed using the method of real axis integration; see e.g. Tsang & Rader 1979 or Cheng & Toksöz 1981.) The upper trace is for a centre frequency of 10 kHz, and the lower for a centre frequency of 5 kHz. In the lower trace one observes the P- and S-refractions and the Stoneley wave, while the pseudo-Rayleigh wave is absent since it is not a propagating mode within the frequency range of the source. In the upper trace the wavetrain is more complicated, since the pseudo-Rayleigh mode is now excited. However, it is still fairly straightforward to pick the P- and S-refractions.

In a slow formation, we have no S-refraction, and the only non-attenuated eigenmode is the Stoneley wave. For the low-frequency case in the lower trace of Fig. 2, a simple wavetrain, with only the P-refraction and the Stoneley wave, is seen. However, going to higher frequencies, as shown in the three upper traces of Fig. 2, the wavetrain becomes more complicated. The additional arrival is due to the lowest pseudo-Rayleigh mode, which becomes clearer as frequency increases. In spite of being geometrically damped, it thus contributes to the wavetrain in this case (wavetrains similar to those in Fig. 2 have previously been discussed by Paillet & Cheng (1986)). Observe that the leaky pseudo-Rayleigh

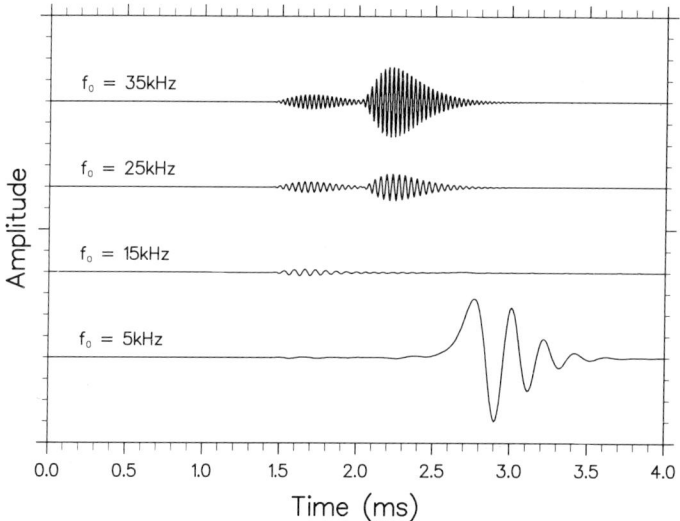

Fig. 2. Examples of acoustic wavetrains in a slow formation. $v_P = 2240$ m/s, $v_P = 1400$ m/s, $v_m = 1500$ m/s, $\rho = 1.63$g/cm^3, $\rho_m = 1$ g/cm^3. Borehole radius is 10 cm. Transmitter–receiver separation is 3 m.

wave is seen above the normal frequency spectrum for acoustic logging, which generally has little energy above 15 kHz. Note also that the Stoneley wave is not seen for the two higher frequencies, and only barely seen at 15 kHz, since it is weakly excited at these higher frequencies.

In the following, we will show that the amplitude of the leaky pseudo-Rayleigh wave in certain cases is strongly sensitive to the shear wave velocity, and less sensitive to other relevant parameters. Figure 3 shows the dependence of the attenuation of the first pseudo-Rayleigh mode at 20 kHz on formation S-wave velocity, borehole radius, mud velocity, mud attenuation, formation P-wave velocity, and formation attenuations. The data in Fig. 3 have been obtained by tracing the pseudo-Rayleigh pole in the complex k_z-plane, while changing the appropriate parameter. The geometric attenuation is computed from the imaginary part of the pole. All computations have been performed disregarding the presence of a logging tool. The logging tool will in practice lead to a reduced effective borehole size. In essence this means that the effects we discuss will be shifted towards higher frequencies.

From the figure it can be seen that, for this case, the attenuation is strongly dependent on shear wave velocity between 1300 and 1500m/s, and quite sensitive for velocities below 1200m/s. Secondly, there is a strong dependence on borehole radius, and on the mud Q below 50. The

attenuation is virtually insensitive to the formation Q-values and the formation compressional velocity, and moderately sensitive to the mud velocity.

This behaviour may be explained in the following way: The attenuation is governed by mud losses and by energy radiated into the formation as shear waves. The mud losses depend on the mud Q, and are important for low mud Q values. The radiation into formation shear waves depends on the matching between the velocity in the mud and the shear velocity in the formation. Thus the formation shear velocity, the borehole radius and the mud velocity are of importance. On the other hand, the radiation into the formation shear waves depends very little on the P-wave velocity and on the attenuations in the formation. Note the little peculiarity that increased shear Q in fact reduces the attenuation a bit. This may be explained by a less effective radiation into the formation for a low shear wave Q.

In Fig. 4, we show synthetic full waveforms as a function of distance for a shear wave velocity of 1300m/s. The strong attenuation of the pseudo-Rayleigh wave due to radiation is apparent when comparing with the amplitude of the P-refraction.

Figure 5 shows the effect on the full waveform at 3.0 m of changing the shear wave velocity. The P velocity is kept at 1.6 times the S velocity. The maximum attenuation around 1200–1300 m/s is seen, and the significant sensitivity

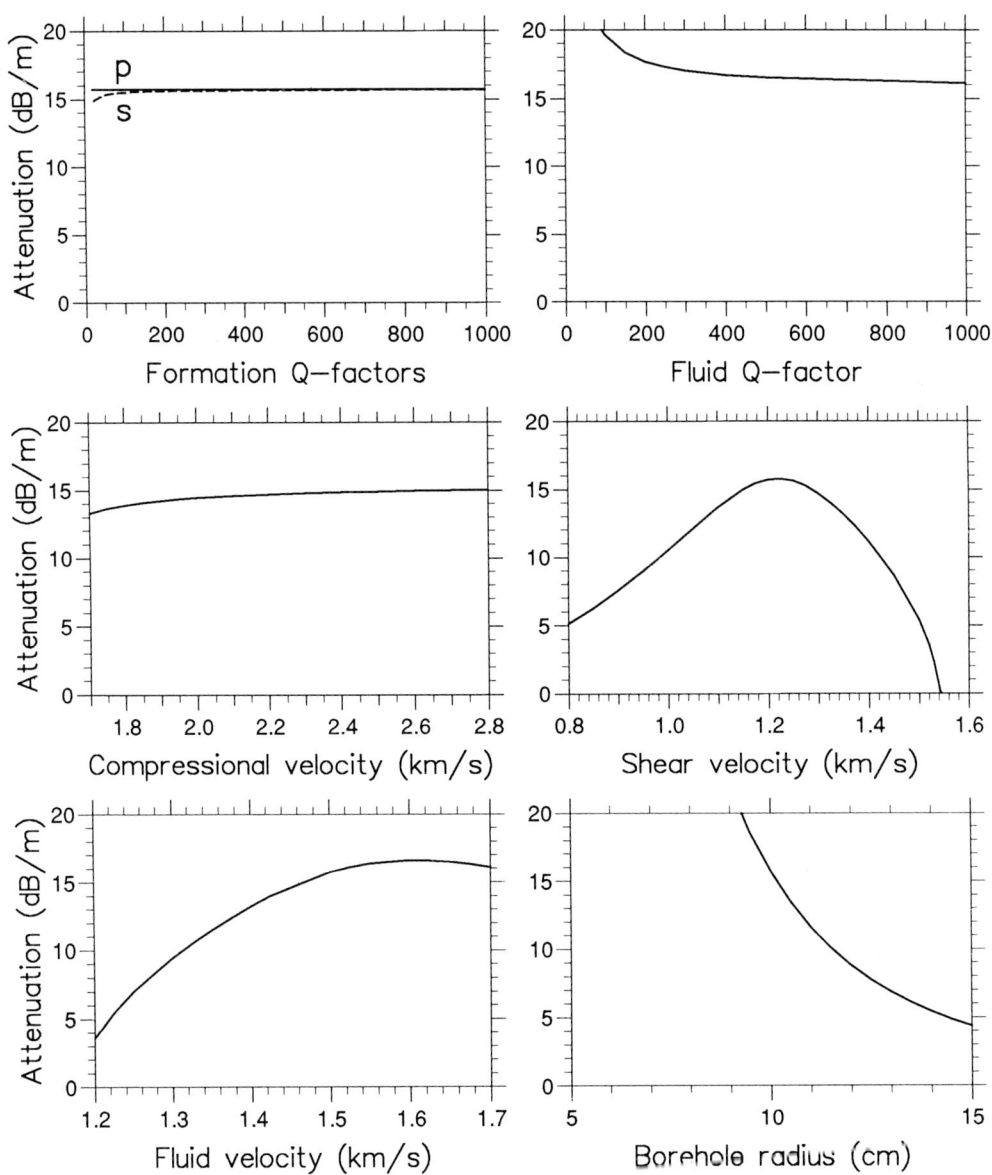

Fig. 3. Dependence of first pseudo-Rayleigh mode attenuation at 20 kHz on various parameters: $v_p = 2200$ m/s, $v_s = 1200$ m/s, $v_m = 1500$ m/s, $\rho = 1.63$ g/cm^3, $\rho_m = 1$ g/cm^3. Borehole radius is 10 cm.

between 1300 and 1500 m/s for this model is clearly illustrated.

In Fig. 6 we show a similar plot, where we keep the P-wave velocity constant at 2200 m/s. The behaviour of the pseudo-Rayleigh train is the same as in Fig. 5, as expected since the sensitivity to the P velocity is small. Further, one should note the sensitivity of the P-wave amplitude to the shear wave velocity, in accordance with the discussion in the Introduction. Note

that in contrast to the leaky pseudo-Rayleigh wave, the amplitude effect on the P-wave refraction is due to the variation of the coupling of energy into the refracted P-waves, and not to an attenuation mechanism.

Experimental results

A model experiment was performed on a hollow cylinder of outcrop chalk. The sample had an

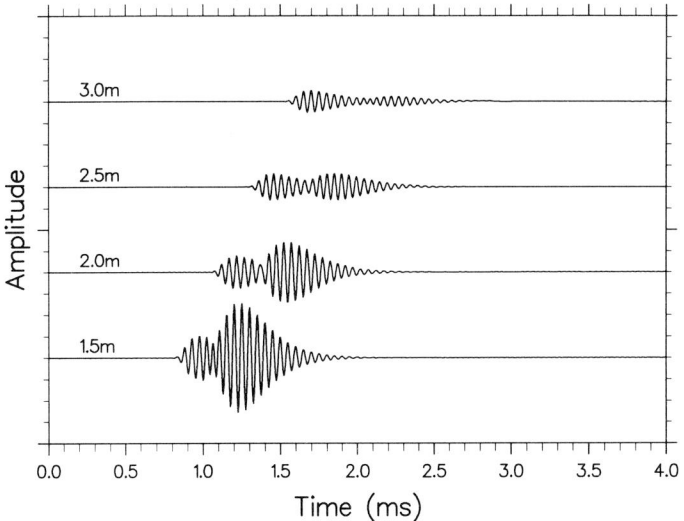

Fig. 4. Acoustic wavetrains as a function of distance for a source centre frequency of 20 kHz: $v_p = 2080$ m/s, $v_s = 1300$ m/s, $v_m = 1500$ m/s, $\rho = 1.63$ g/cm^3, $\rho_m = 1$ g/cm^3. Borehole radius is 10 cm.

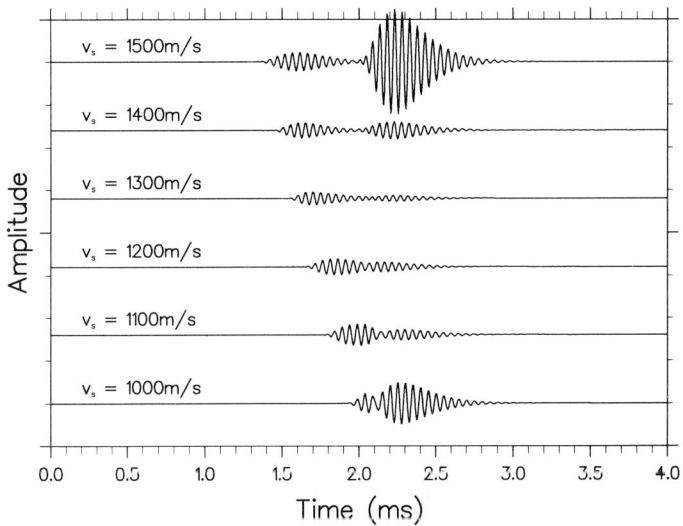

Fig. 5. Acoustic wavetrains at various shear wave velocities. $v_p = 1.6 v_s$, $v_m = 1500$ m/s, $\rho = 1.63$ g/cm^3, $\rho_m = 1$ g/cm^3. Borehole radius is 10 cm. Transmitter–receiver separation is 3 m.

outer diameter of 20 cm and an inner diameter of 2 cm. The sample was saturated with water in chemical equilibrium with the chalk. Full acoustic wavetrains were recorded with a miniature acoustic logging tool with two receivers.

Solid cylinders of this chalk fail under hydrostatic load at about 8–9 MPa (Raaen & Renlie 1990). Unconfined acoustic velocities are typically $v_p = 2100$ m/s and $v_s = 1250$ m/s.

Figure 7 shows the acoustic wavetrains as a function of the applied external stress. The following should be observed from the figure. The velocity of the P-refraction is nearly constant, but increases a little from 3 to 5 MPa and then decreases. The amplitude of the P-refraction is somewhat increased around 5 MPa. The amplitude of the leaky pseudo-Rayleigh waves first decreases a little, and then increases consider-

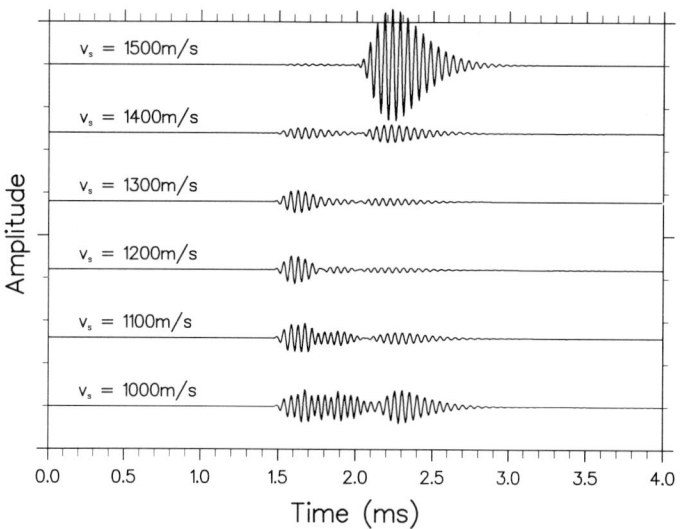

Fig. 6. Acoustic wavetrains at various shear wave velocities. $v_p = 2100 \, \text{m/s}$, $v_m = 1500 \, \text{m/s}$, $\rho = 1.63 \, \text{g/cm}^3$, $\rho_m = 1 \, \text{g/cm}^3$. Borehole radius is 10 cm. Transmitter–receiver separation is 3 m.

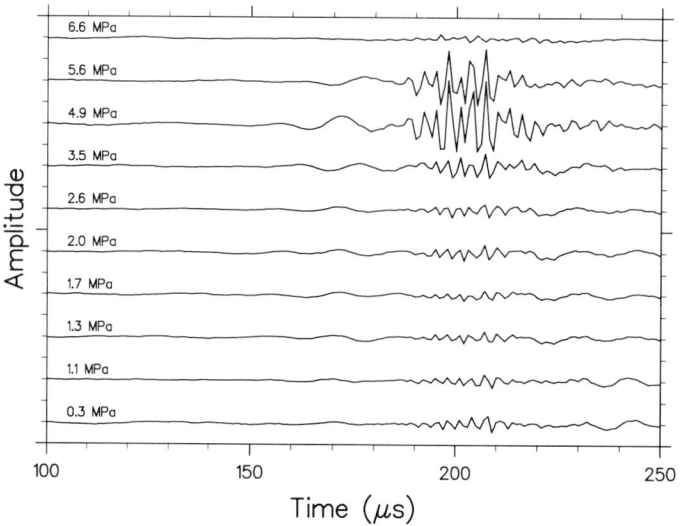

Fig. 7. Observed wavetrains as a function of applied external stress in an outcrop chalk. Transmitter–receiver separation is 30.5 cm.

ably from 3 to 5 MPa, before decreasing at higher stress levels.

During these experiments, one may assume the mud Q-factor and velocity to remain constant. The borehole radius may be expected to decrease, particularly close to failure. As seen from Fig. 3, a decrease in borehole radius will increase the attenuation of the pseudo-Rayleigh wave. The other parameters that may be expected to vary are the formation velocities. From the P refraction it is clear that the P velocity is rather constant. Also, any change in the P velocity would not affect greatly the amplitude of the pseudo-Rayleigh wave, according to

Fig. 3. The variation in the amplitude must therefore be explained in terms of the shear velocity increasing or decreasing in the sensitive region. A possible explanation is that the shear velocity decreases as the load increases from 3 to 5 MPa. This is also consistent with the amplitude increase of the P-wave refraction, as seen from Fig. 6. The decrease above 5 MPa could then be explained by a continued reduction of the shear velocity, or alternatively by a reduction in the borehole radius associated with the borehole failure. The present dataset is, however, not sufficient to allow a firm conclusion to be drawn.

Conclusions

We have shown by computing eigenmode attenuations and synthetic full wavetrains that the high-frequency part of the leaky pseudo-Rayleigh wave in a soft formation is sensitive to the shear velocity, the borehole radius, and the mud Q-value. It is moderately sensitive to the mud velocity, and insensitive to the formation P-wave velocity and the formation Q-factors.

The effects have been illustrated by experiment, although the interpretation of the data is not fully conclusive.

With the advent of the multipole logging tools, and the push towards lower frequencies for the monopole tools, this indirect method is not expected to become of significant value in the field. However, new insight into acoustic propagation has been gained, which should be useful for the interpretation of both old and new high-frequency data. It is expected that the effect could be useful in the interpretation of data from properly designed laboratory experiments. For example, since one senses the shear velocity in the near wellbore region, information on stress dependent elastic moduli (see e.g. Santarelli *et al.* 1986) may be obtainable.

The author would like to thank Fina Exploration Norway as responsible for the Mechanical Properties subproject of the North Sea Chalk Research Programme for the permission to include the experimental data in Fig. 7. The referee is thanked for giving valuable comments.

References

CHENG, C. H. 1989. Full waveform inversion of P waves for v_s and Q_p. *Journal of Geophysical Research*, **94**, 19619–19625.

—— & TOKSÖZ, M. N. 1981. Elastic wave propagation in a fluid-filled borehole and synthetic acoustic logs. *Geophysics*, **46**, 1042–1053.

——, —— & TUBMAN, K. M. 1983. Determination of shear wave velocities in 'soft' formations. *SEG 53rd Annual Meeting and Exposition, Las Vegas, September 11–15*, Paper BHG8.

HARRISON, A. R., RANDALL, C. J., ARON, J. B., MORRIS, C. F., WIGNALL, A. H., DWORAK, R. A., RUTLEDGE, L. L. & PERKINS, J. L. 1990. Acquisition and analysis of sonic waveforms from a borehole monopole and dipole source for the determination of compressional and shear speed and their relation to rock mechanical properties and surface seismic data. *SPE 20557*.

PAILLET, F. L. & CHENG, C. H. 1986. A numerical investigation of head waves and leaky modes in fluid-filled boreholes. *Geophysics*, **51**, 1438–1449.

RAAEN, A. M. & RENLIE, L. 1990. Relations between acoustic and static mechanical properties of chalk.

Third North Sea Chalk Symposium, Copenhagen, June 11–12.

SANTARELLI, F. J., BROWN, E. T. & MAURY, V. 1986. Analysis of borehole stresses using pressure-dependent, linear elasticity. *International Journal of Rock Mechanics and Mining Sciences & Geomechanical Abstracts*, **23**, 445–449.

SCHMITT, D. C., BOUCHON, M. & BONNET, G. 1988. Full wave synthetic acoustic logs in radially semiinfinite saturated porous media. *Geophysics*, **53**, 807–823.

TSANG, L. & RADER, D. 1979. Numerical evaluation of the transient acoustic waveform due to a point source in a fluid-filled borehole. *Geophysics*, **44**, 1706–1720.

WHITE, J. E. 1967. The Hula log: A proposed acoustic tool. *Transactions of SPWLA 8th Annual Logging Symposium*.

—— 1983. *Underground Sound*. Elsevier, New York.

ZEMANEK, J., ANGONA, F. A., WILLIAMS, D. M. & CALDWELL, R. L. 1984. Continuous Acoustic Shear wave logging. *Transactions of SPWLA 25th Annual Logging Symposium*.

Laboratory measurements of the seismic properties of sedimentary rocks

C. McCANN & J. SOTHCOTT

Postgraduate Research Institute for Sedimentology, University of Reading, Whiteknights, Reading RG6 2AB, UK

Abstract. A laboratory technique for the measurement of compressional and shear wave velocities and attenuations (quality factors) of sedimentary rocks at pressures up to 70 MPa and at frequencies up to 1 MHz was published by Winkler and Plona in 1982. In the method the rock sample is enclosed between buffer rods. The arrival times and amplitudes of the signals reflected from its top and base are compared. A similar technique has been used for many years in the non-destructive testing of metals and plastics. This paper describes the use of a single-frequency pulsed sine wave for the velocity and attenuation measurements, in addition to the transient pulse used in the Schlumberger method. Measurements were made on samples with accurately known compressional and shear wave properties and on replicated rock samples. The results demonstrate that the determinations of velocities are accurate to $\pm 0.3\%$, and that the determinations of attenuation are accurate to ± 0.1 dB/cm at 0.85 MHz.

Compressional and shear wave sonic logs form an important part of the suite of logs used in the interpretation of the lithological and petrophysical properties of formations penetrated by boreholes. The purpose of this paper is to bring to the attention of a wider audience of log analysts the availability of a laboratory technique for accurately measuring the compressional and shear wave velocities and attenuations of sedimentary rocks under simulated in situ conditions of pressure. In the method the sample is sandwiched between two buffer rods. A compressional and/or shear wave is input to the top of one buffer rod. Velocity and attenuation are determined by comparing the travel times and the amplitudes of the pulses reflected from the top and base of the rock sample. The method was developed at Schlumberger by Winkler & Plona (1982).

The results are of intrinsic value in supplementing log data as well as contributing towards our understanding of the relationships between the seismic and petrophysical properties of these rocks and hence potentially enhancing the accuracy and reliability of the interpretations. As an example of the former, both shear and compressional velocities are required for the interpretation of amplitude-versus-offset surveys (Ostrander 1984), yet it is still relatively unusual for shear wave log data to be routinely acquired. In this case the laboratory estimates of shear wave velocity can be of considerable importance in placing limits on the in situ wave velocities.

It is important for the accurate interpretation of log data that a substantial body of accurate laboratory seismic velocity data together with the associated sedimentological data should exist, to define and explain the relationships between the compressional and shear wave velocities and the petrophysical properties of sedimentary rocks. At the present time it is not possible to predict *accurately* the seismic properties of a rock from its sedimentological and poro-perm properties or *vice versa*, but much progress has been made in understanding these relationships since the pioneering work of Wyllie *et al.* (1958), who demonstrated an inverse linear relationship between compressional velocity and porosity, summarized in the 'time-average equation' for clean sandstones. Their database was extended to include both compressional and shear wave velocities for shaley sandstones, mudstones, limestones and dolomites by Pickett (1963) and by Domenico (1984). Castagna *et al.* (1985) and Han *et al.* (1986) demonstrated the following linear relationships between compressional wave velocity, V_p(km/s), shear wave velocity, V_s(km/s), fractional porosity, ϕ, and fractional clay content, V_{cl}, for clastic silicate rocks:

$$V_p = 5.81 - 9.42\,\phi - 2.21 V_{cl} \qquad (1)$$

$$V_s = 3.89 - 7.07\,\phi - 2.04 V_{cl}. \qquad (2)$$

Wilkens *et al* (1984) have demonstrated similar relationships for carbonate rocks. However, Hilterman (1990) showed that these equations oversimplify the actual relationships particularly as

From HURST, A., GRIFFITHS, C. M. & WORTHINGTON, P. F. (eds), 1992,
Geological Applications of Wireline Logs II. Geological Society Special Publication No. 65, pp. 285–297.

the velocities appear to depend on the clay type and habit, not just on the clay content. Other factors which cause deviations from the simple relationships include the elastic moduli of the inter-grain cements (Castagna *et al.* 1985) and the nature and degree of saturation of the pore fluid(s) (Murphy 1982). Wang & Nur (1990) showed the effect of temperature on the acoustic velocities in sedimentary rocks saturated with hydrocarbons and the importance of understanding this in following the progress of steam floods for enhanced oil recovery.

A second seismic property which defines the amplitude and quality of the signal observed on compressional and shear wave logs is the attenuation or absorption of the elastic energy. This may be defined as the loss per unit distance (attenuation coefficient, measured in dB/cm or dB/m) or the inverse of the fractional loss per period or wavelength (quality factor, Q, dimensionless). Attenuation has rarely been measured on borehole logs in the past for the following reasons. Amplitude losses due to the complex refraction wave path, mode conversion and thin bed scattering of the high frequency signal often exceed the anelastic losses (Goldberg & Zinszner 1989). Hence very careful analysis of the complete wavefield is necessary to extract the contribution of the anelasticity (Patton 1988). Little effort has been devoted to this as it was not perceived that attenuation could give significant information about the petrophysical or sedimentological properties of the rocks. Recently, however, Klimentos & McCann (1990) have demonstrated that there is a linear relationship between compressional wave attenuation, a_p (dB/cm at 1 MHz and 60 MPa), fractional porosity, ϕ, and fractional clay content, V_{cl} in water-saturated shaley sandstones under simulated in situ conditions in the laboratory:

$$a_p = 0.132 + 3.15\,\phi + 24.1\,V_{cl} \qquad (3)$$

Recent work (unpublished by A. Best and the authors) has demonstrated a similar relationship between shear wave attenuation, porosity and clay content. But how relevant are laboratory measurements of seismic quality factor at frequencies of hundreds of kilohertz to log measurements in situ at frequencies of a few kilohertz or less? The early work of Attewell & Ramana (1966) indicated that the compressional wave quality factor is independent of frequency. Spencer (1981), Murphy (1982) and Tittman *et al.* (1984) showed in the laboratory that in dry sedimentary rocks absorption is low and the quality factor essentially frequency independent. In fluid saturated sedimentary rocks at ambient

pressure the presence of pore fluids increases the absorption and causes the quality factor to display strong frequency dependence. Morig & Burkhardt (1989) showed that the observed peaks of loss consist of two parts: there is a contribution from the rock intrinsic attenuation *and* a contribution from flow of the pore fluid into and out of the surface of the sample (the Biot–Gardner boundary effect). The absorption at ambient pressure is dominated by the effects of the microfractures pervading the sample (Winkler 1985). As the confining pressure applied to a sample is increased the absorption decreases by an order of magnitude (Klimentos & McCann 1990) as the microfractures are closed, to a residual or terminal value which as shown above is related to the sedimentology of the rock. It is the variation of the residual absorption with frequency which is important in comparing high- and low-frequency behaviour. In this context the work of O'Hara (1989) is crucial. He made the following observations on the extensional logarithmic decrement in brine-saturated Berea Sandstone over nearly two decades of frequency. At a confining pressure of 5 MPa the logarithmic decrement increased at the power of frequency of 0.31. The power dependence progressively decreased with increasing confining pressure until at the maximum confining pressure of 40 MPa the logarithmic decrement increased at a power of frequency of 0.12. This would correspond to a change of logarithmic decrement by a factor of only 1.7 from 10 kHz logging frequency to 1 MHz laboratory frequency. Much more work has to be done to establish the true variation of the quality factor with frequency at high confining pressure, but O'Hara's results indicate that it may not be large.

It may thus be concluded that accurate measurements of attenuation should be carried out to define more precisely the relationship of attenuation with the sedimentary properties of the rocks, that these can usefully be made at ultrasonic frequencies in the laboratory and that it is a reasonable *prediction* that the properties observed in the laboratory will be applicable to, or at the least give guidance on, the properties at the lower frequencies used in logging. With the advent of sophisticated multi-receiver, full-waveform sonic logging equipment and powerful full-wave-field processing of the data, it can be expected that attenuation measurements will be added to the suite of physical properties available from the borehole and that these data will give diagnostic information on the sedimentology and petrophysical properties of the rocks. It should be pointed out that Oliver *et al.* (1991)

have successfully measured compressional wave quality factors from standard vertical seismic profiles at frequencies from 20 to 80 Hz, and have demonstrated a relationship between the resulting quality factors and the lithology.

Laboratory measurement of seismic properties

The standard laboratory technique used by many workers to obtain the compressional and shear wave velocities of rocks is the measurement of the travel time of an ultrasonic signal (0.4 MHz to 1.5 MHz) transmitted directly through the rock. This method was used by Wyllie et al. (1958) to measure compressional wave velocity, and by King (1965) to measure both compressional and shear wave velocities. It is a powerful technique because the sample and the transducers may be placed inside a pressure cell and thus the measurements carried out under simulated in situ conditions of temperature and pressure. This is important, as the pervasive fractures which exist inside samples of sedimentary rocks as a result of the stress relief during coring render meaningless measurements of the seismic properties at ambient pressure. In the method cylindrical samples of the rock, 5 cm diameter and 1.5 cm to 3.0 cm long, are drilled from core or hand specimens. The two ends of the sample are ground flat and parallel within 0.01 mm. In the transmission method as normally used, a transient pulse is generated by applying a high voltage spike to the transmitting transducer; the velocity is measured by estimating the onset time of the pulse received by the second transducer at the far end of the sample. There are a number of disadvantages with the standard technique which may lead to systematic errors in the measurements of velocity and attenuation.

Four particular sources of error may degrade the accuracy of the measurement of velocity. Firstly, the travel time of the pulse through the system includes that through the transducer faces and through any coupling material. Consequently, the system must be calibrated by measuring the travel time through a material of known compressional and shear wave velocity. Secondly, the onset time of a transient pulse cannot be measured precisely. Thirdly, the exact frequency of the measurement cannot be measured and fourthly, as a consequence of the third, it is not possible to make corrections for the phase shifts (diffraction corrections) arising from the transducer geometry. The first and second of these are largely overcome by applying sufficient coupling pressure to the transducers

and by using a specimen that is sufficiently long for the travel time error to be negligible, but the third and fourth limitations remain. The length of the sample is ultimately limited by the anelastic attenuation. Furthermore, although unlikely with modern electronic equipment, there exists the possibility of electronic phase shifts and pulse delays, which cannot be observed, within the overall travel time. In summary, the disadvantages of the method are that it relies totally on calibration with a specimen of known compressional and shear velocities, that the frequency of the measurement is not known and that it is then difficult to apply corrections for the diffraction effects of the transducers.

It is well known that it is difficult to make accurate attenuation measurements with the direct transmission system. There is unknown and variable coupling between the transducers and the sample; the transmitted pulse shape and spectrum cannot be determined and there is unknown and unmeasurable change in amplitude of the transmitted pulse arising from the impedance contrast between the transducer and the sample faces. It is not possible to reproduce exactly the conditions of the measurement between the calibration material and the rock sample. In the case of shear waves, the transmitting and receiving transducers must be exactly aligned to obtain accurate attenuation measurements, a procedure of some technical difficulty.

To overcome these problems we use a pulse reflection technique, widely used in the non-destructive testing industry (see, for example, Papadakis 1975), and described with application to sedimentary rocks by Winkler & Plona (1982). Figure 1 shows the arrangement of the equipment. A compressional or shear wave transducer transmits a pulse through a perspex buffer rod into the rock sample and thence into a second perspex buffer rod. The same transducer is used to receive the reflections from the perspex rock interfaces at the top (reflection A) and at the bottom (reflection B) of the sample. The reflected pulses are displayed on the oscilloscope screen for direct measurement of the travel times and amplitudes and stored on the computer for subsequent spectral analysis. The whole system comprising the steel housing of the transducer, the perspex buffer rods and the rock sample is enclosed in a rubber jacket and placed inside a pressure cell, capable of applying up to 70 MPa hydrostatic pressure. A feature of the high pressure cell is a ram which extends from the inside to the outside of the chamber and which remains in contact with the top of the transducer housing. Measurement of the movement of the top of the ram enables the decrease in length of the sample with the increasing pressure to be monitored with a dial micrometer to an accuracy of ±0.01 mm. It also enables a uniaxial pressure to be applied, if required, via the top of the steel transducer housing. The rock can be dry or saturated with an appropriate pore fluid and the pore pressure can be varied through a fluid inlet in the lower buffer rod. After pressurizing the sample to 65 MPa, travel time and amplitude measurements are taken as the pressure is reduced in steps of 10 MPa, allowing the whole system to equilibriate at each pressure for 30 minutes.

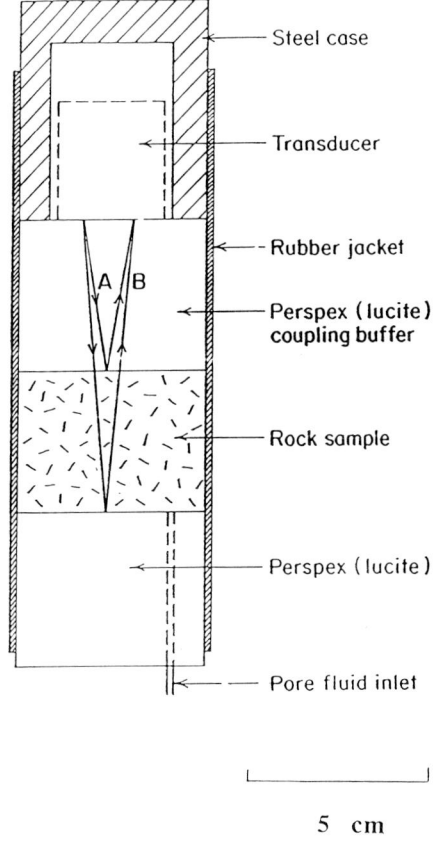

Steel case

Transducer

Rubber jacket

Perspex (lucite) coupling buffer

Rock sample

Perspex (lucite)

Pore fluid inlet

5　cm

Fig. 1. Arrangement of the transducer and sample for the reflection system of measuring velocity and attenuation in sedimentary rock samples.

The acoustic analysis of the rock is based on the comparison of reflections at the top and bottom of the sample; any variations in the amplitude or spectral content of the source pulse due to the transducer couplant, electronic variations or impedance contrast with the buffer rod affect both reflections equally and have no effect on the measurement of velocity and attenuation. It has been observed in our experiments that the perspex buffer rods are effectively in 'welded' contact with the rock specimen at pressures in excess of 10 MPa and thus the amplitudes of the reflected pulses may be accurately predicted from the relative acoustic impedances of the perspex and the rock sample. In many cases multiple reflections within the rock sample can be observed and measured, effectively doubling or tripling the sample length and the travel times, enabling confidence limits to be placed on the resulting velocities and attenuations.

An inconvenience in obtaining well material is that the 5 cm diameter samples are double the size of the core plugs normally used in the industry for petrophysical and sonic measurements. However, it is important that the diameter-to-wavelength ratio and the diameter-to-length ratio of the core should be as large as possible in order that *true* bulk compressional and shear wave velocities and attenuations are measured. If the cylinder of rock is too thin, the Young's modulus wave velocity is determined rather than the true bulk compressional wave velocity.

Winkler & Plona (1982) measured velocity and attenuation as a function of pressure by comparing the phase and amplitude spectra of the reflections from the top and the base of the rock sample. They estimated that their measurements of velocity were accurate to within 1%, and their measurements of attenuation accurate to within 0.5 dB/cm. Certainly their calibration measurements on materials of known velocities and attenuations show discrepancies of these magnitudes. For example, they give the published value for the compressional velocity of fused silica to be 5869 m/s; their measured value (Fig. 3 of Winkler & Plona 1982) is 5900 m/s, as closely as can be determined from the diagram. Similarly, their measured compressional wave attenuation for fused silica was between 0.2 dB/cm and 0.5 dB/cm, whereas the true value is 0 dB/cm. Previous measurements made in our laboratory using the transient pulse technique (Klimentos & McCann 1990) were considered to have similar accuracies.

In trying to investigate the relationships between the seismic and the petrophysical properties of sedimentary rocks it is important that the seismic data should be as accurate as possible in order that observed differences between samples can be related to the sedimentology. The technique of using the difference between the phase spectra to determine velocity is ultimately limited by the sampling interval of the pulses; in our case this is ±0.02 μs. With an

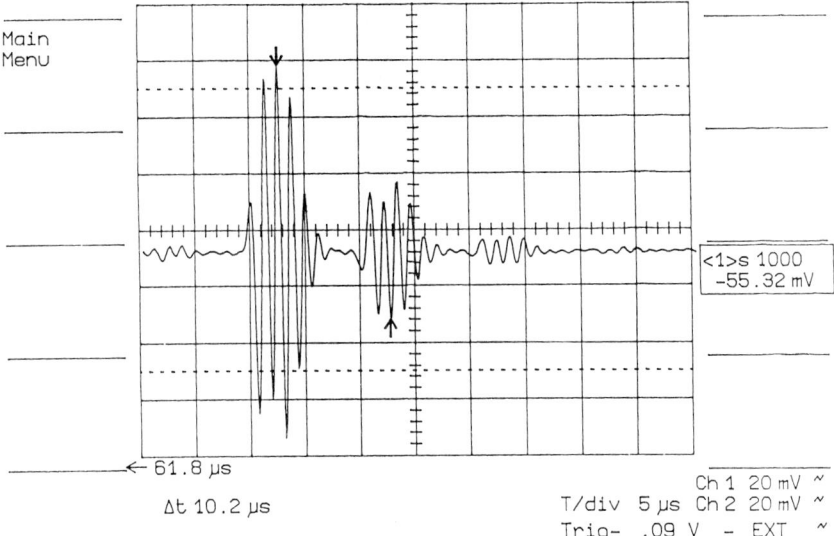

Main
Menu

61.8 μs

Δt 10.2 μs

<1>s 1000
-55.32 mV

Ch 1 20 mV ~
T/div 5 μs Ch 2 20 mV ~
Trig- .09 V - EXT ~

Fig. 2. Example of a shear wave pulsed-sine-wave signal. The first pulse is the reflection from the top of the sample. The second and subsequent pulses are the primary and multiple reflections from the base of the sample. The pulse is displayed at 5 μs/division. Confining pressure is 50 MPa. The sample is water-saturated.

average time difference of 10 μs, the precision cannot be better than ±0.2%. This can be improved by an order of magnitude by making measurements on *tone-burst or flat-topped sine wave pulses* and it is worthwhile explaining in detail how this is achieved using the Lecroy Type 9400 Digital Oscilloscope. Tone-burst or flat-topped sinusoidal-pulses are almost-single frequency pulses of which the duration can be varied from about three cycles upwards; in our implementation of the method they are generated by an Arenberg Type PG650C pulse generator and the transducer is driven through a home-built transmit–receive circuit. A typical shear-wave pulse pair reflected from the top and base of a rock sample is shown in Fig. 2.

For each time and amplitude measurement, 1000 signals are summed and averaged to maximize the signal-to-noise ratio. The arrival times of the third positive cycle of the reflection from the base of the top buffer rod and the third negative peak of the first and subsequent reflections from the base of the rock sample are measured to an accuracy of 1 ns (see below). The amplitudes of the signals from the top of the third peak (trough) to the bottom (top) of the next trough (peak) is measured to an accuracy of 0.01 mV. In many cases as many as four reflections can be observed and measured. The frequency of each separate pulse is measured and recorded. The internal frequency of the sinusoid can be varied, as the compressional and shear

wave transducers are broadband (Panametrics nominal 1 MHz centre frequency). Each acoustic measurement is repeated twice to check for consistency. As each individual pulse is acquired by the digital oscilloscope it is sampled at 100 MHz (10 ns sampling interval).

How can a signal that is digitized at 10 ns sampling interval be measured with a precision of 1 ns? The time measurement between the reflections from the top and the base of the rock sample is made by displaying a particular peak or trough of the sinusoid on the oscilloscope screen (Fig. 2) at a timebase rate of 50 ns/cm. The oscilloscope screen displays the signal with the resolution of 1024 by 1024 samples. Although each individual pulse acquired is only sampled at 10 ns intervals, the random relationship between the digitizer and the pulse generator trigger ensures that all pixels on the screen are filled with samples and a 'smooth' waveform with an equivalent sample interval of 0.5 ns is observed. This is true whether the oscilloscope is in continuous acquisition mode or is digitally summing 1000 pulses to give a stationary, virtually-noise-free waveform. The absolute time of the waveform relative to the trigger is displayed with a precision of 1 ns. It is only if a single 'shot' is acquired that the waveform is displayed on the screen with samples at time separation of 0.01 μs joined by straight lines. Other factors contributing to the precision of the method are the precision of the oscilloscope time base, which is calibrated to 1 part in 10^5 (calibration certificate No 9400/85590), and the ability of the operator to define accurately the position of the actual peak or trough of the cycle. By displaying it with maximum amplification and large offset the peak becomes a sharply defined point on the screen, and can usually be defined to the nearest 1 ns. The potential precision of the time difference between the peaks/troughs using this method is thus about ± 2 ns. We conclude that the precision of determination of a travel time difference of 10 μs, is $\pm 0.02\%$.

To summarize, velocity determination by measurement of the difference between the phase spectra of transients is limited in precision by the width of the sampling interval to about $\pm 0.2\%$. The ability of the oscilloscope screen to display 1024 data points on a timebase of 50 ns/cm, combined with the random relationship between the 100 MHz digitizer and the pulse trigger enable the peaks or troughs of the sinusoidal pulse to be displayed with an observed sampling interval of 0.5 ns. The potential precision of the on-screen measurement of the time difference between peaks of tone-bursts is an order of magnitude better, about $\pm 0.02\%$. Time measurement ceases to be a problem of limitation and the final accuracy of the velocity measurement is determined by other factors. The accuracy of the sample thickness measurement is comparable to that of the time measurement at about $\pm 0.025\%$, but the overall accuracy will be degraded by inaccuracies in the phase corrections for the geometry of the transducer (see below) and by the variability of the rock samples. An important advantage of making the high precision velocity measurement at a single frequency is to provide an absolute value to calibrate the velocity dispersion curve determined with lower precision from the phase spectra of the transient pulses.

As explained before, the amplitudes of the two pulses may be measured directly from the oscilloscope screen for attenuation measurements. There is no equivalent significant improvement in the precision of the attenuation measurements as a result of using a single frequency sinusoid. Both the sinusoid and the transient are digitized at 8-bit resolution by the oscilloscope. The random relationship between the digitizer and the trigger enables the full 1024 vertical display resolution of the screen to be filled, giving an effective resolution of 10 bits. A precision of about 0.2% in the measurement of the amplitude ratio would be expected. The actual accuracy attained in the attenuation measurements on samples is discussed below. There are two advantages in measuring attenuation using the single frequency sinusoid. Firstly, all the power of the pulse is contained within a narrow frequency band, leading to excellent signal-to-noise ratio. Secondly, the presence of interfering pulses, such as arise from scattering, mode conversion etc., distort the characteristic flat top of the pulse and this is a sensitive test of the reliability of the final attenuation results.

Both travel time and amplitude data are analysed to calculate the compressional and shear wave velocity and attenuation of each sample at each specified frequency. The computer program combines the arrival times of all the reflection readings to obtain an approximate velocity; this value is then used to calculate the diffraction corrections. These arise from the interference between the elementary wavefronts transmitted from each point on the 2.54 cm diameter transducer and are a function of the wave frequency, transducer diameter and path length. The net result is a change in the *phase* of the observed signal, which gives rise to an apparent decrease in the travel time and thus an apparently higher velocity in the sample and a decrease in the *amplitude* of the signal resulting in an apparently higher attenuation in the sample. Both effects become more pronounced and significant with decreasing frequency. Papadakis *et al.* (1973) and Papadakis (1976) published graphs of the necessary phase and amplitude corrections; detailed tables of the corrections were published by Benson & Kiyohara (1974). The corrections refer to compressional waves only and to propagation in an infinite medium. We have investigated the applicability of the corrections to shear waves and to specimens of finite size: the results are given in the following section. Using the calculated velocities together with the density of the sample, the computer program corrects the observed amplitudes of the pulses for reflection and transmission losses at the interfaces of the sample with the buffer rods, and for the amplitude diffraction losses. It then calculates the attenuation coefficient of the sample from the primary and subsequent multiple reflections, together with a 95% confidence interval. This measurement may be repeated at several different frequencies and quoted on its own, or used to supplement and calibrate the results of spectral analysis of a transient pulse.

The range of frequencies used in the experimental work is between 0.5 MHz and 1 MHz; for most rocks the compressional wave measurements are carried out at 0.85 MHz and 1 MHz, and the shear wave measurements at 0.85 MHz only. At lower frequencies the diffraction effects of the transducer geometry cannot be properly calculated; at frequencies greater than 1 MHz wave scattering is observed due to the finite size of the sand grains. In measuring compressional wave velocities on rock samples of less than 2 cm length, and on the high velocity calibration aluminium sample (see next section), it is found to be helpful to use both frequencies, as the short time between the end of the reflection from the top of the sample and the beginning of the pulse from the base of the sample sometimes make it difficult to identify the correct cycle in the latter. This is not a problem in making the shear wave measurements; on the other hand the high attenuation of shear waves in the rocks means that the signal at the higher frequency is too small to be measured with reasonable accuracy. Our standard practice is to prepare two samples of different lengths from each core horizon independently and then to measure and compute velocity and attenuation independently. This practice ensures, as far as possible, that significant errors due to incomplete water saturation (see below), measurement errors, and so on, are picked up.

It is worth discussing the problems of fluid saturating sedimentary rocks. As Murphy (1982) comments, this 'is not a trivial procedure' and it is well known that ultrasonic properties, particularly attenuation, are very sensitive to incomplete saturation. Our system is as follows. The cleaned sample is placed under vacuum until a pressure of 1 Pa (1E-5 bar) is achieved. It is then flooded with de-ionized, distilled, de-aired water and pressurized by hand pump, maintained over several days until the pressure stabilizes at 7 MPa. The fluid-saturated sample is then transferred to the ultrasonic pressure rig with the minimum exposure to atmosphere. A confining pressure of 65 MPa is applied and the pore fluid is pressurized to 30 MPa. As the solubility of gases is directly proportional to the pressure, this procedure ensures that any residual gas dissolves and should cause no problem. The following techniques were adopted to establish that full saturation was achieved.

First technique. The total gas remaining in the pressurizing system was determined by carrying out the full saturation procedure on a glass vial (0.05 cm internal diameter and 3 cm length, closed at one end). This was mounted with its open end attached to a large funnel such that all the remaining gas in the system collected in the funnel and was thus trapped in the sealed end of the glass vial. The total volume of gas collected was 3E-4 ml from a total chamber volume of 100 ml, indicating a residual gas pressure in the

system at the time of flooding of 3 Pa. A similar system was published by Worthington (1978).

Second technique. Ultrasonic measurements made on rocks at 60 MPa confining pressure with 30 MPa pore pressure were found to agree within the confidence interval with measurements on the same rocks at 30 MPa confining pressure with 0 MPa pore pressure. In other words the rock properties are measured to be identical at the same effective pressure. It would be an extraordinary coincidence if any residual gas behaved identically at the two different pore pressures (A. Best, pers. comm., 1991).

Third technique. Consistency of the ultrasonic data for separate samples from the same core horizon.

Fourth technique. Agreement within the experimental errors of the fluid and gas determined porosities of the separate samples from the same core horizon.

The disadvantages of the reflection technique compared to the transmission method are as follows. Firstly, the received signals (transient or pulsed sine wave) are of significantly smaller amplitude than the equivalent transmitted signals, particularly where the acoustic contrast between the perspex and the rock is small and the attenuation is high. It is almost always necessary to use a digital oscilloscope with facilities for stacking multiple shots to increase the signal-to-noise ratio to an adequate value. Secondly, if shear wave transducers are used in the transmission method there is usually sufficient compressional wave signal generated for the velocities of both waves to be measured simultaneously. This is not the case in the reflection method where very little mode conversion occurs and both compressional and shear waves are remarkably uncontaminated. This limitation (at least as far as rapid velocity measurements are concerned) has now been overcome in our equipment by the use of a double compressional/shear wave transducer.

Velocity and attenuation calibration

An accurately machined aluminium alloy cylinder of length 2.4976 cm was used to check the compressional-wave and shear-wave velocities and attenuations obtained in the high-pressure rig using the reflection system. Confidence intervals of 95% were estimated from the variability of the results obtained from the primary and subsequent reflections multiply re-

Table 1. *Velocities and quality factors of compressional and shear waves for aluminium calibration sample and for replicate sandstone samples. 95% confidence intervals*

Sample (Note)	Compressional		Shear	
	Velocity (m/s)	Quality factor	Velocity (m/s)	Quality factor
Aluminium (a)	6400 ± 1	150,000	3150 ± 1	150,00
Aluminium (b)	[1]6408 ± 4	>350	3159 ± 6	>700
	[2]6403 ± 4			
Sandstone 1 (c)			2578 ± 5	31 ± 6
Sandstone 1 (c)			2572 ± 8	32 ± 3
Sandstone 2 (d)	4613 ± 2			
Sandstone 2 (d)	4611 ± 2			
Sandstone 3 (e)			3040 ± 4	
Sandstone 3 (e)			3028 ± 2	
			3079 ± 5	
			3078 ± 2	

Notes
All confidence intervals are 95%, i.e. two standard deviations. The data refer to a frequency of 0.85 MHz unless state otherwise in the notes.
(a) True values of aluminium velocities measured at 5 MHz on the bench. Quality factors from Zamanek & Rudnick (1961).
(b) Aluminium velocities and quality factors measured at 60 MPa, 0.85 MHz [1] and 1 MHz [2] using the reflection technique. The quality factors correspond to a 95% confidence interval on the measurement of attenuation coefficient of ±0.1 dB/cm at 0.85 MHz.
(c) Sandstone 1. Shear wave velocity and quality factor measured on two separate occasions.
(d) Sandstone 2. Compressional velocity measured on two separate occasions.
(e) Sandstone 3. Sample showed significant layering. Shear velocity measured on two separate occasions. Second set show measurements at transducer orientations of 0, 45 and 90 degrees respectively to determine the seismic effect of the sample anisotropy.

flected through the sample. Measurements were made at frequencies of about 1 MHz and 0.85 MHz for the compressional waves and at about 0.85 MHz for the shear waves, these being the frequencies at which measurements were made on the rock samples. The absolute values of the velocities in the aluminium alloy sample were obtained on the bench by measuring the times of arrivals of multiple reflections of very high frequency (5 MHz) pulses for which the diffraction corrections were calculated to be negligible. Compressional and shear wave attenuations in aluminium alloy are negligible at these frequencies (Zemanek & Rudnick 1961).

The velocity and attenuation results are given in Table 1. The velocities quoted have been corrected for the diffraction effects of the transducer using the tables published by Benson & Kiyohara (1974). For compressional waves these correspond to a reduction of the measured velocities of 49 m/s at 0.85 MHz and 44 m/s at 1.0 MHz, a systematic change of about 0.8%. The corrected results at the two different frequencies are consistent with each other within

the 95% confidence intervals. The average value of 6405 ±4 m/s measured at 60 MPa agrees with the true velocity of 6400 ±1 m/s measured at 5 MHz on the bench. The diffraction correction for the shear waves corresponds to a reduction of the measured velocity of 10 m/s. The corrected value of 3159 ±6 m/s measured at 60 MPa is slightly, but significantly different from the true value of 3150 ±1 m/s measured on the bench. This systematic discrepancy in the shear wave velocity probably arises from inadequate diffraction corrections. The Benson & Kiyohara tables strictly apply to a compressional-wave piston transducer radiating into an infinite medium. Tang *et al.* (1990) have also analysed the diffraction effects of piston transducers radiating into inifinite media and show that both compressional and shear waves act as acoustic waves in the far field. The far-field of a 0.85 MHz transducer of diameter 2.54 cm starts at a distance of 10 cm from the transducer face. In our equipment the buffer rod system ensures that the first reflection has travelled 10 cm before it returns to the transducer face, and is thus just about in the far field. Hence, the compressional wave diffraction corrections are just about applicable to the shear waves, with appropriate velocities and wavelengths. Neither the Benson and Kiyohara nor the Tang, Toksöz and Cheng diffraction analyses consider the effects of the finite sample size. We conclude that although the precision of our velocity measurements is better than 0.1%, the limitations of the corrections applied to account for the diffraction effects of the transducers can cause the final value of the shear wave velocity to be systematically in error by 0.3% at 0.85 MHz. The systematic error is smaller for the compressional waves and will be smaller for materials with lower shear wave velocities. It will be larger at lower frequencies where the diffraction correction is larger. We therefore conservatively estimate the accuracy of our velocity results to be ±0.3% at 0.85 MHz.

The results of the aluminium attenuation measurements are summarized in Table 1. The true attenuation values for aluminium at frequencies between 0.85 MHz and 1 MHz are negligible for both compressional and shear waves. Zemanek & Rudnick (1961) give quality factors of 150 000. The actual attenuation values determined on the reflection system at 0.85 MHz were −0.04 ± 0.02 dB/cm (compressional waves) and −0.01 ± 0.1 dB/cm (shear waves), after application of the diffraction corrections. A typical diffraction correction for compressional waves at 0.85 MHz in the aluminium sample is 1.85 dB. Failure to apply the correction results in a final attenuation coefficient of 0.37 dB/cm for the aluminium (quality factor of 98), a significant systematic error. It is pointless to give negative quality factor figures in Table 1; rather we conservatively quote the minimum quality factor corresponding to the maximum overall measurement error, ±0.1 dB/cm, observed for the shear wave data. The actual attenuation results are 0 dB/cm (infinite quality factor), within this 95% confidence interval. We conclude that the estimates of the diffraction corrections of the pulse amplitudes are not a significant source of systematic error in the measurement of attenuation at 0.85 MHz. The attenuation values would, however, be significantly and systematically in error if the diffraction corrections were ignored. We estimate that the 95% confidence interval of our attenuation measurements, incorporating all measurement errors and corrections, to be ±0.1 dB/cm at 0.85 MHz (corresponding to a 95% confidence interval of ±4 in a quality factor of 50). The width of this confidence interval increases rapidly with increasing quality factor as it is inversely proportional to the attenuation coefficient. It will also decrease with

decreasing frequency as the diffraction corrections become larger and more difficult to estimate accurately.

Assessment of the reliability and repeatability of the velocity and attenuation measurements

In the previous section velocity and attenuation data for an aluminium sample were used to estimate the overall accuracy of results, including known sources of random and systematic error in the measurement technique. When analysing real rocks, other errors arise due to the smaller signal-to-noise ratio, wave scattering, variations in the applied pressure, effects of incomplete water saturation, anisotropy etc. These effects were investigated by analysing independent repeat measurements of compressional wave and shear wave velocities and attenuations on selected samples with identical processing applied to the results. The results are given in Table 1.

Sandstone 1 was a water-saturated sample for which the shear wave velocity and attenuation were measured at 60 MPa on separate dates by different operators. The average quoted 95% confidence interval in the velocities, determined from the variability of the results between the first and multiple reflections, is ±0.25%, and the independent estimates differ by 0.23%. The quoted 95% confidence interval in the quality factor estimates, determined in the same way, corresponds to an average error in the attenuation coefficients of ±0.37 dB/cm, although the actual attenuation estimates differ by only 0.03 dB/cm.

Sandstone 2 was a dry sample for which the compressional wave velocity was determined on two separate occasions by different operators. The quoted 95% confidence intervals are ±0.04%; the values differ by 0.04%.

Sandstone 3 was a dry sample for which the shear wave velocity was measured on two separate occasions. The sandstone showed significant layering and the purpose of the repeat measurement was to investigate the variation of shear velocity with azimuth of the polarization of the shear wave transducer. The quoted 95% confidence intervals average ±0.01%, and the second set of readings straddles the first reading.

We conclude that the agreement between replicated measurements of velocity and attenuation is within the overall confidence interval deduced from the aluminium data.

It is also necessary to assess the repeatability of the velocity and attenuation measurements on

Table 2. *Sedimentological and ultrasonic data for sandstones and a shale*

Sample number	Rock type (mm)	Thickness	Porosity (%)	Permeability (mD)	Clay content (%)	Compressional		Shear	
						Velocity (m/s)	Quality	Velocity (m/s)	Quality
201A	S'st	13.214	10.4	62	0	4955	82	3163	49
201B	S'st	21.266	10.8	112	0	4950	84	3168	46
202A	S'st	13.653	14.6	144	0	4502	74	2847	46
202B	S'st	23.034	15.1	238	0	4501	77	2834	52
101A	S'st	12.677	12.9	14	9	4498	28	2712	21
101B	S'st	18.464	14.3	4	9	4081	26	2403	10
102A	S'st	14.062	17.4	51	10	4174	42	2578	31
102B	S'st	22.153	17.3	39	10	4072	47	2491	30
108A	S'st	13.214	9.1	0.5	17	4707	51	2834	41
108B	S'st	19.791	11.4	0.5	17	4689	56	2849	28
105A	Shale	13.67	10.8	0	33	3397	24	1906	18

samples of rock taken from the same horizon in a core. This tests, not the precision of the technical method, but the validity of the ultrasonic data in terms of the sedimentological homogeneity of the rocks and the success of our efforts to prepare and fluid saturate the rocks. The data in Table 2 demonstrates the repeatability and stability of the measurements of velocities and quality factors on short- and long-water-saturated sandstone samples taken from the same 5 cm core plug, at 60 MPa in the pressure cell.

All the samples are from North Sea Wells, but they are not identified further for reasons of confidentiality. Table 2 gives data on sample thickness, porosity, gas permeability and clay content as well as the ultrasonic data. In each case the sample marked 'A' is the shorter of the two. The data are tabulated in order of increasing clay content.

Samples 201A and 201B are clean sandstones with similar porosities and permeabilities. The ultrasonic data show excellent agreement, well within the 95% confidence limits previously deduced.

Samples 202A and 202B are clean sandstones with similar porosities and permeabilities. Again the ultrasonic data show excellent agreement.

Samples 101A and 101B are sandstones containing approximately 9% clay. Their porosities are significantly different, which partly accounts for the differences in the compressional and shear wave velocities. Examination of the samples shows them to be vertically heterogeneous. The compressional wave quality factors are similar. The shear wave attenuations in both samples were so high that only the first reflection could be measured and the signal-to-noise ratio was small. It was not possible to estimate 95% confidence intervals for these results.

Samples 102A and 102B are visually homogeneous sandstones containing approximately 10% clay. The porosities and permeabilities are in good agreement.

The compressional and shear wave velocities are significantly different, but the quality factors agree well within the 95% confidence interval. More detailed investigation of the mineralogy of these samples is needed to determine the source of the velocity discrepancy. Measurement of a further sample would be helpful.

Samples 108A and 108B are sandstones containing 17% clay. The porosities are significantly different. The compressional and shear velocities and the compressional wave quality factors are in reasonable agreement, given the difference in porosity between the samples. The measured confidence interval for the shear wave quality

factors are ± 15 for 108A and ± 9 for 108B. Hence, the value of 28 given for 108B is the more reliable of the two.

Sample 105A is a single shale with 33% clay. This is included for comparison with the sandstones as it shows the lowest velocities and quality factors. There are rather few measurements of ultrasonic velocity and attenuation for shales in the literature, because they are difficult to prepare for the high pressure cell.

The gross variations of velocities and quality factors observed from these few samples confirm the results of Han et al. (1986), and of Klimentos & McCann (1990), that low porosity, low clay-content sedimentary rocks show high compressional- and shear-wave velocities, and high compressional-wave quality factors, and vice versa. The shear wave quality factors show the same general trend.

Although the data presented here are a small sub-set of a much larger set which is still being processed, they confirm that the reflection method gives accurate and repeatable values of compressional and shear wave velocities and quality factors for fluid saturated sedimentary rocks. They also confirm that replicated samples from the same core horizon on the whole give duplicate ultrasonic data, unless the samples are obviously heterogeneous. Our practice of performing replicate measurements on samples of slightly different lengths gives confidence in the final results. For high velocity, high quality-factor samples it is best to use long samples in order to maximize the overall travel time and attenuation. On the other hand low-velocity, low-quality factor samples must be short in order that the high attenuation coefficient does not degrade the signal-to-noise ratio excessively.

Conclusions

The pulse reflection method designed by Winkler & Plona (1982) at Schlumberger for measuring compressional and shear wave velocities and attenuations at pressures up to 70 MPa on dry and water saturated rock samples in the laboratory has been implemented. In addition to using a transient pulse for estimating attenuation as a function of frequency and velocity dispersion, we use a pulsed-sine wave for very accurate measurement of velocity and attenuation at specified frequencies. Compressional and shear wave velocities and attenuations were measured at frequencies between 0.85 MHz and 1 MHz at pressures up to 60 MPa on an aluminium alloy sample and on replicate rock samples. The results demonstrate that it is important to apply diffraction corrections to both the velocity and to the attenuation measurements. The results for the aluminium alloy sample also show that the diffraction corrections do not bring the shear velocity into exact agreement with the 'true' value. However, even with this small systematic error, the 95% confidence interval of the velocity results is $\pm 0.3\%$ and that of the attenuation results is ± 0.1 dB/cm, at 0.85 MHz and 60 MPa. The replicate measurements on rock samples agreed within these confidence intervals.

Important consequences of the use of this technique are that both the velocity and the quality factor of a sample may be reported with reference to a particular frequency and with reliable 95% confidence intervals. Differences in the ultrasonic results between samples may be ascribed to sedimentological differences, rather than possible experimental error. The velocity

and quality factor data may be reliably compared with the in situ values, with due regard for the differences in measurement frequency and scale.

We acknowledge with grateful thanks the provision of borehole samples and financial support by Amerada Hess Ltd. Many thanks also to D. M. McCann and D. Entwisle of the British Geological Survey for the loan of our first high-pressure cell and a spare Arenberg Type PG650C pulse generator. Some of the measurements reported here were carried out by A. Best and J. S. Oliver, research students of the Postgraduate Research Institute for Sedimentology at Reading University. We are grateful for very useful comments from two anonymous referees.

This is contribution No. 159 of the Postgraduate Research Institute for Sedimentology, University of Reading.

References

ATTEWELL. P. B. & RAMANA, R. V. 1966. Wave attenuation and internal friction as functions of frequency in rocks. *Geophysics*, **31**, 1049–1056.

BENSON, G. C. & KIYOHARA, O. 1974. Tabulation of some integral functions describing diffraction effects on the ultrasonic field of a circular piston source. *Journal of the Acoustical Society of America*, **55**, 184–185.

CASTAGNA, J. P., BATZLE, M. L. & EASTWOOD, R. L. 1985. Relationship between compressional-wave and shear-wave velocities in clastic silicate rocks. *Geophysics*, **50**, 571–581.

DOMENICO, S. N. 1984. Rock lithology and porosity determination from shear and compressional wave velocity. *Geophysics*, **49**, 1188–1195.

GOLDBERG, D. & ZINSZNER, B. 1989. P-wave attenuation from laboratory resonance and sonic waveform data. *Geophysics*, **54**, 76–81.

HAN, D., NUR, A. & MORGAN, F. D. 1986. Effects of porosity and clay content on wave velocities in sandstones. *Geophysics*, **51**, 2093–2107.

HILTERMAN, F. 1990. Is AVO the seismic signature of lithology? A case history of Ship Shoal-South Addition. *Geophysics, The Leading Edge*, **9**, 15–22.

KING, M. S. 1965. Wave velocities in rocks as a function of changes in overburden pressure and pore fluid saturants. *Geophysics*, **31**, 50–73.

KLIMENTOS, T. & MCCANN, C. 1990. Relationships between compressional wave attenuation, porosity, clay content and permeability of sandstones. *Geophysics*, **55**, 998–1014.

—— & MCCANN, C. 1988. Why is the Biot slow compressional wave not observed in real rocks? *Geophysics*, **53**, 1605–1609.

MORIG, R. & BURKHARDT, H. 1989. Experimental evidence for the Biot–Gardner theory. *Geophysics*, **54**, 524–527.

MURPHY, W. F. 1982. Effects of partial water saturation on attenuation in Massilon Sandstone and

Vycor porous glass. *Journal of the Acoustical Society of America*, **71**, 1458–1468.

NUR, A. 1989. Four dimensional seismology and (true) direct detection of hydrocarbons: the petrophysical basis. *Geophysics, The Leading Edge*, **8**, 30–36.

O'HARA, S. G. 1989. Elastic wave attenuation in fluid saturated Berea Sandstone. *Geophysics*, **54**, 785–788.

OLIVER, J. S., MCCANN, C. & CULLEN, E. 1991. Attenuation measurements from VSPs: considerations for improved results. *Marine and Petroleum Geology*, submitted.

OSTRANDER, W. J. 1984. Plane wave reflection coefficients for gas sands at non-normal angles of incidence. *Geophysics*, **49**, 1637–1648.

PAPADAKIS, E. P. 1975. Ultrasonic diffraction from single apertures with application to pulse measurements and crystal physics. *In*: MASON, W. P. & THURSTON, R. N. (eds) *Physical Acoustics*, Volume 11. Academic, New York.

—— 1976. Ultrasonic diffraction loss and phase change in anisotropic materials. *Journal of the Acoustical Society of America*, **50**, 863–876.

——, FOWLER, K. A. & LYNWORTH, L. C. 1973. Ultrasonic attenuation by spectrum analysis of pulses in buffer rods: method and diffraction corrections. *Journal of the Acoustical Society of America*, **53**, 1336–1343.

PATTON, S. W. 1988. Robust and least-squares estimation of acoustic attenuation from well-log data. *Geophysics*, **53**, 1225–1232.

PFFENHOLZ, J. & BURKHARDT, H. 1989. Attenuation and modulus measurements in the seismic frequency range and strain range on partially saturated sedimentary rocks. *Journal of Geophysical Research*, **94**, B7, 9493–9507.

PICKETT, G. R. 1963. Acoustic character logs and their applications in formation evaluation. *Journal of Petroleum Technology*, June, 659–667.

SPENCER, J. W. 1981. Stress relaxations at low frequencies in fluid saturated rocks: Attenuation and velocity dispersion. *Journal of Geophysical Research*, **86**, 1803–1812.

TANG, X. M., TOKSÖZ, M. N. & CHENG, C. H. 1990. Elastic wave radiation and diffraction of a piston source. *Journal of the Acoustical Society of America*, **87**, 1894–1902.

TITTMAN, B. R., BALAU, J. R. & ABDEL-GAWAD, M. 1984. The role of viscous fluids in the attenuation and velocity of elastic waves in porous rocks. *In*: JOHNSON, D. L. & SEN, P. N. (eds) *Physics and Chemistry of Porous Media*. American Institute of Physics, New York.

TONN, R. 1991. The determination of the seismic quality factor Q from VSP data: a comparison of different computational methods. *Geophysical Prospecting*, **39**, 1–28.

WANG, Z. & NUR, A. 1990. Wave velocities in hydrocarbon-saturated rocks: Experimental results. *Geophysics*, **55**, 723–733.

WILKENS, R., SIMMONS, G. & CARUSO, L. 1984. The ratio V_p/V_s as a discriminant of composition for siliceous limestones. *Geophysics*, **49**, 1850–1860.

WINKLER, K. 1985. Dispersion analysis of velocity and attenuation in Berea sandstone. *Journal of Geophysical Research*, **90**, 6793–6800.

—— & PLONA, T. J. 1982. Techiques for measuring ultrasonic velocity and attenuation spectra in rocks under pressure. *Journal of Geophysical Research*, **87**, 10776–10780.

WORTHINGTON, A. E. 1978. A technique for detecting incomplete saturation of cores. *Journal of Petroleum Technology*, **30**, 1716–1717.

WYLLIE, M. R., GREGORY, A. R. & GARDNER, G. H. F. 1958. An experimental investigation of factors affecting elastic wave velocities in porous media. *Geophysics*, **23**, 459–493.

ZEMANEK, J. & RUDNICK, I. 1961. Attenuation and dispersion of elastic waves in a cylindrical bar. *Journal of the Acoustical Society of America*, **33**, 1238–1288.

Thermal conductivity prediction from petrophysical data: a case study

C. M. GRIFFITHS,[1] N. R. BRERETON,[2] R. BEAUSILLON[3] & D. CASTILLO[4]

[1] *BP-Statoil Alliance, Ranheimsveien 10, N-7004 Trondheim, Norway*
[2] *British Geological Survey, Keyworth, Nottingham, UK*
[3] *CRG Garchy, 58 150 Puilly Sur Loire, France*
[4] *Department of Geophysics, Stanford University, Stanford, California 94305, USA*

Abstract. Thermal conductivity measurements were carried out on over 400 samples from Tertiary and Cretaceous sediments and Jurassic basalts during Ocean Drilling Program Leg 123 in the Argo Abyssal Plain, North West Australia. Ship-board physical property measurements were obtained from cores over the same intervals, and petrophysical wireline logs were run. New algorithms are demonstrated relating thermal conductivity to matrix grain type and porosity, derived only from wireline log data. Comparison with heat flows calculated using high-resolution temperature measurements at Site 765 suggest that thermal conductivities predicted from wireline logs can be used to estimate heat flow. This implies that existing exploration and production wells in which high resolution temperature measurements and thermal conductivities have not been measured could still be used to prepare regional maps of heat flow variation.

The ability to map regional heat flow variation throughout a sedimentary section is of interest in basin modelling, especially if no new measurements are required. In order to achieve this, thermal conductivities must be derived from other physical properties in a way that enables a three-dimensional thermal conductivity model to be constructed. The measurement of thermal conductivity from core material is, however, time consuming, costly, and requires not only the existence of a core, but the presence of measuring facilities in close proximity to the well site. The use of a down-hole needle-probe for thermal conductivity measurement is also time consuming. It is, therefore, an attractive proposition to be able to infer or predict thermal conductivities from other routine measurements such as those provided by wireline logs that do not need core material and, given stable hole conditions, provide a continuous record throughout a well.

Previous work

Using hand specimens, Rzhevsky & Novik (1971), Balling *et al.* (1981), Lovell (1983, 1985*a,b*), Lovell & Ogden (1984) and Brereton (1990) have demonstrated various relationships between thermal conductivity, porosity and matrix mineralogy and/or density. These published equations are discussed in a later section. Although Ratcliffe (1960), Bullard (1954) and others have commented on the weak control that

matrix mineralogy exerts on thermal conductivity and diffusivity, they were only considering the extremely high porosity sediments on the seabed. At moderate depths (a few hundred metres to a kilometre) below seafloor the porosity is reduced to 20 to 30%, and the combination of mineralogy and grain contacts becomes increasingly significant. Lovell (1983, 1985*a, b*) and Lovell & Ogden (1984) examined the relationship between thermal and electrical conductivity/electrical formation factor (which are also partly a function of grain contacts).

The use of wireline logs to predict thermal conductivity is a possibility that has interested several previous workers. Goss *et al.* (1975) used linear regression to derive an empirical relationship between bulk thermal conductivity, porosity and acoustic velocity on a set of core samples. They then applied the same equation to log values of porosity and velocity, and predicted a range of thermal conductivities which they compared to values measured on compacted cuttings samples. Given the fact that the nature of grain contacts appears to be important in thermal conductivity (see below) it is not surprising that the two sets of values did not agree particularly well, and Goss *et al.* considered that 'There seems little hope of more general predictive relationships being successful...' However, they did not resolve the problem of which set of measurements was the more accurate. Vacquier *et al.* (1988) approached the problem by dividing the 'lithological universe'

From HURST, A., GRIFFITHS, C. M. & WORTHINGTON, P. F. (eds), 1992,
Geological Applications of Wireline Logs II. Geological Society Special Publication No. 65, pp. 299–315.

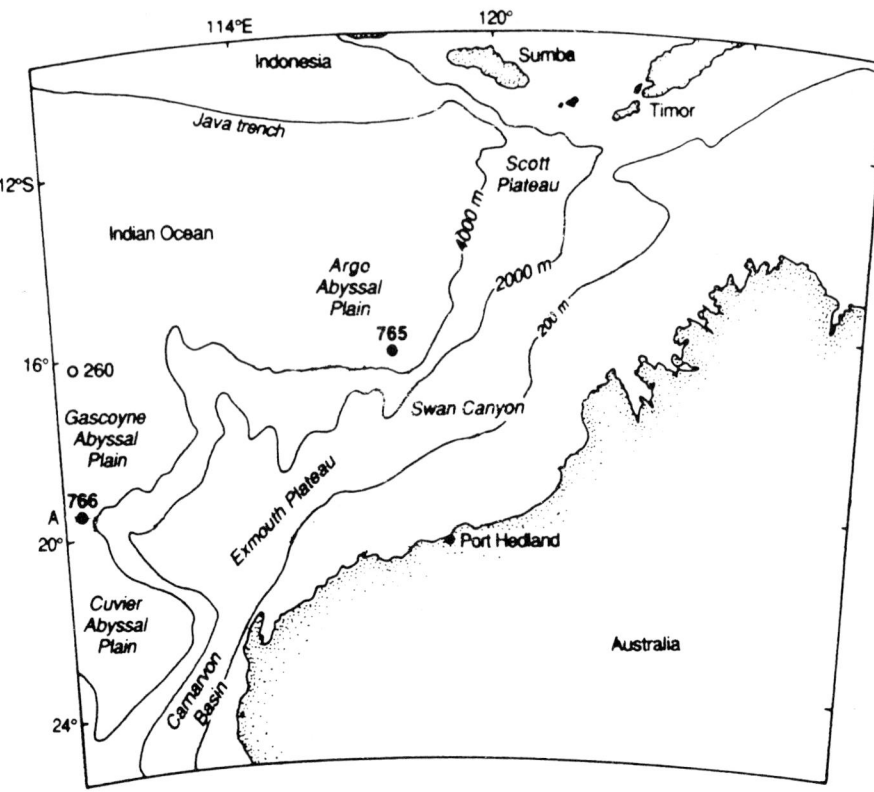

Fig. 1. Location of Sites 765 and 766, Ocean Drilling Program Leg 123. Indian Ocean.

into seven subsets and, based on core measurements, determined a mean bulk thermal conductivity for each lithology. No further subdivision appears to have been made on the basis of porosity. A five-term linear regression was then developed for each lithology, using density, acoustic transit time, porosity and shale volume. Practical application then relied on the identification of the relevant lithotype in the well before selection of the appropriate equation. Correlation coefficients between measured and predicted values ranged from 0.55 (clean sandstones) to 0.89 (argillaceous rocks). The fact that this study did not take into consideration porosity variation within a lithotype is a possible cause of the observed variation in predictability. More recent work by Brigaud *et al.* (1990) has used the long-established geometric mean approach attributed to Woodside & Messmer (1961)

$$K = \prod_{i=1}^{m} k_i^{V_i}$$

where k_i is the thermal conductivity of the ith component, and V_i its fractional volume. The use of this equation will be discussed below. The important point to note here is that there are many ways of calculating components. One can include every mineral or fluid thought to be present, or one can aggregate components (for example 'clastic grains', 'pore fluid', 'clay'). Attempts to use individual components usually fail because of the difficulty of identifying many components from few logs, i.e. we have an underdetermined system. Using a few components involves the problem of explicit aggregation. Brigaud *et al.* (1990) approached this problem by using two stages in the analysis. The percentages of sand, carbonate and shale at each sample location were first determined from cuttings analysis. The thermal conductivity for each endmember was determined by combining the thermal conductivities of the constituent minerals using the geometric mean equation. The wireline logs were then used to identify in situ porosity at the sample location, and the bulk thermal conductivity estimated by means of the geometric mean equation. This approach has the effect of reducing the uncertainty in lithology estimation but suffers from some major problems. The first is the tying of cuttings sample depths to log depths, not an easy task even with perfect hole conditions. The second is the need to carry out lithological analyses on many hundreds of cuttings, and the third is the loss of mineralogical components during cuttings transport.

The study discussed here uses wireline logs exclusively, deriving necessary matrix parameters from the log response.

Details of the study area

Leg 123 of the Ocean Drilling Program (ODP) drilled at two sites in the northeast Indian Ocean, off the northwestern margin of Australia. Site 765 is located in the Argo Abyssal Plain, 60 km north of the Swan Canyon on the northern edge of the Exmouth Plateau (Fig. 1). The Argo Abyssal Plain is bounded to the south by the Exmouth Plateau and is underlain by Cretaceous (to possibly Jurassic) oceanic crust.

Site 766 is located on the western limit of the Exmouth Plateau, at the foot of the continental slope at the boundary between the Cuvier and Gascoyne Abyssal Plains. The sites are described in Gradstein & Ludden *et al.* (1990).

The water depth at Site 765 is around 5.7 km. At this site 931 m of largely Miocene turbiditic carbonates and clays overly tholeiitic basalts capped by a sill of iron–titanium basalt. The basalt has been dated at 155 Ma ± 3 Ma (Gradstein, pers. comm.). At site 766, with a water depth of 4 km, 466 m of late Valanginian to recent sediments are underlain by a series of diabase sills to a depth of 527 mbsf. The true sediment base may not have been reached at this site.

The lithology of Site 765 consists mainly of nannofossil ooze with varying clay and fossil fragment content. Clay is the dominant lithology for the lowest 300 m of the section, changing to nannofossil ooze above (Fig. 2). Wireline logging was possible over much of the Miocene section from 147 metres below sea floor (mbsf) to 460 mbsf, but bad hole conditions prevented open-hole logs being run from 460 mbsf to 940 mbsf. The basalt section from 940 to 1145 mbsf was successfully logged in open hole.

Site 766 also has a Tertiary section that is dominated by nannofossil ooze, but during the Barremian to Albian (136 mbsf to 304 mbsf) the section is composed mostly of clays. From 304 mbsf to 466 mbsf the section consists mostly of Hauterivian siliciclastics and altered volcaniclastics.

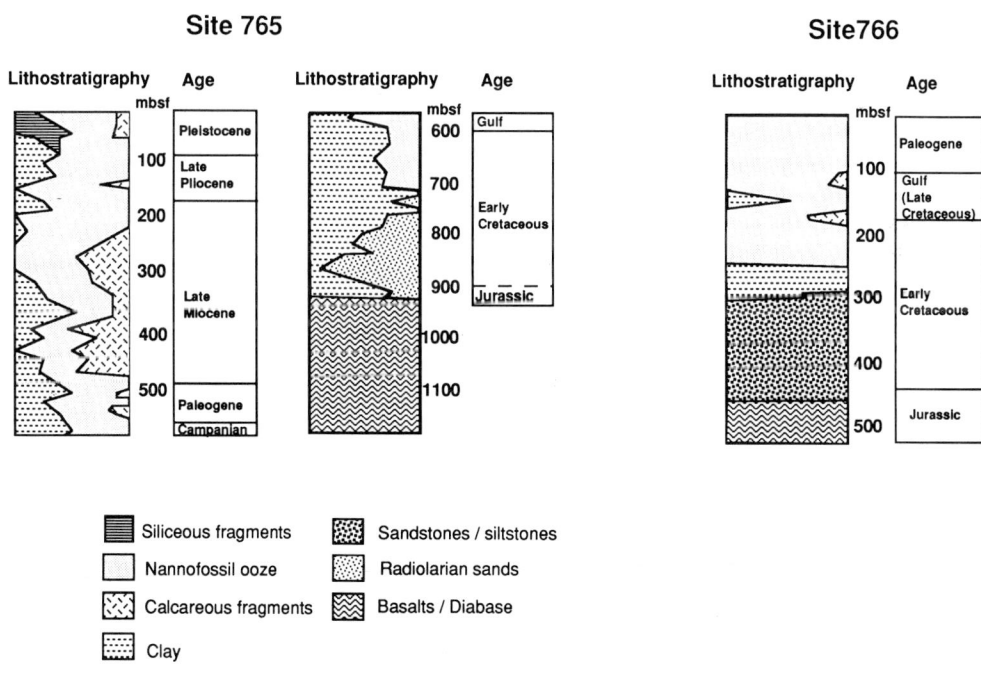

Fig. 2. Lithostratigraphy at Sites 765 and 766.

Table 1. *Lithology and physical property subdivisions for Sites 765 and 766*

Lithological Unit	Lithology	Grain density (gm/cm^3)	Porosity
Site 765			
Unit A (0–80 mbsf)	Unconsolidated calcareous ooze	2.62 ± 0.13	0.74
Unit B (80–350 mbsf) (265–290 mbsf)	Debris flows and turbidites Clay deposits interlayered with thin carbonate cemented sands	2.71 ± 0.11	0.55 ± 0.02
Unit C (350–590 mbsf)	More lithified debris flows, carbonates, and calcareous turbidites	2.70 ± 0.13	0.43 ± 0.01
Unit D (590–896 mbsf)	Dark red claystone with increasing amounts of water-sensitive minerals	2.66 ± 0.14	0.42 ± 0.01
Unit E (936–1176 mbsf)	Pillow basalts and massive basalt flows	2.87 ± 0.05	0.04
Site 766			
Unit A (0–100 mbsf)	Calcareous ooze	2.66 ± 0.10	
Unit B (100–185 mbsf)	Claystones and chalk	2.78 ± 0.14	
Unit C (185–240 mbsf)	Chalk with hard chert layers	2.47 ± 0.19	
Unit D (240–300 mbsf)	Claystone, with water sensitive minerals	2.5 at the top to 2.7 at the base	
Unit E (300–459 mbsf)	Glauconitic siltstones and sandstones with periodic layers of highly lithified bioclastic sandstone. Grey-black clay-rich sandstone at base.	2.71 ± 0.08	
Unit F (459–463 mbsf)	Basalt sill	2.68 at margins 2.84 at centre	
Unit G (463–467 mbsf)	Dark shale similar to base of Unit E		
Unit H (467–516 mbsf)	Diabase	2.91 ± 0.05	

Lithostratigraphy

The section at Site 765 has been subdivided into four sedimentary units and one igneous rock unit (see Fig. 2 and Table 1), whereas the Site 766 section has been subdivided into six sediment and two igneous rock units on the basis of physical properties.

Wireline logging measurements

The holes were logged by Schlumberger using tool combinations developed for the Ocean Drilling Program. The logs are basically the same as for oil-field investigations and give comparable results after environmental corrections have been carried out. Logs of particular use in

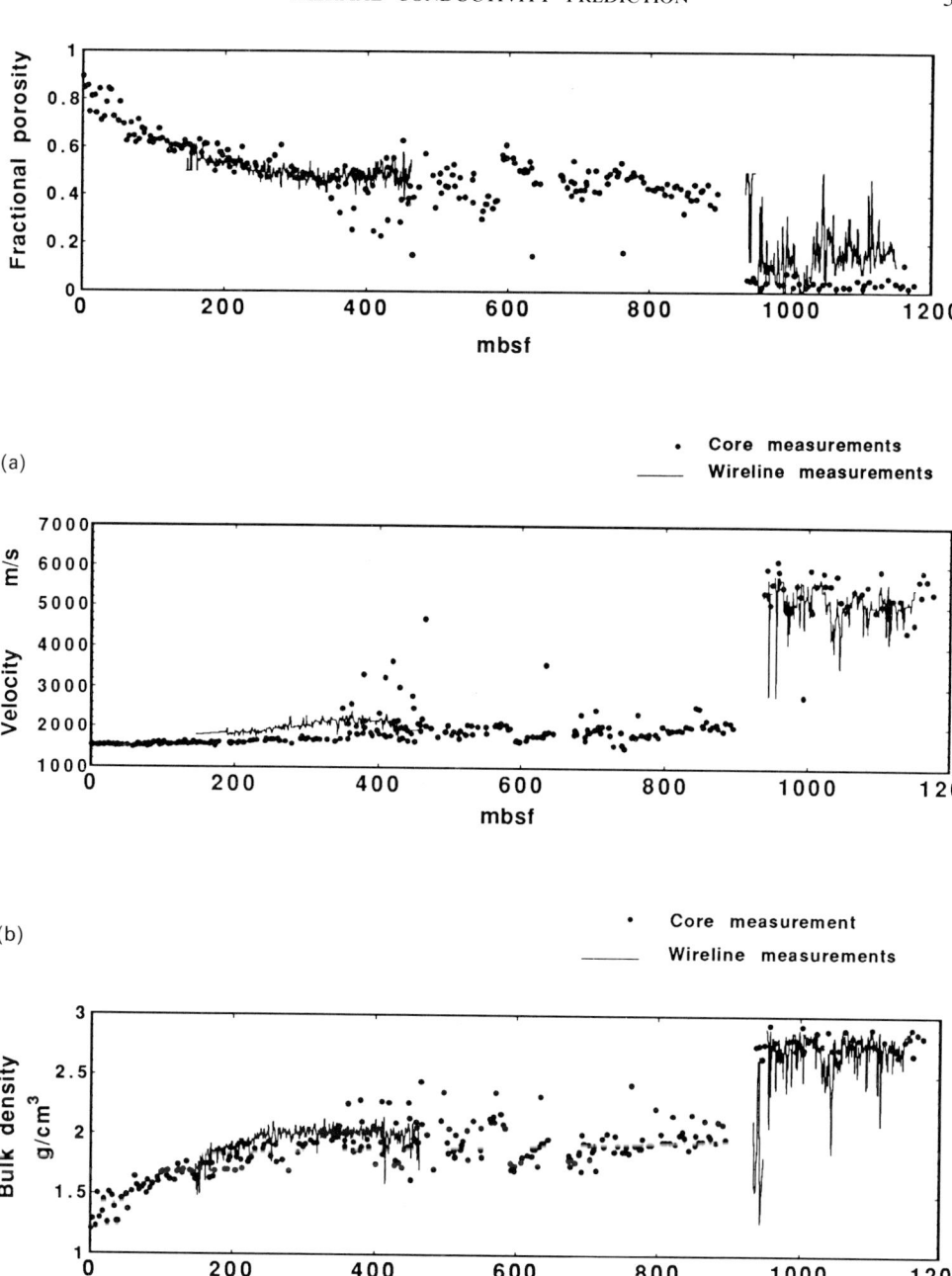

(a)

(b)

(c)

Fig. 3. (a) Comparison of porosity measured both on core samples and by wireline logs at Site 765. (b) Comparison of compressional wave (acoustic) velocity measured both on core samples and by wireline logs at Site 765. (c) Comparison of bulk density measured both on core samples and by wireline logs at Site 765.

the present study are the neutron porosity log, litho-density log (giving rock bulk density), natural gamma log, and sonic log (giving acoustic transit time or reciprocal velocity). For further details on the principles of operation of these tools and their quality control the reader is referred to the appropriate Schlumberger manuals (Schlumberger Log interpretation principles/applications, 1989; Log interpretation charts, 1989; Bateman 1985; Theys 1990). The wireline log data were acquired at the Schlumberger standard 6-inch (0.1524 m) spacing while the transmitter–receiver distance for most tools is 0.3 to 0.6 m. This leads to a smoothing of petrophysical values that many authors have attempted to correct by using so-called 'blocking' algorithms (Griffiths 1982; Schlumberger 1989, p. 108; Rider 1985). The logs used in this study were blocked using the algorithm presented in Griffiths (1982).

The sonic, density and neutron logs do not function particularly well through steel tubing, so the use of wireline log data concerns only those parts of the borehole that were logged in open hole, i.e. the Miocene oozes and Jurassic basalts from Site 765, and the Barremian and Hauterivian clays and siliciclastics at Site 766. The logs were calibrated against shipboard physical properties measurements, and found to be in reasonable agreement given the difference in the nature of sampling (Figs 3a, b, c).

A number of different 'porosities' were then calculated. Porosity is a rather complex concept that perhaps needs some explanation. The neutron 'porosity' logging tool does not in fact measure porosity. It measures the 'hydrogen index' of a body of rock close to the borehole wall. The tool used here (a thermal neutron tool) is sensitive to all neutron absorbers such as hydrogen, boron and chlorine. Most of the response is due to hydrogen and the hydrogen index is proportional to the quantity of hydrogen per unit volume, with the hydrogen index of fresh water at surface conditions taken as unity (Schlumberger 1989, pp 5–20). The 'porosity' recorded by the neutron porosity tool will therefore also record high porosities in shales and hydrated minerals where bound water and water of crystallization are present. It is common practice to control the neutron porosity by also calculating a porosity from the bulk density log (Φ_D) assuming a certain grain density (ρ_g) and fluid density (ρ_f). This density-controlled porosity is referred to here as the total porosity (Φ_T). The equation used here is the simple expression:

$$\Phi_T = (\Phi_N - \Phi_D)/3 + \Phi_D.$$

For further details see Schlumberger *Log Interpretation Principles/Applications* (1989, chapter 6), or Elphick (1987)). There is also an 'effective porosity' (Φ_E) which is that fraction of the rock occupied by moveable fluids. This is less than the total porosity by the amount of shale influencing the total porosity. One estimate of this is provided by:

$$\Phi_E = \Phi_T(1 - V_{sh}) \qquad (0)$$

where V_{sh} is the volume of clay/shale in the rock (which is thus considered to be composed of matrix + clay + effective porosity).

Shipboard physical properties measurements

The following physical properties were determined on board ship: porosity, bulk density, grain density and water content (collectively referred to as index properties); compressional wave velocity; thermal conductivity. The measurement of thermal conductivity is discussed in the following section.

Drilling technology employed by the ODP enables continuous coring from the softest sediment at the seawater–sediment boundary, to the hardest basalts. The cores were recovered in 10 m plastic-sleeved core barrels thus retaining the integrity of the softer materials. Soon after core recovery, the plastic liners were cut into 1.5 m sections and each section split lengthwise. One half was archived while the other was used for sampling. Samples for index property and compressional wave velocity measurements were taken by either cutting parallel sided pieces with a knife (in the softer sediments), or using a double-bladed diamond saw for the more lithified or brittle material. Igneous rock samples were obtained using a 2.5 cm rock corer. In almost all cases these two sets of measurements (index properties and compressional wave velocity) were made on the same sample. For practical reasons the thermal conductivity measurements were made at neighbouring locations. At no time were the samples allowed to dry out prior to measurements being taken, and sample temperatures were allowed to equilibrate to the stabilized laboratory temperature of about 25°C.

For the index property determinations samples were weighted wet using two Scientech 202 electronic balances, interfaced with a microcomputer which compensates for the ship's motion by taking the average of 100 sample weighings over about 1 minute. The wet sample volumes were determined using a Quantachrome helium Penta-Pycnometer (QhPP). Dry sample weights and volumes were determined by the same procedure after freeze-drying the sample for 12 hours. The accuracy of the weight and volume determinations was periodically checked using calibration standards.

Compressional wave velocities were calculated from the determination of the travel time of a 500 kHz compressional wave through a measured thickness of sample using a Hamilton Frame velocimeter and Tektronix DC5010 counter/timer system. Travel distance was measured using an attached variable resistor connected to a Tektronix DM 5010 digital multimeter. The Hamilton frame was calibrated with lucite, aluminium, brass, and water standards at the beginning of the cruise. The variable resistor was calibrated with known lengths of aluminium cylinders.

Calibration of index properties

Brereton (1990, pp. 7–8) has discussed some problems that arose in the estimation of porosity and bulk density from the measurements described above, and a brief outline of the conclusions will be given here in so far as they influence the estimation of thermal conductivity from index properties. It transpires that ODP shipboard measurements of *wet* volume using the QhPP is (and has been for some time) in error. The values of porosity and bulk density quoted in the ODP Volumes A should be multiplied by 0.94 to correct for the error in the pycnometer, i.e. the recorded values are about 6.2% too high. The values reported and plotted in this paper are corrected values after Brereton (1990).

Physical property measurements were taken at approximately 4.5 m intervals throughout the borehole when core recovery permitted. Sampling a representative lithology over those intervals was difficult, and occasionally less representative material (such as basalt pebbles) were sampled leading to some high-frequency noise in the measurements, and poorer relationships between index properties, wireline measurements, and thermal conductivity. In areas of rapidly changing lithology, such as the turbidites in Unit B, and in fractured and altered basalts, the choice of which 2.5 cm to sample is nontrivial. Consistent sampling of a minor lithology (such as only altered basalt, or only the pelagic clay component of a turbidite) could generate misleading interpretations. For instance, the scatter of physical properties values from 380 to 480 mbsf (Fig. 3) is a function of sampling basalt pebbles rather than the turbidite matrix.

Shipboard thermal properties measurements

Thermal conductivity was measured directly on the cores using techniques described in Von Herzen & Maxwell (1959) and Vacquier (1985). The cores were allowed to equilibrate to room temperature for over three hours. Needle probes were connected to a Thermcon-85 unit and inserted into the sediment through small holes drilled in the core liner. An additional probe was inserted in a reference material. Once the temperature had stabilized, the probes were heated and the coefficient of thermal conductivity was calculated as a function of the change in electrical resistance of the probe every 12 seconds over a six-minute interval.

Stiff material was pre-drilled before probe insertion, while lithified rock was split and the measurements carried out on a split surface. The needle probe was partially embedded in a slab of insulating material and covered with a layer of Dow Corning 111 heavy silicone lubricant to provide good thermal contact between the probe and the sample. The slab–probe–sample combination was then placed in a saltwater bath and allowed to reach thermal equilibrium with the water. The probe was heated, and resistance changes in the probe monitored every nine seconds over a six-minute interval.

Core-log matching

There are some problems involved in the matching of discrete core samples with corresponding wireline-log values. They include the following.

1. The precise location of missing core intervals within a cored section has until very recently been unknown. Average core recovery on Leg 123 was 65% in general, but only 31% through the basalts at Site 765. In a cored length of 10 m this leads to a cumulative uncertainty of up to +7 m in the location of any one sample (missing core being attributed by convention to the base of the cored section). Compared to a wireline log sample interval of six inches (0.1524 m) this uncertainty is obviously problematic. When the wireline tool transmitter–receiver distance of around 0.3–0.6 m is also considered, it is clear that there is an uncertainty of between 5 and 45 log values when deciding which log value should be compared with a given core measurement. In the absence of unambiguous marker beds this often leads to a circular argument, where the most similar values are selected within the permissible error range.

2. Drillers depths and log depths can vary by up to a meter due to cable stretch. This is often solved by reference to casing shoe depths, where available, followed by interpolation or use of marker beds.

3. The core samples, plugs and thermal conductivity measurements represent point values, 2.5 cm in the case of index and velocity measurements and a few millimetres around the needle probe in the case of thermal conductivity. The wireline log values on the other hand represent averaged values over 0.3–0.6 m. One must then assume that the core measurement was representative of the whole log interval, an obvious oversimplification in complex lithologies. Sample intervals are also significantly greater in the case of core measurements, between 2 and 5 m for index porosities and thermal conductivity, and up to 25 m for downhole temperature measurements.

4. The probable difference in mechanical strength between the recovered and non-recovered parts of the section may be reflected in other physical properties such as thermal conductivity. This probably means that the true range of thermal conductivities has not been sampled in the recovered portion.

The fairly conventional log processing procedure used in this study was as follows:

(i)   Correct all logs for borehole environment effects.
(ii) Block the gamma, density, sonic and neutron porosity logs.
(iii) Depth match the sonic and Lithodensity/Compensated Neutron *(LDT/CNL) logs using marker beds identified on gamma traces. The edges of the blocked beds are easier to match than slopes or peaks.
(iv) Adjust the combined depths to drillers casing shoe depths, or drillpipe in the case of ODP holes.

We are primarily interested in predicting first-order thermal properties, and in order to do so we have chosen the largest sample interval as the controlling factor. The downhole temperature measurement points used by Castillo (*in* Gradstein & Ludden *et al.* 1990) were used as new sample depths. This meant that the thermal gradient at 25 m intervals could be compared with the mean index properties, velocity, thermal conductivity, and wireline log values at the same location in the well. This smoothing improved the correlation between log and core measurements.

Prediction of thermal conductivity from other petrophysical properties

Ratcliffe (1960) and Bullard & Day (1961)

---

* Mark of Schlumberger

showed that water-filled porosity was the single most important factor influencing thermal conductivity of seafloor sediments. The porosity effects on thermal conductivity are most pronounced in seafloor sediments (50–70% porosity) where the contrast in thermal conductivity between water (0.55 W/m°C to 0.7 W/m°C) and most matrix minerals (3–10 W/m°C) is the greatest. Bullard & Day (1961) suggested the empirical relationship for thermal conductivity:

$$\frac{1}{k_s} = (0.161 \pm 0.014) + (0.651 \pm 0.030)\Phi \quad (1)$$

for a water-filled porosity, where K_s is the bulk sediment thermal conductivity. In this case Φ will be the 'effective' porosity or the volume fraction of water that can be removed by drying the sample.

As with many other aspects of porous media such as bulk density, it is possible that thermal conductivity can be considered an additive function, that is to say it has the general form:

$$f_s = f_p\Phi + f_g(1 - \Phi)$$

where f is an appropriate function. This is justified in the same way as the Wyllie time average equation (Wyllie *et al.* 1956) which is known to be more valid for densities than acoustic velocities at high porosity values. In the case of thermal conductivity the equation would take the form;

$$k_s\rho_s = \Phi(k_p\rho_p) + (1 - \Phi)k_g\rho_g \quad (2)$$

where Φ is the porosity, k_s is the thermal conductivity of the whole sample, k_p is the pore fluid thermal conductivity, and k_g is the grain, or rock matrix, thermal conductivity. The terms ρ_s, ρ_p and ρ_g are respectively the sample, pore fluid, and matrix densities. Thus if we know the porosity of the sample, and the density and thermal conductivities of the matrix and pore fluid, we can estimate the thermal conductivity of the sample.

The pore fluid in this case is water and the thermal conductivity of water varies with temperature according to the equation:

$$k_p = 0.56 + 0.002T - 1.01 \times 10^{-5}T^2 + 6.71 \times 10^{-9}T^3 \text{ W/m°C} \quad (3)$$

where T is the borehole temperature in °C (Kaye & Laby 1968, p. 54). Using the temperature profile from Castillo (1991), values of k_p range from 0.56 at the seabed to 0.62 at 1160 mbsf.

Although in theory the value of k_g will vary from around 4 W/m°C (calcite) to between 7 and 10 W/m°C for quartz, the thermal conductivity of the matrix is dramatically reduced in polycrystalline substances. Rzhevsky & Novik (1971 pp 138–147) discuss this phenomenon and quote thermal conductivities of quartz varying from 11.7 W/m°C (monocrystalline), through 3.6 (polycrystalline), to 1.39 (fused quartz). A similar range is observed for calcite, and it appears that the nature of the inter-grain contacts are almost as influential as the pore space itself in controlling thermal conductivity. This is probably due to thin fluid films on the grain contacts. Rzhevsky & Novik (1971) also found that monocrystalline substances have the highest range of thermal conductivities, whereas polycrystalline substances have a much lower range. Aggregation brings moderation. Grain size affects thermal conductivity by increasing the number of grain contacts per unit volume. Rzhevsky & Novik (1971, p 140) report a reduction of 27% of the monocrystalline thermal conductivity values at a grain size of 0.1 mm, and a 50% reduction at 0.05 mm. Fine-grained silts and the types of calcareous ooze found in the deep ocean would therefore expect to have significantly reduced effective matrix thermal conductivites. Clays have both an increased surface area to mass ratio, and increased water content, both contributing to a low thermal conductivity.

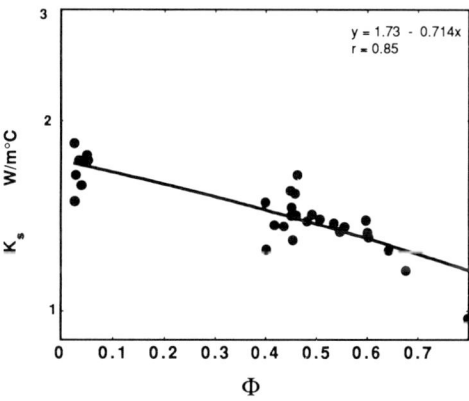

Fig. 4. Log-normal relationship between porosity and thermal conductivity measured on core samples at Site 765.

Brereton (1990, pp 15–16) showed that a rough prediction of measured thermal conductivity values from core porosity could be achieved by using equation (2) with matrix thermal conducti-

vities between 1.65 W/m°C (sediment) and 1.7 W/m°C (basalt). These matrix values were obtained by extrapolating to zero porosity on a porosity–thermal conductivity cross-plot. A log–normal plot of the core measurements for Site 765 alone shows (Fig. 4) that the zero-porosity intercept gives a matrix value of around 1.73 W/m°C for both sediment and basalt. Except for clays, these matrix thermal conductivities are significantly lower than values usually quoted for grain conductivities and suggest that some sort of grain contact mechanism may be a contributory factor. The concept of a matrix conductivity in the case of clays and shales is probably of little use. Of more interest is understanding the rate of change of conductivity with compaction as measured by reduction in water content.

Another approach uses the geometric mean formula (Woodside & Messmer 1961; Nobes et al. 1986);

$$k_s = k_p^{\Phi} k_g^{(1-\Phi)}. \tag{4}$$

A comparison of the results of applying equations (2) and (4) to core data can be seen in Fig. 5. Using a matrix thermal conductivity of 1.75 W/m°C, equation (2) gives a reasonable fit to both sediment and basalts. However, the best fit to sediments using equation (4) is achieved using a matrix value of 2.6 W/m°C, in which case the basalt thermal conductivities are not well described by the equation. The inclusion of some measure of matrix variation therefore seems to be an advantage in modelling thermal conductivity.

Rzhevsky & Novik (1971, p 138) discuss the difference between serial and parallel conduction of heat. In the serial case the mean thermal conductivity of a multicomponent system can be described as;

$$\frac{1}{k_s} = \sum_{i=1}^{n} \frac{V_i}{k_i} \tag{5}$$

where k_s is the thermal conductivity of the sample, V_i is the volume of the ith material component, and k_i is the thermal conductivity of the ith component.

In the case of components arranged in parallel then the resultant equation is;

$$k_s = \sum_{i=1}^{n} V_i k_i \tag{6}$$

In any given porous rock both conditions will be present and equations (5) and (6) describe the extreme values to be expected. If a rock can be considered a statistical mixture of different minerals with equal probability of occurring in series or parallel, then for a bimineralic rock, Rzhevsky & Novik (1971) suggest that the following difference equation may be applied:

$$k_s = \frac{3k_p - 2(1 - \Phi)(k_p - k_g)}{3k_g + (1 - \Phi)(k_p - k_g)} k_g \qquad (7)$$

for $k_g > k_p$.

The results of applying this equation using a value of k_p calculated using equation (3), and a k_g of 1.78 W/m°C, gave too high estimates at low thermal conductivities. The estimate improved with depth as porosity reduced, and thermal conductivity increased.

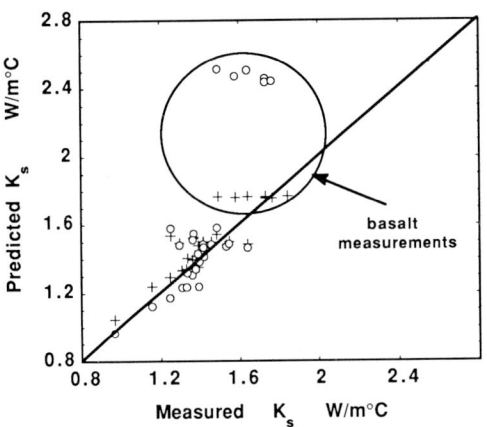

Fig. 5. Comparison of thermal conductivity predictions using both Equation (2), ('time average') and Equation (4) (geometric relationship). Equation (2) uses $k_g = 1.78$ W/m°C and k_p calculated from the thermal gradient. Equation (4) uses $k_g = 2.6$ W/m°C, $k_p = 0.75$ W/m°C. The porosity and density values used are those measured on core samples at Site 765.

Bullard's equation (1) using the measured core porosity, gave moderately good agreement in the upper section, but the estimates became too high as porosity decreased.

Rzhevsky & Novik (1971) mention other controls on thermal conductivity, such as mineral anisotropy and pore shape, suggesting that porosity and bulk density are necessary but not sufficient parameters in the prediction of thermal conductivity.

Assuming a relationship in the form of equation (4), Lovell & Ogden (1984) reported a linear relationship between porosity and logarithmic

thermal conductivity, giving a matrix conductivity of 2.01 W/m°C. They compared this relation with Bullard & Day's (1961) equation and concluded that errors in the needle probe technique of measuring thermal conductivity were probably more important than the differences in the predictive equations. Lovell (1985), describing a similar study on pure quartz sands, found that a matrix value of 8.58 W/m°C fit the data, whereas for carbonate shell sands a value of 3.32 W/m°C was appropriate.

Although Ratliffe (1960), Bullard (1954) and others have commented on the weak control that matrix mineralogy exerts on thermal conductivity and diffusivity, they were only considering the extremely high porosity sediments on the seabed. At moderate depths (a few hundred metres below seafloor to a kilometre) the porosity is reduced to 20 to 30% and the combination of mineralogy and grain contacts becomes increasingly significant. Gretener (1982) reports work by Birch & Clark (1940) showing the steady reduction of thermal conductivity with increasing temperature, a trend that acts to counter the increase in thermal conductivity as a function of reduced porosity with depth.

Balling *et al.* (1981) looked at the thermal conductivities in the three main lithology groups from the Danish Cenozoic and Mesozoic. The thermal conductivities of carbonates, clay/claystones, and sandstones plot as quite distinct curves (Balling *et al.* (1981) Figs 25 to 27). Least-squares fits to their data gave the following sets of equations:

1. Clay, claystone, shale

$$K_{sh} = 0.46^\Phi \times 3.43^{(1 - \Phi)} \text{ W/m°C} \qquad (8)$$

2. Carbonates
$$K_{carb} = 0.54^\Phi \times 3.24^{(1 - \Phi)} \text{ W/m°C} \qquad (9)$$

3. Sandstone
$$K_{sst} = 0.69^\Phi \times 4.88^{(1 - \Phi)} \text{ W/m°C} \qquad (10)$$

In a separate study using a total of 58 Danish (high porosity) and Swedish (low porosity) outcrop samples of water-saturated carbonates, Balling *et al.* (1981) confirmed a geometric relationship of the form:

$$K_{carb} = 0.58^\Phi \times 3.26^{(1 - \Phi)} \text{ W/m°C} \qquad (11)$$

The coefficients here are very similar to those obtained in the previous study. Lovell (1985) published an empirical quartz sand equation:

$$K_{sst} = 0.64^\Phi \times 8.58^{(1 - \Phi)} \text{ W/m°C} \qquad (12)$$

These empirical equations have some interesting features. The estimate of K_g is substantially lower than the monocrystalline values, possibly as a result of grain-contact effects as discussed earlier. The estimate of K_p increases from shale to carbonate to sands, possibly as a function of the difference in the nature of the porosity in the three cases. It is clear that both k_g and k_p should be labelled 'apparent' or 'effective' in these cases, and a knowledge of lithology is necessary before the equations can be used predictively.

To summarize:

(i)   Porosity provides an important but not exclusive control on thermal conductivity.
(ii) The influence of porosity probably decreases with depth (as porosity itself decreases).
(iii) Equation (2), while allowing some aspects of lithology (namely grain density, and matrix thermal conductivity) to be explicitly included, assumes a two-phase (liquid–solid) assemblage. The grain density can therefore be considered to be an 'effective' grain density rather than a true grain density in the same way that the matrix thermal conductivity should be considered an 'effective' value.
(iv) Geometric equations with the form of equation (4) also assume a two-phase assemblage where the porosity-complement coefficient includes all matrix effects as above.
(v)   Different studies have derived different empirical coefficients in different lithology groups. It would seem useful therefore to include lithological information explicitly in any attempt to predict thermal conductivity from wireline log data.

It is worth noting that most of the empirical studies mentioned here have subdivided the lithological universe into three main components; carbonate, clay/shale and sandstone. This is obviously a limitation when crystalline rocks such as basalts and granites are concerned. In the present study we will begin with a three-lithology universe, but extend the approach to basalts.

Assuming a basic clastic lithology, then estimates of shale (clay) volume (V_{sh}) can be derived from a combination of 'shale indicators' such as relative gamma values, relative spontaneous potential, neutron-density cross-plot, etc. Estimates of carbonate content can be obtained from the slope of the neutron-density crossplot (the N-value published by Burke $et\ al.$ (1969)), and from the apparent grain density (ρ_{ga}). Including this extra information in either

equation (2) or (4) requires the use of at least one, possibly two, extra sets of 'effective' thermal conductivities and densities. For example, equation (2) with three components appears as:

$$k_s\rho_s = \Phi_T(K_p\rho_p) + (1 - \Phi_T - V_{sh})K_g\rho_g + V_{sh}K_{sh}\rho_{sh}. \tag{13}$$

Since the monocrystalline values are of little practical use in clastic aggregates, and the empirical effective matrix values are related to specific equations and studies, there appears to be little theoretical basis for choosing relevant values for the coefficients k_g, k_{sh}, ρ_g and ρ_{sh}. Acceptable thermal conductivity values can be derived using an empirical k_{sh}, ρ_{sh} product of ~ 5.0 (using wireline log values), but the physical interpretation of this is rather vague. The results are shown in Fig. 6.

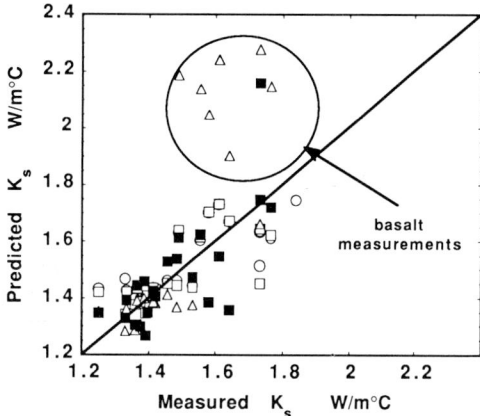

O Equation 2a (including V_{sh})
□ Equation 2b (single matrix component)
△ Equation 4
■ Equation 16

Fig. 6. Comparison of thermal conductivity predictions using both 'time average' and geometric relationships. Equation (2a) uses $k_g = 1.78\ \text{W/m}^\circ\text{C}$, a k_p calculated from the thermal gradient and, a $k_{sh}\rho_{sh}$ product of 5.0 Equation (2b) does not have any shale component. Equation (4) uses $k_g = 2.6\ \text{W/m}^\circ\text{C}$, $= 0.75\ \text{W/m}^\circ\text{C}$. Equation (16) uses the algorithm discussed in the text. The porosity and density values used are those measured on core samples at Site 765.

However, the inclusion of multiple matrix components in equation (4) can be achieved by choosing appropriate combinations of equations (8), (9) and (10) for different matrix component combinations throughout the well. This is best explained algorithmically as follows. Given the total porosity (Φ_T), the effective porosity (Φ_E), and the shale volume (V_{sh}), then

$$\Phi_{(8),(9),(10)} = \Phi_T$$

where $\Phi_{(8)}$ is the porosity to be used in equation (8) etc.

In other words we are considering that the total water volume rather than the effective water volume influences thermal conductivity.

Given that the V_{sh} complement

$$V_{scom} = (1 - V_{sh})$$

then, after the definition of effective porosity (equation (0)),

$$V_{scom} = \Phi_E/\Phi_T \qquad (14)$$

The fractional thermal conductivities for a sediment composed of three components will be:

$$K_{shf} = V_{sh}/K_{sh} \qquad (15a)$$

$$K_{carbf} = V_{scom}/K_{carb} \qquad (15b)$$

$$K_{sstf} = V_{scom}/K_{sst} \qquad (15c)$$

where K_{shf} is the fractional component of the thermal conductivity associated with the observed shale volume, and K_{sh} is calculated from equation (8) using Φ_T. Using equation (14), equations (15b) and (15c) can be rewritten in the form:

$$K_f = \frac{\Phi_T}{\Phi_E} K_{matrix}.$$

This is in effect a weighting function for the 'non-clay' thermal conductivity component. When Φ_T equals Φ_E, (i.e. the rock is 'clean' with no clay) then the matrix thermal conductivity is unchanged. However, as the amount of clay increases, then V_{sh} influences both the log-derived total porosity, and the combined fraction conductivities.

The weighting function operates as shown below:

$$K_s = \frac{K_{sh} K_{matrix}}{K_{sh} (\Phi_E/\Phi_T) + K_{matrix} [1 - \Phi_E/\Phi_T]} \qquad (16)$$

where K_s is the combined sample thermal conductivity, and K_{matrix} is calculated according to one of equations (9) or (10) according to the dominant matrix type, carbonate or quartz. As can be appreciated from equation (16), for clean, clay-free sediments, K_s is equal to K_{matrix}.

The dominant matrix is assumed to be quartz unless:

$$N > 0.64 \text{ and } 2.71 < \rho_{ga} < 2.8$$

where

$$N = \frac{1.0 - \Phi_N}{\rho_b - \rho_f}$$

and ρ_{ga} is the apparent grain density, in which case carbonate matrix values are used (Fig. 6).

The response of equations (8), (9) and (10) in the basalts is not satisfactory due to the lack of an appropriate basalt relationship. If, however, we use the relation expressed in Fig. 4, i.e.

$$K_{basalt} = 0.85^\Phi \times 1.78^{(1 - \Phi)} \qquad (17)$$

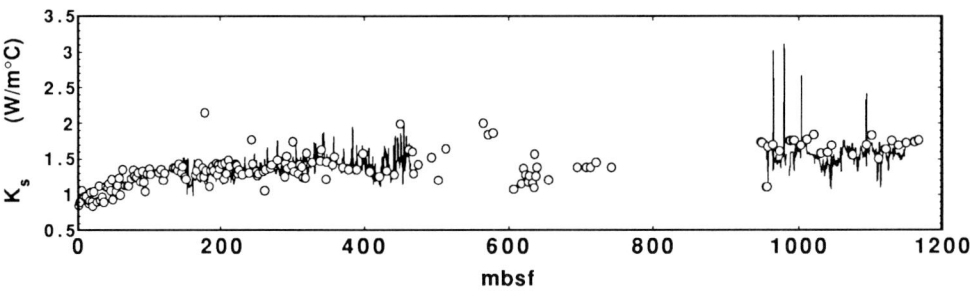

Fig. 7. Comparison of thermal conductivities measured on core samples, and thermal conductivities predicted at log resolution using the algorithm explained in the text (Site 765).

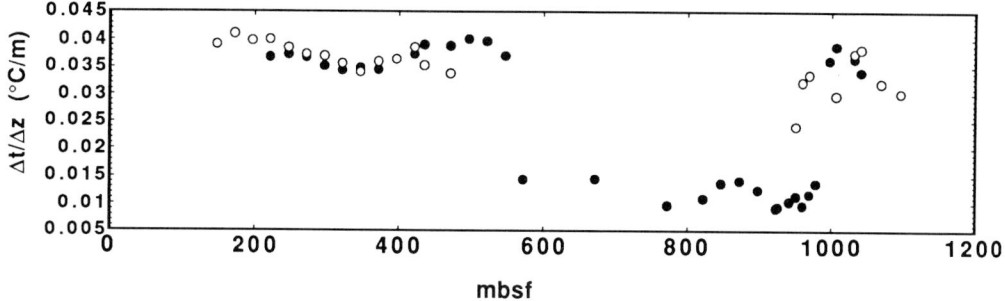

Fig. 8. Comparison of predicted first-order $\Delta t/\Delta z$ from log thermal conductivity estimation from wireline logs, and the nine-term smoothed $\Delta t/\Delta z$ from high-resolution borehole temperature measurements for Site 765.

as K_{matrix} for all samples where $N > 0.7$ and the apparent grain density exceeds 2.8 g/cm^3, then a satisfactory fit is achieved. The continuous thermal conductivity log for Site 765 is shown in Fig. 7 alongside the core measurements. The higher-frequency detail in the log measurements is to be expected from the difference in sample interval.

Thermal gradient and heat flow prediction from log data

Thermal conductivity would remain something of an academic topic were it not for its use in the prediction of local variations in thermal gradient and heat flow.

Predicting local thermal gradients from log data is of use in looking for small-scale variation in vertical heat flow patterns that may indicate fluid flow, and local disturbance. Figure 8 shows the result of using the relationship:

$$\Delta t/\Delta z = Q/K_s \qquad (18)$$

where Q is the heat flow in mW/m^2, to estimate the local downhole thermal gradient.

In Fig. 8 the inverted conductivity predictions from wireline logs, using the algorithm described above, are multiplied by a constant and plotted as a function of depth alongside the nine-term smoothed temperature gradient derived from the high-resolution temperature survey (Castillo 1991). Using a constant of 52 mW/m^2 gives reasonable agreement in the sediment column. Although this heat flow value is also reasonable for the Jurassic crust (140 Ma), it does not appear to be satisfactory in the basalt section

because high-frequency temperature profile perturbations suggest that the basement section is experiencing subsurface fluid flow, and the system has not reached equilibrium.

Using the thermal interval method (Powell et al. 1988), Castillo (1991) derived a conductive heat flow value for the sediment of between 42 ± 4 mW/m^2 to 60 ± 8 mW/m^2 with a weighted average value of 51.02 ± 0.12 mW/m^2. Using a method proposed by Bullard (1939), a figure of 51.21 mW/m^2 was obtained. Considering several techniques, Castillo (1991) concluded that the mean present day heat flow for Hole 765D is 52.8 ± 5.7 mW/m^2. In the Bullard method a conductive steady-state heat flow model is assumed. Thermal resistance $(\partial z/k)$ is summed over discrete intervals of temperature measurement. The cumulative sum is plotted against the corresponding temperature at the base of the interval. The intervals may be lithology intervals or logging runs. If a conductive heat flow model applies then the plot should form a straight line with a slope Q (heat flow) and an intercept T_0 (surface temperature) (Castillo 1991), i.e.

$$T_z = T_0 + Q \sum (\partial z/k) \qquad (19)$$

where T_z is the temperature at depth z.

In the present study, with log and core measurements at small, relatively regular, intervals along the hole it is more convenient to integrate the thermal resistivity $(1/k_s)$ curves. Figs 9–12 show the results of such an analysis using core- and log-derived thermal resistivities.

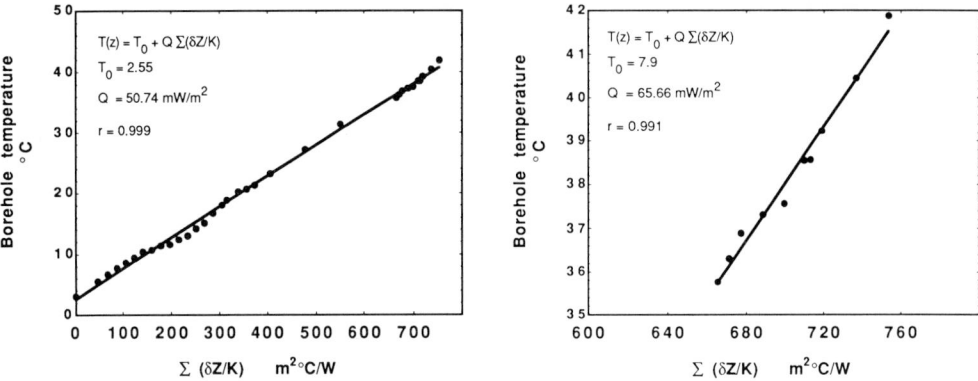

Fig. 9. (a) Prediction of heat flow from thermal conductivities measured on cores using Bullard's method; all measurements (sediments and basalts). (b) Prediction of heat flow from thermal conductivities measured on cores using Bullard's method; basalts only.

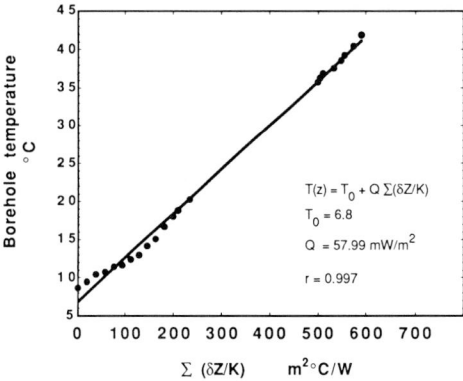

Fig. 10. Prediction of heat flow from thermal conductivities estimated from logs using Bullard's method (sediments and basalts).

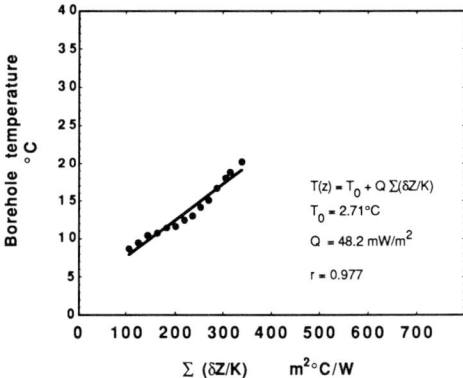

Fig. 11. Prediction of heat flow from thermal conductivities estimated from logs using Bullard's method; sediments only.

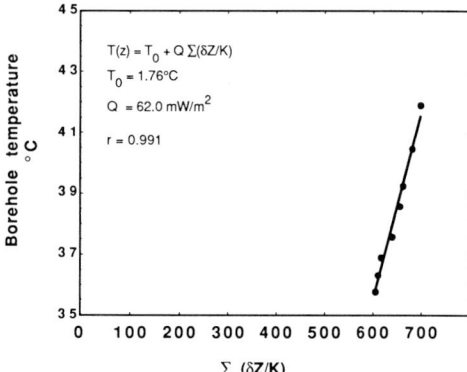

Fig. 12. Prediction of heat flow from thermal conductivities estimated from logs using Bullard's method; basalts only.

It is clear from these figures that a conductive heat flow model is probably appropriate for most of the section. The non-linearity observed in the sediment section (observed with both log- and core-derived data) suggests that some convective heat flow may be occurring. The values for heat flow using core measurements vary from $50.74 \, \mathrm{mW/m^2}$ (complete section) to $65.66 \, \mathrm{mW/m^2}$ (basalt only), whereas using log-derived measurements the range is from $48.2 \, \mathrm{mW/m^2}$ to $62 \, \mathrm{mW/m^2}$. These values approximate the range reported by Castillo (1991) using the interval method. The surface temperature estimates range from 0.63 to $7.9°\mathrm{C}$ compared to a measured value of $2.5°\mathrm{C}$.

These results were obtained with the help of a high-resolution temperature survey, something that is not run routinely, and may not be available for many older wells. In the case of Site 766 the temperature logging equipment was not used, and we are left with the borehole logs, core measurements of thermal conductivity and the bottom hole temperatures from logging runs. If a first approximation to a heat flow can be derived from these measurements then hopefully the method may be more generally useful for exploration drilling or a re-evaluation of old wells.

A reliable bottom-hole temperature was not recorded at Site 766, and the best estimate based on the maximum recorded temperature is around 11 to 12°C. The surface (seabed) temperature is estimated to be around 2°C. Using these two values plotted against the integral of the thermal resistance yielded a heat flow of 37 mW/m² for both the core and log derived data. This is probably too low a value, but is comparable to a published value (36 mW/m²) (Stein *et al.* 1988) for the neighbouring abyssal plain.

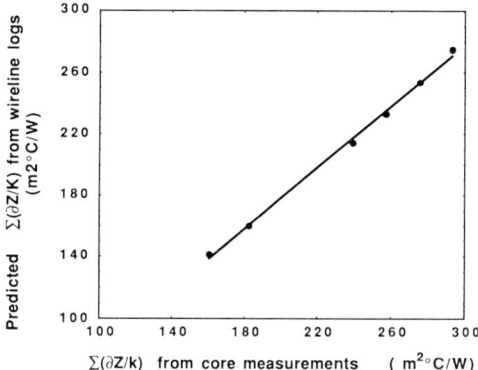

Fig. 13. Comparison of cumulative thermal resistance from both core measurements and log predictions.

Figure 13 shows the comparison of integrated thermal conductivity values for core and log data, while Fig. 14 shows the estimate of heat flow (using only two data points).

Discussion

It is clear from the above results that porosity does indeed dominate the thermal conductivity in sediments, in so far as an acceptable empirical correlation between porosity and thermal conductivity can be achieved using most of the porosity-dominated equations discussed above, in shallow sediments with porosities in excess of 30%. The critical value as far as prediction is concerned is the value of k_g used. A single value is not applicable throughout the section,

especially if igneous rocks, evaporites, and carbonates are encountered. The empirical estimates of k_p are also seen to change, possibly as a function of porosity type although this has not been investigated here. The algorithmic approach detailed in the present article is one possible way of combining matrix and porosity information from petrophysical data with empirically derived porosity-thermal conductivity relationships. Lovell & Ogden (1984) and Lovell (1985) presented a relationship between thermal conductivity and formation factor that may be used in a similar way, in that the nature of the pore space and grain contacts certainly appear to influence the correlation. We did not have the possibility to test a prediction based on formation resistivity factor here due to the poor quality of the resistivity logs obtained on Leg 123, but a comparison of the different approaches on the same data should be informative. Tests using equation (2) for multiple component lithologies have shown that reasonable empirical agreement can be obtained using a $k_{sh}.\rho_{sh}$ product of 5.0, a k_g of 1.78, and a value of ρ_g calculated in the usual way from log values. This raises the question of the relationship between ρ_g and the shale density, as the measured effective grain density should in theory include the shale component. The physical basis of this equation is thus at this stage unclear.

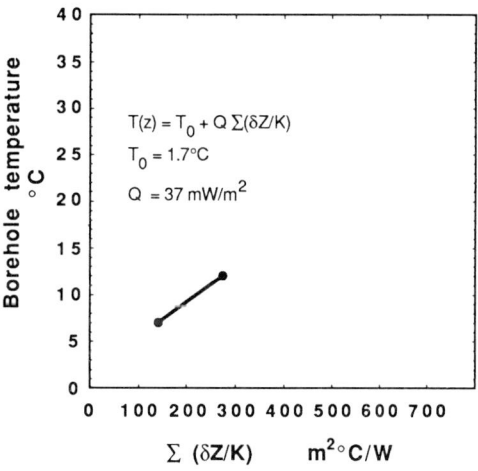

Fig. 14. Estimate of heat flow at Site 766 using only bottom hole temperature, surface temperature, and log-derived thermal resistance.

Another point that should be noted concerns the use of V_{sh} estimates from logs. There are many different methods of calculating V_{sh} using natural gamma, spontaneous potential, neutron, density, sonic and resistivity logs. They will all give different values for clay volume, and the algorithm used should preferably be calibrated against core, side-wall core or cuttings estimates before being used in thermal conductivity prediction.

Thermal conductivity measurements of well lithified, fractured material also pose problems. Core measurements are usually restricted to the lithified fragments, whereas log measurements average over both fractures and lithified blocks. It could be argued in such circumstances that the log values are more representative of the bulk rock response over the transmitter–receiver distance. If this is so, then calibration measurements on core should ideally use the same integration interval as the logs to be of any value in calibration. This will need a measurement technique other than the needle probe currently used on ODP cruises.

Conclusion

A method has been presented that allows heat flow estimates to be prepared from old well data in the absence of detailed temperature logs. Any combination of logs giving an estimate of matrix type, clay content and porosity may be combined with bottom-hole temperature information to give rough vertical heat flow estimates

throughout a basin. Better temperature and lithological information will improve the heat flow estimates.

Both additive and geometric relationships between thermal conductivity and porosity–mineralogy combinations may be used, but the physical basis of the coefficients becomes more diffuse as extra terms are added.

One practical goal of such studies is to enable the prediction of heat flow in three dimensions, at high resolution, throughout a basin. In order to do this it will be necessary to derive not only porosity but some measure of lithology from the seismic record. Although this appears to be beyond the present state of the art of seismic inversion, it may be that a combination of seismic inversion and sequence stratigraphy can provide sufficient lithological and porosity information to attempt a three-dimensional heat-flow map, at least in shallow sections with few tectonic disturbances.

More work is needed concerning the possibility of predicting the change in effective matrix thermal conductivities for real mixtures of mineral components.

This work was supported in Norway by the Nordic Council. We are grateful for valuable suggestions made by Andrew Hurst and an anonymous referee.

References

BALLING, N., KRISTIANSEN, J., BREINER, N., POULSEN, K. D. & RASMUSSEN, R. 1981: Geothermal measurements and subsurface temperature modelling in Denmark, *Geologiske Skrifter*, **16**.

BATEMAN, R. 1985. *Log Quality Control*. IHRDC, Boston.

BIRCH, F. & CLARK, H. 1940. The thermal conductivity of rocks and its dependence upon temperature and composition. *American Journal of Science*, **238**, 529–558.

BRERETON, N. R. 1990. *Physical Properties of Sediments and Basalts from the Argo and Gascoyne Abyssal Plains in the Indian Ocean*, British Geological Survey Technical Report WK/90/11.

BRIGAUD, F., CHAPMAN, D. S. & LE DOUARAN, S. 1990. Estimating thermal conductivity in sedimentary basins using lithological data and geophysical well logs. *AAPG Bulletin*, **74**, 1459–1477.

BULLARD, E. C. 1954. The flow of heat through the floor of the Atlantic Ocean. *Proceedings of the Royal Society of London* A., **222**, 408–429.

—— & DAY, A. A. 1961. The flow of heat through the floor of the Atlantic Ocean. *Geophysical Journal of the Royal Astronomical Society*, **4**, 282–292.

BURKE, J. A. CAMPBELL, R. L. Jr., SCHMIDT, A. W. 1969. The litho-porosity crossplot. *The Log Analyst*, Nov-Dec.

CASTILLO, D. A. 1991 (in press). Thermal and hydrologic properties of old oceanic crust in Hole 765D, Argo Abyssal Plain, Indian Ocean. Submitted to ODP Leg 123 Science Volume.

ELPHICK, R. Y. 1987. Petrophysical corner: Neutron/Density–GR interpretation in shaley sands. *Geobyte*, **2**, 2, 51–55.

GOSS, R., COMBS, J. & TIMUR, A. 1975. Prediction of thermal conductivity in rocks from other physical parameters and from standard geophysical well logs, *Transactions of SPWLA 16th Annual Logging Symposium*, Paper MM.

GRADSTEIN, F. M. & LUDDEN, J. *et al.* 1990. *Proceedings of the Ocean Drilling Program, Initial Reports*. Ocean Drilling Program, College Station, TX, **123**.

GRETENER, P. E. 1982. *Geothermics: Using Temperature in Hydrocarbon Exploration*. AAPG Education Course Notes Series, **17**, American Association of Petroleum Geologists, Tulsa.

GRIFFITHS, C. M. 1982. A proposed geologically consistent segmentation and reassignment algorithm for petrophysical borehole logs. *In*: CUBITT, J. M. & REYMENT, R. A. (eds) *Quantitative Stratigraphic Correlation*. Wiley, Chichester, 287–290.

KAYE, G. W. C. & LABY, T. H. 1968. *Tables of Physical and Chemical Constants*. Longmans, London.

LOVELL, M. A. 1983. *Resistivity–Thermal Conductivity–Porosity Relationships for Marine Sediments*. PhD Thesis, University of Wales, UK.

—— 1985a. Thermal conductivity and permeability assessment by electrical resistivity measurements in marine sediments. *Marine Geotechnology*, **6**, 2, 205–240.

—— 1985b. Thermal conductivities of marine sediments. *Quarterly Journal of Engineering Geology*, **18**, 3, 437–441.

—— & OGDEN, P. 1984. Remote assessment of permeability/thermal diffusivity of consolidated clay sediments. EUR report 9206 EN (contract 263-81-7 WAS UK).

NOBES, D. C., VILLINGER, H., DAVIS, E. E. & LAW, L. K., 1986. Estimation of marine sediment bulk physical properties at depth from seafloor geophysical measurements. *Journal of Geophysical Research*, **91**, 14033–14043.

POWELL, W. G., CHAPMAN, D. S., BALLING, N. & BECK, A. E. 1988. Continental heat-flow density. *In*: HEANEL, R., RYBACH, L. & STEGENA, L. (eds) *Handbook of Terrestrial Heat Flow Density Determination* (1st edn), Kluwer, Massachusetts, 167–222.

RATCLIFFE, E. H. 1960. The thermal conductivity of ocean sediments. *Journal of Geophysical Research*, **65**, 1535–1541.

RIDER, M. H. 1985. *The Geological Interpretation of Wireline Logs*, Blackie, Glasgow.

RZHEVSKY, V. & NOVIK, G. 1971. *The Physics of Rocks*, Mir, Moscow (translated from the Russian by A. K. Chatterjee).

SCHLUMBERGER, 1989. *Log Interpretation Charts*, Schlumberger Well Services, Houston.

—— 1989. *Log Interpretation Principles/Applications*, Schlumberger Well Services, Houston.

STEIN, C. A., HOBART, M. A. & ABBOTT, D. H. 1988. Has the Wharton Basin's heat flow been perturbed by the formation of a diffuse plate boundary in the Indian Ocean? *Geophysical Research Letters*, **15**, 455–458.

THEYS, P. 1990. *Log Data Acquisition and Quality Control*, Editions Technip, Paris.

VACQUIER, V. 1985. The measurement of thermal conductivity of solids with a transient linear heat source on the plane surface of poorly conducting body. *Earth and Planetary Science Letters*, **74**, 275–279.

——, MATHIEU, Y., LEGENDRE, E. & BLONDIN, E. 1988. An experiment on estimating the thermal conductivity of sedimentary rocks from oil well logging. *AAPG Bulletin*, **72**, 758–764.

VON HERZEN, R. P. & MAXWELL, A. E. 1959. The measurement of thermal conductivity of deep-sea sediments by a needle probe method. *Journal of Geophysical Research*, **64**, 1557–1563.

WOODSIDE, W. & MESSMER, J. H. 1961. Thermal conductivity of porous media. *Journal of Applied Physics*, **32**, 1688–1706.

WYLLIE, M. R. J., GREGORY, A. R. & GARDNER, L. W. 1956. Elastic wave velocities in heterogeneous and porous media. *Geophysics*, **21**, 41.

Determination of Young's modulus of the rock mass from geophysical well logs

D. M. McCANN & D. C. ENTWISLE

Engineering Geology and Geophysics Group, British Geological Survey, Nicker Hill, Keyworth, Nottingham, NG12 5GG, UK

Abstract. Geophysical logging of site investigation boreholes can provide a wide range of useful geotechnical information ranging from the formation density (gamma/gamma log) to the dynamic elastic moduli (full wave train sonic log). In this paper, the use of Young's modulus as a measure of the deformation characteristics of the rock mass is examined. A comparison is made between the dynamic value obtained from geophysical borehole logging and the static value obtained directly from the laboratory testing of samples collected from the borehole. The data are obtained from a number of BGS boreholes. It is shown that although the prediction of the static value of Young's modulus from the dynamic value is dependent on a knowledge of the borehole lithology, the use of the dynamic elastic moduli can yield considerable useful engineering information in the assessment of the deformation properties of the rock mass.

The design and construction of most structures requires a knowlege of the engineering properties of the rock mass. Thus, an essential part of the site investigation is the determination of the deformational characteristics of the rock types involved. This is achieved by laboratory tests on intact rock specimens, such as the simple measurement of uniaxial compressive strength or the more complex measurement of the static moduli.

The specimens required for laboratory testing are prepared from samples collected from the site investigation boreholes. However, borehole samples are often highly disturbed and even the most carfully taken sample is subject to stress relief during the drilling process. The question, therefore, arises as to how representative the laboratory tests are of the in situ properties of the rock mass and, indeed, how much reliability can be placed on them without invoking large factors of safety in the design calculations. Hence, in situ loading tests, such as the plate load (Anon 1981; Marsland 1971; Lane 1964; Burland & Lord 1970; Hobbs 1973) and Goodman Jack tests (ISRM 1981; Goodman 1980) are often applied in the field to increase the overall knowledge of the engineering performance of the rock mass.

A borehole is also an access path to the geological structures of interest to the civil engineer and can be used for further in situ engineering tests by the deployment of geophysical logging tools. The geophysical properties of the rock mass are directly related to its lithological and geotechnical properties (Wyllie *et al.* 1956;

Buchan *et al.* 1972). Of particular interest in this context is the determination of the compressional and shear wave velocities from the full wave train sonic log and the formation density from the gamma/gamma log, since these parameters can be used to compute the dynamic elastic moduli of the rock mass. While the velocity of propagation of stress waves in a material is related to its deformational characteristics, the dynamic elastic moduli are rarely used for design purposes in preference to the static moduli in civil engineering projects. The exception to this is the situation where the structure is likely to be subjected to dynamic loading effects, such as wave action on a marine structure, and in this case the dynamic elastic moduli will be used in the design of its foundations.

The major problem which arises in trying to relate static and dynamic moduli is that most geological materials do not behave in a perfectly linear elastic, homogeneous, isotropic manner when they are subjected to static loading. As a result of this in most cases, there is a difference between the static and dynamic moduli which is thought to be related to the difference in strain levels at which the two sets of moduli are measured. Thus, while the strain levels of 10^{-6} or less involved in the measurement of the dynamic elastic moduli still allow the material to behave in an elastic manner the strain levels of 10^{-3} or greater involved in the static testing of the material usually result in permanent deformation of its internal structure and a non-linear stress/strain relationship will be observed.

The determination of dynamic elastic moduli

From HURST, A., GRIFFITHS, C. M. & WORTHINGTON, P. F. (eds), 1992,
Geological Applications of Wireline Logs II. Geological Society Special Publication No. 65, pp. 317–325.

317

is quick and inexpensive to carry out in both the field and the laboratory. Hence, many attempts have been made to establish a relationship between the static and dynamic elastic moduli (Ide 1936; Sutherland 1962). The results of many of these studies are summarized by Lama & Vutukuri (1978), who showed that the dynamic value of Young's modulus (E_d) can be greater than the corresponding static value (E_s) by up to 300 per cent.

The specimen used in laboratory testing will be selected from relatively unfractured sections of the borehole core and the values of the static moduli obtained will be a maximum for a particular lithological unit. Similarly, the dynamic elastic moduli obtained from the geophysical logging of the borehole will have been measured in the same section of borehole and will also represent a maximum value. Thus, by establishing a relationship between the static and dynamic elastic moduli for each lithological unit present in the borehole, it is possible to assess the effects of fracturing in other parts of this borehole on the engineering properties of the rock mass for any lithological unit present in the borehole. A continuous log of the overall engineering performance of the rock mass, which includes the effects of fracturing, can thus be obtained from the geophysical borehole logging programme (McCann et al. 1990).

The main objective of this present study is to examine the relationship between the static and dynamic values of Young's modulus for a wide range of rock types. The data are obtained from a number of BGS boreholes, which had been comprehensively sampled and geophysically logged.

Dynamic elastic moduli

The dynamic elastic moduli of a material are expressed as follows:

Young's modulus (E_d)

$$E_d = 2\rho V_s^2 (1 - v_d) \tag{1}$$

Poisson's ratio (v_d)

$$v_d = \frac{0.5(V_p/V_s)^2 - 1}{(V_p/v_s) - 1} \tag{2}$$

Shear modulus (μ_d)

$$\mu_d = \rho V_s^2 \tag{3}$$

Bulk modulus (K_d)

$$K_d = \rho(V_p^2 - \tfrac{4}{3} V_s^2) \tag{4}$$

where ρ is the bulk density, V_p is the compressional wave velocity and V_s is the shear wave velocity

In a borehole, V_p and V_s can be determined from the full wave train sonic log while the formation density is obtained from the gamma/gamma log. In both cases, the recorded logs are adversely affected by caving in the borehole wall and, hence, it is a normal procedure to run a caliper log to measure the variation in borehole diameter so that some correction can be made to the two logs.

The full wave train sonic log is based on a normal sonic logging tool but use is made of the full wave train observed at the receiver to determine both the compressional P-wave arrival and the later shear S-wave arrival (Fig. 1). The latter arrival results from mode conversion of the P-wave energy from the transmitter to S-wave energy at the borehole wall. Cheng & Toksöz (1980) concluded from model studies that the S-wave arrival may be a pseudo-Rayleigh wave propagating near its cut-off frequency. Siefried & Castagua (1982) suggested that if this is the case then the phase velocity of the pseudo-Rayleigh wave will be within about 10% of the shear wave velocity of the formation in most cases. Lingle & Jones (1977) showed that the S-wave velocity values obtained on laboratory samples from the borehole core were consistently higher by 10 to 15 per cent than those obtained from the full wave sonic log, which seems to confirm the conclusions of Cheng & Toksöz (1980). Use of this log in engineering studies is described in Geyer & Myung (1970) and Lingle & Jones (1977).

The timing of the S-wave arrival is complicated by the problem of identifying the precise point in the wave train where it commences. Siefried & Castagua (1982) have applied signal processing techniques to the extraction of the shear wave arrival from the recorded wave train and report significant improvement in the signal-to-noise ratio and, hence, the shear wave velocity.

When the shear wave velocity in the formation is lower than the compressional wave velocity, no significant shear wave energy is transmitted in the formation. Cheng & Toksöz (1980) indicated that in this case it might be possible to calculate the shear wave velocity in the formation from analysis of the Stoneley wave which is also observed in the sonic wave train. Where it is not possible to identify the shear wave in the

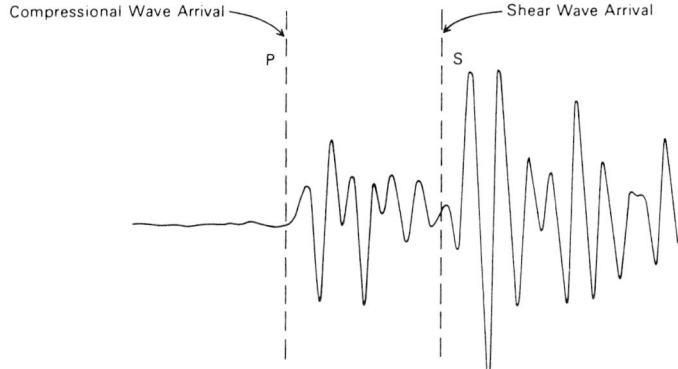

Fig. 1. Typical received pulse in the full wave train sonic log.

sonic wave train, the shear wave velocity can be calculated from Christensen's equation:

$$V_s = V_p \left[1 - 1.15 \left(\frac{1/\rho + 1/\rho^3}{e^{1/\rho}} \right) \right]^{3/2} \qquad (5)$$

Entwisle & McCann (1990) considered this equation in some detail and showed that it gives an over-estimate of the shear wave velocity in mudrocks and soft sediments. This will give rise to a corresponding increase in the computed values of the dynamic elastic moduli.

Using the data from a number of published papers Castagna *et al.* (1985) derived the relationship

$$V_p = 1.16V_s + 1.36 \qquad (6)$$

for mudrocks, where V_p and V_s are in km/s. Castagna *et al.* (1985) concluded that the V_p/V_s ratio is highly variable in mudrocks, since it ranges from less than 1.8 in quartz-rich rocks to over 5 in loose, water-saturated sediments.

It is also important to record the frequency at which the compressional and shear wave velocities are determined. While it has been generally considered (O'Brien 1973) that the velocity of propagation of both compressional and shear waves in natural materials is independent of frequency, recent studies have shown that, in many cases, frequency dependent velocities may be observed (Winkler 1986). This is normally associated with rocks with a high attenuation coefficient (low-Q) and in this case the dynamic elastic moduli will also vary with frequency.

Static elastic moduli

During strength testing for soils, axial deformation is usually measured using externally mounted dial gauges; this may result in greater errors associated with 'end effects'. However, the recent introduction of local strain measurement such as internal water proof transducers or Hall Effect apparatus has improved the measurement of axial strain in soils testing. Axial deformation in rocks is usually carried out using strain gauges which measure strain in the central portion of the sample.

In the case of purely elastic materials, the stress-deformation line shown in Fig. 2(a) is linear and the slope of the line is

$$E = \delta\sigma/\delta\varepsilon \qquad (7)$$

where σ = deviator stress and ε = change in length/the initial length. However, many materials are non-linear in their behaviour and this has resulted in a number of methods to measure E.

(a) Tangent Young's modulus, E_t, is measured at a stress level which is some fixed percentage of the ultimate strength. It is generally taken at a stress equal to 50% of the ultimate uniaxial compressive strength (Fig. 2(b)).

(b) Average Young's modulus, E_{av}, is determined from the average slope of the straighter portion of the stress–strain curve.

(c) Secant Young's modulus, E_s, is taken as the gradient of the line joining the origin to some fixed percentage of the ultimate uniaxial compressive strength on the stress/strain curve, generally 50%.

If stress is measured in kPa, and strain expressed as a percentage, then

$$E = \frac{\delta\sigma}{\delta\varepsilon(\%)} \times \frac{1}{10} \quad \text{MN/m}^2 \text{ (or MPa)}. \qquad (8)$$

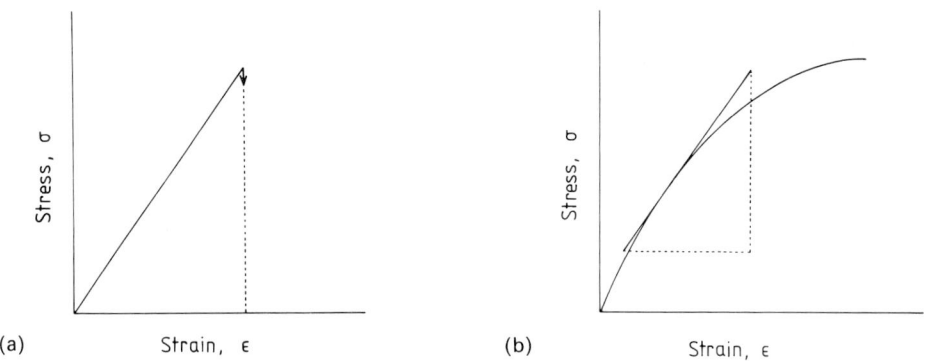

Fig. 2. Stress/strain curve for the computation of Young's modulus: (a) linear case; (b) tangent modulus.

Fig. 3. Positions of BGS boreholes used in the study.

The static Young's modulus results presented here were calculated using the tangent method.

A number of factors may affect the static Young's modulus; these include strain rate, yielding at low strains and lack of initial alignment of the test specimen. The static Young's modulus increases with increasing rate of strain.

Conventional soils or weak rock testing usually measure Young's modulus at about 0.3 to 1% strain or more. Research into the stiffness of heavily over-consolidated soils and soft rocks has demonstrated that the stress–strain behaviour of these materials is usually non-linear. Laboratory tests in which the stiffness reduces by a factor of at least ten within the range of practical interest is common (Atkinson *et al.* 1990). Yielding at axial strain of 0.1% or less reduces Young's modulus and makes accurate measurement of small strains important, but this is not normally undertaken. Modern techniques of measurement of low axial deformation are becoming more common and the effect of axial strain on material stiffness is now better understood (Anderson & Stokoe 1978; Baldi *et al.* 1988; Jardine *et al.* 1986).

The measurement of static elastic moduli in the laboratory should take into consideration effects such as stress relief to be more comparable with in situ tests. This effect is most noticeable in weak rocks and soils.

Comparison of the dynamic and static values of Young's modulus

The geotechnical and geophysical information used in this study is derived from boreholes drilled by the British Geological Survey. The boreholes were mainly of Jurassic age and included:

(a) Sea Barn Farm
(b) Harwell 3
(c) Tattenhoe

As a contrast, information from the three boreholes drilled at Altnabreac in Caithness, Scotland through granites and Monian metasediments were also included in the study. All the boreholes mentioned above were continuously cored and geophysically logged; their locations are shown in Fig. 3.

The static values of Young's modulus were determined using this appropriate procedure given in the section 'Static elastic moduli'. The dynamic value of Young's modulus for the rockmass in the equivalent sample positions in the borehole was obtained from the geophysical logs using equation (1). This required the calculation of the compressional and shear wave velocities from the full wave train sonic log and the formation density from the gamma/gamma log.

A comparison of the static and dynamic values of Young's modulus for all the specimens tested is shown in Fig. 4 and the relationship is expressed as:

$$E_{ST} = 0.69\,E_{DY} + 6.40 \qquad (9)$$

with a correlation coefficient of 0.75.

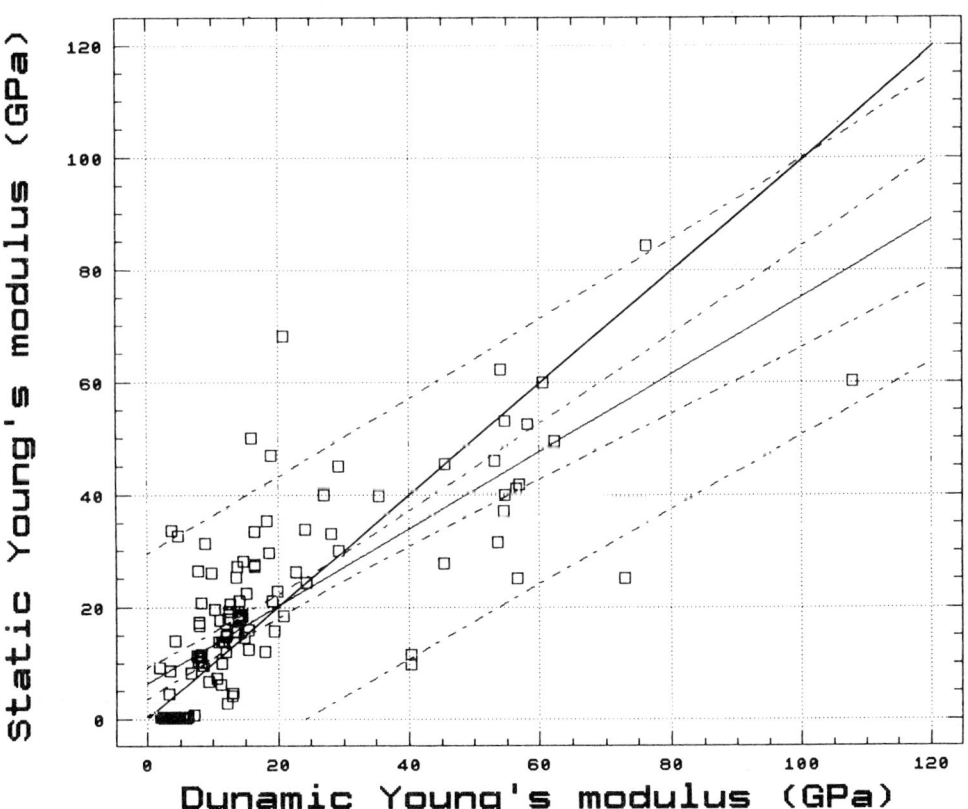

Fig. 4. Plot of static and dynamic values of Young's modulus showing 5 and 95 per cent confidence limits.

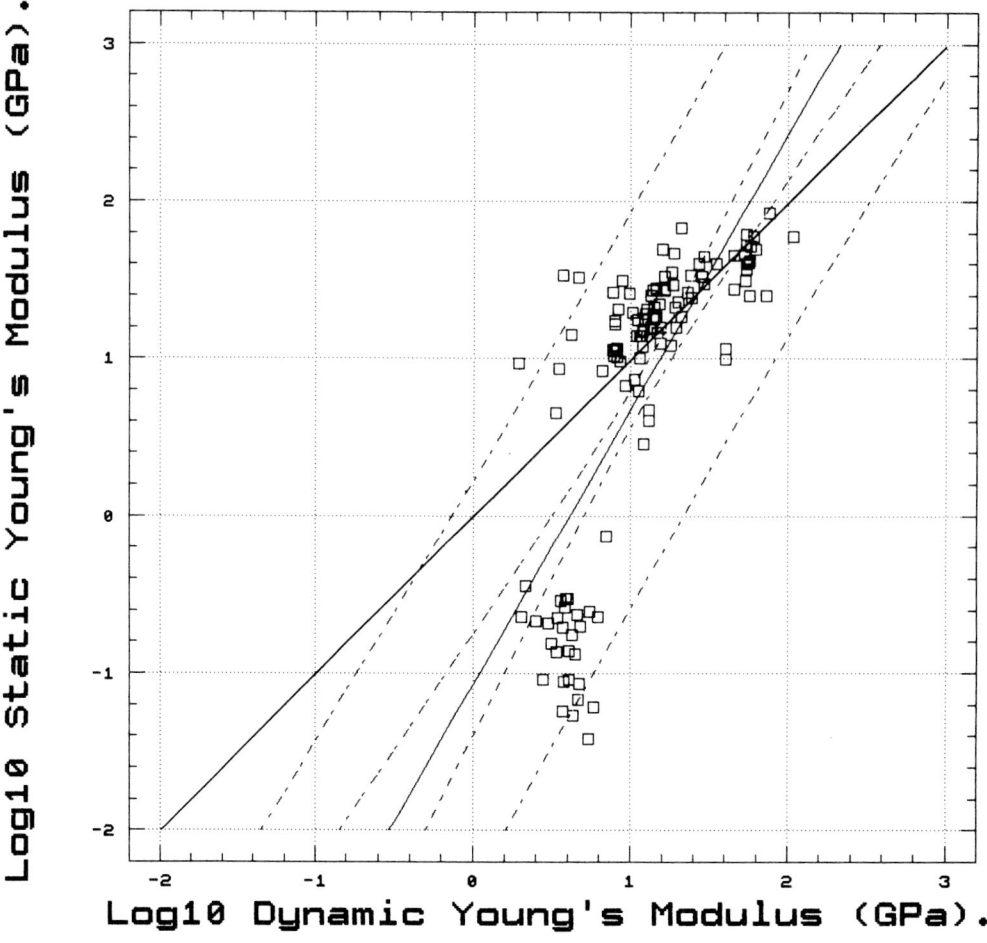

Fig. 5. Log_{10} plot of static and dynamic values of Young's modulus showing 5 and 95 per cent confidence limits.

A logarithmic plot of the same data is shown in Fig. 5 and the relationship is expressed as:

$$\log_{10} E_{ST} = 1.749 \log_{10} E_{DY} - 1.075 \quad (10)$$

with a correlation coefficient of 0.75.

A similar laboratory study for rocks from the Canadian Shield is reported by King (1983) where the relationship between the static and dynamic moduli is expressed by:

$$E_{ST} = 1.26 E_{DY} - 29.5 \quad (11)$$

with a correlation coefficient of 0.9.

In this equation if E_{DY} is less than 23.4 GPa then negative values of E_{ST} will be obtained.

Eissa & Kazi (1988) considered a wide range of rock types taken from the literature and developed the following relationship between the static and dynamic elastic moduli:

$$E_{ST} = 0.64 E_{DY} - 0.32 \quad (12)$$

with a correlation coefficient of 0.84.

The similarity between equations (9) and (12) which are based on the results of the present study and those of Eissa & Kazi (1988) respectively should be noted. In Fig. 5 the majority of the data points are grouped around the 1:1 line with the exception of a group of weak mudrocks. For these rocks the E_{DY}/E_{ST} ratio is around 100:1, which compares favourably with the observations of a ratio of 200:1 observed in soft alluvial materials by Imai & Yoshimura (1975). The difference between E_{DY} and E_{ST} is controlled by the low strain levels (10^{-5}) at which the dynamic values for Young's modulus are measured compared with strain levels of 10^{-3} to 10^{-2} measured in the static test. The different strain levels have little effect on the stronger, well cemented rocks since they may be considered to perform in a quasi-elastic manner and return to their original state after the loading is applied during the static test. The softer clays and mudrocks are subjected to plastic deformation on loading and clearly the strain level is a significant factor in the static test since the samples do not return to their original state. There will be some deformation of most rocks during static testing and only the very high strength rocks will approach the ideal elastic state so that the 1:1 ratio between the dynamic and static values of Young's modulus is observed.

It is also important to consider which value of Young's modulus is derived from the stress/strain curve shown in Fig. 2(b) following the definitions quoted above. The closer the stress/strain curve is to a linear relationship then the closer the various derived values of static Young's modulus will be to each other. Again this is very dependent on how closely the material under test approaches the elastic state and the quality of the laboratory testing programme; possible sources of error which might arise during static testing of rock samples are considered in some detail in Baldi *et al.* (1988).

In the determination of the dynamic values of Young's modulus, Christiansen's equation (5) was used for the computation of shear velocity in the softer clays and mudrocks, since no detectable shear wave arrival was observed in the sonic wave train. As mentioned above, Entwisle & McCann (1990) showed that, for these material types, Christensen's equation gives an over-estimate of the shear wave velocity and, hence, the computed value of Young's modulus will be correspondingly larger. As Entwisle & McCann (1990) pointed out, the equation was probably derived empirically from measurements on medium to high strength rocks commonly sampled in hydrocarbon exploration boreholes, rather than the weaker mudrocks.

The computation of the dynamic value of Young's modulus does require a value for the formation density. This information is derived from the gamma/gamma log and provided the correct calibration procedures are adopted the formation density can be determined to within $\pm 0.05 \text{ gm/cm}^3$.

The difference between the static and dynamic values of the elastic moduli have been examined by a number of authors. Simmons & Brace (1965) showed that the presence of cracks and cavities in the rock fabric, which tend to close up under pressure, has a marked effect on the static curves. King (1970) carried out both static and dynamic laboratory tests under pressure and at high stress levels and showed fairly close agreement between the two values of Young's modulus. Myung & Helander (1972) showed that good agreement was obtained between the dynamic elastic moduli computed from measurements of the compressional and shear wave velocity, both on samples under simulated formation pressure in the laboratory, and from the borehole sonic log in the field. As was noted above, the results from King (1983) indicate that only in the case of rocks with a Young's modulus in excess of 100 GPa will there be a 1:1 ratio between static and dynamic values. For weaker rocks his equation suggests that the static value of Young's modulus will be very much lower than the corresponding dynamic value, an observation that is also confirmed by the results of Eissa & Kazi (1988) and those of the present study.

Conclusion

The results of this study indicate that the dynamic values of Young's modulus derived from the full wave train sonic and formation density borehole logs are a reasonable estimate of the engineering properties of the rock mass provided that due regard is given to its lithology. The 1:1 relationship between the dynamic and static values observed in the study for harder, denser rocks, confirm the conclusions of King (1983), Eissa & Kazi (1988) and Van Heerden (1987). The results of the present study and other authors including Eissa & Kazi (1988) and Imai & Yoshimura (1978) indicate that in the softer mudrocks and alluvial materials the E_{DY}/E_{ST} ratio can be as high as 200:1 where the materials are subject to plastic rather than elastic deformation. The E_{DY}/E_{ST} ratio does vary between these two extreme values and, thus, the effects of lithology must be considered if the dynamic values are used in the place of the static values, since it would be desirable to know the E_{DY}/E_{ST} ratio for the range of lithologies present in a particular site investigation.

The key parameter in the calculation of the dynamic Young's modulus is the shear wave velocity. As was noted above, the identification of the shear wave pulse in the full wave train sonic log can be difficult particularly in the soft mudrocks. While empirical relationships between the shear wave velocity V_s and the compressional wave velocity V_p, such as Christensen's equation, may be used to predict V_s, such an approach may be subject to considerable error and will result in corresponding errors in the computation of the dynamic value of Young's modulus.

Similar problems can arise in the measurement of the static value of Young's modulus, which is dependent on the point on the stress/strain curve at which the value is measured. It is also necessary to consider the effects of sampling, particularly in mudrocks, where stress relief during drilling and swelling may result in significant changes in their geotechnical properties.

However, once a relationship has been established between the dynamic and static values of Young's modulus for each lithological unit within the borehole being logged, the dynamic values derived from well logging can be used in the following ways.

1. In situ static values of Young's modulus can be estimated from the dynamic values from unfractured sections of the rock mass.

2. An assessment can be made of the relevance of the laboratory testing programme on a suite of borehole core samples to the in situ rock mass engineering properties by comparing the in situ and laboratory dynamic properties.

3. Using the standard geophysical well logging methods an assessment can be made of the deformation characteristics of the rock mass on a continuous basis and the effects of fracturing can also be estimated, since the elastic moduli determined in unfractured sections of the borehole will be a maximum.

This paper is published by permission of Director, British Geological Survey (NERC).

References

ANDERSON, D. G. & STOKOE, K. H. II 1978. Shear modulus: A time-dependent soil property. *Dynamic Geotechnical Testing*. ASTM STP 654, American Society for Testing and Materials, 66–90.

ANON 1981. *Code of Practice for Site Investigation: BS 5930*. British Standards Institution.

ATKINSON, J. H., COOP, M. R., STALLEBRASS, S. E. & VIGGIANI, G. 1992. Measurement of stiffness of soils and weak rocks in laboratory tests. *In*: CRIPPS, J. C. & MOON, C. F. (eds) *Engineering Geology of Weak Rock*. Balkema, Rotterdam.

BALDI, G., HIGHT, D. W. & THOMAS, G. E. 1988. *Advanced Triaxial Testing of Soil and Rock*. ASTM STP 977. DONOGHE, R. T., CHANEY, R. C. & SILVER, M. L. (eds). American Society for Testing and Materials, 219–263.

BUCHAN, S., MCCANN, D. M. & TAYLOR SMITH, W. 1972. Relations between the acoustic and geotechnical properties of marine sediments. *Quarterly Journal of Engineering Geology*, **5**, 265–284.

BURLAND, J. B. & LORD, J. A. 1970. The load deformation behaviour of Middle Chalk at Mundford, Norfolk: a comparison between full-scale performance and in situ and laboratory measurements. *Proceedings of a Conference on In Situ Investigations in Soils and Rocks*. British Geotechnical Society, London, 3–15.

CASTAGNA, J. P., BATZLE, M. L. & EASTWOOD, R. L. 1985. Relationship between Compressional Wave and Shear-Wave Velocities in Elastic Silicate Rocks. *Geophysics*, **50**, 571–581.

CHENG, C. H. & TOKSÖZ, M. N. 1980. Modelling of Full Wave Acoustic Logs. *Transactions of SPWLA 21st Logging Symposium*, Paper J.

EISSA, E. A. & KAZI, A. 1988. Relation between Static and Dynamic Young's Moduli of Rocks. *International Journal of Rock Mechanics, Mining Science and Geomechanical Abstracts*. **25**, 479–482.

ENTWISLE, D. C. & MCCANN, D. M. 1990. An assessment of the use of Christensen's equation for the prediction of shear wave velocity and engineering parameters. *In*: HURST, A., LOVELL, M. A. & MORTON, A. C. (eds) *Geological Applications of Wireline Logs*. Geological Society, London, Special Publication, **48**, 347–354.

GEYER, R. L. & MYUNG, J. I. 1970. The 3D velocity log. A tool for in situ determination of the elastic moduli of rocks. *Proceedings of the 12th Symposium on Rock Mechanics, University of Missouri-Rolla, Missouri*. Society of Mining Engineers, AIME.

GOODMAN, R. E. 1980. *Introduction to Rock Mechanics*. Wiley, New York.

HOBBS, N. E. 1973. Effects of non-linearity on the prediction of settlements of foundations on rock. *Quarterly Journal of Engineering Geology*, **6**, 153–158.

IDE, J. M. 1936. Comparison of statically and dynamically determined Young's modulus of rocks. *Proceedings of the National Academy of Science*, A. **22**, 2, 81–92.

IMAI, T. & YOSHIMURA, M. 1975. *The Relationship of Mechanical Properties of Solid to P- and S-Wave Velocities for Soil Ground in Japan*. Oyo Technical Note TN-07, Oyo Corporation of Japan.

INTERNATIONAL SOCIETY OF ROCK MECHANICS 1981. *Rock Characterisation, Testing and Monitoring: ISRM Suggested Methods*. BROWN, E. T. (ed.) Pergamon, Oxford.

JARDINE, R. J., POTTS, D. M., FOURIE, A. B. & BURLAND, J. B. 1986. Studies of the influence of non-linear stress–strain characteristics in soil-structure interaction. *Geotechnique*, **36**, 377–396.

KING, M. S. 1970. Static and dynamic moduli of rocks under pressure. *In*: SOMERTON, W. H. (ed.) *Proceedings of the 11th Symposium on Rock Mechanics, California*. American Institute of Mining Engineers, New York, 329–351.

—— 1983. Static and Dynamic Elastic Properties of Rocks from the Canadian Shield. *International*

Journal of Rock Mechanics, Mining Science and Geomechanical Abstracts, **20**, 237–241.

LANE, R. G. T. 1964. Rock foundations: Diagnosis of mechanical properties and treatment. *Transactions of the International Conference on Large Dams.* **1**, 141–146.

LAMA, R. D. & VUTUKURI, V. S. 1978. *Handbook on Mechanical Properties of Rocks. Trans. Tech. Publications*, 196–220.

LINGLE, R. & JONES, A. H. 1977. Comparison of Log and Laboratory Measured P-Wave and S-Wave Velocities. *Transactions of SPWLA 18th Annual Logging Symposium*, Paper N.

MARSLAND, A. 1971. Large in situ tests to measure the properties of stiff fissured clays. *In: Proceedings of the 1st Australian–New Zealand Conference on Geomechanics, Melbourne.* Vol. 1, 180–189.

McCANN, D. M., CULSHAW, M. G. & NORTHMORE, K. J. 1990. Rock mass assessment from seismic measurements. *In:* BELL, F. G., CULSHAW, M. G., CRIPPS, J. C. & COFFEY, J. R. (eds) *Field Testing in Engineering Geology.* Geological Society, London, Engineering Geology Special Publication, **6**, 257–266.

MYUNG, J. I. & HELANDER, D. P. 1972. Correlation of elastic moduli dynamically measured by in situ and laboratory techniques. *Transactions of*

SPWLA 13th Annual Logging Symposium, Paper H.

O'BRIEN, P. N. S. 1953. Velocity dispersion of seismic waves. *Geophysical Prospecting*, **19**, 1–12.

SIEFRIED, R. W. & CASTAGNA, J. P. 1982. Full Waveform Sonic Logging Techniques. *Transactions of SPWLA 23rd Annual Logging Symposium*, Paper I.

SIMMONS, G. & BRACE, W. F. 1965. Comparison of static and dynamic measurements of the compressibility of rocks. *Journal of Geophysical Research*, **70**, 5649–5656.

SUTHERLAND, R. B. 1962. Some dynamic and static properties of rocks. *In:* FAIRHURST, C. (ed.) *Proceedings of the 5th Symposium on Rock Mechanics, Minnesota*, Pergamon, New York, 473–490.

VAN HEERDEN, W. L. 1987. General Relations between Static and Dynamic Moduli of Rocks. *International Journal of Rock Mechanics, Mining Science and Geomechanical Abstracts*, **24**, 381–385.

WINKLER, K. W. 1986. Estimates of Velocity Dispersion between Seismic and Ultrasonic Frequencies. *Geophysics*, **51**, 183–189.

WYLLIE, M. R. J., GREGORY, A. R. & GARDNER, L. W. 1956. Elastic wave velocities in heterogeneous and porous media, *Geophysics*, **21**, 41–70.

Fluid salinity and dynamics in the North Sea and Haltenbanken basins derived from well log data

KJETIL GRAN, KNUT BJØRLYKKE & PER AAGAARD

Department of Geology, University of Oslo, P.O. Box 1047, Blindern, 0316 Oslo 3, Norway

Abstract. The salinities of porewaters in sandstones and carbonate reservoir rocks from the North Sea and Haltenbanken basins have been calculated using resistivity logs. Applying Archie's equation, hydrocarbon containing intervals were avoided and a 100% water saturation assumed in the calculations of the formation water resistivity. The salinities obtained vary from 20 000 to 300 000 ppm NaCl equivalents. The highest salinities, which approach saturation with halite, are found in the Southern North Sea, close to the Upper Permian evaporite sediments. There is a marked upwards decrease in the salinity values away from the evaporites. The salinities seem to be a function of the vertical distance to the evaporites and are independent of stratigraphic horizon. In the Northern North Sea, where Permian salt is absent, the salinity values are much lower (20 000–50 000 ppm NaCl eq.). This salinity stratification strongly indicates that large scale vertical mixing of porewater by convection or compaction driven pore water flow has not taken place. We can therefore conclude that saline pore water has not been transported from the deeper parts of the basin to the shallower parts. This also implies that there is no transport of carbonates or silicates in solution over the same distances. Calculation of salinities from well logs is subject to considerable uncertainties. However, the values obtained in this study agree well with reported values from formation water analyses. Mapping the distribution of salinity in the pore waters of sedimentary basins is very important and the information obtained can be used to constrain our models for porewater flow.

The degree of vertical mixing of pore waters and transport of solids in solution are important unknown factors pertinent to modelling of diagenetic reactions in sedimentary basins. Leaching of minerals or precipitation of cement may be used as indicators of fluid flow. However, such evidence is often not conclusive because the origin of undersaturated or supersaturated pore water can rarely be uniquely determined. At elevated temperatures ($>100°C$) relatively rapid reaction rates limit the distance pore water can flow before reaching equilibrium with the mineral phases (Giles 1987). Chlorides, however, which usually represent the main proportion of solids in pore waters, are not in equilibrium with the carbonate and silicate minerals, as there is no sink for chloride outside the evaporite deposits. In general, saline pore waters overlying evaporites subside almost at the same rate as the basin itself (de Caritat 1989), creating a layering in the pore water chemistry. The concentration of chlorides therefore record mass transport away from the evaporites, either by advection or diffusion. Large scale convective pore water flow would cause a mixing of pore water and a homogenization of the salinities. The presence of distinct salinity versus depth trends therefore indicates that convective pore water flow has not

taken place. Increasing salinities with depth cause an increase in pore water density with increasing depth, reducing or eliminating the driving force for convection. A salinity gradient of 30 000 ppm NaCl eq./km is sufficient to compensate for the thermal expansion of water (Bjørlykke *et al.* 1988), so that the fluid density is not reduced with increasing temperature.

Analyses of samples of formation water from a large number of wells in the North Sea and Haltenbanken basins (Egeberg & Aagaard 1989) show salinity versus depth trends, in addition to a crude stratification of the pore water with respect to other components. They showed that the low salinities often observed in shallow reservoirs were frequently associated with low $\delta^{18}O$ values, suggesting that the pore water represented modified meteoric water. The number of reliable samples of formation waters is however limited and there are rarely several samples from different depths in the same well.

The present investigation covers the North Sea and Haltenbanken basins (Fig. 1), with data from more than 90 wells. The majority of the wells examined are located in the Southern North Sea (56–59°N), penetrating the Permian Zechstein Group sediments (Fig. 2). In the Haltenbanken basin, very few wells reach the Trias-

From HURST, A., GRIFFITHS, C. M. & WORTHINGTON, P. F. (eds), 1992,
Geological Applications of Wireline Logs II. Geological Society Special Publication No. 65, pp. 327–338.

327

sic evaporites. Wells in the Northern North Sea (59–62°N) have been chosen based on the total depth, independently of stratigraphy.

The purpose of the present paper is to demonstrate that log derived salinity estimates can represent an important supplement to formation water analyses, and that the salinity trends obtained are diagnostic of pore water flow within sedimentary basins.

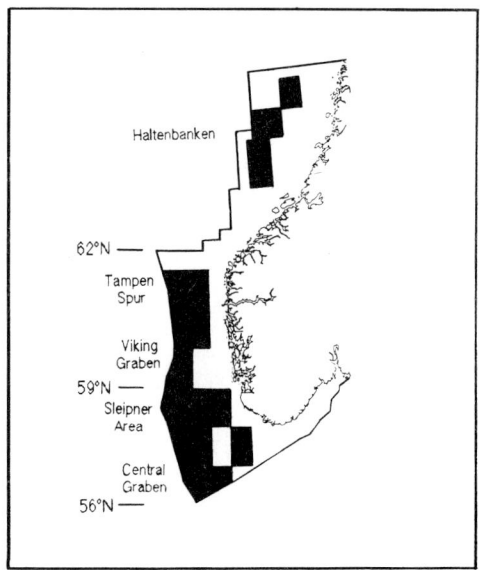

Fig. 1. Outline of the North Sea and Haltenbanken study area.

Method of analysis

The formation water resistivity is estimated using Archie's equation for water saturation:

$$S_w^n = FR_w/R_t \qquad (1)$$

where S_w is the water saturation of uninvaded zone, n is the saturation exponent, F is the formation factor, R_w is the formation water resistivity and R_t is the formation resistivity (uninvaded zone).

Sequences containing hydrocarbons have been avoided and a 100% water saturation assumed in the intervals of interest. R_t, true formation resistivity, is taken directly from the deep-reading induction log. With saline formation fluid and saline drilling mud, R_t should ideally be taken from the deep laterolog. However, the laterolog is generally run over a very limited depth interval, usually the anticipated reservoir.

A comparison between R_t values from both the laterolog and induction log shows that the use of the induction log alone is justified in formations with minimum mud invasion.

The general relationship for the formation factor is:

$$F = a/\phi^m \qquad (2)$$

where ϕ is the effective porosity and 'a' and 'm' are constants which vary with lithology. In this investigation the most common coefficients and exponents have been applied (Asquith 1982; Rider 1986). According to Winsauer *et al.* (1952) these values will yield acceptable formation factors within the range of porosities normally encountered in sedimentary rocks.

Lithology	Carbonates	Unconsoli-dated sandstones	Consolidated sandstones
a	1	0.62	0.81
m	2	2.15	2

Porosity is calculated from the density logs;

$$\phi_{den} = \frac{\rho_{ma} - \rho_b}{\rho_{ma} - \rho_f} \qquad (3)$$

where: ϕ_{den} is the porosity, ρ_{ma} is the matrix density, ρ_b is the bulk density (from log) and ρ_f is the fluid density.

Extensively caved sections have been avoided and corrections due to variable borehole size have not been made in the porosity calculations. Standard values have been chosen for matrix density and fluid density (Asquith 1982; Rider 1986). Avoiding shaly formations, a matrix density of 2.65 g/cm³ for sandstones and 2.71 g/cm³ for limestones has been applied. Assuming saline formation waters, a fluid density of 1.10 g/cm³ is used in the calculations, although this varies as a function of salinity, temperature and pressure (Schlumberger 1989). Data on fluid densities from the North Sea and Haltenbanken (Egeberg & Aagaard 1989) show a range of 1.01 to 1.19 g/cm³, with a salinity variation from less than 20 000 ppm to more than 250 000 ppm NaCl equivalents. Using 1.10 g/cm³ throughout will therefore not lead to significant errors in the calculations. Archie's equation is only valid for clean, shale-free formations. Manheim & Paull (1981) suggested a formula for direct determination of salinity from calculated R_w values in shaly sandstones. However, in the North Sea

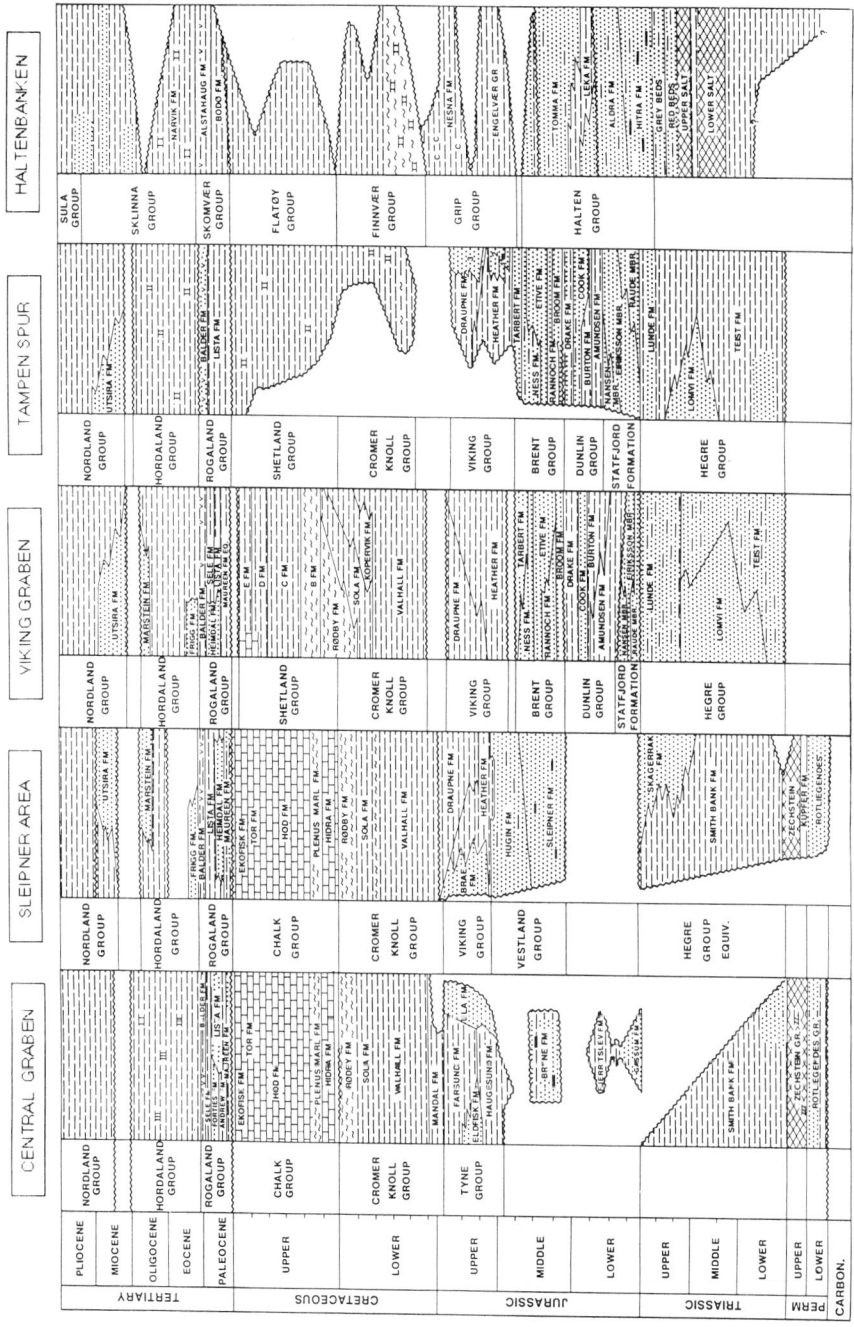

Fig. 2. Lithostratigraphic setting in the North Sea and Haltenbanken basins. Compiled from Campbell & Ormaasen (1987).

and Haltenbanken basins, the suggested equation gives unrealistically low formation water resistivities.

Formation temperatures have been estimated from bottom hole temperature measurements reported on log headings, using the corrections suggested by Carstens & Finstad (1981). The temperature gradients used in this study are 36.0°C/km in the Southern North Sea and 31.0°C/km in the Northern North Sea and Haltenbanken.

The salinity of the formation waters, given as ppm NaCl equivalents, can then be determined directly from charts (e.g. Schlumberger 1986), using corrected formation temperatures and R_w values averaged over the interval of interest.

In situ salinity gradients

The investigated basins have been separated into three areas: Southern North Sea (56–59°N), Northern North Sea (59–62°N) and Haltenbanken (Fig. 1). The estimated formation water salinities are shown both as individual well profiles and as a function of burial depth for all the wells within each area. A salinity value is plotted

for each stratigraphic interval investigated. This represents an average value for the whole sequence, providing the lithology is generally uniform throughout the interval and not several hundred metres in thickness. For very thick intervals, the formation water salinity is represented by two or more values. In sequences with interbedded shales and sandstones, the formation water resistivity has been estimated only in the shale-free sandstone beds. The salinity is then given either as an average over the whole interbedded sequence, or if the sandstone beds are sufficiently thick, as individual values representing each unit. Regression lines are shown on all the plots, and will be discussed later. Additional salinity data are taken from formation water anlayses reported by Egeberg & Aagaard (1989).

Southern North Sea (56–59°N)

Well profiles from the Southern North Sea (Fig. 3) show a distinct salinity increase towards the Upper Permian Zechstein Group evaporites. In well A, the salinity concentration approaches halite saturation close to the Permian deposits.

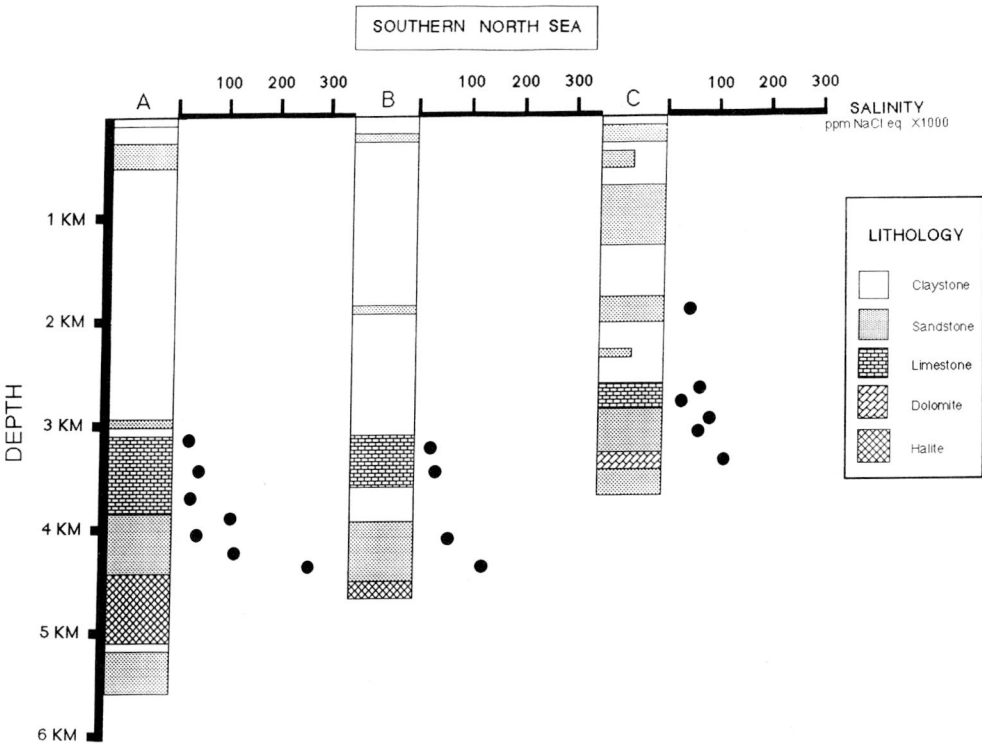

Fig. 3. Well profiles showing salinity versus burial gradients in the Southern North Sea. Positions are outlined in Fig. 1. (Well A: 2/1–7; well B: 7/12–5; well C: 16/1–3.)

This tendency decreases towards the Sleipner area (Well C). Here the Zechstein Group sediments consist mainly of anhydrite and dolomite, with only minor halite.

Formation water salinity as a function of burial depth for all the Southern North Sea wells (Fig. 4) shows a very large scatter. The values range from less than 25 000 ppm to 300 000 ppm NaCl equivalents.

Northern North Sea (59–62° N)

The formation water salinities in the Northern North Sea indicate a totally different geochemical development than observed in the Southern North Sea. Well profiles from the Viking Graben and the Tampen Spur area (Fig. 5) show only minor compositional changes with increasing depth. Salinity values versus depth from all the studied wells in the Northern North Sea (Fig. 6) confirm this pattern for the whole area. The results give salinities of about 50 000 ppm NaCl

equivalents or less. This observation is further strengthened by the formation water data from Egeberg & Aagaard (1989).

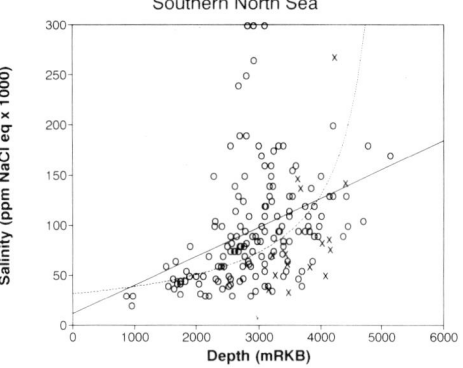

Fig. 4. Formation water salinity versus burial depth for the Southern North Sea. Salinity values from this study are shown as circles, while data from Egeberg & Aagaard (1989) are indicated with ×. Regression data are given in Table 1. Linear regression line: ——. Multiplicative regression line: . . . . .

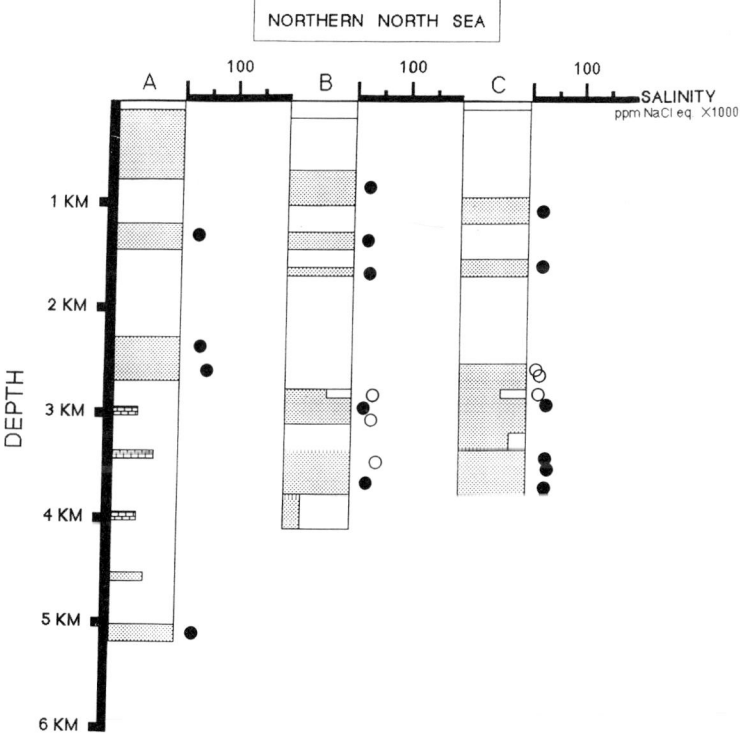

Fig. 5. Well profiles showing salinity gradients in the Northern North Sea. Salinity values from this study are shown as infilled circles while data from Egeberg & Aagaard (1989) are given as open circles. Lithology explanation is shown in Fig. 3. Positions are indicated in Fig. 1. (Well A: 24/12–2; well B: 30/6–15; well C: 34/7–16.)

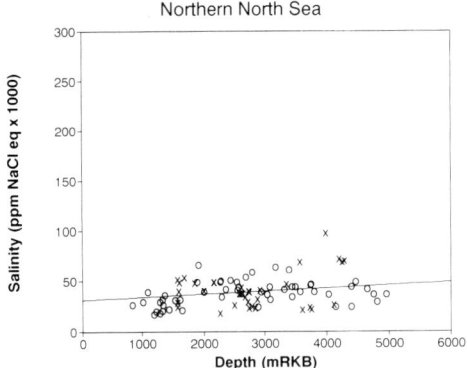

Fig. 6. Formation water salinities versus burial depth for the Northern North Sea. Salinity values from this study are shown as circles, while data from Egeberg & Aagaard (1989) are indicated with ×. Regression data are given in Table 1. Linear regression line: ——.

Haltenbanken

In the Haltenbanken area, concentration gradients again seem to increase with burial depth (Fig. 7), reflecting a similar development as seen in the Southern North Sea. The general lithostratigraphic column for the Haltenbanken sediments (Fig. 2) shows two periods of evaporite

deposition in the Triassic. Unfortunately, only a few wells in this area do actually penetrate the evaporites, and the salinity never approaches halite saturation.

The development of the formation water salinity versus burial depth in all the Haltenbanken wells (Fig. 8) also resembles that observed in the Southern North Sea. Salinity values here are much less scattered indicating a more gradual increase in formation water salinity with burial depth.

Discussion

The salinity distribution observed in the North Sea and Haltenbanken basins has implications for both the evolution of formation waters and fluid flow. Although log-derived salinity values are generally viewed with scepticism (Society of Professional Well Log Analysts 1989), extensive regional investigations such as the present one can provide valuable information on the pore water geochemistry and set constraints on pore water flow models in relation to diagenetic mineral reactions. Depth profiles have been presented by Dickey (1969) and Hanor (1979, 1984) for several U.S. sedimentary basins, showing generally increasing salinities with increasing burial. In the Southern North Sea and Halten-

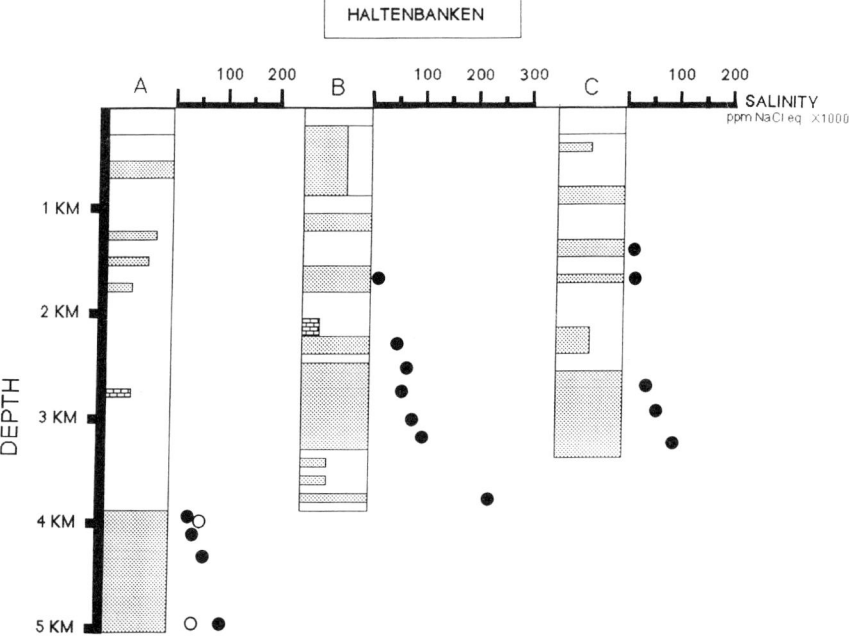

Fig. 7. Well profiles showing salinity gradients in the Haltenbanken area. Salinity values from this study are shown as infilled circles, while data from Egeberg & Aagaard (1989) are given as open circles. Lithology explanation is shown in Fig. 3. (Well A: 6507/12–1; well B: 6610/7–1; well C: 6406/3–1.)

Table 1. *Regression analysis on salinity values versus depth*

Area	Regression type	Number of samples	Correlation coefficient	r^2
Southern North Sea	Linear	156	0.39	15.0%
	Multiplicative	156	0.54	29.3%
Northern North Sea	Linear	51	0.29	8.8%
Haltenbanken	Linear	61	0.78	60.8%
	Multiplicative	61	0.79	62.3%

banken, well profiles (Figs 3 & 7) also show increasing salinity with depth. However, regression analysis (Table 1) reveals significant differences between these areas. In the Haltenbanken area (Fig. 8), both linear and multiplicative regression give correlation coefficients of about 0.8, suggesting a close relationship between salinity and burial depth. Similar analyses for the Southern North Sea (Fig. 4) give much lower correlation coefficients. Rather than totally dismissing a relationship between salinity and depth, the reduced correlation coefficients probably reflect differences in thickness and movement of the Upper Permian deposits of the Southern North Sea compared to the Triassic evaporites of the Haltenbanken area. In the Southern North Sea, the thickness of the Zechstein Group evaporites can exceed 2000 metres in the Central Graben (Taylor 1984). With extensive halokinesis, lateral transport of dissolved halite from nearby salt domes can explain the large scatter in salinity values in the Southern North Sea. A similar salinity development has been reported from the Gulf Coast (Hanor & Sassen 1990). In the Haltenbanken area, the thickness of the evaporites is much smaller and diapirs have not been reported, giving a more distinct salinity versus depth relationship in these sediments.

The difference in salinity distribution in the Southern and Northern North Sea (Figs 4 & 6) is striking. It indicates a direct relation between the presence of evaporite deposits and the degree of change in the concentration of total dissolved solids with increased burial depth. In the Northern North Sea, significant increases in salinity have not been observed. Instead, a slight decrease in the salinity gradient is sometimes observed in the deepest parts. Linear regression (Table 1) shows clearly that the salinity development in the Northern North Sea is totally independent of burial depth. With the inherent uncertainties in log evaluation (Theys 1988), only a moderate confidence level would be given to each of the salinity values separately. However, the salinities reported from formation water analyses (Egeberg & Aagaard 1989) show an excellent correlation with the log data.

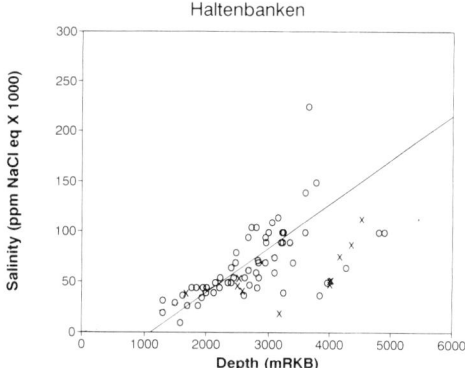

Fig. 8. Formation water salinities versus burial depth for the Haltenbanken area. Salinity values from this study are shown as circles, while data from Egeberg & Aagaard (1989) are indicated with ×. Regression data are given in Table 1. Linear regression line: ——. Multiplicative regression line: .....

The use of oxygen and hydrogen isotope geochemistry (Clayton *et al.* 1966; Hitchon & Friedman 1969; Kharaka & Carothers 1986; Knauth & Beeunas 1986) has revealed that formation brines are often mixed with meteoric water. Egeberg & Aagaard (1989) found that these formation waters were mixtures of an original formation brine and meteoric water recharged at a much later stage. A comparison of the reported formation water salinities and $\delta^{18}O$ values are shown in Fig. 9. The oxygen isotope data are from sandstone reservoirs covering formation temperatures from 55 to 160°C, over a depth interval of 1550 to 4400 metres of burial. The plot shows that the oxygen isotope signatures gradually get more positive with increasing salinity. Evidence of mixing of the original brine with meteoric water in both the Southern and Northern North Sea indicates that the low salinities in the Northern North Sea are not a product of late stage influx of meteoric water. Instead the low salinities are believed to be the result of absence of an evaporite source to provide the high concentration of chlorides seen in the Southern North Sea and the Haltenbanken areas.

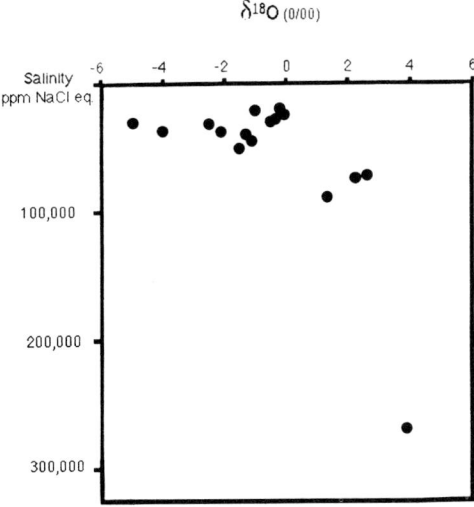

Fig. 9. $\delta^{18}O$ versus formation water salinity in the North Sea, indicating heavier oxygen isotopes with increasing salinity. Data compiled from Egeberg & Aagaard (1989).

Kharaka *et al.* (1985) and Morton & Land (1987) suggested that local areas of low pore water salinities in the U.S. Gulf Coast could be caused by water–rock interactions during mineral diagenesis, together with overpressuring. In this study, the results indicate lowest salinities in the shallow parts (<2000–2500 metres), where diagenesis involving water rich minerals (e.g. smectite → illite) are least extensive. Secondly, the observed independence of salinity with respect to specific lithostratigraphic levels suggests that overpressured sequences do not control the salinity development in these sediments. Dewatering of clay minerals may be a significant factor in the deeply buried sediments in the Viking Graben, where already low salinities seem to be further reduced compared to the overlying sequences (Fig. 5, Well A). In the Southern North Sea or Haltenbanken areas, patterns like this have not been observed. In addition to providing information on the regional changes in the salinity pattern, the distinct concentration differences between the Southern and Northern North Sea could also be used in

an attempt to limit the northern extent of the evaporite deposits of the Permian Zechstein Group. According to Taylor (1984), the distribution of the Zechstein Group sediments north of latitude 59°N is not well known, but Ziegler (1982) and Doré & Gage (1987) reported that Permian deposits were found well into the Northern North Sea. Wells in the Sleipner Field area (Fig. 1) have only very thin sequences of Zechstein Group sediments, generally consisting of anhydrite and/or dolomite, with halite mostly absent or present only in very small amounts. The lack of any major salinity increases between 59°N to 62°N suggests that the northern limit of the Zechstein evaporites is close to 59°N.

The actual origin and evolution of highly saline formation waters has been the topic of much debate. Early this century, several theories were suggested for the origin of formation brines, such as burial of marine waters (Lane 1908), subsurface dissolution and diffusion (Richardson 1917), and subsurface evaporation (Mills & Wells 1919). Russel (1933) dismissed evaporation as a brine-forming mechanism and instead introduced infiltration of subaerial brines and reverse osmosis as possible processes. According to Hanor (1987a), membrane filtration, burial of subaerially produced brines, and subsurface dissolution of evaporites are currently viewed as the possible mechanisms for producing the salinity concentrations found in several of the sedimentary basins around the world. Well documented evidence for dissolution of evaporites as a mechanism for this increase has been presented for the U.S. Atlantic Coast (Manheim & Horn 1968), the Gulf Coast area (Carpenter 1978; Land & Prezbindowski 1981; Hanor & Bailey 1983) and the Western Canada Basin in Alberta (Hitchon *et al.* 1971; Spencer 1987). With the solubility of NaCl being only slightly affected by increasing pressure and temperature (Braitsch 1971), suitable conditions for halite dissolution should therefore also exist in the North Sea and Haltenbanken basins. Ranganathan & Hanor (1987), using numerical modelling, found that diffusion of NaCl could contribute to the salinity pattern found in the Gulf Coast area. However, according to their model, diffusion alone could not be the only transport mechanism for solids. Convection systems, either set up by thermal or density differences, have often been suggested to induce fluid flow (Wood & Hewett 1982, 1984; Haszeldine *et al.* 1984; Blanchard & Sharp 1985; Rabinowicz *et al.* 1985; Evans & Nunn 1989). Hanor (1987b) and Ranganathan & Hanor (1988) proposed local convection associated with salt diapirs. However, according to Bjørlykke *et al.* (1988),

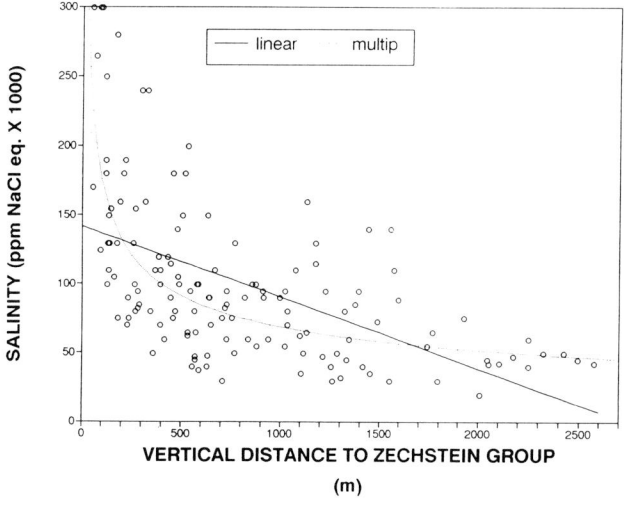

DEVIATION FROM MULTIPLICATIVE MODEL

Fig. 10. (a) Salinity versus vertical distance to the Permian Zechstein Group evaporite sediments in the Southern North Sea. Regression data are given in Table 2. (b) Deviation ($\pm$) from multiplicative regression line.

convection will not be a major factor in long-distance transport of dissolved solids. The observed salinity gradients produce a density increase with depth which removes the driving force for convective flow. In addition, the thickness of a uniformly porous layer needed to set up a convection cell will have to exceed 100–300 metres, which is rarely observed in any sedimentary basin. The presence of even very thin layers of low permeability will totally block the convection system. If large-scale convection were active in the fluid flow in the North Sea and Haltenbanken, gradients originally present in the salinity concentration would be erased and the for-

mation waters would attain a uniform geochemistry. The large regional variations seen in the salinity in the North Sea indicate that fluid transport in the form of convection cells cannot be of great importance in this area. The presence of distinct concentration gradients in the Haltenbanken basin suggests that similar limits to fluid flow conditions exist in this area also.

Compaction driven advective flow probably does not control the salinity distribution either. From Figs 4 & 8, the formation waters in sediments buried shallower than 2000–2500 metres generally show low salinity values (< 50 000 ppm NaCl eq.). This is similar to the values generally

Table 2. *Regression analysis on salinity versus vertical distance to Zechstein Group evaporites*

Area	Regression type	Number of samples	Correlation coefficient	r^2
Southern North Sea	Linear	161	−0.54	29.2%
	Multiplicative	161	−0.67	44.9%

found throughout the Northern North Sea (Fig. 6), which suggests that highly saline waters have not been transported upwards through advection. In addition, compactional water available from the Permian sediments is limited and will not significantly influence the salinities in the overlying sequences. This also limits the transport of other dissolved solids, e.g. calcium and silica, important for mineral diagenesis and cementation of the reservoir rocks.

Instead, it is proposed that diffusion is the main mechanism for transport of solids in these sediments. In the Southern North Sea, formation water salinity is seen as a function of vertical distance to the Zechstein Group (Fig. 10a) and is rapidly reduced in the sediments immediately overlying the evaporites. With some exceptions, sequences with a vertical distance of more than 500 metres from the evaporites generally have fairly low salinities. Regression analysis (Table 2) gives, at best, a correlation coefficient of −0.67, explaining 44.9% of the data. However, the validity of the diffusion model can be demonstrated by a plot of the deviation from the multiplicative regression line (Fig. 10b). Negative Δ-salinity values are generally low and decrease with increasing vertical distance (this is a function of measuring the deviation vertically from the regression line). Deviations indicating possible influence of advective porewater flow are found at distances of 500 metres and more, where a few Δ-salinity values of 50 000 to 100 000 ppm NaCl equivalents are observed. These deviation peaks are most likely the result of lateral transport from local salt domes rather than flow from deeply buried evaporites. The presence of significant upward fluid flow would result in considerably higher salinity values in the shallow sequences and larger positive deviations from the regression line.

With diffusion generally being a very slow process, distinct concentration gradients will occur in the sediments containing evaporite deposits, as found in the Southern North Sea and Haltenbanken. The absence of homogenization of formation water geochemistry or long distance transport of highly saline pore water, either vertically or horizontally, explain the

marked differences in salinity development between the Southern and Northern North Sea.

Conclusions

The distribution of salinity in pore waters of sedimentary basins is very important in terms of constraining pore water flow. Usually, the number of available formation water samples is too small to provide a detailed regional control of saline pore waters. This study presents log derived salinities from wells located in the North Sea and Haltenbanken basins. Salinity values estimated from well logs are generally considered to be inaccurate. However, the log derived salinities agree well with results from formation water samples from these basins. Log evaluation appears therefore to be a valuable supplement to formation water samples in analysing the pore water salinity.

Salinity values are very widely scattered in the Southern North Sea, ranging from 20 000 to 300 000 ppm NaCl equivalents, and having only a weak relationship with burial depth. In the Haltenbanken area, salinity values are less scattered, and show a good correlation with depth. Fewer high salinity values are recorded, however, since only a few wells reach the Triassic evaporites. The difference in salinity distribution between these areas is probably caused by the extensive halokinesis in the Upper Permian evaporites of the Southern North Sea, giving a greater range in salinity.

In the Northern North Sea, the salinities do not show any increase with burial depth. Generally, the salinity values are less than 50 000 ppm NaCl equivalents overall. This may be explained by the absence of evaporites in this region. The low salinities suggest that membrane filtration cannot produce salinity trends similar to those observed where evaporites are present.

The salinity trends in the Southern North Sea and Haltenbanken are evidence against large-scale vertical mixing of pore fluids in these basins. Both large scale convective pore water flow and compaction driven flow from the deeply buried sediments would have transported saline pore water into the shallower parts of the basins. Since there is no sink for Cl^- outside the evaporites, the salinity distribution does not only reflect the present flow pattern, but can also be used to constrain earlier patterns of pore water flow. Salinity shows an inverse correlation with vertical distance to the underlying Permian evaporites in the Southern North Sea. Within 300–400 metres of the evaporites the salinity is controlled mainly by diffusion. The advective component is small in the deepest part of the basin, and the salinity distribution demonstrates that upward pore water flow driven by compaction is slower than the basin subsidence, as argued by Bjørlykke *et al.* (1988) and de Caritat (1989).

This study is part of the German–Norwegian Research and Development Programme on Basin Analysis and Reservoir Studies. It is supported by VISTA, a research cooperation between the Norwegian Academy of Science and Letters and Den Norske Stats Oljeselskap (Statoil). We thank Conoco, Norsk Hydro, Saga, Shell and Statoil for providing the well logs. K. Gran wishes to thank T. Nedkvitne and M. Ramm of the Department of Geology for discussions and assistance during the preparation of the manuscript. C. M. Griffiths and two anonymous reviewers are also thanked for their constructive criticism and comments.

References

ASQUITH, G. B. 1982. *Basic Well Log Analysis for Geologists.* Methods in Exploration Series No. 3. American Association of Petroleum Geologists, Tulsa.

BJØRLYKKE, K., MO, A. & PALM, E. 1988. Modelling of thermal convection in sedimentary basins and its relevance to diagenetic reactions. *Marine and Petroleum Geology*, **5**, 338–351.

BLANCHARD, P. E. & SHARP, J. M. 1985. Possible free convection in thick Gulf Coast sandstone sequences. *Transactions of the Southwest Section of American Association of Petroleum Geologists*, 6–12.

BRAITSCH, O. 1971. *Salt Deposits. Their Origin and Composition.* Springer, Berlin.

CAMPBELL, C. J. & ORMAASEN, E. 1987. The discovery of oil and gas in Norway: an historical synopsis. *In*: SPENCER A. M. *et al.* (eds) *Geology of the Norwegian Oil and Gas Fields.* Graham & Trotman, London, 1–37.

CARPENTER, A. B. 1978. Origin and chemical evolution of brines in sedimentary basins. *Oklahoma Geological Survey Circular*, **79**, 60–77.

CARSTENS, H. & FINSTAD, K. G. 1981. Geothermal gradients of the Northern North Sea Basin, 59–62°N. *In*: ILLING, L. V. & HOBSON, G. D. (eds) *Petroleum Geology of the Continental Shelf of North-West Europe.* Heyden, London, 152–161.

CLAYTON, R. N., FRIEDMAN, I., GRAF, D. L., MAYEDA, T. K., MEENTS, W. F. & SHIMP, N. F. 1966. The origin of saline formation waters. 1. Isotopic composition. *Journal of Geophysical Research*, **71**, 3869–3882

de CARITAT, P., 1989. Note on the maximum upward migration of pore water in response to sediment compaction. *Sedimentary Geology*, **65**, 371–377.

DICKEY, P. A. 1969. Increasing concentration of subsurface brines with depth. *Chemical Geology*, **4**, 361–370.

DORÉ, A. G. & GAGE, M. S. 1987. Crustal alignments and sedimentary domains in the evolution of the North Sea, Northeast Atlantic Margin and Barents Shelf. *In*: BROOKS, J. & GLENNIE, K. (eds) *Petroleum Geology of North West Europe.* Graham & Trotman, London, 1131–1148.

EGEBERG, P. K. & AAGAARD, P. 1989. Origin and evolution of formation waters from oil fields on the Norwegian shelf. *Applied Geochemistry*, **4**, 131–142.

EVANS, D. G. & NUNN, J. A. 1989. Free thermohaline convection in sediments surrounding a salt column. *Journal of Geophysical Research*, **94**, 12413–12422.

GILES, M. R. 1987. Mass transfer and problems of secondary porosity creation in deeply buried hydrocarbon reservoirs. *Marine and Petroleum Geology*, **4**, 188–204.

HANOR, J. S. 1979. The sedimentary genesis of hydrothermal fluids. *In*: BARNES, H. L. (ed.) *Geochemistry of Hydrothermal Ore Deposits.* Wiley, New York, 137–168.

—— 1984. Variation in the chemical composition of oilfield brines with depth in Northern Louisiana and Southern Arkansas: implications for mechanisms and rates of mass transport and diagenetic reaction. *Transactions of the Gulf Coast Association of Geological Societies*, **34**, 55–61.

—— 1987a. *Origin and Migration of Subsurface Sedimentary Brines.* Society of Economic Paleontologists and Mineralogists, Short Course No. 21.

—— 1987b. Kilometre-scale thermohaline overturn of pore waters in the Louisiana Gulf Coast. *Nature*, **327**, 501–503

—— & BAILEY, J. E. 1983. Use of hydraulic head and hydraulic gradient to characterize geopressured sediments and the direction of fluid migration in the Louisiana Gulf Coast. *Transactions of the Gulf Coast Association of Geological Societies*, **23**, 115–122.

—— & SASSEN, R. 1990. Evidence for large-scale vertical and lateral migration of formation waters, dissolved salt, and crude oil in the Louisiana Gulf Coast. *Gulf Coast Society of the Society of Economic Paleontologists and Mineralogists Foundation Ninth Annual Research Conference, Proceedings*, 283–296.

HASZELDINE, R. S., SAMSON, I. M. & CORNFORD, C. 1984. Quartz diagenesis and convective fluid movement: Beatrice oil field, U.K. North Sea. *In*: MORGAN, D. J. *et al.* (eds) Patterns of mineral diagenesis on the northwest European continental shelf and their relation to facies and hydrocarbon accumulation. *Clay Minerals*, **19**, 391–402.

HITCHON, B., BILLINGS, G. K. & KLOVAN, J. E. 1971. Geochemistry and origin of formation waters in the western Canada sedimentary basin—III. Factors controlling the chemical composition. *Geochimica et Cosmochimica Acta*, **35**, 567–598.

—— & FRIEDMAN, I. 1969. Geochemistry and origin of formation waters in the western Canada sedimentary basin—I. Stable isotopes of hydrogen and oxygen. *Geochimica et Cosmochimica Acta*, **33**, 1321–1349.

KHARAKA, Y. K. & CAROTHERS, W. W. 1986. Oxygen and hydrogen isotope geochemistry of deep basin brines. *In*: FRITZ, P. & FRAPE, J. Ch. (eds) *Handbook of Environmental Isotope Geochemistry, Vol. 2. The Terrestrial Environment*. Elsevier, Amsterdam, 305–360.

——, HULL, R. W. & CAROTHERS, W. W. 1985. Water–rock interactions in sedimentary basins. *In*: GAULTIER, D. L. *et al.* (eds) *Relationship of Organic Matter and Mineral Diagenesis*. Society of Economic Paleontologists and Mineralogists, Short Course No. 17, 79–176.

KNAUTH, L. P. & BEEUNAS, M. A. 1986. Isotope geochemistry of fluid inclusions in Permian halite with implications for the history of ocean water and the origin of saline formation waters. *Geochimica et Cosmochimica Acta*, **50**, 419–433.

LAND, L. S. & PREZBINDOWSKI, D. R. 1981. The origin and evolution of saline formation water, Lower Cretaceous carbonates, South-Central Texas, U.S.A. *Journal of Hydrology*, **54**, 51–74.

LANE, A. C. 1908. Mine waters and their field assay. *Geological Society of America*, **19**, 501–512.

MANHEIM, F. T. & HORN, M. K. 1968. Composition of deeper subsurface waters along the Atlantic continental margin. *Southeastern Geology*, **9**, 215–236.

—— & PAULL, C. K. 1981. Patterns of groundwater salinity changes in a deep continental-oceanic transect off the southeastern Atlantic coast of the U.S.A. *Journal of Hydrology*, **54**, 95–105.

MILLS, R. V. & WELLS, R. C. 1919. The evaporation and concentration of waters associated with petroleum and natural gas. *Bulletin, United States Geological Survey*.

MORTON, R. A. & LAND, L. S. 1987. Regional variations in formation water chemistry, Frio Formation (Oligocene), Texas Gulf Coast. *AAPG Bulletin*, **71**, 191–206.

RABINOWICZ, M., DANDURAND, J.-L., JACUBOWSKI, M., SCHOTT, J. & CASSAN, J.-P. 1985. Convection

and hydrocarbon migration. *Earth and Planetary Science Letters*, **74**, 387–404.

RANGANATHAN, V. & HANOR, J. S. 1987. A numerical model for the formation of saline waters due to diffusion of dissolved NaCl in subsiding sedimentary basins with evaporites. *Journal of Hydrology*, **92**, 97–120.

—— & —— 1988. Density-driven groundwater flow near salt domes. *Chemical Geology*, **74**, 173–188.

RICHARDSON, G. B. 1917. Note on diffusion of sodium chloride in Appalachian oil-field waters, Washington (D.C.). *Academy of Science Journal*, 73–75.

RIDER, M. H. 1986. *The Geological Interpretation of Well Logs*. Blackie, Glasgow.

RUSSEL, W. L. 1933. Subsurface concentration of chloride brines. *AAPG Bulletin*, **17**, 1213–1228.

SCHLUMBERGER 1986. *Log Interpretation Charts*. Schlumberger Well Services, Houston.

—— 1989. *Log Interpretation. Principles/Applications*. Schlumberger Well Services, Houston.

SPENCER, R. J. 1987. Origin of CaCl brines in Devonian formations, western Canada sedimentary basin. *Applied Geochemistry*, **2**, 373–384.

Society of Professional Well Logging Analysts 1989. *North Sea Rw Catalogue 1989 Edition*. Society of Professional Well Logging Analysts, London Chapter.

TAYLOR, J. C. M. 1984. Late Permian–Zechstein. *In*: GLENNIE, K. W. (ed.) *Introduction to the Petroleum Geology of the North Sea*. Blackwell, Oxford, 61–83.

THEYS, P. 1988. Log quality control and error analysis: a prerequisite to accurate formation evaluation. *Transactions of the SPWLA Eleventh European Formation Evaluation Symposium*, paper V.

WINSAUER, W. O., SHEARIN, H. M., MASSON, P. H. & WILLIAMS, M. 1952. Resistivity of brine-saturated sands in relation to pore geometry. *AAPG Bulletin*, **36**, 253–277.

WOOD, J. R. & HEWETT, T. A. 1982. Fluid convection and mass transfer in porous sandstones—theoretical model. *Geochimica et Cosmochimica Acta*, **46**, 1707–1713.

—— & —— 1984. Reservoir diagenesis and convective fluid flow. *In*: MCDONALD, D. A. & SURDAM, R. C. (eds) *Clastic Diagenesis*. American Association of Petroleum Geologists Memoir, **37**, 99–111.

ZIEGLER, P. A. 1982. *Geological Atlas of Western and Central Europe*. Shell International Petroleum Maatschappij B. V., Elsevier, Amsterdam.

The Temperature Decay Log: a different approach to presenting a temperature survey

PER-GUNNAR ALM

*Department of Engineering Geology, University of Lund, P.O. Box 118, S-221 00 Lund,
Sweden*

Abstract. In connection with 'shut-in' temperature surveys numerous temperature logs are
run over a period of time. When it comes to presenting the results, several of the recorded
temperature logs are often shown in the same figure. In order to increase the readability of
the figure, some of the recorded logs might have to be excluded. This can cause vital
information concerning the formation surrounding the borehole to be excluded. The paper
gives an example of a different way to present the recorded temperature logs. Normally the
presentation shows temperature versus depth. A different approach is to study the change of
temperature versus time, after shut-in. In this case, the temperature behaviour at every
depth is studied throughout the survey. Thereafter all the temperature logs are included in
one final temperature decay log. This means that all the temperature logs run during the
survey are used in the final presentation. The method gives as a result one single curve
showing the temperature recovery after shut-in. The curve shows the time of recovery versus
depth. In the figure, the different zones of the borehole are easily discerned. By use of several
of the above mentioned temperature decay logs it is possible to distinguish parts of the
borehole with uniform temperature behaviour. The paper gives an example from a fractured
formation where the two methods of presenting temperature logs are compared. (The
conventional one and the one suggested in the paper.)

The temperature log can be used in many different types of temperature survey. The normal use
of the log is to determine the geothermal gradient in a certain area. Temperature measurements
are also carried out in connection with gas production. Another application may be to locate
permeable and impermeable zones or fractures
in a formation. In this case, temperature logging
is performed before and after 'shut-in'. A well is
shut-in when production or injection, of gas or
fluid, are deliberately stopped. This paper will
show a different way of presenting the temperature logs in connection with such an application.

When permeable zones or fractures are to be
located with temperature logs, the temperature
survey consists of two phases; temperature
measurements prior to, and after, shut-in. The
first phase is a flow phase, either produced or
injected fluid is present in the borehole. During
this time it is preferable that the flow rate and
fluid temperature are constant. This phase continues until steady state occurs in the borehole,
i.e. the temperature in the borehole does not
change with time. This of course means that the
fluid, which moves in both the borehole and the
surrounding formations, has reached the same
temperature, at least close to the borehole.

When steady state is reached, phase two starts
with a shut-in. During phase two the temperature in the borehole and the surrounding forma-

tions is allowed to return to the undisturbed
temperature. The undisturbed temperature is
recorded in the borehole before any injection or
production is started. It is during this second
phase that most temperature logs are run. The
temperature distribution in the borehole is
recorded at various times after shut-in. This
enables a presentation of the temperature recovery. When all the logs are presented in the same
figure different zones of a formation can be
discerned.

In a borehole filled with a fluid, with a temperature different from the borehole surroundings, two major kinds of heat transfer can occur.
They are conduction and convection. There will
always be heat exchange by conduction but
when moving fluid is present, heat exchange by
convection will also take place. Heat transfer by
conduction is the slower process of the two.

An injection temperature survey will here be
described in greater detail. Assume that a fluid at
a constant flow rate is injected into a borehole
with an uncased section. Assume also that the
temperature of the injected fluid is constant and
higher than the borehole surroundings.

During the first phase, the fluid is entering the
permeable zones or fractures. The fluid is losing
heat to the surroundings, in the borehole as well
as in the permeable formations. This process
continues until both the fluid and the formation

From HURST, A., GRIFFITHS, C. M. & WORTHINGTON, P. F. (eds), 1992,
Geological Applications of Wireline Logs II. Geological Society Special Publication No. 65, pp. 339–348.

339

reaches the same temperature. A 'heat front' is created, and this front gradually moves outwards from the borehole as the injection continues. A temperature log recorded during injection, just before shut-in, will record a temperature that is valid both for the fluid and the close vicinity of the borehole.

The formation close to the borehole wall will achieve a temperature that differs from the temperature within the undisturbed surrounding formation. Since the temperature of the fluid is higher than that of the formation, the heat is transported into the permeable zones with the injected fluid. The heat exchange in the permeable zones is governed by both convection and conduction. The impermeable zones, however, are heated by conduction only.

As heat transfer by conduction is a slow process, it means that during the same amount of time, a much smaller volume of an impermeable zone is heated, compared to a permeable one where convection also takes place (see Fig. 1). The figure shows, in general, how the heated volume of some zones around a borehole can vary due to different permeability. The figure shows the situation at shut-in.

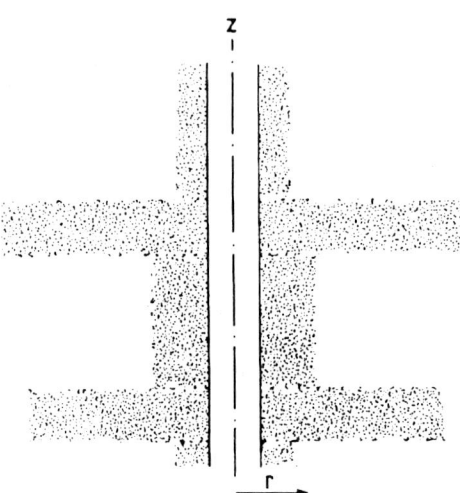

Fig. 1. A model of a hypothetical temperature distribution at shut-in. The shaded areas indicate heated parts of the formation (Persson 1985).

When phase two starts, at shut-in, the heat in the formation declines toward the undisturbed temperature, i.e. the geothermal temperature for that specific area. The heat loss after shut-in is caused by conduction only if it is assumed that there is no movement of fluid in the formation. If only conduction is present and if the thermal

diffusivity is the same, the heat exchange will be the same in permeable zones as well as in impermeable ones. The time it takes to decrease to the undisturbed temperature is only dependent on the volume which has been heated.

The temperature in front of the impermeable zones will reach the geothermal temperature much faster than in the permeable zones. The heat has not intruded those zones as deeply as in the permeable ones.

Presentation of temperature curves

The temperature logs recorded in a temperature survey, such as the one described above, are often presented as shown in Fig. 2. When presenting temperature logs in this way, it is of vital importance to present the curves such that the temperature anomalies are easily discerned. To be able to identify the temperature variations due to the survey, an 'undisturbed' temperature curve is always presented. The curve is recorded in the borehole prior to the survey.

When temperature logs are run the time intervals are of great importance. If a curve is recorded too late after the previous one, temperature anomalies might be missed. If the curve is recorded too soon, it might have to be excluded from the presentation due to overlapping of the curves.

Sometimes it is necessary to make a temperature curve selection, and therefore some important sections of the temperature behaviour can be missed. This does not often influence the overall picture of the temperature behaviour. However, if some curves have to be excluded, then only part of the total temperature decline is presented. There is, however, a way of using all the recorded temperature values and take into account the total temperature recovery in the borehole.

The Temperature Decay Log

As mentioned earlier, the most common way of presenting the temperature is as a plot versus depth. This paper suggests that the temperature may also be studied versus time. Instead of studying the temperature in the complete borehole at one single time, much information can be yielded by studying *one* single depth during one entire temperature survey.

The method is described by the following example. Figure 3 shows the temperature at one single depth, during a complete temperature survey. The curve has been produced by plotting the temperature values, recorded at a certain depth, throughout the complete survey. The

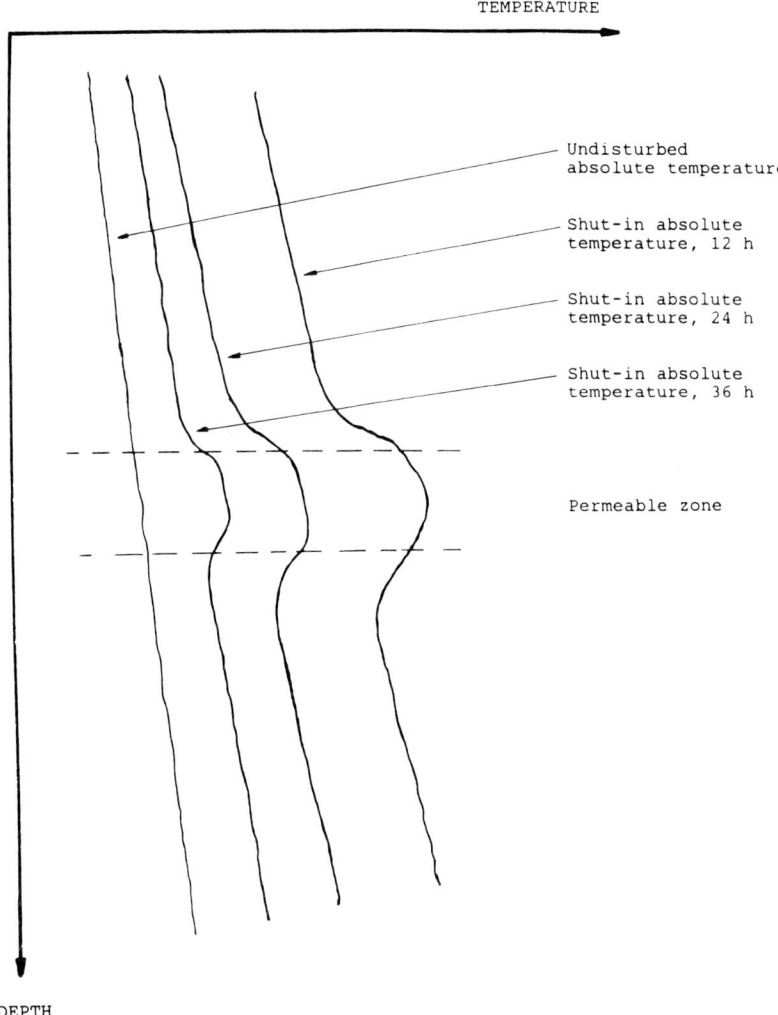

Fig. 2. Schematic temperature curves showing the temperature decline at various times after shut-in.

plotted values are all collected from the ordinary temperature logs that were run after shut-in.

The figure shows the temperature declining after shut-in. In this example the injected fluid had a higher temperature than the initial temperature at that specific depth. Zero time is at shut-in. The curve in the figure contains all the recorded temperature values at that specific depth. The temperature in the example was raised from the initial temperature of 18°C to approximately 27°C. The initial temperature, at that depth, is represented by a dasehd line in the figure. As can be seen, the temperature decays to this temperature level. It is also possible to see at what time any given temperature decrease occurs at that depth and in that specific borehole.

The curve in the figure shows how the formation at a certain depth is losing heat with respect to time. The time it takes to reach initial temperature is dependent on how great a volume of the formation has been heated. The smaller the volume the faster the recovery and vice versa.

In Fig. 3, a line representing a temperature decrease of 50% has been included. The temperature decrease is calculated relative to the increased part of the temperature, *at that depth*, at shut-in. In the example, the temperature increase was 9°C (27 − 18). In this case, the temperature decrease of 50% occurs when the temperature has declined to 22.5°C. From the figure it can be seen that this occurs 7–8 hours after shut-in.

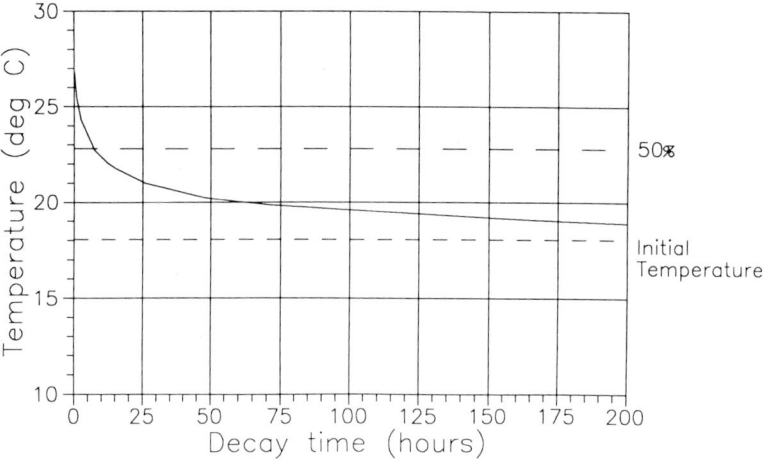

Fig. 3. Temperature decay at *one* specific depth in a borehole during an injection temperature survey.

If every depth in a borehole is studied as described in Fig. 3, it is possible to create a log showing how many hours it takes for a certain temperature decrease to occur. This type of log can be called a 'Temperature Decay Log', and shows temperature decay time versus depth. See Fig. 4 for an example of the log. The curve in Fig. 4 does not show the temperature in the borehole. The curve is showing how the borehole, i.e. the borehole surrounding, is responding to the temperature increase created during phase one. In this case the curve shows how long it takes for the various parts of the borehole to lose a certain amount of the increased temperature at shut-in. The temperature increase is referred to the initial undisturbed temperature.

A case study—an example from a fractured formation

To illustrate the advantage of this method, an example from a borehole with some fractured zones is described. The purpose of the survey was to try to locate, with the help of temperature logging, two fractured zones in a borehole. The total depth of the borehole was 87 m and the casing shoe was placed at 52 m.

The temperature survey was performed with an injection phase prior to shut-in. Water with a temperature slightly over 19°C was injected in the borehole over a nine hour period (see Fig. 5). The temperature recordings of the water were measured in the injection pipe at surface. As can be seen, the temperature varies at the beginning of the injection phase. The variations are rather

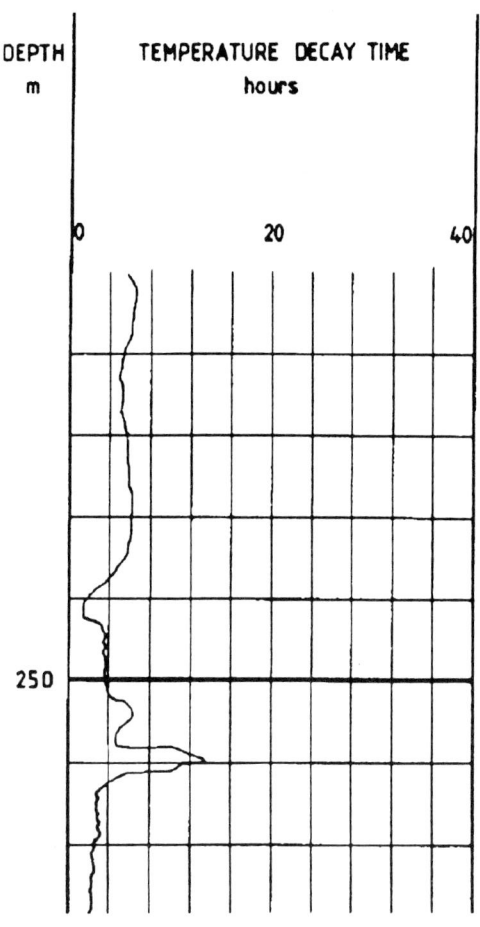

Fig. 4. The Temperature Decay Log.

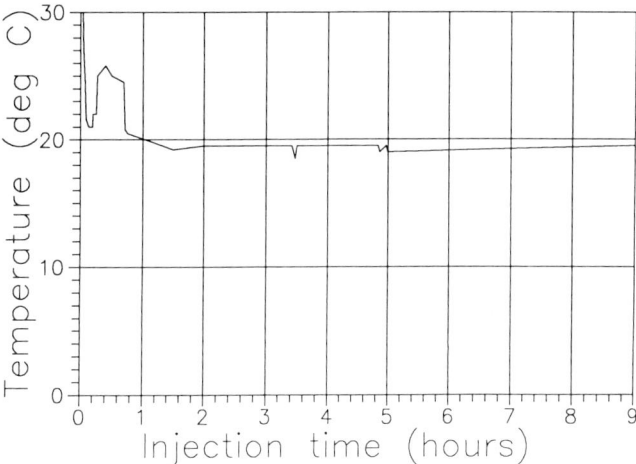

Fig. 5. Temperature variation of injection water.

small and since no variations occur in the end of the injection phase, it will have no significant effect on the temperature increase in the bore-hole. The water was injected through a pipe which was perforated in the open section of the borehole. A total of 11.2 m³ was injected.

During the injection phase, several tempera-ture logs were run to monitor the temperature increase in the borehole. These logs are presented in Fig. 6 where the injection points can also be seen. A lithology profile for the borehole is also presented in the figure. This profile has been constructed from cuttings and other logs carried out during and after the drilling. A caliper log was run and it is presented in track one. The caliper log shows two major zones where the diameter varies. The upper one is between 58–60 m and the second one occurs around 67.5 m. For reference purposes, a temperature log was recorded before any injection was started. This is shown as curve 1.

To see the difference in presentation of the temperature recordings during phase two, both methods are presented. First of all, the 'normal' way of presenting the temperature curves is given. This is done in Fig. 7. The undisturbed temperature curve is the reference curve. See curve 1. The figure presents the results from six temperature logs recorded at different times after shut-in. Even though the complete temperature survey lasted for more than 1200 hours, only a few of the 28 recorded logs could be presented.

The reason for not presenting any curves recorded later than 72 hours after shut-in is due to the smoothing effect on the temperature

curves as time after shut-in increases. This is one disadvantage with this method. Only a few curves can be presented at the same time, other-wise the figure can be unclear and anomalies difficult to discern.

The second method, plotting temperature decay versus time, is now presented. Even though it is not necessary for the final presentation, the temperature decline at one depth is also shown. Figure 8 shows the temperature decline at 60 m. In the figure are also included lines representing a temperature decrease of 50% and the initial temperature at 60 m. At that depth, a 50% temperature decrease will occur after 20 hours.

The figure also shows the recorded tempera-ture during phase one prior to shut-in, hence the reason for the negative time values on the time axis. Zero time is at shut-in. As mentioned before, the temperature survey lasted for more than 1200 hours but only the first 10% of the survey is presented in Fig. 9. The temperature in the borehole was raised from the initial tempera-ture of 8°C to approximately 17.5°C. The initial temperature, at 60 m, is represented by a dashed line in the figure.

Figure 9 shows the temperature decay log. In this figure no reference curve is needed. Still, a reference temperature log has to be recorded since the temperature increase at every depth is calculated from that measurement. In the figure it can easily be seen that there exist two major permeable zones in the borehole. The two frac-tured zones identified from the caliper log are clearly shown on the log. They are located at the depths of 60 m and 67.5 m. Furthermore it can

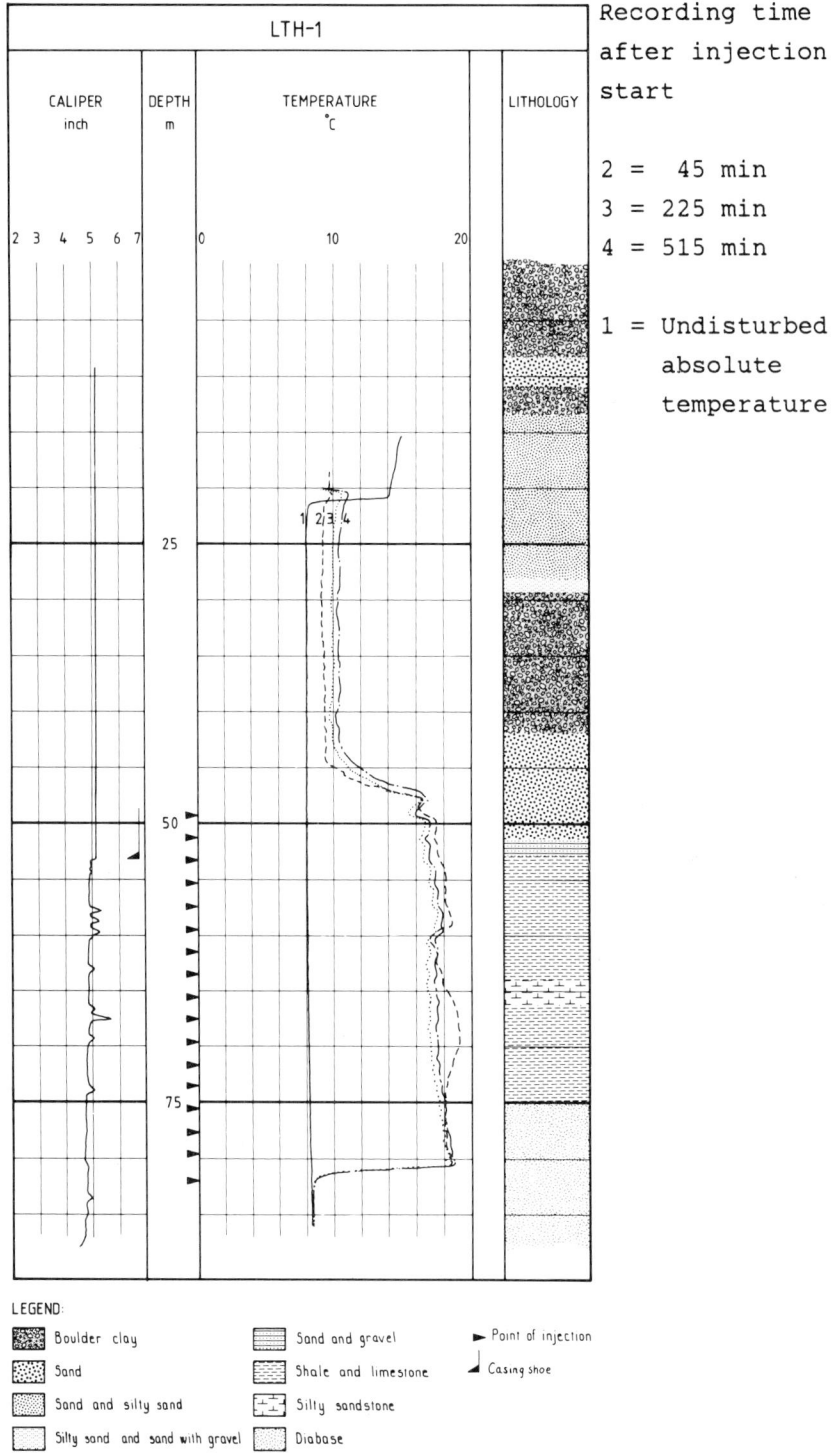

Fig. 6. Temperature logs run before shut-in (Persson 1985).

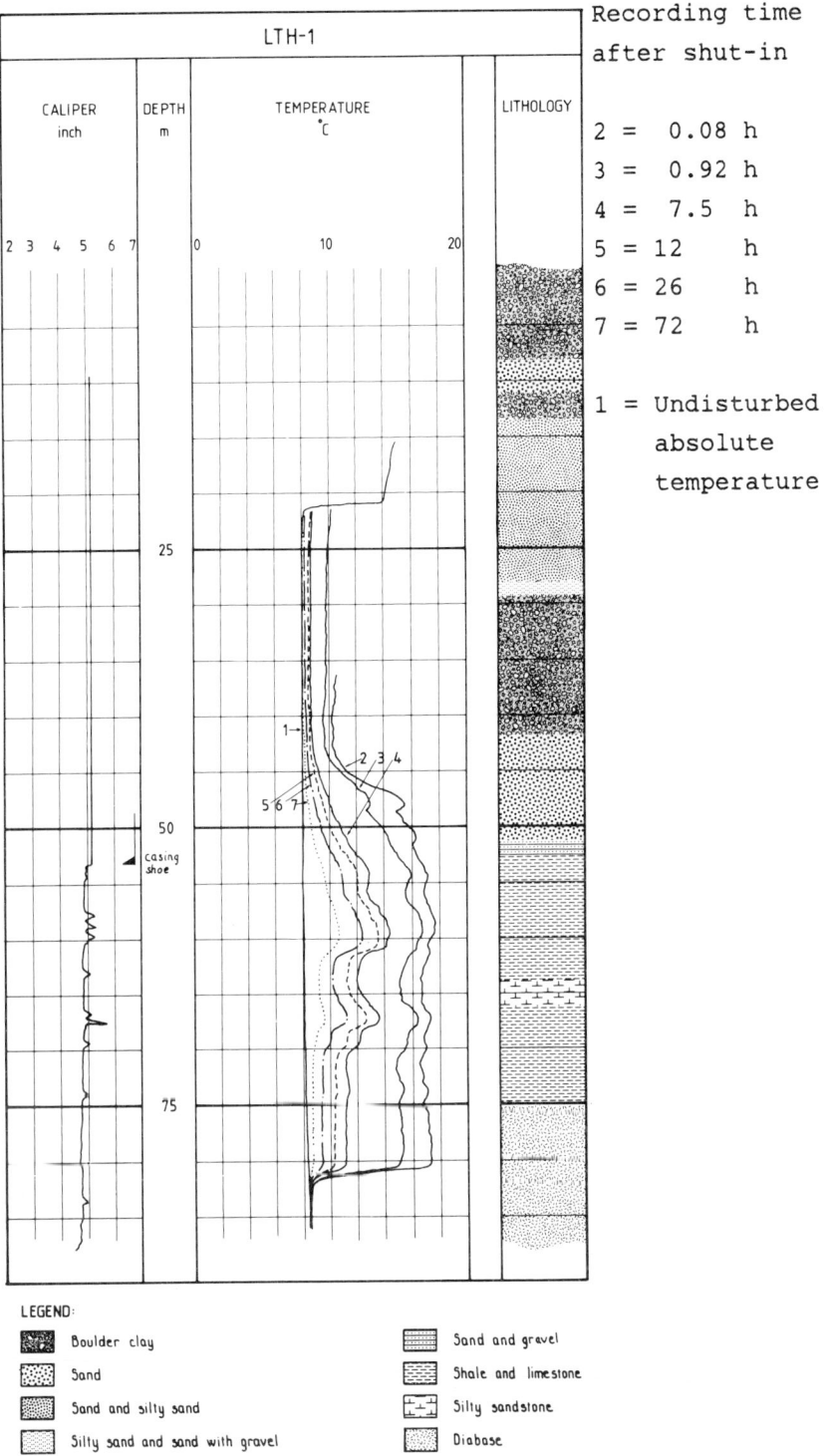

Fig. 7. Normal presentation of a shut-in temperature survey (Persson 1985).

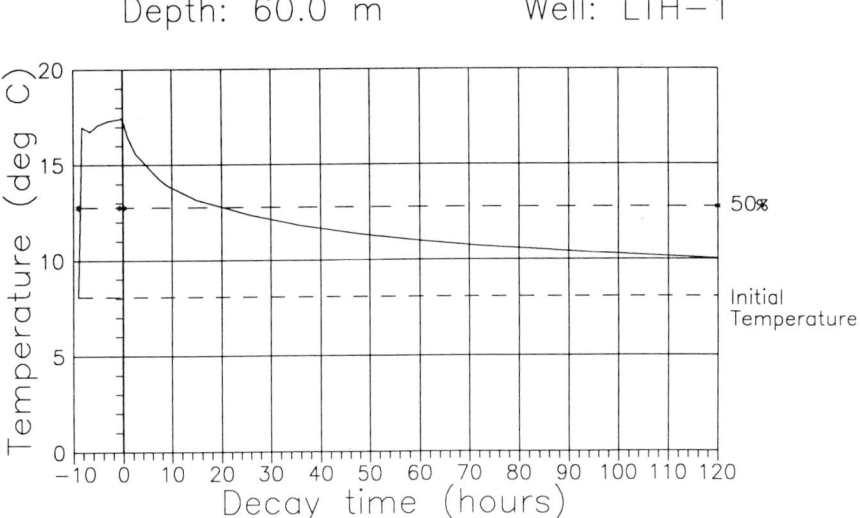

Fig. 8. Temperature decay at 60 m in LTH-1. A dashed line representing a temperature decrease of 50% is included.

be seen that the heated water has penetrated the upper zone to a greater extent than the lower one. This is shown by the slower temperature recovery in the upper zone. For the same reason it is possible to see that the lower part of the upper zone has received more water than the upper part of that zone.

The figure shows very clearly that the borehole surroundings respond very differently to the temperature increase. The time it takes to lose 50% of the temperature increase at shut-in, varies between 3 and 20 hours. It is not only possible to discern where there are fractures, but also variations within the impermeable zones can be seen. These variations are due to different heat capacity and thermal conductivity within the formation. This can be seen above the casing shoe. The formations behind the casing have only been heated by conduction. If the formation had been uniform the temperature decline would also have been uniform.

Several temperature decay logs can naturally be presented together as has been done in Fig. 10. With the help of these curves it is easier to discern different parts of the formation with uniform temperature behaviour, not just the permeable and impermeable zones, but also parts with different heat capacity.

As is shown in Fig. 10, it takes a very long time for the fractured zones to lose all the temperature increase, even though the injection lasted for only nine hours. It took more than 320 hours for the upper zone to lose 90% of the increased temperature, for example. The

influence of the injection was seen more than 1200 hours after shut-in.

Conclusion

This paper has shown a different method of presenting the recorded temperature curves in a shut-in temperature survey. Often an injection temperature survey is presented with several temperature recordings in one figure. This can result in figures that are difficult to read.

The temperature decay log may occasionally be preferred. The presentation is clearer, only one curve is needed to show differences between permeable and impermeable zones. The log is therefore very useful for locating permeable zones. One of the advantages of the method is that information from all the recorded temperature logs is included in the final curve. There is no risk that vital information concerning the temperature behaviour has been excluded.

The use of the Temperature Decay Log also makes it possible to discern parts of the formation with homogeneous temperature behaviour.

The temperature decay time log can be used to present the time it takes for a borehole to recover a certain percentage of a temperature increase.

Reference

PERSSON (ALM), P. G. 1985. *The Temperature Log as an Indicator of Permeable Zones and Fractures in Wells.* Dept. of Engineering Geology, Lund, Sweden, LUTVDG/(TVTG-3009)/1–94 (1985)

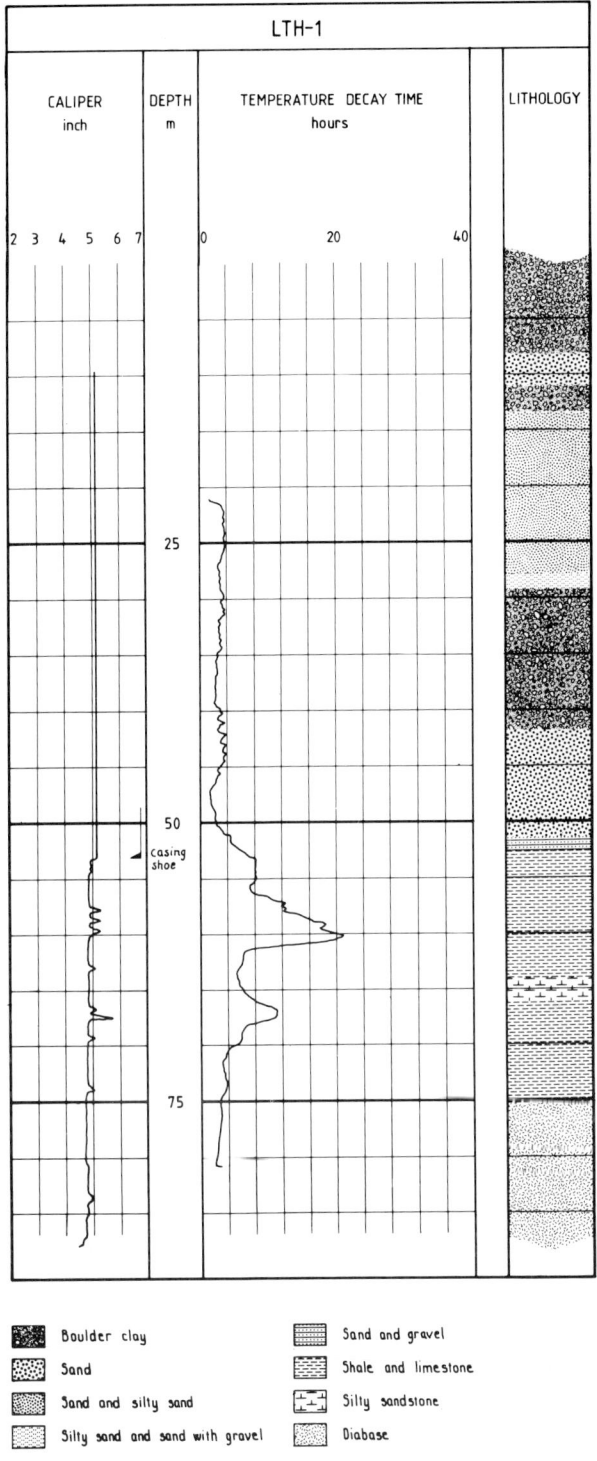

Fig. 9. The Temperature Decay Log for the well LTH-1 (Persson 1985).

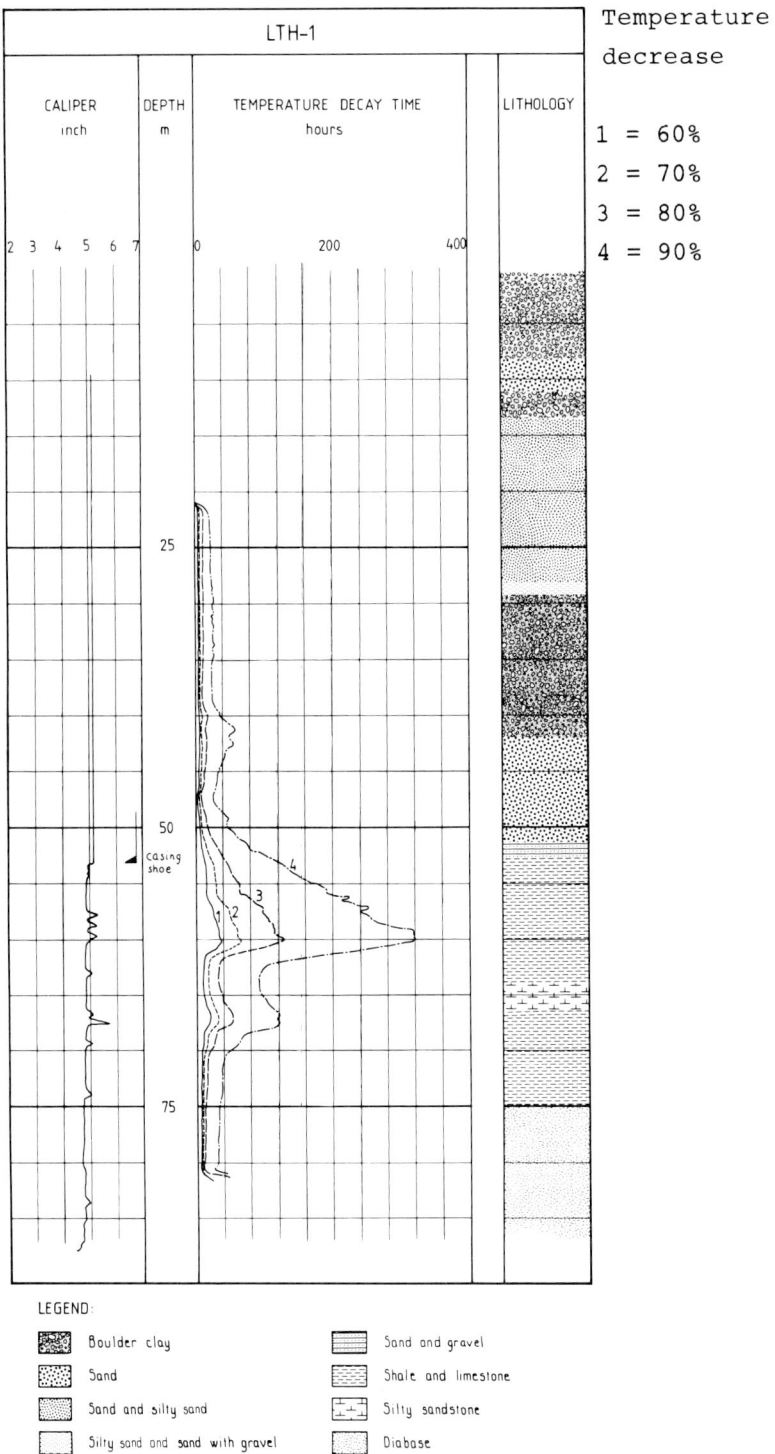

Fig. 10. Temperature decay time curves for various temperature decreases (Persson 1985).

A new approach to the interpretation of nuclear borehole logs

P. L. ØLGAARD

*Department of Electrophysics, Technical University of Denmark, DK-2800 Lyngby,
Denmark*

Abstract. When interpreting nuclear borehole logs it is usually assumed for simplicity that
the tool signal and a formation property, e.g. its density, are uniquely related. However,
while the tool signal will usually depend primarily on one formation property, it will also
be affected by other formation properties. To take this multi-parameter dependence of all
the nuclear tools into account, a new approach to nuclear log interpretation has been
developed. In this approach four nuclear tools are considered: the gamma density tool, the
neutron porosity tool, the pulsed neutron tool and the natural gamma tool. The responses of
these four tools depend on formation density (ρ), porosity (ϕ), neutron absorption (Σ_a), and
geological type (including shale content). For each of the four tools a physical model has
been developed. The signals measured with the four tools and the four models represent
together four equations with four unknowns: ρ, ϕ, Σ_a and geological type. These four
equations may be solved by iteration and a consistent set of formation properties obtained.
The interpretation approach has been implemented on a personal computer. Examples of
results of such interpretations are presented.

When interpreting borehole logs obtained by use
of nuclear tools, it is usually assumed that there
exists a unique relation between the tool signal
and a specified formation property, e.g. the
density or the porosity of the formation. This
relation may be an empirical or semi-empirical
calibration curve or expressed as an algorithm.
By use of such relations the corresponding formation properties may be obtained and used in
the log interpretation. The tool signals are count
rates or count-rate ratios. A closer investigation
of the physics of the nuclear tools reveals that
while the signal of a given tool may primarily
depend on one formation property, it will also
depend on other properties. This means that if a
consistent interpretation is to be made, the
multi-property dependence of the signals has to
be taken into account.

In the investigations reported here, the neutron porosity tool, the gamma density tool, the
pulsed neutron tool and the natural gamma tool
were considered. The properties on which these
four tools depend are listed in Table 1. For
example, the signal of the thermal neutron porosity tool is primarily determined by the volume
fraction of water or oil, but it is also affected by
the bulk density, the neutron absorption, and
the geological type of the formation.

The new approach

From Table 1 it is noted that there are four
different secondary properties, and that these are
identical to the four primary properties. Consequently there are four unknown properties.
Further there are four tools. This means that if
theoretical models which relate the tool signal to
the formation properties could be developed for
each of the four tools, then one would have four
equations with four unknowns. Solving these
four equations will yield the four unknown
properties.

The idea behind the present approach is to do
exactly that. It should be noticed that while the
porosity, ϕ, (i.e. the volume fraction of water or
oil), the bulk density, ρ, and the neutron absorption (i.e. the neutron absorption cross section),
Σ_a, are all characterized by one figure which will
come out of a solution of the equations, the
same is not the case for the formation type. This
poses an additional problem which will be discussed later.

When developing the four models a compromise has to be made. On the one hand, reasonably accurate models have to be derived so that
the interpretation is reasonably reliable. On the
other hand, a short computation time is desirable in order to permit real-time or near-real-
time interpretation of the logs. This means that a
number of effects cannot be taken into account.
These effects included the mud cake correction
and—for all but the natural gamma tool—the
corrections for borehole diameter, mud composition, and tool stand-off. This could be a
serious limitation of the approach discussed
here, but corrections of these effects could be
introduced using the empirical correction of the
logging companies.

From HURST, A., GRIFFITHS, C. M. & WORTHINGTON, P. F. (eds), 1992,
Geological Applications of Wireline Logs II. Geological Society Special Publication No. 65, pp. 349–358.

Table 1. *Nuclear tools and formation properties on which the tool signals depend*

Tool type	Primary formation properties	Secondary formation properties
Thermal neutron porosity	H_2O/oil vol. fraction	Bulk density Neutron absorption Formation type
Gamma density	Bulk density	H_2O/oil vol. fraction Formation type
Pulsed neutron	Neutron absorption	H_2O/oil vol. fraction Bulk density Formation type
Natural gamma	Formation type	Bulk density H_2O/oil vol. fraction

Model for the neutron porosity tool

The neutron porosity tool consists of a fast neutron source, usually [241]Am–Be, and two thermal neutron detectors, usually [3]He proportional counters. The two detectors are placed at different distances from the source: the near detector closer to and the far detector farther from the source. By use of a bow spring the tool is in mechanical contact with the borehole wall.

In the model the actual geometry is approximated by a point source of fast neutrons, situated in an infinite, homogeneous medium, and two line detectors. This means that the geometry considered is one-dimensional. The neutron flux distribution of the medium is determined by use of three-group diffusion theory. With the assumptions made, an analytical expression for the thermal neutron flux, ϕ_3, is obtained:

$$\phi_3(r) = \frac{q\,\kappa_1^2\,\kappa_2^2}{4\pi D_3(\kappa_2^2 - \kappa_1^2)\,r}$$

$$\times \left(\frac{\exp(-\kappa_1 r) - \exp(-\kappa_3 r)}{\kappa_3^2 - \kappa_1^2} \right.$$

$$\left. - \frac{\exp(-\kappa_2^2) - \exp(-\kappa_3 r)}{\kappa_3^2 - \kappa_2^2} \right)$$

Here r is the distance to the source and q the source strength. D_3, κ_1, κ_2, and κ_3 are nuclear constants, determined by the formation composition. By use of $\phi_3(r)$ the count rate of the two detectors and the count-rate ratio of the near to the far detector is calculated. This count-rate ratio is the tool signal.

Satisfactory results with this type of model have earlier been experienced (Ølgaard & Haahr 1967). It represents an approximation since the effect of the tool itself and the borehole mud is not taken into account. However, the perturbation of the thermal neutron flux, caused by the tool and the borehole mud, will be roughly the same for both detectors. Since the ratio of the count rate of the two detectors is used as the tool signal, the effect of the borehole mud and the tool will be greatly reduced.

In Fig. 1, two calibration curves for neutron porosity tools of the same type are shown. They originate from the same logging company and were obtained through analysis of field logging data (Ølgaard & Petersen 1988). The reason for the two different calibration curves of the logging company is presumably that the two tools are not completely identical, e.g. due to unintended differences in the detectors.

In Fig. 1 calibration curves, calculated by use of the model, are also shown. It is seen that in general the agreement is reasonable, but there are differences. The calibration curves of the logging company are used for any type of geological formation, irrespective of geological type, density and neutron absorption. The calculated calibration curves, which are valid for water-saturated limestone with three different neutron absorption cross sections, demonstrate that this is not quite permissible. The model assumes that the near and far detectors are identical. This may in practice not quite be the case, but could be corrected for by normalization. Also the logging companies usually prefer linear calibration algorithms while the curves in practice are non-linear. Approximation in the model may of course also contribute to the disagreement.

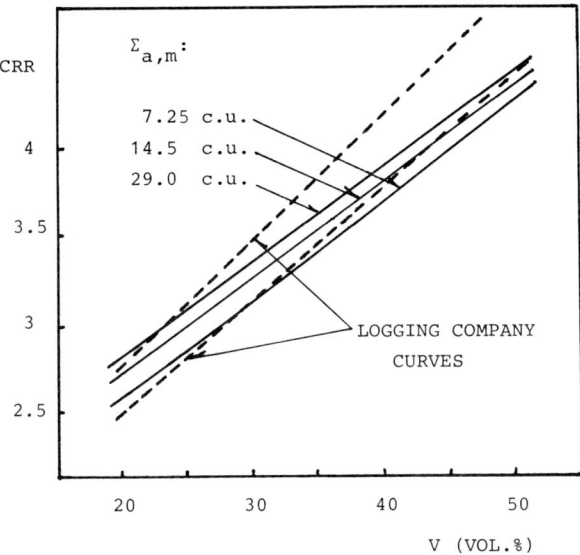

Fig. 1. Count rate ratio (*CRR*) versus volume per cent of H_2O/oil (*V*) for a neutron porosity tool. Dashed curves: Calibration curves of the logging company. Solid curves: Calibration curves calculated by use of the model. Formation: Water-saturated limestone with different matrix absorption cross sections ($\Sigma_{a,m}$, in capture units).

Model for the gamma density tool

The gamma density tool consists of a gamma source, usually a ^{137}Cs, 0.662 MeV source, and two gamma detectors, usually NaI scintillators. Here too there is a far and a near detector. The near detector is strongly collimated and used for mudcake corrections. It is not considered by the model. The source and the far detector are both somewhat collimated. The source and the detector are placed on a pad which, by use of a tool arm, is pressed towards the borehole wall. The source emits gamma quanta into the formation and the far detector detects quanta coming from the formation.

In the model the tool geometry is approximated by a gamma point source, situated in an infinite, homogeneous medium, and a point detector. Here too the geometry is one-dimensional. The gamma flux at the point of the far detector is calculated by use of five-group diffusion theory. Usually diffusion theory is not well-suited for gamma calculations due to strong forward scattering of the quanta. However, by modifying the diffusion coefficient of the highest energy group, good agreement with the more accurate moment method used by Goldstein & Wilkins (1954) was obtained.

Here too analytic expressions for the gamma flux of the five energy groups were derived. The energy range covered by the five groups ranges from 0.662 MeV down to 0.131 MeV. The value of 0.131 MeV was selected due to lack of better data. Later information indicates that the value should be 0.18 to 0.2 MeV. Investigations show that this will not have much effect on the SiO_2 and Al curves, but may affect the $CaCO_3$ curves. The advantage of analytical expressions is that the computing time is substantially decreased. The count rate of the far detector, which is the tool signal, is assumed proportional to the gamma flux at the detector.

The model represents an approximation of the actual geometry of the tool. It may be added that Monte Carlo calculations are under way to investigate the importance of the geometrical approximations. However, as shown in Fig. 2, the model agrees reasonably well with the calibration curve used by the logging company. This calibration curve was obtained through an analysis of field logging data (Ølgaard & Petersen 1988). Since the tool signal is the count rate, not a count-rate ratio, it is necessary to normalize the calculated curves to the calibration of the logging company. This is done in Fig. 2 by normalizing, more or less arbitrarily, the count rate of the calculated SiO_2 curve to be equal to

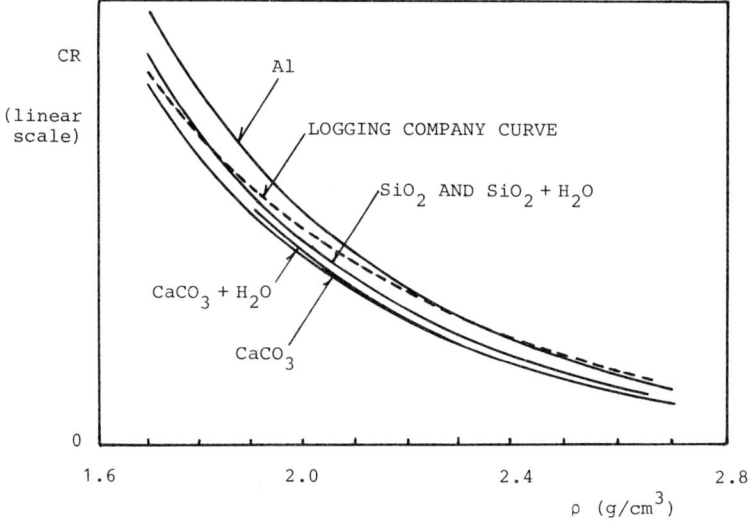

Fig. 2. Count rate of the far detector of a gamma density tool (CR, arbitrary scale) versus formation density (ρ). Dashed curve: Calibration curve of the logging company. Solid curves: Calibration curves calculated by use of the model for different media.

that of the logging company curve at $\rho = 1.8$ g/cm^3.

It is seen that the slope of the logging company curve is somewhat lower than that of the calculated curves. This may be due to approximations of the model or to the fact that the upper part of the logging company curve is normalized to fit measurements in aluminium which will give too high a count rate. It is also seen that the model gives different curves for different media—as was to be expected due to the fact that Z/A is not exactly equal to 1/2 for all elements considered—while the logging company uses only one curve. As mentioned above, the difference between the SiO_2 and the $CaCO_3$ curves may be due to the too low cut-off energy used. In Fig. 2, SiO_2 stands for SiO_2–vacuum mixtures, $CaCO_3$ for limestone–vacuum mixtures, and Al for Al–vacuum mixtures. $SiO_2 + H_2O$ and $CaCO_3 + H_2O$ stand for water saturated sandstone and limestone mixtures.

Model for the pulsed neutron tool

The fast neutron source of the pulsed neutron tool is a mini-accelerator which by use of a D + T nuclear reaction produces 14 MeV neutrons. The tool also contains two gamma detectors (NaI) which measure the gamma flux originating from the capture of the thermal neutrons produced by slowing-down of the fast source neu-

trons. The two gamma detectors are placed different distances from the source.

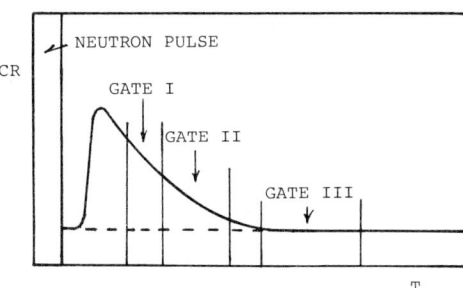

Fig. 3. Count rate of a gamma detector of a pulsed neutron tool (CR) versus time (T) after emission of neutron pulse and gate arrangement.

The pulsed neutron tool measures the time variation of the capture gamma flux after the emission of each fast neutron pulse. This is, in older tools, done as shown in Fig. 3, where the gates during which the gamma detector pulses are recorded, are shown. Gate III is used for background correction of gate I and II. The ratio of the background corrected counts of gate I and II is, summed over a number of pulses, the signal of the tool. In newer tools the gate arrangement is more elaborate, but the principle of the measurements is the same.

Logging companies usually assume that the decrease of the gamma count-rate curve is exponential and may be expressed by

$$\exp(-v\Sigma_a t)$$

where v is the thermal neutron velocity, Σ_a the neutron absorption cross section, and t the time. By use of this exponential and the background corrected counts of the gates Σ_a may be calculated.

In the model the tool geometry is approximated by a pulsed point source, situated in an infinite, homogeneous medium, and a point gamma detector. Thus once again the tool geometry is assumed one-dimensional. The time dependent capture gamma flux at the point detector is calculated by use of Fermi-age theory for the fast neutrons, diffusion theory for the thermal neutrons and build-up factor theory for the capture gamma quanta. (Glasstone & Edlund 1960, Ølgaard 1985). This results in the following expression for the gamma flux ϕ_g at the detector:

$$\phi_g(a,t) = \frac{S \, v \, \Sigma_a \exp(-v\Sigma_a t)}{4\pi^{3/2} \, a \, [4(\tau + Dvt)]^{1/2}}$$

$$\times \int_0^\infty \frac{e^{-\mu\rho}}{\rho} B(\mu\rho) \left[\exp\left(\frac{(\rho - a)^2}{4(\tau + Dvt)}\right) \right.$$

$$\left. - \exp\left(\frac{(\rho + a)^2}{4(\tau + Dvt)}\right) \right] d\rho$$

where

$$B(\mu\rho) = 1 + \beta_1\mu\rho + \beta_2(\mu\rho)^2 + \beta_3(\mu\rho)^3.$$

Here a is the source-detector distance, t the time, S the number of neutrons per pulse, v the thermal neutron velocity, Σ_a the neutron absorption cross section, τ the Fermi-age to thermal energy for the fast neutrons, D the diffusion coefficient of the thermal neutrons, and μ the gamma attenuation coefficient. $B(\mu\rho)$ is the build-up factor and β_1, β_2, and β_3 are constants. The build-up factor corrects for the scattered gamma quanta.

The integral given above has to be determined by numerical integration and the count rate of the two gates has also to be calculated by numerical integration. The count rate of the gamma detector is assumed proportional to ϕ_g.

When comparing the exponential decrease of the count rate, assumed by logging companies, with that obtained from the model, it is seen that the latter is much more complex. This is due to

the fact that the logging companies tacitly assume that the thermal neutron flux distribution with respect to space does not change with time while the model takes this change into account. This change, which consists in a flattening of the neutron flux distribution with time, is called the diffusion effect.

The correctness of the model was tested by comparison with measurements. Wahl *et al.* (1970) have reported the measurements presented in Fig. 4. They comprised pulsed neutron tool measurements in quartz–saltwater and limestone–fresh water mixtures. The source–detector distance a was also varied during the measurements. In Fig. 4 the measured results are presented as dashed lines and the calculated results as solid curves. It is seen that the agreement between the model and the measurements is quite acceptable except for pure limestone. The reason for the disagreement is likely to be that the measurements on pure limestone were made in too small a limestone block. This would result in substantial neutron leakage which has the same effect as an increase in Σ_a or a decrease in $1/(v\Sigma_a)$. Calculations indicate that in order to simulate 'an infinite medium', a dry limestone block should have a diameter of more than 2 m. It should be noted that the measurements show that the $1/(v\Sigma_a)$ value is affected by the source–detector distance. The model results in the same effect while the method used by logging companies includes no such effect.

Model for the natural gamma tool

The natural gamma tool contains a gamma detector (NaI), and it measures the gamma quanta emitted by the decay of naturally occurring radioactive nuclei in the formation. These nuclei consist of potassium, thorium and uranium isotopes and their daughter nuclei. The basic idea behind the use of the natural gamma tool is that there exists some degree of correlation between the type of formation and its content of radioactive elements. Limestone and dolomite have in general a low content, sandstone a higher content, and shale the highest content.

The count rate of the natural gamma tool is not only dependent on the type of formation but also on the tool diameter, the borehole diameter, the composition of the mud, in particular its density, and whether the tool is run centered or eccentered. For this reason the measured count rates are always corrected so as to correspond to a standard borehole–tool geometry and mud composition (density and chemical composition).

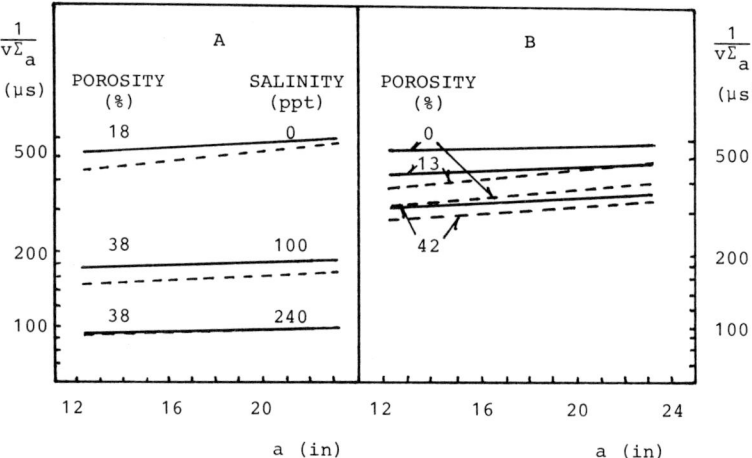

Fig. 4. Comparison between measured (dashed lines) and calculated (solid lines) $1/(v\Sigma_a)$-values, obtained with a pulsed neutron tool, versus source-detector distance, a. A: Quartz–salt water mixtures. B: Limestone–fresh water mixtures.

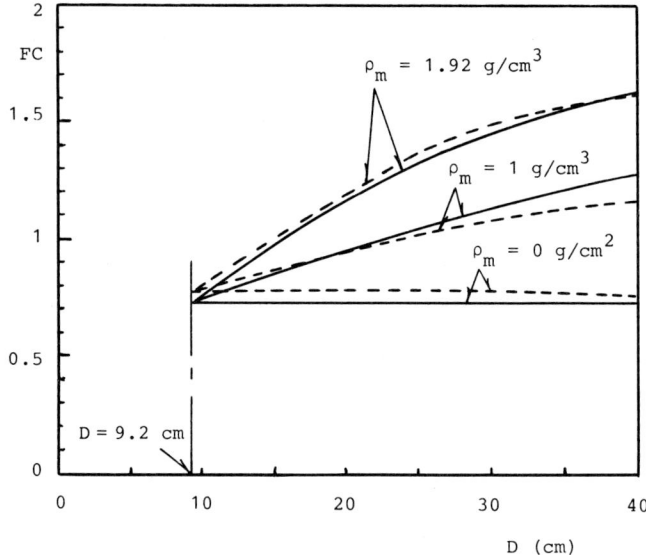

Fig. 5. The correction factor (*FC*) by which the measured count rate must be multiplied to give the count rate of the reference borehole, as a function of borehole diameter (*D*). Eccentered, 92 mm diameter tool. Solid curves: Model calculations. Dashed curves: Logging company. ρ_m: mud density.

The purpose of the model for the natural gamma tool is to make this correction. Initially the model was based on a two- or three-dimensional numerical integration over the formation. Later the model has been improved so that it only involves one or two numerical integrations (Hovgaard 1990, pers. comm.). The model is based on geometrical considerations and on the build-up factor concept. This concept corrects for the scattered gamma quanta. While the models discussed earlier are one-dimensional, the models for the natural gamma tool are two-(centered) or three-dimensional (eccentered).

In Fig. 5 the correction factor FC with which the measured count rate must be multiplied in order to correspond to the count rate of the reference borehole is given. The curves of Fig. 5 apply to an eccentered tool with three different mud densities. It is seen that the calculated curves agree quite well with those of the logging company (Schlumberger Well Services 1985). The model may also take into account the presence of barite in the mud.

Log interpretation

The logs considered so far have been limited to rather pure limestone and mixtures of sandstone and shale.

The natural gamma tool is used to determine whether a formation layer is limestone or a shale–sandstone mixture. If the corrected count rate of the natural gamma tool is below around 25 API, the formation is assumed to be limestone. If it is higher, the formation is assumed to be a shale–sandstone mixture. Initially it had been hoped that the natural gamma count rate could be used to determine the amount of shale in the mixtures (cf. Heslop 1974). However, this approach seemed to give too high uncertainty. Thus another approach was used.

It is well known that the bulk density, ρ, may be written in the following form for a formation containing water and gas:

$$\rho = (1 - \phi)\rho_m + \phi\, S_w\rho_w + \phi(1 - S_w)\rho_g.$$

Here ϕ is the total porosity, S_w the volume fraction of the porosity filled with water, and ρ_m, ρ_w and ρ_g the densities of matrix (including shale), water and gas, respectively. The value of ρ is obtained by use of the gamma density tool, and the value of ϕS_w is obtained from the neutron porosity measurement. If ρ_m, ρ_w and ρ_g are known, ϕ and the volume fraction of gas, $\phi(1 - S_w)$, can readily be obtained.

During the first runs with the interpretation programme it turned out that the volume frac-

tion of gas in the shale–sandstone mixtures was always negative where it should have been very close to zero. The reason for this was that shale contains bound water. Since the neutron porosity tool measures the total concentration of water whether bound or in the pores, the value of ϕS_w obtained is too large. Assuming that the weight fraction of bound water in the shale is known it is possible to calculate the volume fraction of shale and of sandstone, provided the volume fraction of gas is very small. This is the procedure used to calculate the volume fractions of shale and sandstone. If the bound water content of the shale is not known it may be obtained by use of the cross plot technique (Hovgaard & Ølgaard 1990).

NULIP, a nuclear log interpretation programme

By use of the tool models and the log interpretation procedure described above it is possible to obtain a consistent log interpretation. This may be achieved by use of the NULIP programme which by use of the measured tool signals and the four models (or equations) calculates ϕ, ρ, Σ_a, and the geological type by an iterative procedure.

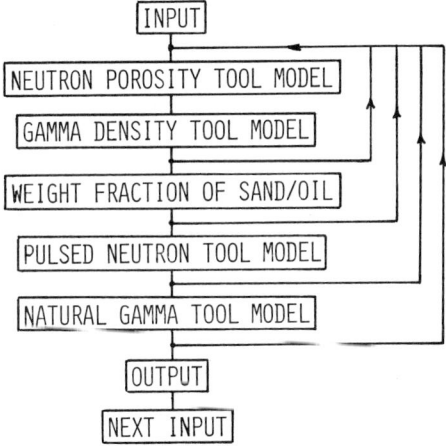

Fig. 6. Simplified block diagram of the nuclear log interpretation programme NULIP.

A simplified version of the block diagram for NULIP is shown in Fig. 6. For each measurement depth the signals of the four tools considered are read into the computer together with trial values of the four unknowns. These trial values are—except for the very first measure-

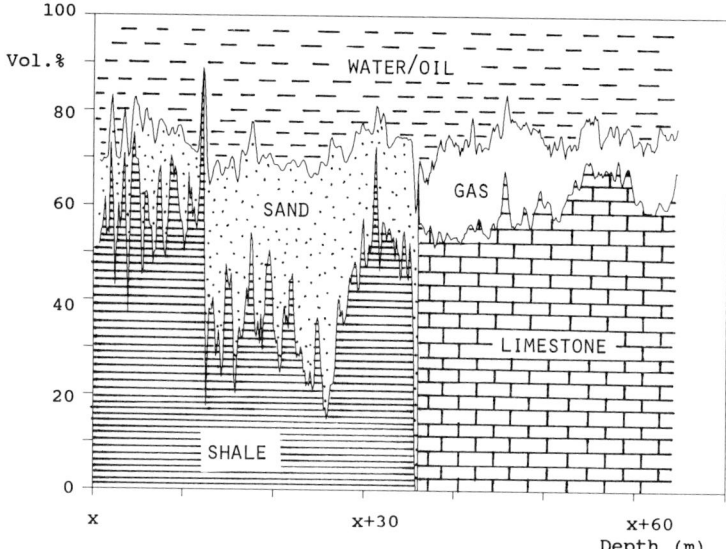

Fig. 7. Interpretation of logs from the four nuclear tools. The measurements were made in a borehole in the North Sea.

ment depth—equal to the values determined for the preceding measurement depth.

The first model considered is that of the neutron porosity tool. Using the trial values of ρ, Σ_a and geological type the value of the porosity ϕ is varied until the neutron porosity tool signal calculated by use of the model is equal to that measured.

The second model considered is that of the gamma density tool. Using the just determined value of ϕ and the inferred geological type, the density ρ is varied until the calculated tool signal is equal to that measured with the gamma density tool. Since the new ρ value is rarely consistent with the new ϕ value, calculations with the two models are repeated until convergence.

Provided the formation is a sand/shale mixture, the new values of ϕ and ρ are used to calculate a new sand and shale weight percent of the matrix. If the weight percentages change a new ϕ–ρ iteration is performed.

Upon convergence of ϕ and ρ the model of the pulsed neutron tool is considered using the new values of ϕ and ρ. The value of Σ_a is varied until the calculated tool signal is equal to that measured with the pulsed neutron tool. Since the new Σ_a is not necessarily consistent with the ϕ value used, calculations are repeated until convergence.

Next the measured count rate of the natural gamma tool is normalized so that it corresponds to the reference borehole geometry. Also, it is determined whether the normalized count rate is on the same side of a predefined value, normally

25 API, as that of the preceding calculation. If this is not the case, the matrix material is changed and the whole calculation is repeated. If it is the case, the calculation is finished and the programme proceeds to the next measurement depth.

By use of the calculated values of ϕ and ρ as well as the assumed values of ρ_m, ρ_w and ρ_g it is possible to determine the volume fractions of matrix, water/oil and gas. If the matrix is a shale/sand mixture, the volume fraction of these two components may also be determined, provided no gas is present and the bound water content of the shale is known.

The computing time per depth value is 1.65 seconds for a 25 MHz 80386 personal computer with math-processor.

Examples of applications of the NULIP programme

In Fig. 7 an example of a NULIP interpretation of logs of the four tools is presented. The logs were obtained in a borehole in the North Sea. An important shortcoming of NULIP is that it cannot distinguish between water and oil. It was hoped that Σ_a could be used to determine the content of water and oil since the water contains NaCl which has a high neutron absorption cross section while oil does not. But attempts to determine the oil and the water volume fractions through this method have not been successful. One reason is that neither the Σ_a of the matrix material nor that of the formation liquid has a

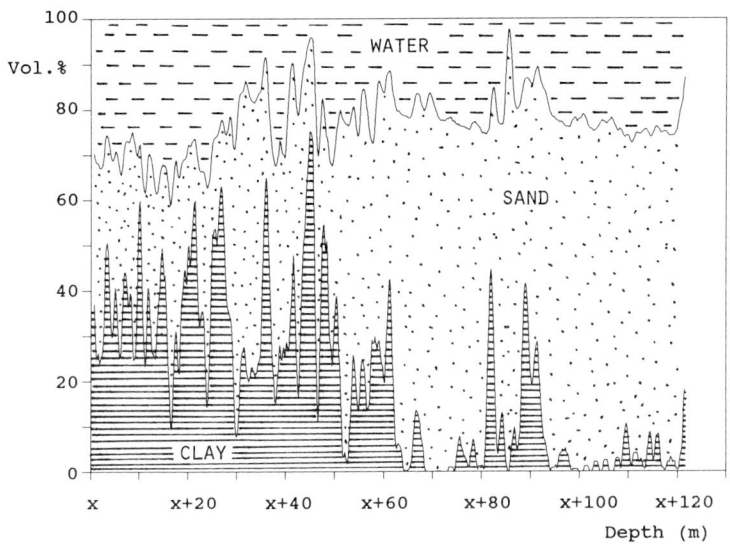

Fig. 8. Interpretation of logs from a borehole at Zealand. The raw logs of the gamma density, the neutron porosity, and the natural gamma tools were used together with Σ_a value from a pulsed neutron tool.

well defined value. It is the intention to investigate whether this important shortcoming of NULIP can be overcome by including, for example, a resistivity log in the interpretation.

A similar interpretation was made based on the ρ and ϕ values obtained by the logging company. The matrix volume fraction obtained in the two cases agreed very well, but the interpretation of the logging company data gave a larger volume fraction for gas and a smaller one for oil/water. Unfortunately, available information did not permit an assessment of which of the two interpretations was more correct.

In Fig. 8 another example of the use of NULIP is given. Here the logs originate from a borehole connected to investigations of a gas storage facility in a sand layer. In this case raw tool signals were not available for the pulsed neutron tool, but the Σ_a data were, so they were used in the interpretation. This means that the pulsed neutron tool model was not used, whereby the computing time for each depth value was reduced to one second.

Concluding remarks

From the results presented above it is seen that the NULIP programme can be used to perform a consistent interpretation of the logs of the four nuclear tools considered.

The NULIP interpretation seems more consistent than that of the logging companies since it takes into account the multi-property dependence of the tool signals, and since the models used in the interpretation give—in principle—a more correct physical treatment of the logs than the empirical relations of the logging companies. On the other hand the models described above do contain a number of approximations which will introduce inaccuracies in the results. It could be feared that the model approximations could lead to a situation where convergence in the iteration would not be achieved. So far this has not been the case.

One shortcoming of the NULIP interpretation is that it only takes into account borehole effects to a very limited extent. This is due to the requirement of near real-time interpretation. The logging companies have developed a number of borehole corrections which could be included in the programme.

In principle NULIP needs as input the signals from all the four nuclear tools. In practice NULIP may work with fewer logs provided estimates of the value of the missing properties are available and supplied to the computer. While the natural gamma, the gamma density, and the neutron porosity tool are almost always used in borehole logging, the use of the pulsed neutron tool is the exception. And, if it is used, the logging with this tool is undertaken several months after the use of the other tools. So if the pulsed neutron logs are not available they may be replaced by an estimate of Σ_a of the formations. The same may be done for the natural gamma tool.

NULIP interpretations have only been performed on logs from a limited number of boreholes, and there is definitely a need for more tests. The present version of the NULIP can undoubtedly be improved, e.g. by taking more logs into accounts, by taking more formation types into account, and by improving the models used.

The most important—and most difficult—point is to assess the degree of correctness of the NULIP interpretation. It is hoped that such an assessment will ultimately be possible, e.g. by comparisons with core measurements. An assessment may well indicate the need for further improvements of NULIP, but it is nevertheless felt that the basic approach behind NULIP—the use of physical tool models which takes into account the multi-property dependence of the tool signals—is worthwhile pursuing.

It is a great pleasure to acknowledge the financial support received from the Commission of the European Communities and the Danish Energistyrelsen.

We are greatly indebted to the Mærsk Olie og Gas company and to Dansk Olie- og Gasproduktion for making field logging data available to us.

Finally the programming work performed by Rudi Petersen and the assistance rendered by J. Hovgaard is greatly appreciated.

References

GLASSTONE, S. & EDLUND, M. C. 1960. *The Elements of Nuclear Reactor Theory*. D. Van Nostrand.

GOLDSTEIN, H. & WILKINS, J. E. 1954. Calculations of the penetration of gamma rays. Final report. Nuclear Development Associated, Inc. United States Atomic Energy Commission. NYO-3075.

HESLOP, A. 1974. Gamma-ray log response of shaly sandstone. *The Log Analyst*, **15**, 16–21.

HOVGAARD, J. & ØLGAARD, P. L. 1990. *Investigations of the Stenlille-4 Borehole Logs*. Department of Electrophysics, Technical University of Denmark. BHR-55.

ØLGAARD, P. L. 1985. An improved model for a pulsed-neutron borehole gauge. Department of Electrophysics, Technical University of Denmark.

BHR-31.

—— & HAAHR, V. 1967. Comparative experimental and theoretical investigations of the DM neutron moisture probe. *Nuclear Engineering and Design*, **5**, 311–324.

—— & PETERSEN, R. 1988. *Processing of Logging Data from Nuclear Tools*. Department of Electrophysics, Technical University of Denmark. BHR-45.

SCHLUMBERGER 1985. Log interpretation charts. Schlumberger Well Services.

WAHL, J. S., NELLIGAN, W. B., FRENTROP, A. H., JOHNSTONE, C. W. & SCHWARTZ, R. J. 1970. The thermal neutron decay time log. *Society of Petroleum Engineers Journal*, Dec. 1970, 365–379.

Mineralogy and geochemistry

Downhole mineralogy logs: mineral inversion methods and the problem of compositional colinearity

P. K. HARVEY & M. A. LOVELL

Borehole Research, Department of Geology, University of Leicester, Leicester, LE1 7RH, UK

Abstract. The inversion of chemical data derived from geochemical and related nuclear spectroscopy tools to provide mineralogy logs is now a widespread approach to the interpretation of these data. The most valuable of such transforms attempt to provide the percentages of actual phases present (minerals and fluids) at each depth interval rather than the ideal minerals occurring in simple theoretical models. Of the numerous problems involved in the inversion for a particular phase assemblage, the most serious is probably that of compositional colinearity in which three or more of the phases sought lie on, or close to, the same compositional plane. Depending upon the algorithm used for the inversion the effects of such compositional constraints may vary between a failure to find any (numerical) solution, failure to find a unique solution, or a solution which may be significantly in error for only a very small deviation in the chemical log or phase compositions. These effects are illustrated using geochemical logs from sedimentary environments together with examples from laboratory derived and numerically simulated geochemical data.

Nuclear logging tools are now capable of measuring most major and some minor elements which occur in rocks, thus producing continuous downhole geochemical logs. Schlumberger's Geochemical Logging Tool (GLT), for example, which comprises four separate tools run in series, measures absolute abundances of Si, Al, Ti, Fe, Ca, K, S, the minor elements Gd, Th and U together with H and Cl (Herzog *et al.* 1987). Measurement of the photoelectric absorption cross section also allows estimation of the subtotal of (Mg + Na) by assuming that these are the only major elements not directly measured. This component is usually treated as Mg on the grounds that Mg is much greater than Na in most sediments, though there are some obvious exceptions to this assumption. With appropriate calibration the initial elemental yields may be recast into the more conventional oxide form to give a virtually complete major element oxide analysis at each measured depth level, typically every 15 cm down the borehole. The use of such data in the study of sediments, and particularly their application in reservoir characterization and formation evaluation, is at a very early stage. Potentially a geochemical log should be able to provide significantly better lithological information than has hitherto been possible, in addition to providing estimates for certain physical parameters, such as grain density or thermal conductivity, that are difficult to obtain from other logs.

Faced with six or more log curves, each representing one of the major oxides, an immediate problem with geochemical logs is one of presentation, and a useful approach is to convert the oxide data into quantitative mineral assemblages. These may then be portrayed as mineralogy logs, in exactly the same manner as lithology logs, or converted through an appropriate mineralogical classification to provide geochemically derived lithology logs (Herron 1988).

The conversion of the geochemical data to mineral abundances (mineral transformation or mineral inversion) has long been employed as an investigative tool by geologists. One possibility is to recompute the oxides to a fixed set of (usually ideal or theoretical) minerals to produce a norm. The CIPW norm, still widely used by igneous petrologists, is probably the best known procedure of this type. This approach is also employed by Herron & Herron (1990) in the context of geochemical logging. While the normative approach is useful for comparative purposes, a mineralogy log is of much greater value if expressed in terms of the actual minerals present, that is, as a chemical mode. In the discussion that follows the aim is to generate mineral modes from the chemical data.

Despite the apparent simplicity of converting a chemical analysis to a set of minerals, there are a large number of pitfalls if an accurate 'chemical mode' is to be attempted. There are several basic assumptions which may be made about the

From HURST, A., GRIFFITHS, C. M. & WORTHINGTON, P. F. (eds), 1992, *Geological Applications of Wireline Logs II.* Geological Society Special Publication No. 65, pp. 361–368.

361

relationship between the chemistry of a rock and its mineralogy:

(a) the phase assemblage in a rock (its mineralogy plus other phases such as water) is a linear function of the chemistry;
(b) for a sequence of rocks, perhaps encountered sequentially in a borehole, the mineral assemblage may vary from one rock to the next;
(c) even for two rocks with the same mineral assemblage the compositions of some of the minerals involved may vary from one rock to the next;
(d) measurement errors expected in the chemical estimates, which in the case of GLT data can be quite significant, may take the form of random errors of measurement or systematic biases from the true values.

With these basic constraints it is clear that in order to generate an 'accurate' mineralogy log from sequential borehole measurements, data for each analysis (depth level) must be considered as a separate mineral inversion problem.

Compositional colinearity can arise when three or more mineral phases included within a model lie on, or close to, the same compositional plane or vector. In the limit this can prevent a numerical solution, or at least a unique solution, being obtained from the chemical data alone. More typically the solution determined may be significantly in error. In the former cases the problem is usually obvious (algorithm failure), while in the latter, which is probably more serious, it may not be noticed. This contribution demonstrates the effects of compositional colinearity, and indicates situations in which these effects should be considered when interpreting the final mineralogy logs.

Setting up a model

For each depth interval, then, it is necessary to choose the appropriate mineral assemblage and the compositions of the model minerals that are to be employed. In our experience, the very minimum that should be known at this stage is a list of minerals which actually occur in the borehole under study, obtained from mineralogical analysis of core samples. The same data should also yield a list of possible mineral parageneses. The strategy for selecting the correct mineral assemblage is still very much open for discussion; some simple strategies are described by Harvey et al. (1990).

A knowledge of the compositions of the minerals to be used for modelling is usually much more difficult to obtain and it is common

practice to use ideal or 'typical' compositions. This practice is usually safe for minerals like quartz, more difficult for, say, calcite ($CaCO_3$ + Mg? + Fe? etc.) and almost impossible for minerals like biotite, chlorite or the clay minerals, unless appropriate core data are available for guidance. Once a model is chosen there is the problem of its solution. There are several algorithms available for deriving a mineral assemblage from a list of minerals and their allotted compositions. These are briefly summarized in the following sections.

(a) Subtractive algorithms

The CIPW norm represents the concept of a subtractive algorithm where fractions are subtracted successively from an original chemical analysis in proportion to the amount and composition of the mineral involved. Typically one of the chemical components acts as a limiting concentration. In this way if ilmenite were to be extracted, then on the assumption that all the titanium present was in ilmenite, the percentage of normative ilmenite becomes a simple function of the $TiO_2\%$, and a corresponding stoichiometric amount is subtracted from the Fe_2O_3 concentration, TiO_2 being set to zero. With successive extraction of minerals in a precisely defined order, the original chemical analysis eventually becomes exhausted. The method has proved to be of considerable use in igneous petrology, and to some extent in metamorphic studies, where there is generally a high degree of chemical equilibrium and control over the minerals present, but it has been less successful in sediments (Harvey et al. 1990). Compared with some of the other methods discussed below, subtractive algorithms always produce a stable solution but there is no way in which the 'quality' of the norm or mode produced can be quantified. In that the chemical analysis taken as the input is exactly partitioned into the minerals, any errors in the original analysis must be passed directly to the derived mineral assemblage.

(b) Error minimization (least squares) algorithms

Mineralogy logs are usually generated by the solution to sets of linear simultaneous equations. The approach is well established in log analysis and has been the basis of lithology modelling for some time (Savre 1963; Doveton 1986). In these applications various physical logs (sonic, neutron, density, etc.) are related to lithological logs through a matrix containing the (physical) properties for each lithology of interest, so that:

$$Xp = c \qquad (1)$$

where p is the vector of phase (lithology) concentrations, c the vector of component or log responses (physical measurements, or oxide data in a mineralogical application), and X the physical constants matrix (see Harvey *et al.* 1990 for a more complete discussion). In a typical log analysis application the physical properties for each lithology (X) are defined and X inverted once; p is then determined at each depth interval as the log is processed. Similar treatment with geochemical data carries the assumption that the mineral assemblage is the same at each depth interval, as are the compositions of the minerals. Equation (1) is appropriate when the number of equations is equal to the number of phases; that is, for a fully determined system of equations, and is not in itself an error minimization model. Once, however, the number of equations exceeds the number of phases, the system becomes over-determined and a solution is sought by an error minimization technique, usually least squares. The latter performs a 'best fit' between the given geochemistry and the compositions of the minerals included in the model. If the analytical precision of the individual elements is known, a weighted least-squares solution could be preferable using the inverse elemental precisions as weights (Herron & Herron 1990). There can be severe problems in obtaining a stable solution using this approach and negative mineral proportions in a solution are common.

(c) *Optimization algorithms*

Another approach is offered through a range of optimization techniques of which linear programming has already been used for solving petrological mixing problems and computing chemical modes (Wright & Doherty 1970; Banks 1979). Similar in certain respects to the least squares algorithms, a set of constrained equations are established, groups of which are solved in turn as a minimum or maximum is sought in a defined objective linear function. In the mineral transform model the objective function is the sum of the modal minerals (the maximum being sought). For each oxide, the constrained equation requires that the amount of oxide needed to 'make' the modal concentrations of all the minerals does not exceed the total amount available in the original (input) analysis. Other constraints may be included to ensure that the Fe/Mg ratio between coexisting minerals, for example, is held constant, and that the derived mineral proportions must be greater or equal to zero.

One effect of these constraints is that the mineral proportions are constrained to be positive, the observed (tool response) compositions cannot be exceeded by that required by the optimized final estimates of the proportions, while the sum of the proportions will be maximized towards, but not exceed, unity. This means that negative mineral concentrations cannot occur; if a mineral does 'not fit' it is determined as a zero proportion, though by the same constraints the effects of a low chemical value will be propagated through the whole analysis. The simplex method is the most commonly employed optimization algorithm in this context.

(d) *Euclidian distance algorithm*

Another algorithm considers the multidimensional geometry inherent in the X matrix and makes use of the euclidian distances between the (end-member) model mineral compositions and the (input) chemical composition. As an example, consider the (ideal) minerals quartz, sillimanite and K-feldspar, which may be represented completely in a three-dimensional compositional space defined by the orthogonal axes: SiO_2, Al_2O_3 and K_2O (Fig. 1). Any rock composed only of these three minerals (point +, for example, in Fig. 1) will lie on the compositional plane that contains these minerals and within the compositional triangle defined by them. The composition of +, in terms of the three minerals, is directly related to is normalized euclidian distances from the apices of the compositional triangle. While this illustration is three dimensional, the same idea may be extended to hyper-polygonal or multi-component compositional systems and hence has application as a general solution to the mineral inversion problem.

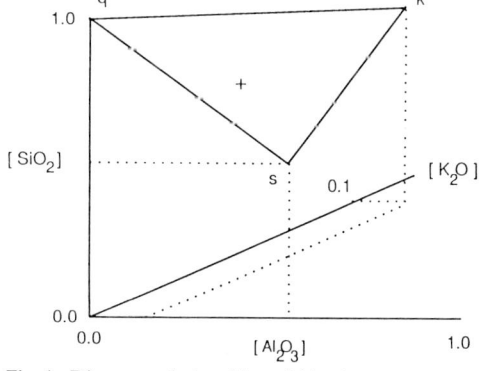

Fig. 1. Distance relationships within the system $[SiO_2]–[Al_2O_3]–[K_2O]$. q: quartz, s: sillimanite, k: potash-feldspar. Axes are in units of molecular proportions. See text for explanation.

This method allows up to the same number of minerals (phases) as components, with an implied constraint that the derived mineral assemblage will sum to 100%. Sample compositions which lie outside the compositional simplex defined by the model minerals will give negative results for those minerals. For example, in Fig. 1, if there were sufficient silica in a particular sample to form some sillimanite and some k-feldspar, but insufficient remaining in excess to allocate to quartz then that sample would plot outside the QSK triangle, beyond the SK join, and away from the Q corner. Any compositional solution based on quartz, sillimanite and k-feldspar for this sample would give a negative quartz abundance.

Comment on algorithms

All the algorithms can give apparently good or sensible results provided appropriate minerals and model mineral compositions are used, and will, under ideal conditions, yield identical solutions. Serious problems can, however, begin to arise when several of the minerals included in the models have similar chemistry. This raises the problem of compositional colinearity.

Compositional colinearity

Compositional colinearity can arise when three or more mineral phases included within a model lie on, or close to, the same compositional plane or vector. A limiting situation, unlikely ever to be encountered in a real log, is shown in Fig. 2. Consider the three minerals quartz, kyanite and corundum. Given the analysis of a sample consisting only of SiO_2 and Al_2O_3 (all other components being zero), such as S, in Fig. 2, then $[SiO_2] = 0.25$, $[Al_2O_3] = 0.75$. (Note square brackets refer throughout to molecular proportions, rather than weight percentages, which have been used for reasons of simplicity.) In terms of the three minerals, there is no unique solution for S; extreme possibilities in molecular units, include:

quartz: 25% kyanite: 0% corundum 75%
or quartz: 0% kyanite: 50% corundum 50%.

A related situation has arisen in the mineral modelling of quartz- and opal-bearing sediments encountered by the Ocean Drilling Program (J. F. Bristow, pers. comm.). The two minerals have the same effective chemistry (SiO_2) for modelling purposes and their relative proportions cannot be estimated from the geochemical logs alone. The mineralogical distinction

between these two forms of silica is very important if a thermal conductivity log were subsequently to be derived from the mineral log, the thermal conductivity of quartz being significantly higher than that of opal.

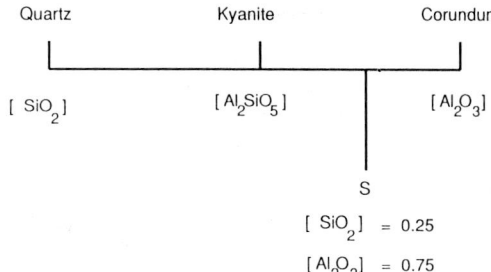

Fig. 2. Compositional colinearity in the system $[SiO_2]$–$[Al_2O_3]$. The mineralogical composition of sample 'S' is indeterminate. See text for a full explanation. Axis is in units of molecular proportions.

A more common and important case can be seen in the mineral modelling of detrital sediments with the possible occurrence together of K-feldspar, illite, kaolinite and muscovite. The 'ideal' compositions of all these minerals consist of at least three of the oxides: SiO_2, Al_2O_3, K_2O and H_2O. These compositional relationships are compared in Fig. 3 where the molecular proportions are shown normalized to one molecule of Al_2O_3. Hence, K-feldspar for instance, can be pictured as having three molecules of SiO_2 to one each of Al_2O_3 and K_2O (no H_2O). Between the four minerals the compositional range is limited: two or three molecules of SiO_2 each together with zero or one molecule of each of the other oxides. These relationships are well known by implication in metamorphic petrology where, for instance:

muscovite   = K-feldspar + Al_2SiO_5 + H_2O (a)

K-feldspar + H_2O = kaolinite + SiO_2 + K_2O (b).

Reaction (a) represents the high temperature break-down of muscovite (Al_2SiO_5 = sillimanite/kyanite/andalusite) while reaction (b) is one reaction involved in the alteration (kaolinization) of granites and compositionally related rocks. Any real (or imaginary) reaction which can be written in this manner implies a compositional colinearity, just as the assemblage quartz, kyanite and corundum (Fig. 2) could never be a chemically stable mineralogical assemblage:

quartz + corundum = kyanite

Table 1. *Comparsion of mineral inversion algorithms on a synthetic sample (Target comp.) which contains a significant degree of compositional colinearity. lsq: least squares algorithm, solved by matrix inversion (inv), or by direct solution (gsj: gauss-jordan, maximum pivot; chl: choleski decomposition); simplex opt.: linear optimization, simplex method, eucd, dist.: euclidian distance algorithm. MSE: mean square error against the target composition.*

	Target comp.	lsq (inv)	lsq (gsj)	lsq (chol)	simplex opt.	eucl. dist.
Quartz	20.00	23.54	20.34	19.82	19.83	19.96
K-feldspar	30.00	21.48	29.14	30.47	30.47	30.14
Muscovite	10.00	21.68	11.15	9.38	9.40	9.85
Kaolinite	30.00	24.02	29.30	30.38	30.38	30.15
Illite	10.00	10.71	10.07	9.96	10.00	9.91
MSE		7.181	0.731	0.396	0.390	0.115

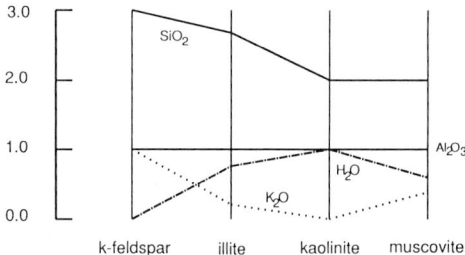

Fig. 3. Comparison of the composition of common aluminous ($+/-$ potassic) silicates normalized to one molecule of Al_2O_3 (Y-axis).

An additional problem which accentuates these effects is that no reliable estimate for H_2O is available from the geochemical logging tools. The determination of bound water from other log measurements offers a possible alternative but, for the benefit of this discussion and with the present state of processing nuclear logging data, the conversion from measured elemental yields to oxides, through an oxide closure model (Herzog *et al.* 1987), produces weight percent oxides which sum essentially to 100%, without water (or CO_2 etc).

A simple simulation in Table 1 illustrates this effect, where a 'synthetic' rock composed of quartz, K-feldspar, muscovite, kaolinite and illite is re-evaluated using a selection of algorithms. The mineral compositions used both to 'make' the simulated rock and in the algorithms are identical. Ideal compositions were employed for quartz, K-feldspar and muscovite while actual analyses were used for illite and kaolinite. The direct result of the built-in compositional colinearity (Table 1) shows, not surprisingly, that solving the least squares equations by matrix inversion produces almost totally erroneous re-

sults and that direct solution of the simultaneous equations is far superior. Also that good results, under the circumstances, are obtained by linear optimization and the euclidian distance algorithm. The latter in this simulation, and frequently in the majority of such simulations, has the lowest mean square error. Apart from a knowledge of the correct mineralogy and mineral compositions to use in this simulation, the reason that meaningful results can be obtained at all is because ideal ('pure') compositions were not used for two of the minerals. The illite in particular had respectable concentrations of TiO_2 (0.9), Fe_2O_3 (5.8), MgO (2.3), CaO (1.8) and Na_2O (0.5). Had ideal compositions been used then, in the absence of H_2O, all five minerals could be represented in the system $K_2O–Al_2O_3–Si_2O$. The latter is shown in Fig. 4, from which the full extent of the compositional colinearity can be seen with four possible reaction relationships being represented:

quartz + muscovite + kaolinite	= illite
K-feldspar + kaolinite	– illite
K-feldspar + kaolinite	= muscovite + quartz
quartz + muscovite	= K-feldspar + illite

In natural rocks, 'full' compositional colinearity of this type fortunately occurs rarely. In clastic sediments, where the situations discussed above are likely to be found, the clays and micas may differ significantly from their ideal compositions and potentially offer the extra degrees of freedom which enable a sensible solution to be reached. Muscovite, for example, is typically phengitic, and may contain several per cent of iron oxide. The important feature is that at least some idea of the mineral compositions is needed.

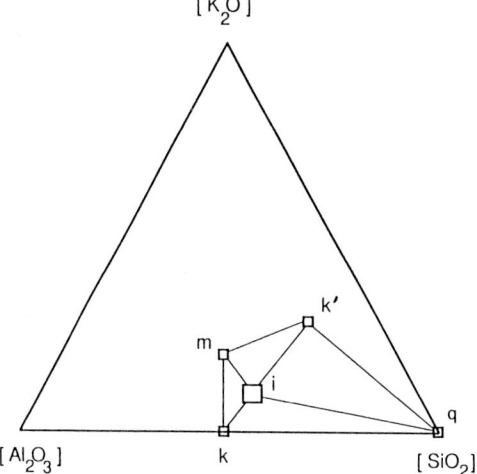

Fig. 4. Compositional colinearity in the system [SiO₂]–[Al₂O₃]–[K₂O]. m: muscovite, k: kaolinite, k′: potash-feldspar, i: illite, q: quartz. See text for explanation. Axes are in units of molecular proportions.

Table 2. *Illite compositions. See text for explanation.*

	Illite-1	Illite-2	Illite-3
SiO_2	56.91	54.13	54.79
Al_2O_3	18.50	23.76	20.67
TiO_2	0.81	0.61	0.82
Fe_2O_3	5.28	4.07	5.83
MgO	2.07	1.59	1.55
CaO	1.59	1.29	1.19
Na_2O	0.43	0.32	0.41
K_2O	5.10	4.83	6.63

top 20%, of a sandy (quartz rich) lithology (1), followed by some 35% of the log of a shaly horizon with less than about 50% quartz and illite the dominant clay mineral (2). Beneath this (3) there are dominantly sandstones with a significant content of feldspar together with clay minerals. These features are consistent with the known mineralogy.

A further example, with additional clay phases, can be introduced with the inclusion of smectite group minerals which vary in composition from the Si/Al bearing smectite, pyrophyllite, through montmorillonite and other minerals which contain small amounts of Ca, Na, Fe and Mg. The effects of mineral composition can be demonstrated here with a short section of log from a North Sea borehole. The core mineralogy (XRD) indicates the presence of quartz, K-feldspar (orthoclase), albite, kaolinite, montmorillonite and illite as the essential phases through the section. For modelling purposes, ideal compositions were taken for the first three minerals. We have no information about the compositions of the three clays but for illustrative purposes have used mineral compositions from core data elsewhere in the same stratigraphic unit. In the latter the main variation is seen to occur in the composition of illite. For this reason three models are presented in which fixed compositions have been used for all minerals except illite, a different illite composition being employed for each model. The compositions of the three illites are given in Table 2.

Figures 5, 6 and 7 show the mineral logs derived for these three models. All three are good 'fits' to the original data with the standard error (SErr log in Figs 5–7) between the chemistry of the derived mineralogy and the input chemistry rising at worst to about 2% but being under 1% for most depth intervals. The log can be broken roughly into three sections:

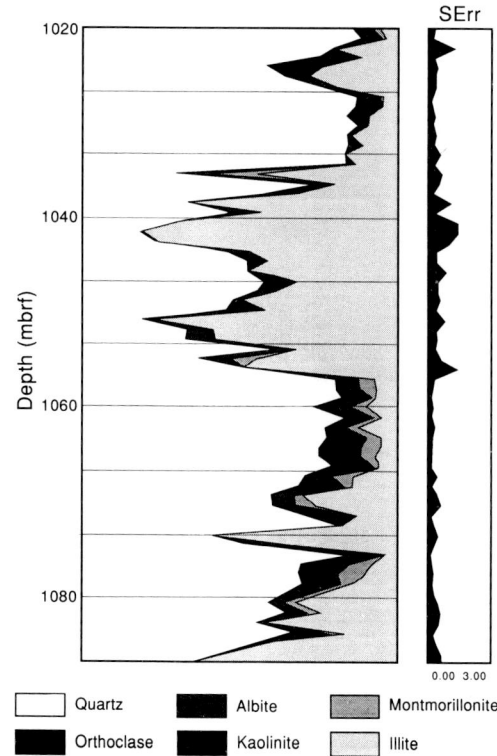

| Quartz | Albite | Montmorillonite |
| Orthoclase | Kaolinite | Illite |

Fig. 5. Model 1 mineralogy log from a test section in North sea clastic sediments. Modelled using Illite-1 (Table 2). SErr: standard error of fit.

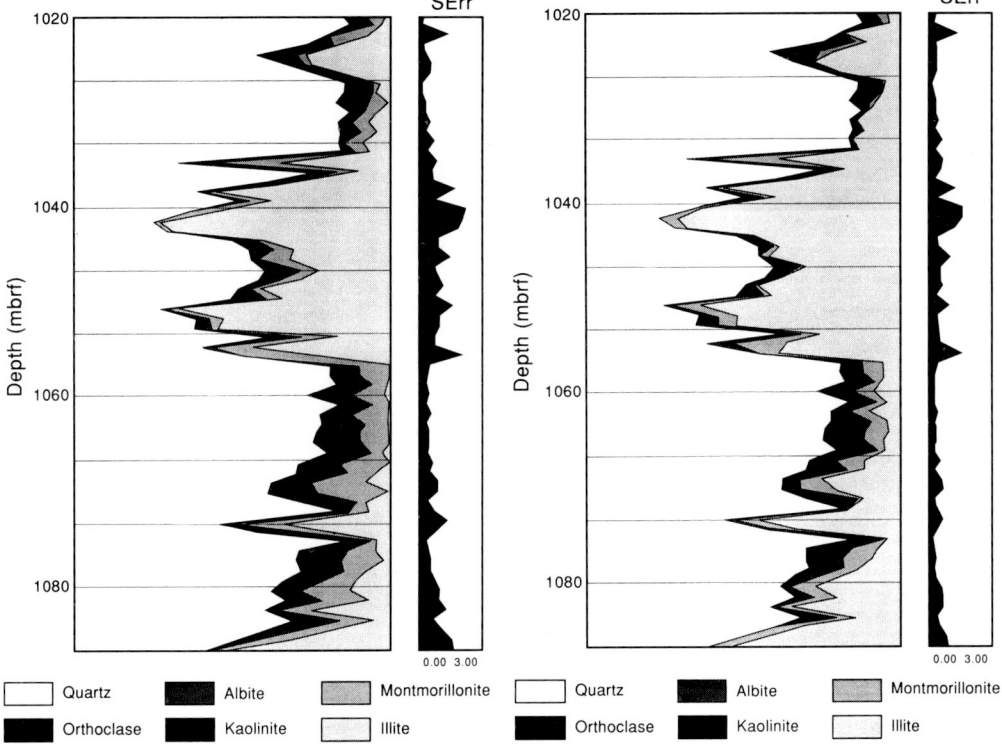

Fig. 6. As Fig. 5 but modelled using Illite-2 (Table 2).

Fig. 7. As Fig. 5 but modelled using Illite-3 (Table 2).

Table 3. *Summary statistics for the comparison of three mineralogy logs using different illite compositions. See text for explanation. Mean: arithmetic mean, S.D: standard deviation, Min.: minimum, Max.: maximum, Orth.: orthoclase, Kaol.: kaolinite, Mont.: montmorillonite. Number of depth intervals within the analysed section: 66.*

	Quartz	Orth	Albite	Kaol.	Mont.	Illite
Model 1 (Fig. 5)						
Mean	62.48	2.47	4.26	0.41	1.95	28.43
S.D.	19.15	2.52	2.42	1.29	3.06	21.39
Min.	17.81	0.00	0.00	0.00	0.00	3.73
Max.	87.83	9.53	9.90	6.82	16.97	81.53
Model 2 (Fig. 6)						
Mean	64.85	5.01	2.97	0.04	8.09	19.03
S.D.	17.16	2.83	2.38	0.33	5.47	19.88
Min.	25.37	0.00	0.00	0.00	0.36	0.00
Max.	88.14	10.39	8.78	2.69	25.49	72.88
Model 3 (Fig. 7)						
Mean	64.40	2.00	3.93	0.21	4.76	24.71
S.D.	17.90	2.51	2.46	0.94	4.17	18.69
Min.	23.07	0.00	0.00	0.00	0.00	3.27
Max.	88.16	9.02	9.53	6.16	23.91	70.98

Comparing Figs 5, 6 and 7 there are some obvious differences which would not have been seen if a single illite composition had been chosen. In Section 2 the lower illite contents are in the first model (Fig. 5) where there is virtually no orthoclase. The latter, however, is distinct in parts of Section 2 in the second model (Fig. 6). In Section 3 the presence of feldspars is quite distinct in all three models though significantly more feldspar and montmorillonite are present at the expense of illite.

Taken over the three models the summary statistics for the mineral proportions are given in Table 3. With the variation in illite composition imposed on the models significant variation occurs in all minerals except quartz. Despite significant compositional colinearity within this mineral assemblage, realistic mineralogy logs have been produced, though in the absence of reliable quantitative mineralogy from core measurements we cannot as yet distinguish between or test the accuracy of these models.

Conclusions

1. Compositional colinearity and the additional associated consequences of important oxides being absent from the oxide closure models are serious potential sources of error in the production of meaningful mineral logs from geochemical data.

2. These problems can probably be virtually eliminated if the mineral assemblage is known, the compositions of the mineral phases are known, and an appropriate inversion algorithm is applied.

3. Details of the actual assemblage require a combination of core derived mineralogy and, in an automated data analysis system, a reliable strategy to try and validate the choice of minerals.

4. The compositions of the minerals should also be obtained from core studies. Such information is far less forthcoming at present from core analysis than is desirable and the measurement of mineral chemistry should be encouraged. In addition, simulation studies should be undertaken to establish a predictable magnitude of error arising from errors in the mineralogy. Such errors should be directed as much towards parameters that might be derived from the mineralogy logs (such as grain density, cation exchange capacity, etc.) as to the quantitative mineralogy itself.

5. With regard to the choice of algorithm, matrix inversion techniques should be avoided; there is little to choose between the other models we have used, but both linear optimization and euclidian distance algorithms have some useful properties, not discussed in this contribution, which may make them more suitable in determining the validity of a mineral assemblage.

References

BANKS, R. 1979. The use of linear programming in the analysis of petrological mixing problems. *Contributions to Mineralogy and Petrology*, **70**, 237–244.

DOVETON, J. 1986. *Log analysis of Subsurface Geology. Concepts and Computer Methods*. Wiley, New York.

HARVEY, P. K., BRISTOW, J. F. & LOVELL, M. A. 1990. Mineral transforms and downhole geochemical measurements. *Scientific Drilling*, **1**, 163–176.

HERRON, M. M. 1988. Geochemical classification of terrigenous sands and shales from core or log data. *Journal of Sedimentary Petrology*, **58**, 820–829.

—— & HERRON, S. L. 1990. Geological application of geochemical well logging. *In*: HURST, A., LOVELL, M. A. & MORTON, A. C. (eds) *Geological Applications of Wireline Logs*. Geological Society, London, Special Publication, **48**, 165–176.

HERZOG, R., COLSON, L., SEEMAN, B., O'BRIEN, M., SCOTT, H., MCKEON, D., WRAIGHT, P., GRAU, J., SCHWEITZER, J. & HERRON, M. 1987. Geochemical logging with spectrometry tools. Society of Petroleum Engineers, Paper SPE No. 16792.

SAVRE, W. C. 1963. Determination of a more accurate porosity and mineral composition in complex lithologies with the use of sonic, neutron and density surveys. *Journal of Petroleum Technology*, **15**, 945–959.

WRIGHT, T. L. & DOHERTY, P. C. 1970. A Linear Programming and Least Squares Computer Method for Solving Petrologic Mixing Problems. *Geological Society of America Bulletin*, **81**, 1995–2008.

Determining total organic carbon contents from well logs: an intercomparison of GST* data and a new density log method

K. J. MYERS[1] & K. F. JENKYNS[2]

[1] BP Exploration Operating Company Limited, 4/5 Long Walk, Stockley Park, Middlesex UB11 1BP, UK

[2] Department of Geology, University of Manchester, Manchester M13 9PL, UK

Abstract. A detailed case study shows that more accurate in situ organic carbon (TOC) estimates can be made by the interpretation of conventional density logs than by the carbon/oxygen log method, and at a fraction of the cost. The performance of the two methods has been compared in a cored section of the Schiste Carton source-rock in the Paris Basin. The density log method described in this paper method is relatively quick and simple to use, whereas the C/O method involves the interpretation of a specially run log.

Results show that with the density log method 80% of the TOC estimates fall within $\pm 20\%$ of the core measured values in a TOC range 2–6.5 wt%. The C/O log method, which involves 30 ten minute stationary measurements or three passes at 600 ft/hour, achieves comparable accuracies at higher TOCs (> 5 wt%), but is considerably less accurate at lower TOCs with only 50% of values within $\pm 20\%$ of the core measured values.

Many papers have proposed methods for the quantification of the organic carbon contents of hydrocarbon source rocks using wireline logs. Most use empirical correlations between various wireline log responses and total organic carbon (TOC) contents, e.g. gamma-ray and uranium contents (Schmoker 1981), bulk density (Schmoker 1979), sonic and resistivity (Dellenbach et al. 1983; Passey et al. 1990), porosity and resistivity (Stocks & Lawrence 1990). These methods have been successful, to varying degrees, in source-rocks with organic carbon contents > 2–3 wt%, but suffer from the need for calibration with core data for specific source rocks.

Schlumberger have developed a method for the quantification of total organic carbon (TOC) using the carbon/oxygen (C/O) log (Herron 1986). High accuracy at low levels of TOC has been obtained by stationary borehole measurements, but at an uneconomically low data recovery rate, while a continuous method involving three repeat passes has produced more rapid results only at the expense of lower accuracy (Herron & Le Tendre 1990).

Here we have modified the Schmoker (1979) method for TOC quantification using conventional density logs, modelling the source-rock as a three component system (matrix, kerogen and porosity) using equations adapted from those of Mendelson & Toksöz (1985). The method we describe is used routinely in-house and has been tested on source-rocks from many different

depositional environments in basins worldwide. It has the advantages of speed and economy, as no special logging tools are required.

In this paper we compare the performance of the C/O and the density log methods in estimation of TOC contents in the lower Toarcian Schiste Carton source-rock in the Paris Basin. Eighteen metres of source rock within the Schiste Carton were cored and logged using the Schlumberger C/O tool. The core was sampled at both 60 centimetre and 20 centimetre intervals and analysed for TOC. The TOC profile for the 60 centimetre sampling interval (Fig. 1, adapted from Herron & Le Tendre (1990), is comparable with logging tool resolution, and so serves as an ideal calibrant for both the C/O log and the density log methods. The density log published by Herron and Le Tendre will be used to calculate TOCs and the results will be compared statistically to those of the C/O log method.

Principle of the Schlumberger C/O log method

The measurement of C/O ratios is made by inelastic gamma ray spectroscopy (GST) with nuclear logging tools (Roscoe & Grau 1985). The method of TOC calculation involves multiplying the measured C/O ratio obtained with a Schlumberger GST tool by an estimated oxygen content to obtain the total amount of carbon, both organic and inorganic. The inorganic car-

---

* Mark of Schlumberger

From HURST, A., GRIFFITHS, C. M. & WORTHINGTON, P. F. (eds), 1992,
Geological Applications of Wireline Logs II. Geological Society Special Publication No. 65, pp. 369–376.

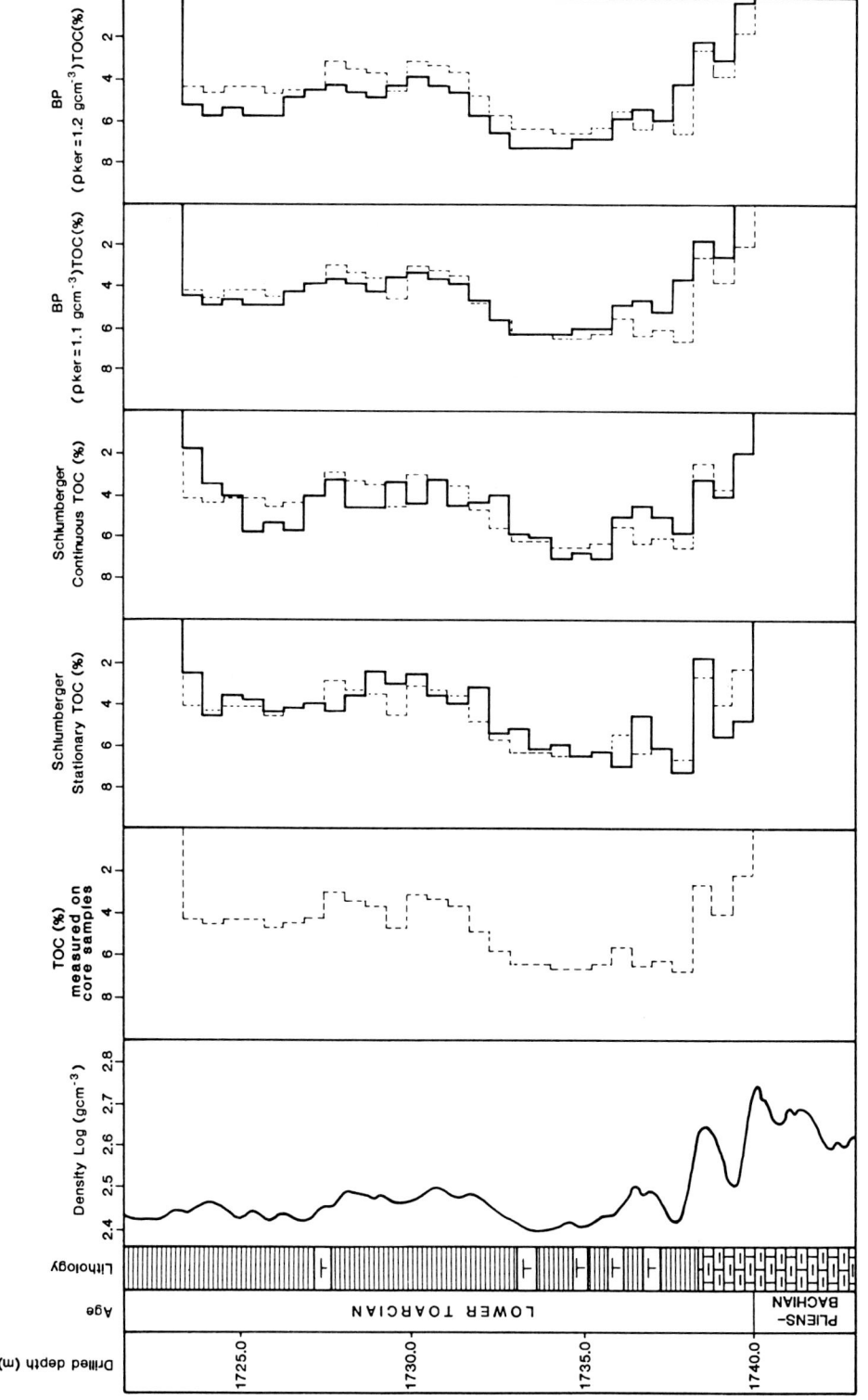

Fig. 1. Lithology, density log and core measured TOC profile for the Lower Toarcian Schiste Carton source rock in a Paris Basin well (after Herron & Le Tendre 1990), compared with estimated TOC profiles obtained by the density log and C/O methods.

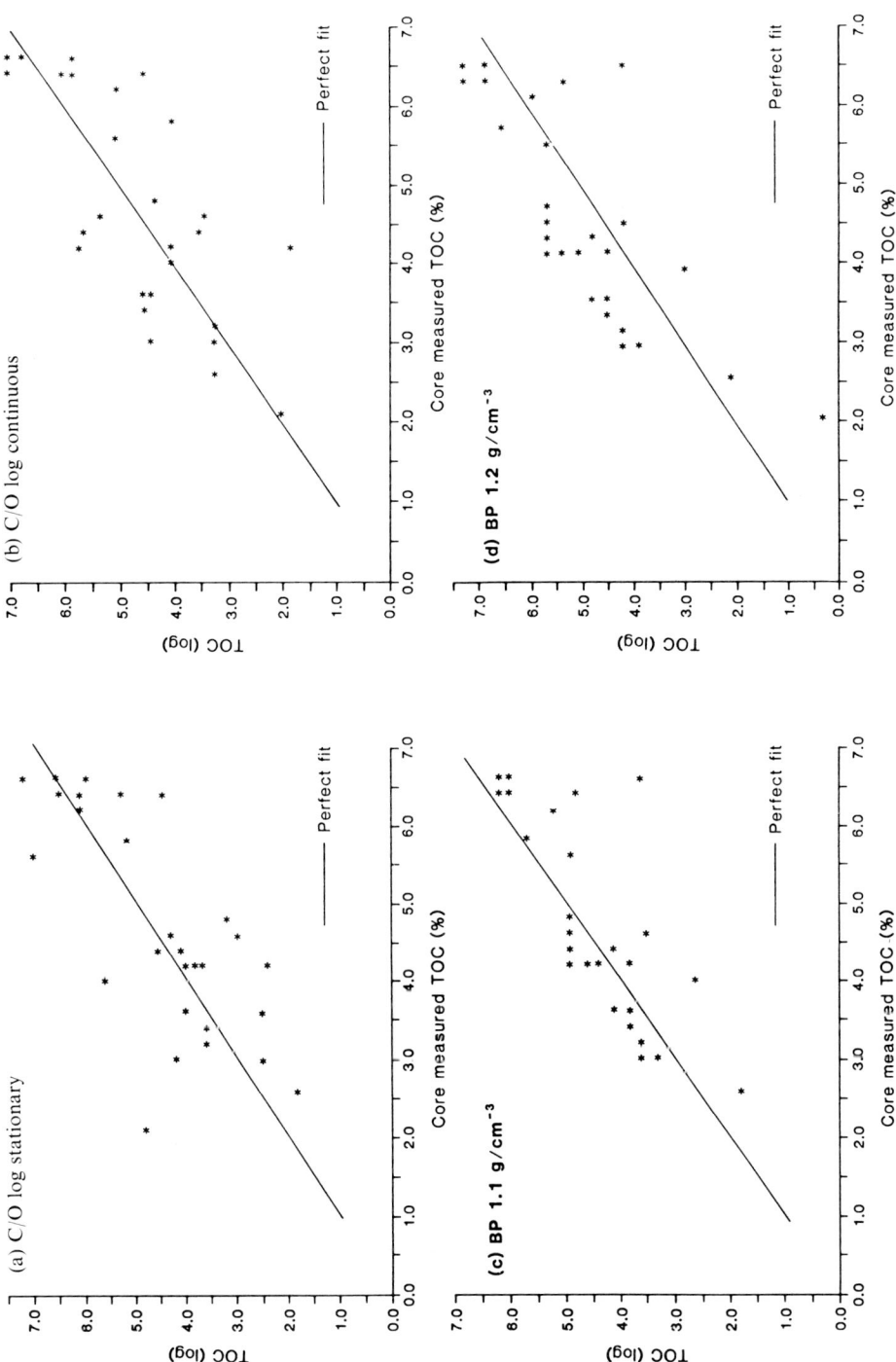

Fig. 2. TOC estimates versus core-measured TOC values.

bon is then estimated using calcium concentration logs (where carbon is present in calcite) or both calcium and magnesium logs (where dolomite also occurs), and subtracted from the carbon to obtain TOC. The method is described in more detail in Herron & Le Tendre (1990).

Principle of the density log method

A simple three-component petrophysical model of the source-rock is used to interpret measured bulk density in terms of weight% TOC. It represents a development of original work by Schmoker (1979) and Mendelson & Toksöz (1985). In this method, the potential source-rock is modelled in terms of a volume of matrix, a water-filled porosity and a volume of kerogen. Knowing the true porosity and matrix density of the organic-rich rock, the method calculates the weight% of organic matter (of an assumed density) required to effect the observed decrease in bulk density (see appendix).

To make the calculation the following parameters are required: the true porosity (ϕ_t), the kerogen density (ρ_{ker}), and the matrix density (ρ_{ma}). For the purposes of calculation, the true porosity of the mudrock is assumed to be independent of increasing organic carbon content and can therefore be calculated from the bulk density in an adjacent organic-poor interval or from a porosity–depth curve. This is the important development from the Mendelson & Toksöz (1985) method.

Kerogen densities vary but, in our experience, values of 1.1 to 1.2 g cm^{-3} are appropriate for many organic-rich mudrocks. The matrix density varies as a function of mineralogy but for modelling purposes is assumed to be constant. Laboratory analyses (unpublished internal data) have confirmed that matrix density values of 2.7 g cm^{-3} are appropriate for many mudrocks (see also Schmoker 1979).

The compressibility of mudrocks is assumed to be independent of organic carbon content and so the true porosity of the source interval can be calculated in an adjacent organic-poor interval.

In this case the true porosity was calculated in the Pliensbachian calcareous shales underlying the Schiste Carton source-rock (Fig. 1) using equation 1 (appendix). The true porosity was estimated to be 6 p.u., equivalent to a bulk density of 2.6 g cm^{-3}.

The source-rock shales are divided into 60 centimetre sampling units. The volume of kerogen is then calculated from the density log, using the calculated true porosity figures, and assuming kerogen densities of 1.1 and 1.2 g cm^{-3} and a matrix density of 2.7 g cm^{-3} (equation (2)). The kerogen volume is converted into weight, and finally into percent TOC using a wt% kerogen to wt% carbon conversion factor of 0.85 (equation (3)).

Results

The results of both the C/O and density log estimates of TOC are presented in Table 1. The C/O log results using both stationary measurements at 60 centimetre intervals and three passes of continuous measurements at 600 feet/hour are compared with the density log method using the same borehole interval and kerogen densities of 1.1 and 1.2 g cm^{-3}. The C/O log results were read from Herron & Le Tendre (1990, figs 13 and 16). The data are illustrated graphically in Figs 1–4.

The four sets of results are compared with TOC measurements made on core samples (i.e. core-measured) in Fig. 1. The same data are presented graphically as cross-plots in Fig. 2. The deviation of each estimated TOC value from its core-measured equivalent is recalculated as a percentage of the core measured value and plotted against the core-measured equivalents in Fig. 3. These percentage deviations are also illustrated as frequency histograms in Fig. 4.

Three statistical tests were run on each of the four sets of results, to compare them in accuracy with each other and with the results of TOC sampling at 60 centimetre intervals (Table 1). The three indices of accuracy thus produced are correlation coefficients, root-mean-square deviatons and bias indices. In general, correlation

Table 1. *C/O and density log estimates of TOC*

	C/O log (Stationary)	C/O log (Continuous)	BP (ρ_{ker} = 1.2 g cm^{-3})	BP (ρ_{ker} = 1.1 g cm^{-3})
Correlation coefficient	0.732	0.762	0.794	0.799
r.m.s. deviation	1.052	0.998	1.030	0.933
Bias	−1.700	−3.300	+10.800	−9.700

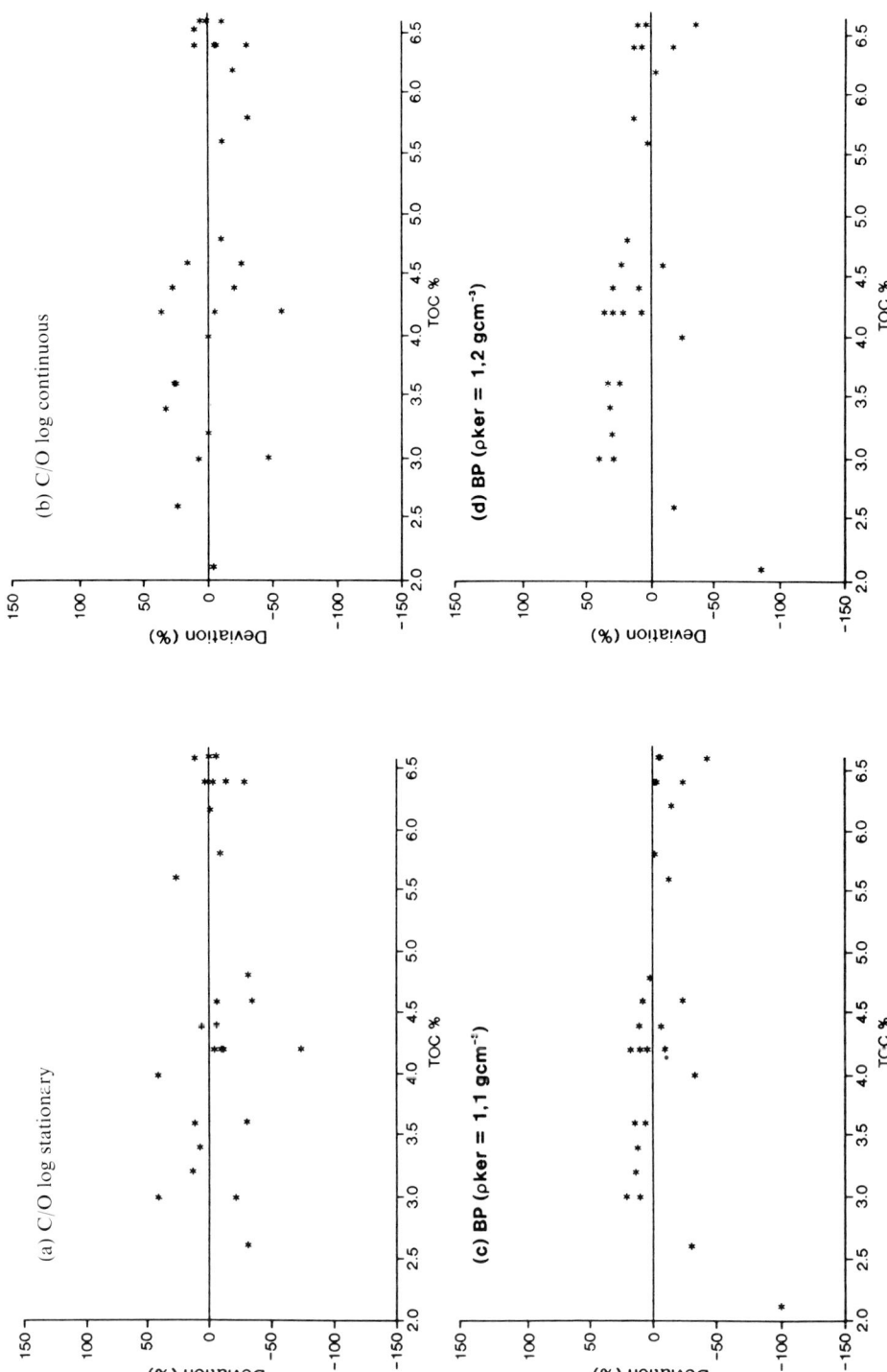

Fig. 3. Percentage deviation of log-derived TOC estimates from core-measured values as a function of increasing TOC.

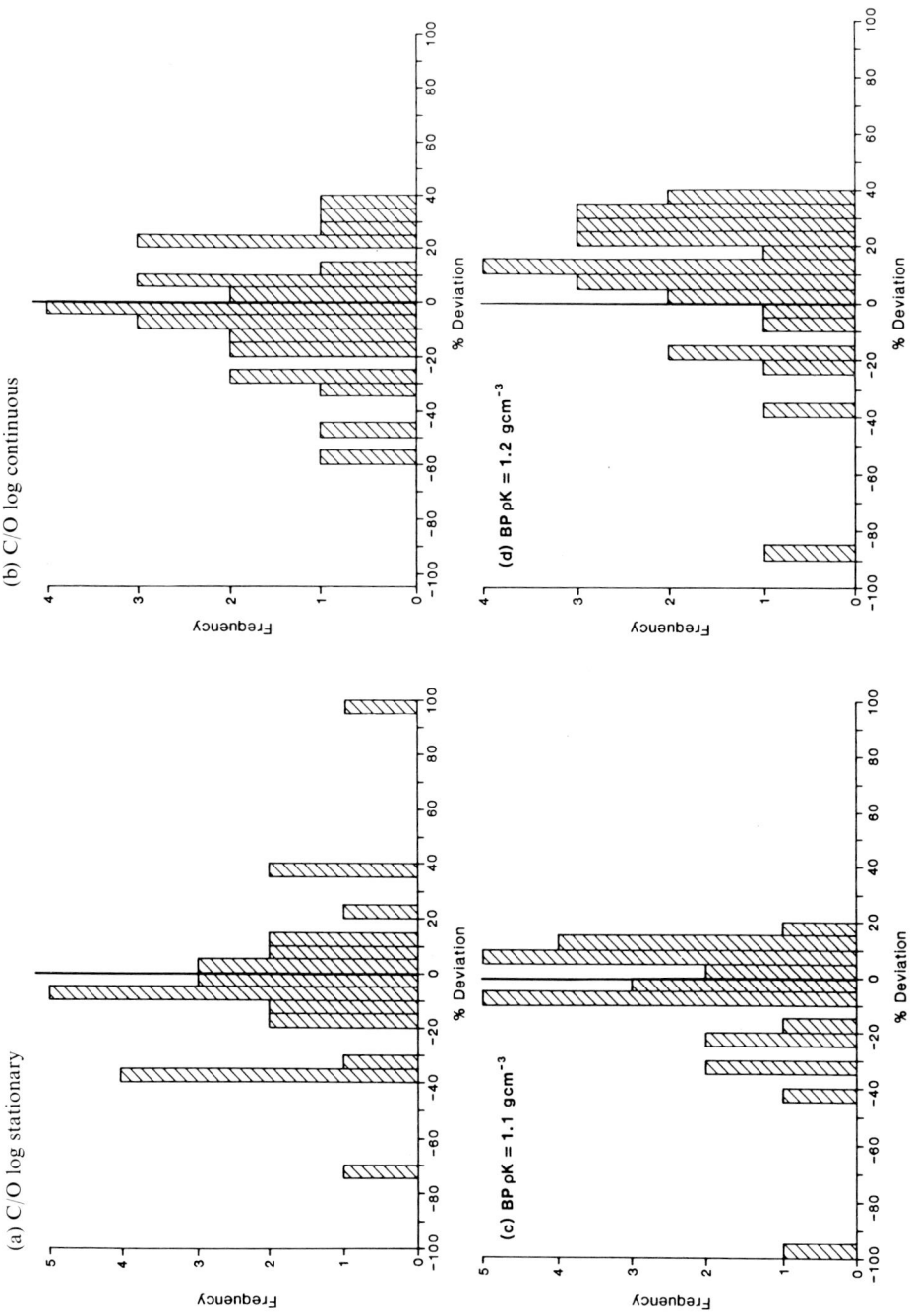

Fig. 4. Frequency histogram of percentage deviation of log-derived TOC estimates from core-measured values.

coefficients indicate the degree of correlation between the estimated and sampled TOCs, but will not reveal any systematic bias in the estimates. Root mean square devation is an index of the degree of departure of the estimates from perfect agreement with the sample data. This deviation may be the result of wide random scatter, in which case the bias index is low, or it may result from systematic over- or under-estimation, in which case the bias index is high and respectively either positive or negative.

Table 1 summarizes the results. As can be seen, the four methods give broadly similar results but with the density log method using a kerogen density of $1.1 \, \mathrm{g \, cm^{-3}}$ having the highest correlation coefficient and lowest root mean square deviation.

Interpretation and discussion

Figs 3 & 4 illustrate the comparative accuracy of the four methods. The density log estimate using a kerogen density of $1.1 \, \mathrm{g \, cm^{-3}}$ offers the highest accuracy, with 80% of the estimated TOCs deviating no more than 20% from core-measured values over a range of 2.5 to 6.5% TOC (core-measured TOCs themselves are considered to have accuracies of ±10%). With a kerogen density of $1.2 \, \mathrm{g \, cm^{-3}}$, 66% of the TOC estimates fall within 20% of the core-measured values, but with a bias towards positive deviations (i.e. over-estimates of TOC). The stationary and continuous C/O log techniques do not differ significantly in accuracy. Over a range of 5 to 6.5% TOC the C/O log results are comparable in accuracy with those of the density log, but at lower TOCs accuracy deteriorates, with only 50% of the estimates falling within 20% of core-measured values.

However, despite the higher overall agreement produced by the density log method, our TOC estimates show a tendency to be consistently slightly biased to one side or the other of the core measured TOCs, depending on the assumed kerogen density. The C/O log results show a greater scatter, but the scatter is more random. This difference is illustrated in the bias indices calculated for the four methods.

One possible consequence of the bias in the density log estimates might be an over- or under-estimation of the overall potential in the organically rich interval. Nonetheless, this bias can be minimized with a reasonable assumption of kerogen density.

Limitation of the density log method

Our experience is that this density log method

can be used to estimate the TOCs of organic-rich mudrocks in a variety of depositional settings in basins worldwide. The use of the technique in conjunction with conventional geochemical sampling can greatly enhance estimates of resource potential.

There are limitations of course. The calculation of the true porosity of the source-rock interval from an adjacent organic-lean interval may not be valid where overpressuring occurs or in organic-rich carbonate lithologies. Kerogen densities can vary with organic facies, generally increasing from type-I to type-IV kerogen. The presence of high-density minerals such as pyrite can affect the matrix density and be a further source of error.

Conclusions

1. Detailed calibration of TOCs interpreted from the density log using the BP methodology can predict TOCs to within ±20% of measured values.

2. The accuracies of the conventional density log and C/O log methods are comparable, but in this case study the density log offers more precise results.

3. Accurate estimates of TOC can be obtained without recourse to specially run geochemical logging tools.

We thank the many colleagues at BP who helped develop and test the techniques described here, particularly David Mann and Thomas Schwarzkopf. We also acknowledge the helpful reviews of the paper and thank BP Exploration for permission to publish.

Appendix. The density log method

Assumption 1: The source rock comprises mineral matrix (density = $2.7 \, \mathrm{g \, cm^{-3}}$), water-filled porosity (density = $1.05 \, \mathrm{g \, cm^{-3}}$) and kerogen (density = 1.1 or $1.2 \, \mathrm{g \, cm^{-3}}$).

Assumption 2: The non-source interval has the same mineral matrix and water density, and the same water-filled porosity, as the source interval.

Using:

density of non-source interval (taken from log) = ρ_{ns}
density of source interval (taken from log) = ρ_s
density of matrix = ρ_{ma} = $2.7 \, \mathrm{g \, cm^{-3}}$
water-filled porosity = ϕ_{fl}
density of kerogen = ρ_{ker} = 1.1 or $1.2 \, \mathrm{g \, cm^{-3}}$
density of water = ρ_{fl} = $1.05 \, \mathrm{g \, cm^{-3}}$
mass of kerogen = M_{ker}

then:

$$\rho_{ns} = \rho_{ma}*(1 - \phi_{fl}) + \rho_{fl}*\phi_{fl}$$

and

$$\phi_{fl} = \frac{\rho_{ns} - \rho_{ma}}{\rho_{fl} - \rho_{ma}} \qquad (A1)$$

and

$$\rho_s = \rho_{ma}*(1 - \phi_{fl} - \phi_{ker}) + \rho_{fl}*\phi_{fl} + \rho_{ker}*\phi_{ker}$$

and

$$\phi_{ker} = \frac{\rho_s - \rho_{ns}}{\rho_{ker} - \rho_{ma}} \qquad (A2)$$

and

$$TOC = 0.85*M_{ker} \qquad (A3)$$

From (A1), (A2) and (A3):

$$TOC = \frac{0.85*\rho_{ker}*\phi_{ker}}{\rho_{ker}*\phi_{ker} + \rho_{ma}*(1 - \phi_{fl} - \phi_{ker})}.$$

References

DELLENBACH, J., ESPITAILE, J. & LEBRETON, F. 1983. Source-rock logging. *Eighth European Formation Evaluation Symposium*, Paper D.

HERRON, S. L. 1986. A total organic carbon log for source rock evaluation. *Transactions of SPWLA 27th Annual Logging Symposium*, paper HH. (1986, revised 1987, *Log Analyst*, **28**, 520–527).

—— & LE TENDRE, L. 1990. Wireline source rock evaluation in the Paris Basin. *In*: HUC, A. Y. (ed.) *Deposition of Organic Facies*. American Association of Petroleum Geologists Studies in Geology **30**, 57–72.

MENDELSON, J. D. & TOKSÖZ, M. N. 1985. Source-rock characterization using multivariate analysis of log data. *Transactions of SPWLA 26th Annual Logging Symposium*, Paper UU.

PASSEY, Q. R., CREANEY, S., KULLA, J. B., MORETTI, F. J. & STROUD, J. D. 1990. A practical method for organic richness from porosity and resistivity logs. *AAPG Bulletin*, **74**, 1777–1794.

ROSCOE, B. A. & GRAU, J. A. 1985. Response of the Carbon/Oxygen measurement for the inelastic gamma ray spectroscopy tool. SPE 14460.

SCHMOKER, J. W. 1979. Determination of organic content of Appalachian Devonian shales from formation-density logs. *AAPG Bulletin*, **63**, 1504–1537.

—— 1981. Determination of organic-matter content of Appalachian Devonian shales from gamma-ray logs. *AAPG Bulletin*, **65**, 1285–1298.

STOCKS, A. E. & LAWRENCE, S. R. 1990. Identification of source rocks from wireline logs. *In*: HURST, A., LOVELL, M. A. & MORTON, A. C. (eds) *Geological Applications of Wireline Logs*. Geological Society, London, Special Publication, **48**, 241–252.

The third age of wireline log analysis: application to reservoir diagenesis

R. C. SELLEY

Department of Geology, Royal School of Mines, Imperial College of Science, Technology and Medicine, Prince Consort Road, London SW7 2BP, UK.

Abstract. Geophysical well logs were first developed to identify the porosity and fluid composition of potential petroleum reservoirs. Subsequently their use was extended to include diagnosing the depositional environment of reservoirs, as an aid to predicting their geometry. Both these techniques are now routine.

A new age of the application of geophysical logs now dawns due to the enhanced understanding of the diagenetic processes that affect petroleum reservoirs, including cementation which destroys porosity and solution which enhances it. These ideas can be used to interpret both conventional porosity logs and the new types of log that, when carefully interpreted, allow vertical variations in mineralogy to be measured.

Three main diagenetic effects can be recognized in geophysical well logs: facies-related diagenetic phenomena; cemented and leached zones related to migrating diagenetic fronts of acid meteoric or deep connate fluids; and diagenetic effects related to petroleum migration and entrapment.

The first geophysical log was run in a well at Pechelbronn in Alsace on 5 September 1927. Over the next half-century geophysical well logging became a standard method of evaluating the fluid content of potential petroleum reservoir formations internationally. The accuracy of these interpretations is generally confirmed by subsequent production tests and core analyses.

The second age of log analysis dawned in the 1960s. This was the application of geophysical well logs to the elucidation of the depositional environment of formations, as an aid to predicting the geometry and trend of petroleum reservoirs. The study of modern sedimentary environments enabled diagnostic environmental parameters, such as the suites and sequences of sedimentary structures, to be applied to subsurface formations. This is commonly inhibited due to the scarcity of coring (geologists who have only worked in the North Sea do not appreciate that extensive coring is not ubiquitous internationally). It was also recognized at this time that vertical grain size profiles could be used to identify channels, barrier bars and other sand bodies. Furthermore, upward-coarsening and upward-fining motifs could sometimes be identified on geophysical well logs, notably the gamma and spontaneous potential logs. The dipmeter, and other microresistivity and sonic devices for imaging borehole walls, have still further aided the diagnosis of the depositional environment and the prediction of the extent of reservoirs. This is now a standard technique on which whole textbooks have been written (e.g. Selley 1985; Rider 1986).

A third age of log analysis now dawns. This is the application of logs to the understanding and prediction of porosity variations caused by leaching and cementation within petroleum reservoirs. This has been brought about by the concatenation of two circumstances, one geological, the other geophysical. Firstly, geologists are developing a much clearer understanding of the processes that control the cementation and solution of minerals in the subsurface in general, and in petroleum reservoirs in particular (e.g. McDonald & Surdam 1984). Secondly, while conventional porosity logs, notably the sonic, have long been used to delineate cemented horizons within reservoirs, a new range of logging tools has now been developed whose data can be interpreted to display vertical variations of the mineralogy of reservoir formations (Harvey *et al.* 1990; Heron *et al.* 1990; Wendlandt & Bhuyan 1990). This combination of new science and new technology will have a major impact on petroleum production in the years to come.

The object of this paper is to outline some of the diagenetic motifs that may be identified both in conventional logs and in the newer geochemical logging tools. Three main motifs will be discussed: facies-controlled diagenetic responses; envelopes and diagenetic fronts; and petroleum–water contact effects.

From HURST, A., GRIFFITHS, C. M. & WORTHINGTON, P. F. (eds), 1992,
Geological Applications of Wireline Logs II. Geological Society Special Publication No. 65, pp. 377–387.

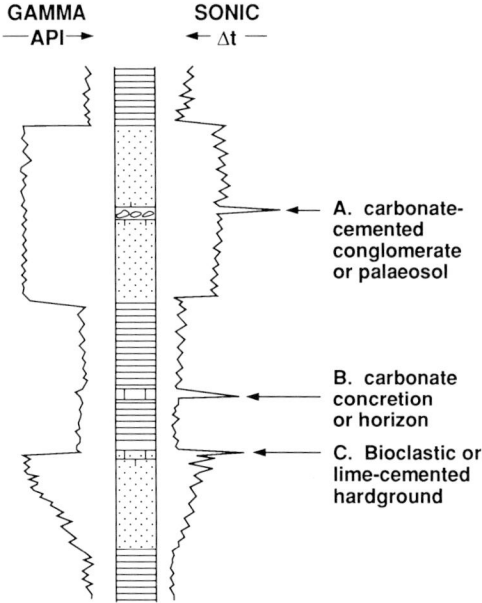

Fig. 1. Hand-crafted gamma-sonic log to show the many geological causes of the ubiquitous sonic spike. A, carbonate cemented conglomerate in sandstones; B, bioclastic horizon and/or hardground; C, carbonate concretion or horizon.

Facies controlled diagenetic phenomena

At their very simplest, geophysical logs have long been used to recognize diagenetic effects that are closely related to facies. They will be dealt with first as they provide a convenient starting point for the subsequent analysis of more complex motifs.

Sharp spikes on the sonic log are common in clastic sequences. They are not susceptible to a single simple interpretation because they are caused by diverse geological phenomena (Fig. 1). Sometimes they indicate bioclastic enriched horizons, often associated with early cementation (hardgrounds). Examples are commonly seen in the Jurassic formations of the Wessex basin, notably in the Bridport Sand (Bryant *et al.* 1988). In the Montrose Group of the northern North Sea, by contrast, sonic spikes are caused by calcite-cemented polymictic conglomerate horizons within multistorey grain flow sands. Sonic spikes are also produced by isolated concretions and continuous conformable early calcite-cemented horizons. Notable examples occur in the eponymous Dogger sands of the Jurassic and within Jurassic shales. It is important to be able to differentiate carbonate concretions from continuously cemented horizons because, while

the former are unlikely to be permeability barriers, the latter may well be (Maher 1981; Hurst 1987). Bjorkum & Walderhaug (1990*a, b*) have described how it may be possible to differentiate concretions from continuously cemented horizons using both cores and geophysical logs. They discuss how continuously cemented horizons may show as tight intervals on sonic, density and neutron logs. Nodules, however, give a less dramatic response on density and neutron logs because of their limited lateral extent. Though tight zones will show on resistivity logs, the dipmeter log may be used to differentiate continuously cemented beds from discontinuous nodules. This is because, in the latter case, resistivity readings will vary around the borehole, indicating the lateral discontinuity of the feature. Resistivity and acoustic borehole imaging tools may also be able to differentiate nodules from continuously cemented layers.

From these relatively simple diagenetic log responses it is now appropriate to turn to more equivocal ones.

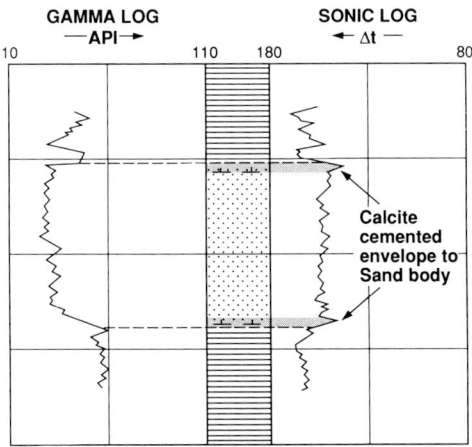

Velocity difference between tight & porous = 10m/s = 7.3% φ

A Gamma - Sonic Log.

Fig. 2. Gamma-sonic logs of wells illustrating cemented envelopes at sand : shale contacts.

Envelopes and diagenetic fronts

For many years geologists have noted that sandstones sometimes have envelopes of quartz and carbonate cement at shale boundaries. This was first described from Venezuela by Fothergill (1955) who attributed it to the retention of anions by filtration at the sands : shale contact. Fuchtbauer (1967) published detailed petrographic accounts of envelopes from Jurassic

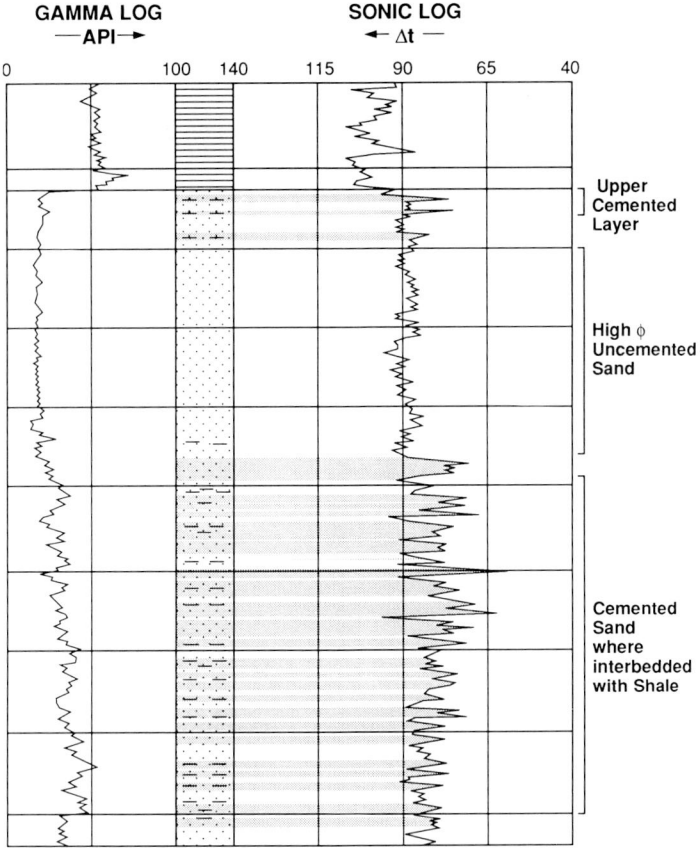

A Gamma - Sonic Log.

Fig. 3. Gamma-sonic log of a well with a clean sand body with well-defined envelopes overlying a ratty section in which the thin sands are tightly cemented.

sands in northern Germany and calibrated them with the spontaneous potential log. He noted that cemented envelopes occurred at the margins of both water-saturated and oil-saturated sands. He thus believed that cementation took place after petroleum invasion, and must be due to precipitation from waters moving into the sands from adjacent compacting shales.

Cemented envelopes can be identified from geophysical logs (Fig. 2). They commonly occur where the contact between a sand and a shale is abrupt rather than gradational. Of all the porosity logs, experience shows that the sonic log most clearly delineates envelopes, especially when they are caused by calcite cement. It has also been noted that thick sands tend to retain porosity, with or without cemented envelopes, whereas thin sands tend to be tightly cemented. This is shown by the log illustrated in Fig. 3.

Combining these two observations leads to the suggestion that sands are often cemented adjacent to shales irrespective of their thickness (Fig. 4). This phenomenon may be attributed to two mechanisms. It is possible that the cements are, and always have been, of only local extent, reflecting the precipitation of minerals from fluids expelled from adjacent shales. It is also possible, however, that the envelopes are relicts of what were once much more pervasive cements (Fig. 5).

The idea of the diagenetic front is a very old one in geology, though known in various guises such as Liesegang rings and roll-fronts (Stansfield 1917). Liesegang rings and related diagenetic front phenomena are more familiar to mining geologists than to petroleum geologists. Roll-front uranium ore bodies have been well documented for many years (e.g. Guilbert & Park

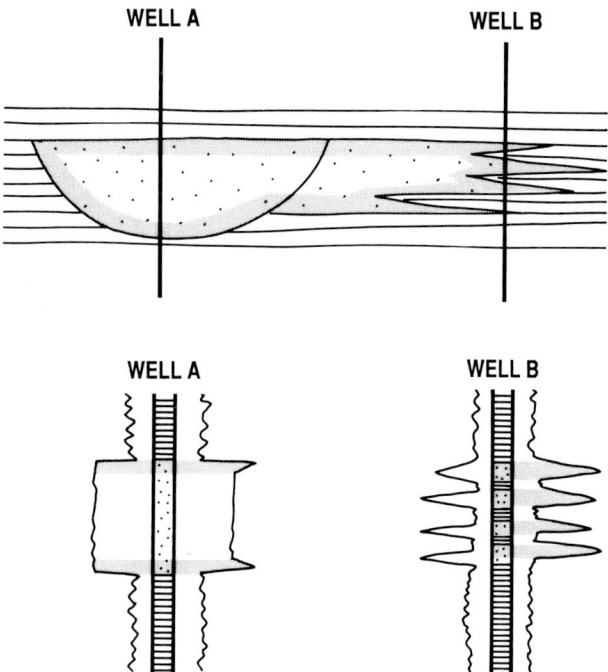

Fig. 4. Cartoon with hand-crafted gamma-sonic logs to show how envelopes in clean sands feather out into tightly cemented thin sands.

1986). Figure 6 illustrates the characteristic morphological and mineralogical features of a uranium roll-front ore body. This is really a caricature of the sort of diagenetic front that commonly appears to have moved through permeable sandstone beds resulting in the precipitation of iron or carbonate cements. Ferruginous diagenetic fronts are common in diverse sedimentary formations in southern England. Examples include the Permo-Triassic New Red Sandstone of Devon, the Cretaceous Lower Greensand, where the fronts are colloquially termed 'Carstones', and in the Tertiary sands of the London and Hampshire basins of southern England (Stoneley & Selley 1991).

Geologists have now established the chemical and physical processes by which diagenetic fronts move (Sultan *et al.* 1990) and have simulated their growth and geometry by computer modelling (Chen & Ortoleva 1990). These studies correlate well with the data observed in nature and provide insight into the spatial distribution of cemented zones seen in well logs of sandstone reservoirs. Shawe (1956) describes how there are three main types of roll-front, that he terms 'c', 's' and 'socket' ore bodies (Fig. 7A).

THIS:

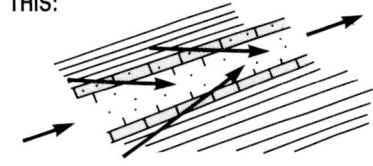

OR THIS:

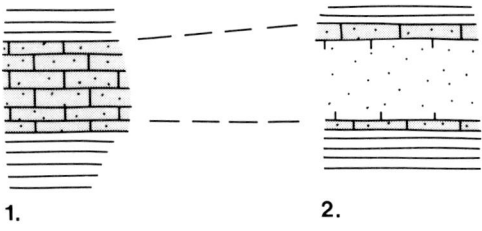

1.   2.

Fig. 5. Cartoons to show how cemented envelopes may either reflect the original extent of cement or be relics of more pervasive cements leached out by migrating diagenetic fronts of acid connate fluids.

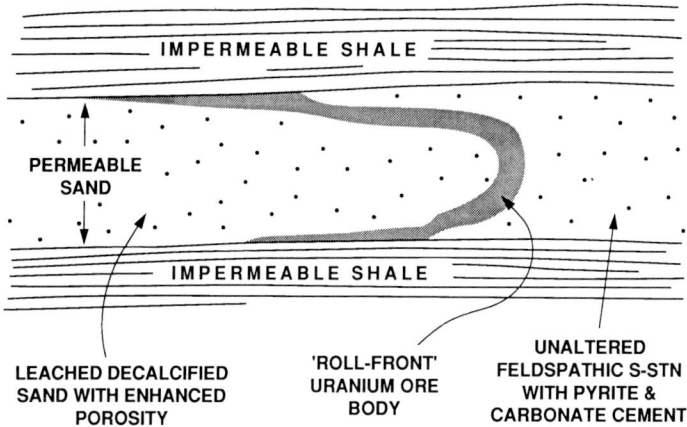

Fig. 6. Cartoon to illustrate the main features of a roll-front uranium ore body. This is a well documented caricature of the type of ferruginous or carbonate diagenetic front commonly seen in permeable sandstones.

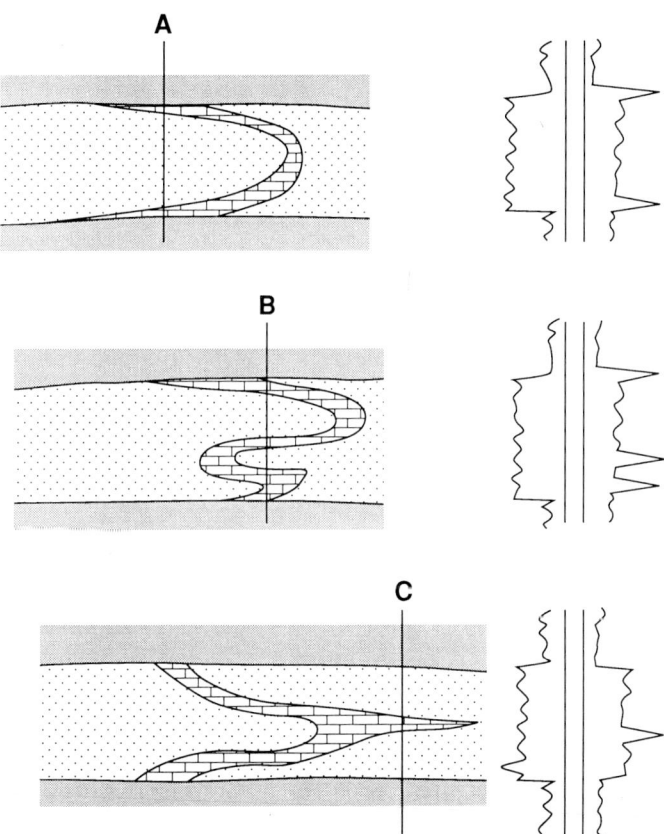

Fig. 7. Cartoons of the 'c', 's' and 'socket' diagenetic fronts described from uranium roll-front ore bodies by Shawe (1956) and hand-crafted gamma-sonic logs for wells drilled in carbonate fronts of analogous geometry. Note how a cemented envelope, illustrated in Fig. 2, may be an indicator that a diagenetic front has passed through a particular well location.

These match the computer simulated geometries described as 'fingers' and 'scallops' in carbonate fronts modelled by Chen & Ortoleva (1990).

A diagenetic front consists of three zones. On the up current concave side of the front (i.e. within the zone that was swept) carbonate grains and cement are leached out, feldspars are kaolinitized or leached out completely, iron is oxidized or leached out. There is normally therefore an overall increase in porosity. The front itself is commonly a tight zone of enhanced cementation, be it of carnotite, limonite or carbonate. The sandstone on the convex down current side of the front (i.e. not yet swept by the advancing diagenetic front) is unaltered, with fresh feldspar and preserved early carbonate cement and bioclasts. Iron may still be in a reduced state.

Sometimes the leaching of glauconite and organic matter shows that the advancing fluid was oxidizing. The ubiquitous leaching of feldspars, carbonate grains and cements shows that the fluid was invariably acidic. Diagenetic fronts are best known due to meteoric flow adjacent to modern land surfaces, and these provide analogues for the diagenetic zones to be encountered in petroleum reservoirs beneath unconformities (Selley 1990). There is now widespread acceptance that diagenetic fronts not only occur where meteoric fluids meet with deeper connate fluids, but also occur in the deep subsurface. These deeper diagenetic fronts are initiated by the compactionally driven flow of acidic connate waters expelled from compacting organic-rich shales (Schmidt & McDonald 1979). As the fronts move through permeable sands they cause the mineralogical reactions outlined above and likewise generate secondary porosity (Giles 1987).

With these concepts in mind it is interesting to hand craft the geophysical logs that would be produced by wells penetrating the three different types of diagenetic front recognized by Shawe (1956) (Fig. 7). Note that Well A, which penetrates the concave side of a 'c' type of front, will exhibit the envelope effect described earlier. A casual inspection of geophysical logs of interbedded sands and shales from wells drilled in many basins reveal not only cemented envelopes at sand:shale boundaries, but also erratically cemented intervals that lend themselves open to interpretation as diagenetic fronts. Figure 8 shows three sandstone beds from UKCS well 16/23/1 that simulate the 'c', 's' and 'socket' roll-front geometries defined by Shawe (1956). The foregoing review supports the thesis that the cemented envelopes of sandstone bodies predate petroleum migration and may serve as an indicator that acidic fluids have driven a diagenetic front through a sand bed.

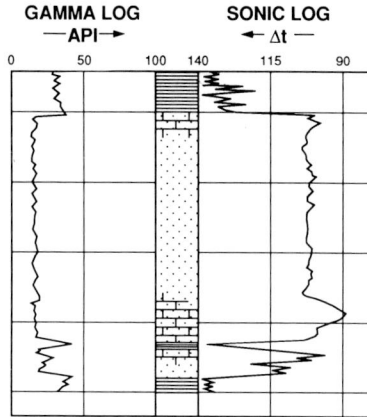

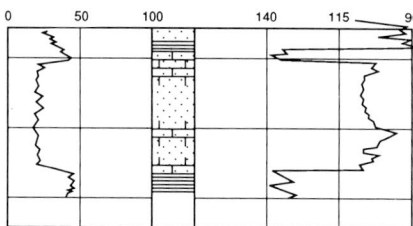

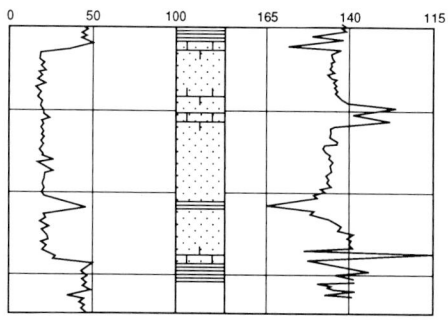

Fig. 8. Gamma-sonic logs from UKCS well 16/23-1 showing carbonate cement motifs that may be attributed to 'c', 's' and 'socket' diagenetic fronts analogous to those of uranium roll-front ore bodies illustrated in Fig. 7.

Careful studies of well logs may make it possible to map the distribution of the diagenetic fronts that mark the boundaries of secondary leached zones. This has an important corollary; it may be also possible, therefore, to delineate cemented zones that may serve as seals to petroleum migration, and may thus give rise to diagenetic traps (Cant 1986).

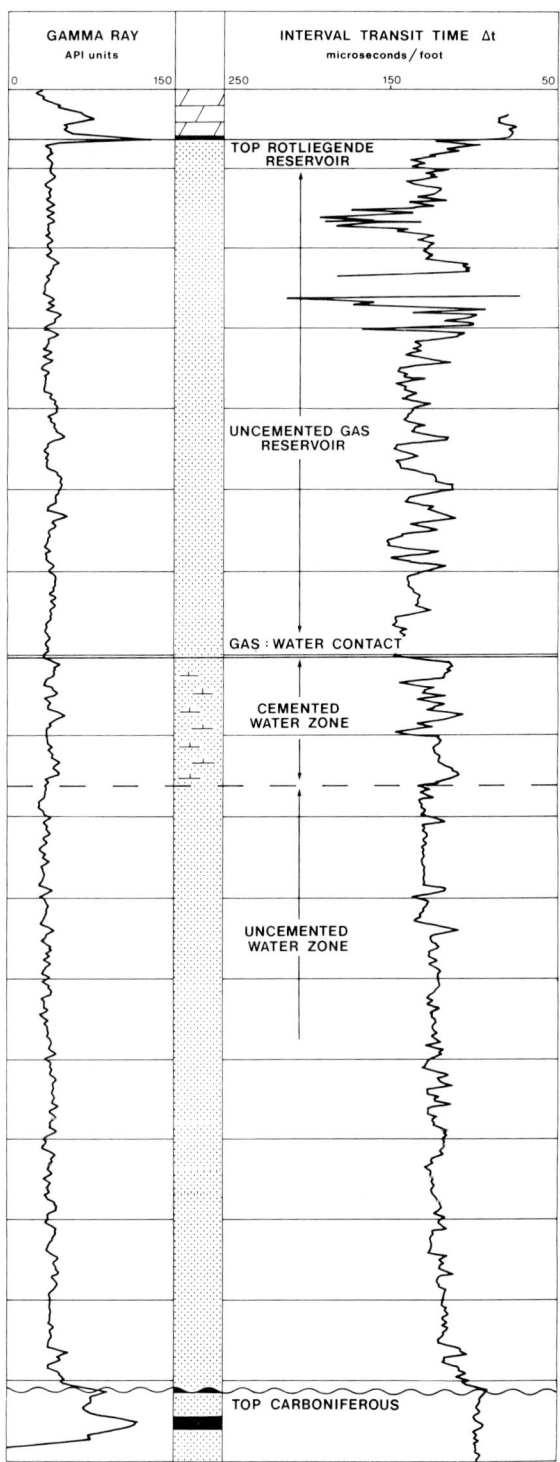

Fig. 9. Gamma-sonic log of a well in the southern North Sea showing a cemented zone beneath the gas : water contact of a Rotliegende (Lower Permian) sandstone reservoir.

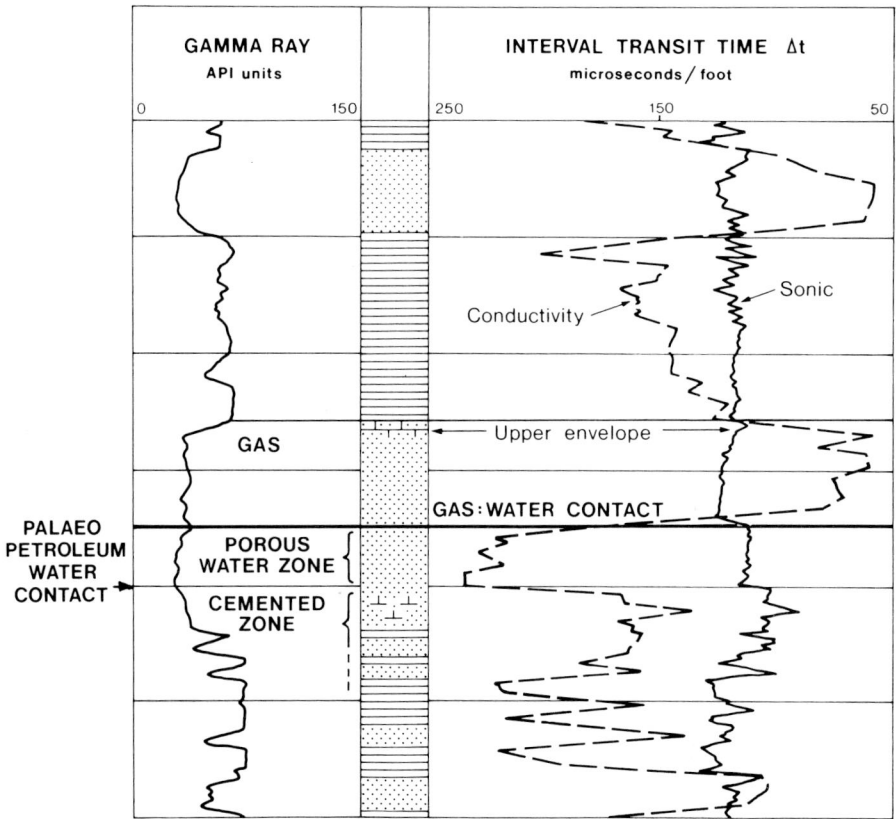

Fig. 10. Well showing a cemented zone some 7 m below the petroleum : water contact. This is open to interpretation as a palaeo-petroleum : water contact in a trap that has leaked or undergone tilting some considerable while after the original time of petroleum migration.

Log responses due to diagenesis at petroleum : water contacts

It has been known for many years that when petroleum invades a reservoir it expels connate fluids and may thus inhibit further porosity loss, at least by cementation. Below the petroleum : water contact, however, the pores are still saturated with connate water. As this continues to flow it may lead to further diagenesis and precipitation of mineral cements. Many fields are known where there are significant diagenetic changes adjacent to the petroleum : water contact. These take several forms. Normally they

include the development of post-migration cements of carbonate, silica, anhydrite, and, sometimes, pyrite. Celebrated carbonate examples have been described from the Arabian Gulf (Dunnington 1967) and good sandstone examples have been documented from the Denver basin by Levandowski *et al.* (1973). Post-migration diagenetic effects in water zones beneath petroleum reservoirs vary from the subtle to the extreme. Conventional geophysical logs may detect the more dramatic cases, but newer geochemical logging tools may detect subtler ones. Fig. 9 illustrates how a sonic log may be used to delineate a possible carbonate

Fig. 11. Gamma-sonic log of a well in the Frigg field of the North Sea. Conventionally the upward decrease in acoustic velocity at the petroleum : water contact would be attributed to the gas effect. But note that there is a 10 m oil column between the water and gas zones, and that Brewster & Jeangeot (1987) have shown that there is a palimpsest 'flat spot' at the original petroleum : water contact, even though another 'flat spot' has developed and moved gradually upward as the field has depleted.

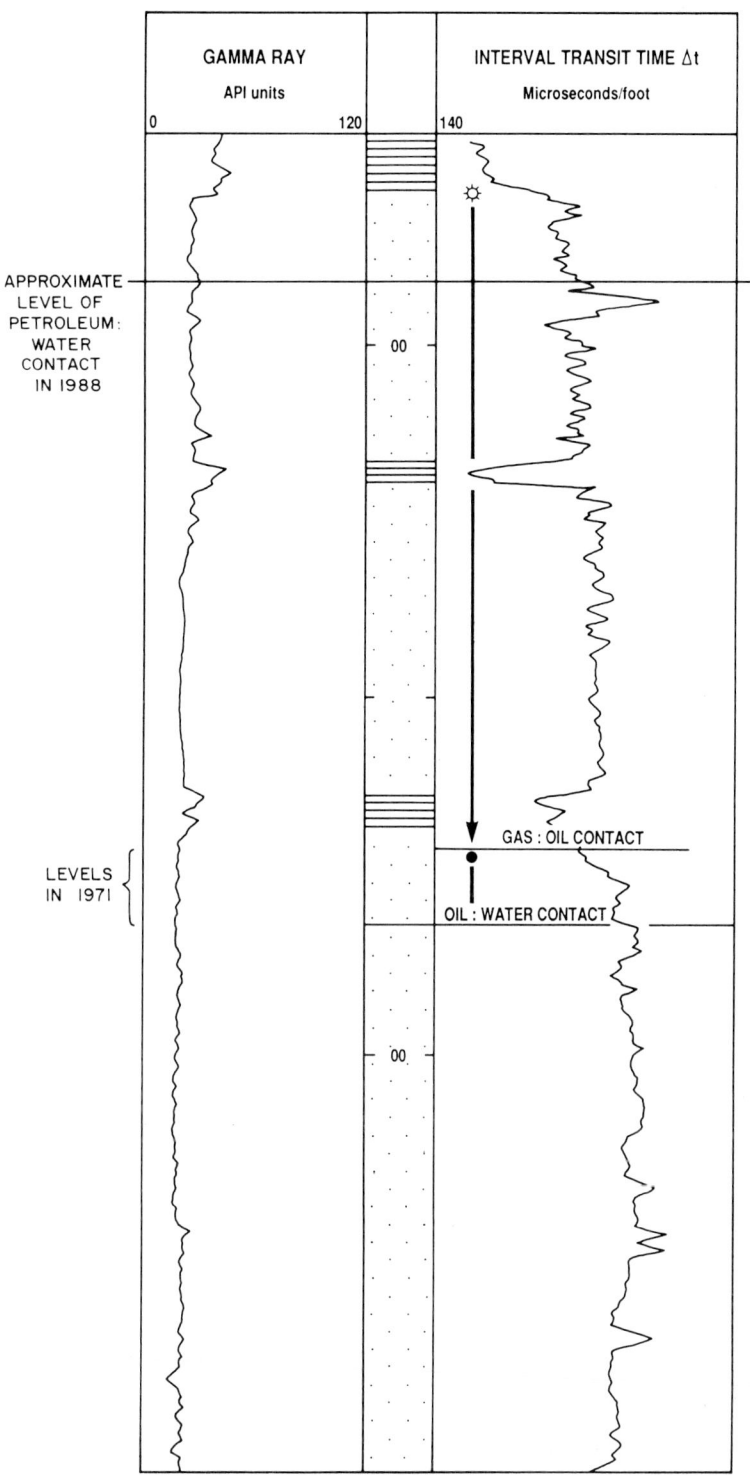

cement seat seal developed beneath a Rotliegende gas reservoir in the southern North Sea.

Sudden changes in porosity and diagenesis
sometimes occur in reservoirs at some depth
above or below petroleum : water contacts.
These are open to interpretation as palaeo-
petroleum : water contacts. The phenomenon
suggests that the trap in question has partially
leaked, or been tilted, leading to remigration and
an adjustment of the petroleum : water contact.
Figure 10 illustrates a well where there is evidence of a cemented zone below the present
petroleum : water contact, suggesting that the
trap may have leaked or undergone tilting in the
past.

The Frigg gas field of the North Sea provides
another interesting case of the relationship
between well logs and diagenesis. When the
Frigg field was first discovered and drilled, the
sonic logs commonly showed a sharp decrease in
acoustic velocity moving up from the water zone
to the petroleum zone (Fig. 11). This change on
the log initially correlated with a dramatic seismic 'flat spot'. Conventionally the break in the
sonic log would be interpreted as the hydrocarbon effect caused by the low velocity of the gas-
saturated sand. It is worth noting though, that
there is a 10 m oil leg separating the main gas
reservoir from the water zone (Heritier et al.
1981). The Frigg field was discovered in 1971
and production began shortly after. Seismic shot
in the mid-1980s revealed that a 'flat spot' has
moved upward as the field has depleted, but
there is still a weak 'flat spot' at the original
horizon (Brewster & Jeangeot 1987). This supports the hypothesis that the original 'flat spot',
and the deflection on the sonic log, was caused
as much by variation in reservoir cementation as
by the vertical change in fluid velocity. Numerous phantom 'flat spots' have been drilled
around the world that have been found to under-
lie water wet reservoirs. A possible explanation
is that the phenomenon marks an old cemented
seat seal beneath a petroleum accumulation that
has since leaked away.

These examples are fairly clear, and can be
recognized on conventional porosity logs. It is

noted in many fields that there are significant
changes in clay mineralogy adjacent to the
petroleum : water contact. Commonly kaolinite
is found in the petroleum leg and illite in the
water zone. This is a very important distinction
because of the way in which illite significantly
diminishes permeability. The Middle Jurassic
fields of the Brent province commonly exhibit
these changes. There are many accounts that
describe an upper zone characterized by secondary porosity and kaolinite produced by the
breakdown of feldspar, and a lower zone where
feldspar is still preserved, carbonate and pyrite
are abundant, and illite is the dominant clay
mineral (e.g. Hancock 1978). The present petroleum : water contact sometimes coincides with
the mineralogical change, sometimes occurs
above, and sometimes below it. This is open to
two interpretations. The boundary may mark a
palaeo-petroleum : water contact similar to those
discussed above. Alternatively the boundary
marks the lower limit of diagenetic fronts due
to meteoric leaching beneath the Cimmerian
unconformity (Selley 1984; Shanmugam 1990).
This is not the occasion to debate these issues.
What is clear, however, is that geophysical logs,
not only new tools, but also the older ones, may
be used to resolve the diagenetic history of
sandstones. Because of an improved understanding of diagenetic processes, it is possible to use
logs, corroborated by cores wherever possible, to
identify zones where porosity has been enhanced
by leaching, or diminished by cementation. Geophysical logs may thus aid the prediction of
porosity and permeability variation within
sandstone petroleum reservoirs. One must never
forget, however, that it has often been remarked
that geologists sometimes use well logs in the
same way that a drunk uses a lamp-post, more
for support than illumination.

The ideas presented in this paper have been based on
observations of well logs made over many years,
tempered by discussions with many students and oil
industry colleagues. I owe a particular debt, however,
to the staff of Schlumberger, to Dr H. S. Shaw of
Imperial College, and to the anonymous referees of
this paper.

References

BJORKUM, P. A. & WALDERHAUG, A. 1990a. Geometrical arrangement of calcite cementation within
shallow marine sandstones. Earth Science Reviews, **29**, 145–161.
—— & —— 1990b. Lateral extent of calcite-cemented
zones in shallow marine sandstones. In: BULLER,
A. T., BJERG, E., HJELMELAND, O., KLEPPE, J.,
TORSAETER, O. & AASEN, J. O. (eds) North Sea Oil
and Gas Reservoirs II. Graham & Trotman,
London, 331–336.
BREWSTER, J. & JEANGEOT, G., 1987. The production
geology of the Frigg Field. In: KLEPPE, J., BERG,
E. W., BULLER, A. T. HJELMELAND, O. and TOR-
SAETER, O. (eds) North Sea Oil & Gas Reservoirs.
Graham & Trotman, London, 75–88.
BRYANT, I. D., KANTOROWICZ, J. D. & LOVE, C. F.

1988. The origin and recognition of laterally continuous carbonate-cemented horizons in the Upper Lias sands of southern England. *Marine & Petroleum Geology*, **5**, 108–133.

CANT, D. J. 1986. Diagenetic traps in sandstone. *AAPG Bulletin*, **70**, 155–160.

CHEN, W. & ORTOLEVA, P. 1990. Reaction front fingering in carbonate-cemented sandstones. *Earth Science Reviews*, **29**, 183–198.

DUNNINGTON, H. V. 1967. Aspects of diagenesis and shape change in stylolitic limestone reservoirs. *Proceedings 7th World Petroleum Congress*, 339–352.

FOTHERGILL, C. A., 1955. The cementation of oil reservoir sands and its origin. *Proceedings 4th World Petroleum Congress*. Section 1, 300–312.

FUCHTBAUER, H. 1967. Influence of different types of diagenesis on sandstone porosity. *Proceedings 7th World Petroleum Congress*. Section 3, 353–369.

GILES, M. R. 1987. Mass transfer and problems of secondary porosity in deeply buried hydrocarbon reservoirs. *Marine and Petroleum Geology*, **4**, 188–204.

GUILBERT, J. M. & PARK, C. F. 1986. *The Geology of Ore Deposits*, 4th edn. Freeman, New York.

HANCOCK, N. J. 1978. Diagenetic modelling in the middle North Sea Brent sand of the northern North Sea. *European Offshore Petroleum Conference, London*. Paper No 92, 275–280.

HARVEY, P. K., BRISTOW, J. F. & LOVELL, M. A. 1990. Mineral transforms and downhole geochemical measurements. *Scientific Drilling*, **1**, 163–176.

HERITIER, F. E., LOSSEL, P. & WATHNE, E. 1981. The Frigg Gas Field. *In*: ILLING, L. V. & HOBSON, G. D. (eds) *Petroleum Geology of the Continental Shelf of North-West Europe*. Heyden, London, 380–394.

HERRON, M. M., HERRON, S. L. EVERETT R. V. & McDONALD, J. E. 1990. Enhanced resistivity interpretation in three wells using geochemical log data. *Transactions of SPWLA 31st Annual Logging Symposium*, Paper U.

HURST, A. 1987. Problems of reservoir characterization in some North Sea sandstone reservoirs solved by the application of microscale geological data. *In*: KLEPPE, J. (ed) *North Sea Oil and Gas Reservoirs*. Graham & Trotman, London, 153–168.

LEVANDOWSKI, D., KALEY, M. F., SILVERMAN, S. R. & SMALLEY, R. G. 1973. Cementation in Lyons sandstone and its role in oil accumulation, Lyons Sand, Denver Basin. *AAPG Bulletin*, **57**, 2217–2244.

McDONALD, D. A. & SURDAM, R. C. (eds) 1984. *Clastic Diagenesis*. American Association of Petroleum Geologists, Memoir **37**.

MAHER, C. E. 1981. The Piper Oilfield. *In*: ILLING L. V. & HOBSON G. D. (eds) *Petroleum Geology of the Continental Shelf of North-West Europe*. Heyden, London. 358–370.

RIDER, M. H. 1986. *The Geological Interpretation of Well Logs*. Blackie, Glasgow.

SCHMIDT, V. & McDONALD, D. A. 1979. *The Role of Secondary Porosity in the Course of Sandstone Diagenesis*. Society of Economic Paleontologists and Mineralogists, Special Publication, **26**, 175–207.

SELLEY, R. C. 1984. Porosity evolution of truncation traps: diagenetic models and log responses. *Proceedings of Norwegian Offshore North Sea Conference, Stavanger*. Norwegian Petroleum Society, Oslo, Paper G3.

—— 1985. *Ancient Sedimentary Environments*. 3rd edn. Associated Book Publishers, London.

—— 1990. Porosity evolution of truncated sandstone reservoirs. *In*: ALA, M., HATAMIAN, H., HOBSON, G. D., KING, M. S. & WILLIAMSON, I. (eds) *Seventy-five Years of Progress in Oil Field Science & Technology*. Balkema, Rotterdam, 103–111.

SHANMUGAM, G. 1990. Porosity prediction in sandstones using erosional unconformities. *In*: MESHRI, I. D. & ORTOLEVA, P. J. (eds) *Prediction of Reservoir Quality through Chemical Modelling*. American Association of Petroleum Geologists, Memoir **49**, 1–23.

SHAWE, D. R. 1956. *United States Geological Survey Professional Paper No. 300*.

STANSFIELD, J. 1917. Retarded diffusion and rhythmic precipitation. *American Journal of Science*, **43**, 1–27.

STONELEY, R. & SELLEY, R. C. 1991. *A Field Guide to the Petroleum Geology of the Wessex Basin*. 3rd edn. Imperial College, London.

SULTAN, R., ORTOLEVA, P., DEPASQUALE, F. & TARTAGLIA, P. 1990. Bifurcation of the Ostwald-Liesegang supersaturation nucleation-depletion cycle. *Earth Science Reviews*, **29**, 163–173.

WENDTLAND, R. F. & BHUYAN, K. 1990. Estimation of mineralogy and lithology from geochemical log measurements. *AAPG Bulletin*, **74**, 837–856.

Sandstone diagenesis: framework of a forward modelling approach by integrating wireline and other geological data

M. J. CHESHIRE & B. W. SELLWOOD

Postgraduate Research Institute for Sedimentology, The University, P.O. Box 227, Whiteknights, Reading RG6 2AB, UK

Abstract. The integration of wireline log information with other forms of geological data allows the recognition of certain diagenetic effects in sedimentary rocks. Diagenetic modification of a sandstone causes changes in the rock properties that will produce subtle shifts in the wireline log characteristics. Such shifts are often difficult to interpret. However, by reference to a now extensive diagenetic information base it is possible to construct models for processes controlling some types of diagenetic change. So it is possible both to model and predict the spatial distribution of diagenetic effects. To interpret diagenetic effects on wireline logs it is necessary to integrate existing geological databases with information derived from wireline logs in a form that can be handled by computer. Interactive modelling involves the iterative refinement of modelled features. This is done by comparing synthetic logs with actual logs over intervals from which core-derived information is available (Rotliegend and Sherwood sandstones provide ideal test-beds for this approach). Predictive models may then be generated by an interactive process of confirming or denying modelled features. This approach allows refinement of the interpretations made. Eventually we expect that a means of directly interpreting certain important diagenetic effects will be possible from wireline log data.

Diagenesis is an important factor controlling reservoir rock properties such as porosity and permeability and thus influences the production of hydrocarbons. Information about the diagenetic characteristics of reservoir rocks is conventionally obtained from core samples. However, cores are usually taken over limited depth ranges through reservoir intervals, and non-reservoir intervals are generally cored in only special circumstances. 'Wireline logs', on the other hand, are conventionally run throughout boreholes, not merely through reservoir zones, and may potentially provide information on rock properties (including diagenetic parmaters) which is not accessible from other methods of investigation Despite this potential, little progress has been made on the interpretation of diagenetic features from wireline data alone, largely because the subtle variations in log response promoted by diagenesis are difficult to distinguish from those generated by other factors (Serra 1984).

The approach that has been used in this study is to use the results from petrographic and laboratory analysis of core samples, integrating these results with wireline log data (which have been environmentally corrected and depth matched) over cored intervals, and then extrapolate these data in hypothetical (but realistic) rock models.

The method discussed here provides the potential to predict more about diagenetic paramaters in a given succession than can be obtained from individual data sources alone (Cheshire & Sellwood 1992; King 1990). Due to the complexity of this integrated approach, extensive computing is required. However, the approach represents a first step towards the development of a forward modelling program for diagenetic modelling from downhole logs.

The models are being both generated and tested against well studied sequences of continental, redbed siliciclastics: the Permian Rotliegend sandstone of the southern North Sea and the Triassic Sherwood sandstone of the English Channel Basin. Extensive literature exists for these successions (e.g. Burley 1984, Glennie 1978; Nagtegaal 1978; Woodward & Curtis 1987) and research programmes are continuing. The diagenetic changes suffered by such redbeds are controlled by processes which are, to some extent, predictable and are thus amenable to modelling (e.g. in terms of fluid flow and burial history; Glennie 1978; Purvis 1989; Wood *et al.* 1985). They would therefore appear to be ideal candidates for the type of approach adopted here. It should be stressed, however, that while accurate modelling of the responses of some wireline logs is possible, others cannot be

From HURST, A., GRIFFITHS, C. M. & WORTHINGTON, P. F. (eds), 1992,
Geological Applications of Wireline Logs II. Geological Society Special Publication No. 65, pp. 389–394.

389

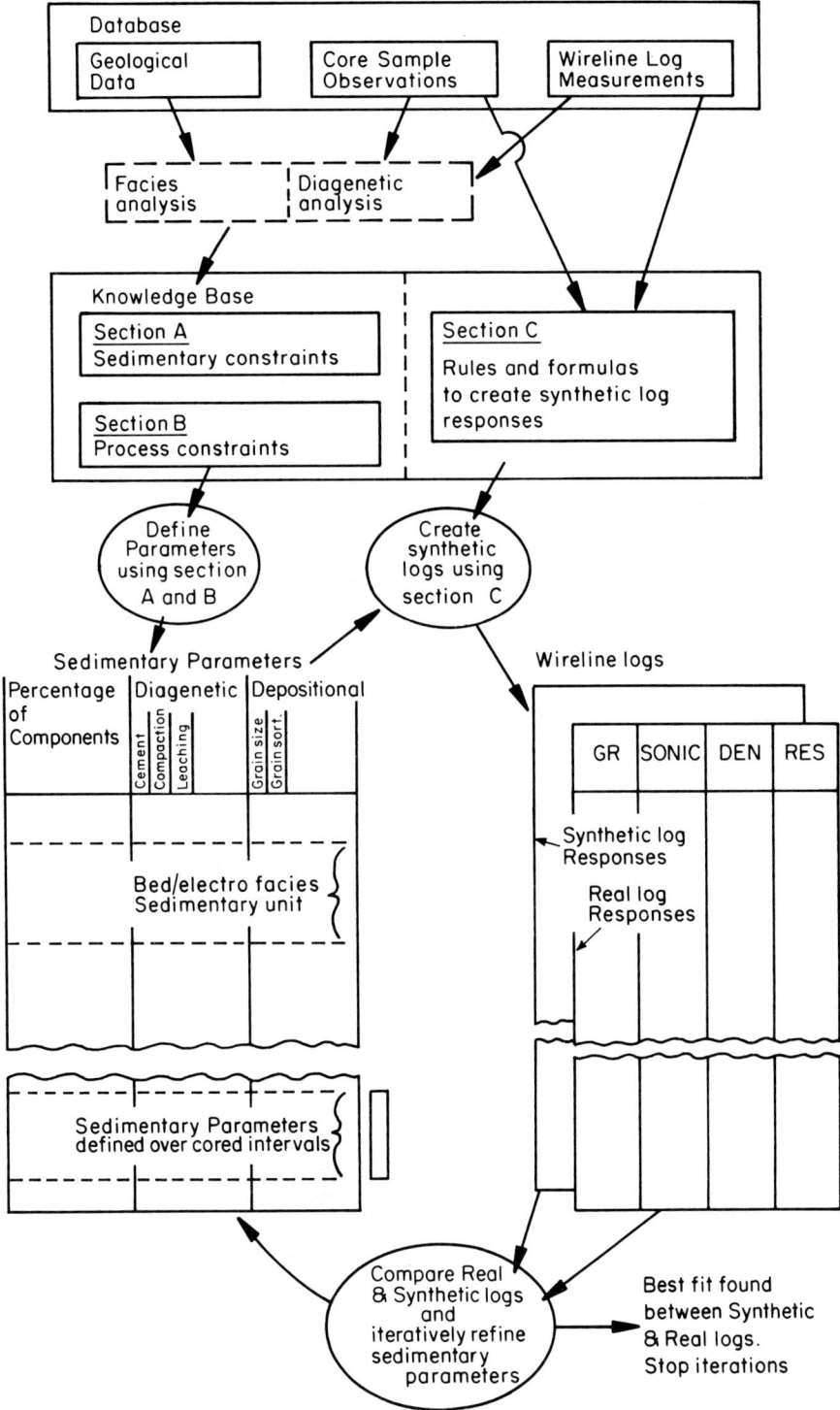

Fig. 1. The elements of a forward-modelling computer program for the interpretation of wireline logging data.

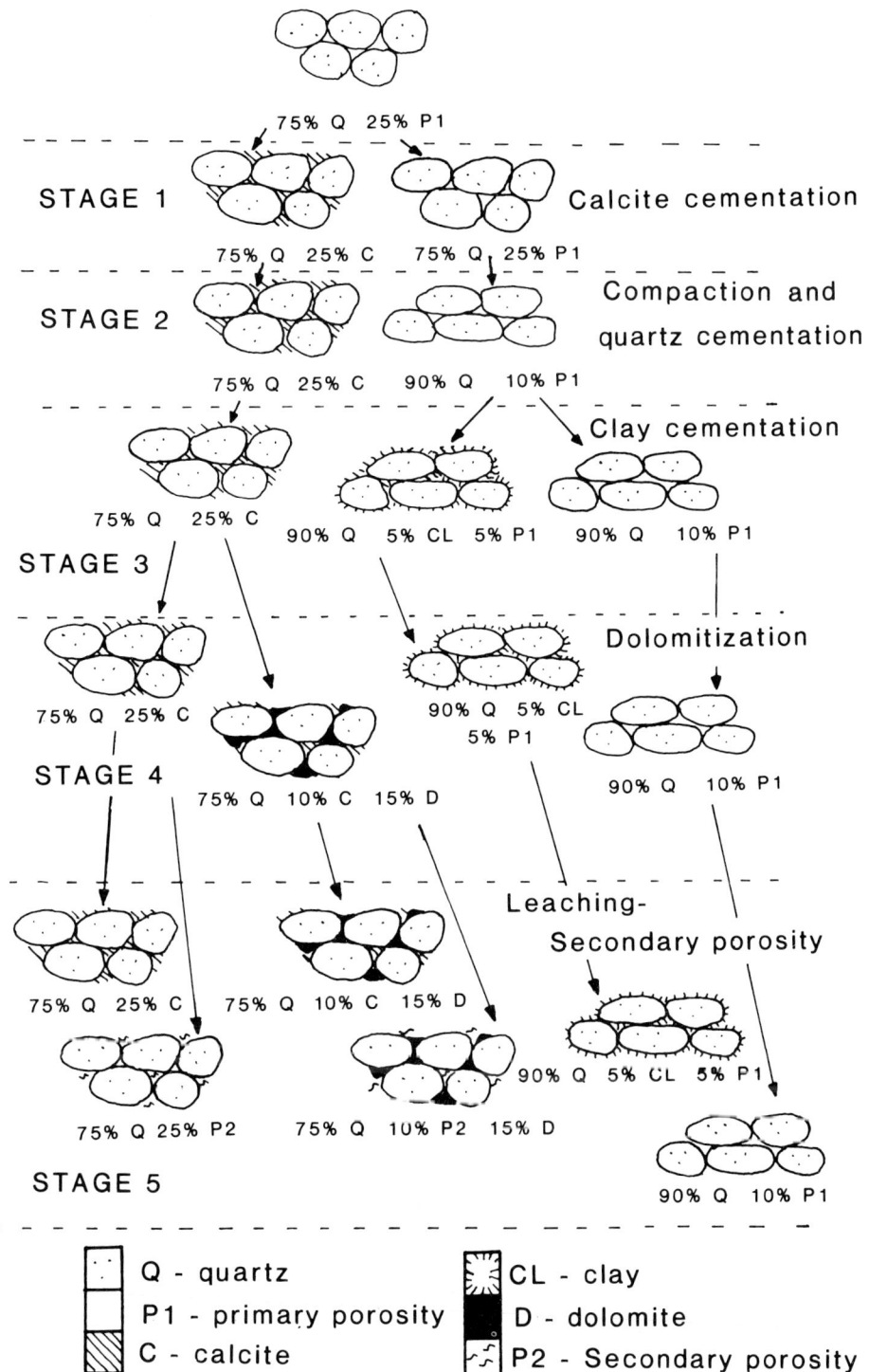

Fig. 2. A hypothetical diagenetic history for a clean quartz dune sand.

modelled. Models are limited at present by the inaccuracy of many forms of geodata currently gathered.

Framework of the computer program

Before any modelling can be attempted, a database must be established which contains all the necessary information (Fig. 1). The database includes all the wireline log measurements and observations from available core material. As has been argued in an earlier work (Cheshire & Sellwood 1992), wherever possible the petrographic parameters available from core materials should be quantified, or at least presented in a numerical form. The data should include mineralogical information, diagenetic parameters (e.g. degree of quartz cementation on a scale of 1 to 10 etc.), grain-size, sorting, compaction (again all semi-quantitatively evaluated). Observations on core samples may have been derived from thin-section, SEM, XRD or other methods (Tucker 1988; Welton 1984). The database also includes other geological data such as structural maps, maturation and palaeoburial histories.

After establishing the database the knowledge base can be constructed (Fig. 1). The knowledge base contains the data in a form that can be processed by a computer program. The information in the knowledge base comprises three types of knowledge (knowledge types A, B and C; Fig. 1).

Sedimentary constraints (knowledge base Section A)

This part of the knowledge base contains the constraints that control the variability of the sedimentary parameters. For example, for a given rock type, such as a fluvial sandstone from the Rotliegend, the porosity would be expected to be in the range of 0% to 20%. However, if such a sandstone received a carbonate cement its porosity would be expected to fall in the range of 0% to 10%. Figure 2 shows a possible diagenetic history for a clean quartzose dune sand. This simplified version is based upon parageneses commonly seen in the Rotliegend. As may be seen, after several phases of diagenetic alteration (e.g. by Stage 5 in Fig. 2), a range of outcomes is possible with the parent rock evolving in a number of potential directions. Each end-member represents a rock with different component compositions. The knowledge base contains rules that help to constrain the sedimentary parameters illustrated in Fig. 2.

Process constraints (knowledge base Section B)

The spatial distribution of diagenetic features in a sedimentary sequence can be related to processes such as fluid flow and palaeoburial. For example, in the Rotliegend sandstone the influx of fluids from Zechstein evaporites effects the distribution of carbonate minerals such as dolomite and siderite. By gaining an understanding of these processes formulae may be derived to model the spatial distribution of diagenetic features.

Rules and empirical formulae to create synthetic wireline logs (knowledge base Section C)

For each core sample analysed a synthetic wireline log response can be generated by using the formula:

$$L = \sum_{i=1}^{n} V_i L_i \qquad (1)$$

where

$$\sum_{i=1}^{n} V_i = 1.$$

Here L is the wireline log response for a rock containing n components, V_i is the proportional volume of component i, and L_i the pure log response for component i. The empirical equation (1) only gives an approximation to the true log response. Differences between the true log response and the synthetic log response (using equation (1)) may be due to textural factors (e.g. grain sorting, packing anisotropy and cementation).

If we consider a two-component rock (Fig. 3), then using equation (1) a straight line relationship would exist in the predicted log response (i.e. between L_i and L_j in Fig. 3) as the proportion of the two components changes. However, because of the textural factors referred to above, it is normally the case that the true relationship is better modelled by a curved line (Fig. 3). The curve may also lie below the straight line. The curved line can be expressed using the quadratic equation (equation (2)) and the straight line using the linear equation (equation (3)):

$$Y_q(x) = -4D^2 + x^2 + (L_j - L_i + 4D)\, x + L_i \quad (2)$$

$$Y_L(x) = (L_j - L_i)\, x + L_i \qquad (3)$$

L_i and L_j are the pure log responses as shown in

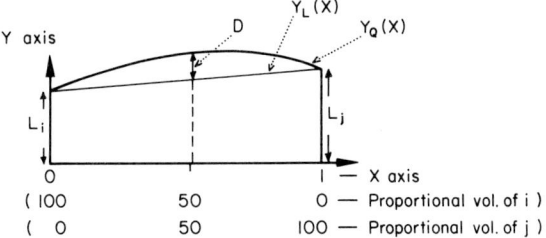

Fig. 3. The determination of a correction factor for a two-component rock.

Fig. 3. The maximum of the curve from the straight line is expressed as D (Fig. 3). It is assumed that this maximum deviation (D-factor) will occur when the two components have a 50 : 50 ratio. The difference between equations (2) and (3) is a correction that must be applied to equation (1) for a two-component rock. The correction, as a function of x (see Fig. 3), is:

$$D_{ij}(x) = Y_Q(x) + Y_L(x) \qquad (4)$$

For a system with more than two components, a more complex correction would be necessary. However, because of the problem of calculating a D-factor, it is sufficient to apply a linear correction (a correction for two components) for each pair of components within a rock. The corrected formula is

$$L_c = L + \frac{1}{(n-1)} \sum_{i=2}^{n} \sum_{j=1}^{i-1} D_{ij}(V_i + V_j) \qquad (5)$$

L is the synthetic log response generated by using equation (1). D_{ij} is the correction for the pair of components i and j. V_i and V_j are the proportional volumes of components i and j. Because more than n corrections are applied, the overall addition of the corrections needs to be scaled down. Each correction should be proportional to the volume of the two components to which it applies.

To use equation (5) it is necessary to calculate D-factors for each pair of components in a rock, and for each type of wireline logging tool. In reality it is only worth calculating the D-factors for those relationships with extreme deviations from the straight line. Slight modifications to the formula are necessary if this is done. Data generated in laboratory and wireline log measurements provide the information necessary for the calculation of actual D-factors. Because rocks usually contain more than two components, the D-factors that are calculated will be summations of the linear D-factors. Of course, in such cases it is necessary to use more than one wireline log measurement, and to solve simultaneous

equations, in order that linear D-factors may be calculated. In reality it is often difficult to establish D-factors in this way and certain assumptions about them need to be made.

It can sometimes be established that the D-factors relate to certain sedimentary parameters. For example, the D-factor between quartz (grains and quartz cement) and water-filled porosity is a function of grain-related parameters (e.g. size, sorting, alignment, etc.). Where such relationships exist, or can be established, they can be used to refine the magnitude of the D-values. Such information about D-factors is incorporated in section C of the knowledge base.

The use of a correction factor, as described earlier, prevents the equation (5) from becoming unnecessarily complicated and only a relatively small number of additional parameters (the D-factors) need to be established.

Once an adequate knowledge base has been acquired the validity of the empirical relationships, and rules, can be checked by comparing specific core samples with actual wireline responses. The knowledge base will need to be refined until a satisfactory confidence level has been achieved. In practice, it will take time before the confidence level becomes commercially acceptable because of practical limitations on the size of the knowledge base.

Sedimentary parameters

The final step in the processing technique is illustrated in the lower part of Fig. 1. This involves, for a given borehole, defining sedimentary parameters for rock units on a bed, or electro-facies scale. These parameters can be defined by using the models incorporated into Section B of the knowledge base. These parameters must satisfy the constraints defined in Section A of the knowledge base. Synthetic logs are then generated using the empirical formulae and rules in Section C of the knowledge base. The synthetic traces can then be compared with the actual traces obtained from the borehole. Then, the sedimentary parameters are modified iteratively until the best fit between the traces is obtained. This should lead to an accurate and representative assessment of the true borehole geology. This method does not give a unique interpretation of the data but it provides an interpretation which is geologically realistic.

Conclusion

At present, and with only a limited but growing knowledge base, only major diagenetic effects can be recognized in the Rotliegend and Sherwood sandstone pilot studies. In the Rotliegend the emplacement of dense carbonate minerals such as dolomite and siderite can be recognized.

On a regional scale the spatial distribution of these phases can be related to two major processes: fluid invasion from the overlying Zechstein, and mineral authigenesis during later burial.

With the implementation of a more extensive knowledge base we confidently expect to be able to detect more subtle diagenetic effects such as clay mineralization and the development of quartz cementation.

The help and cooperation of Intera Information Technologies, the NERC, Shell Exploration and Production, and British Gas plc, is gratefully appreciated. This is PRIS contribution number 164.

References

BURLEY, S. D. 1984. Patterns of diagenesis in the Sherwood Sandstone Group (Triassic), United Kingdom. *Clay Mineralogy*, **19**, 403–440.

CHESHIRE, M. J. & SELLWOOD, B. W. 1992. Diagenetic modelling: An integrated approach using wireline logs and geological data. *Marine and Petroleum Geology*, **9**, 128–138.

GLENNIE, K. W., MUDD, G. R. & NAGTEGAAL, P. J. C. 1978. Depositional environment and diagenesis of Permian Rotliegendes sandstones in the Leman Bank and Sole Pit areas of the southern North Sea. *Journal of the Geological Society, London*, **135**, 25–34.

KING, D. E. 1990. Incorporating geological data in well log interpretation. *In*: HURST, A., LOVELL, M. A. & MORTON, A. C. (eds) *Geological Applications of Wireline Logs*. Geological Society, London, Special Publication, **48**, 45–55.

NAGTEGAAL, P. J. C. 1979. Relationship of facies and reservoir quality in Rotliegendes desert sandstone,

southern North Sea region, *Journal of Petroleum Geology*, **2**, 145–158.

PURVIS, K. 1989. Zoned authigenic magnesites in the Rotliegend, Lower Permian, southern North Sea. *In*: SELLWOOD, B. W. (ed.) *Zoned Carbonate Cements: Techniques, Applications and Implications. Sedimentary Geology*, **65**, 307–318.

SERRA, O. 1984. *Fundamentals of Well-Log Interpretation-2, the interpretation of logging data*. The interpretation of logging data. Elsevier, Oxford.

TUCKER, M. (ed.) 1988. *Techniques in Sedimentology*. Blackwell, Oxford.

WELTON, J. E. 1984. *SEM Petrology Atlas*. American Association of Petroleum Geologists, Methods in Exploration 4.

WOODWARD, K. & CURTIS, C. D. 1987. Predictive modelling for the distribution of production-constraining illites, Morecambe Gas Field, Irish Sea Offshore, UK. *In*: BROOKS, J. & GLENNIE, K. W. (eds) *Petroleum Geology of North West Europe*, Graham & Trotman, London, 205–215.

Index

Page numbers in *italic* refer to figures; those in **bold** refer to tables.

Geology of the Brent Group

Edited by A.C. Morton (British Geological Survey, UK), R.S. Haszeldine (University of Glasgow, UK), M.R.Giles (Shell, The Netherlands) and S. Brown (Petroleum Science & Technology Institute, UK)

The Middle Jurassic Brent Group sediments, and their correlatives on the Norwegian Shelf are in economic terms, the most important hydrocarbon reservoir in NW Europe. In 1971 the Brent Field was discovered by Shell/Esso and tested in 1972 with 1.8 billion barrels of recoverable oil. By 1988 discovered Brent hydrocarbons comprised some 49% of the UK's recoverable reserves, totalling 22.5 billion barrels of oil equivalent.

Now that the UK Brent Province has reached maturity, this book provides a timely review of the geology and petroleum geology of one of the worlds major petroleum reservoirs.

Principal Authors

A.C. Morton (British Geological Survey, UK)
J.M. Bowen (Enterprise Oil, UK)
P. Richards (British Geological Survey, UK)
G. Yielding (Badley Ashton & Associates, UK)
B. Mitchener (BP, UK)
S.J.C. Cannon (Geochem Group, UK)
W. Helland-Hansen (Norsk Hydro, Norway)
E. Scott (University of Oxford, UK)
J. Alexander (University of Wales, UK)
M.F. Whitaker (Geochem Group, UK)
G. Williams (Geostrat Ltd, UK)
E.W. Mearns (Isotopic Analytical Services, UK)
K. Stattegger (University of Kiel, Germany)
K. Bjorlykke (University of Oslo, Norway)
M.R. Giles (Shell, Netherlands)
J.R. Glasmann (Unocal, USA)
N.B. Harris (Conoco, USA)
P.J. Hamilton (CSIRO, Australia)
R.S. Haszeldine (Unversity of Glasgow, UK)
A.J.C. Hogg (Unversity of Aberdeen, UK)
S. Larter (University of Newcastle, UK)
J.D. Kantorowicz (Conoco, UK)
B. Moss (Moss Petrophysical, UK)

Outline of Contents

Introduction · Exploration of the Brent Province · An introduction to the Brent Group: a literature review · The structural evolution of the Brent Province · Brent Group: sequence stratigraphy and regional implications · A regional reassessment of the Brent Group, UK Sector North Sea · Advance and retreat of the Brent delta: recent contributions to the depositional model · The palaeoenvironments and dynamics of the Rannoch-Etive nearshore and coastal successions, Brent group, northern North Sea · A discussion of alluvial sandstone body characteristics related to variations in marine influence, Middle Jurassic of the Cleveland Basin, UK, and the implications for analogous Brent Group strata in the North Sea Basin · Palynological review of the Brent Group, UK Sector, North Sea · Palynology as a palaeoenvironmental indicator in the Brent Group, northern North Sea · Samarium-neodymium isotopic constraints on the provenance of the Brent Group · Provenance of Brent Group sandstones: heavy mineral constraints · Statistical analysis of garnet compositions and lithostratigraphic correlation: Brent Group sandstones of the Oseberg Field, northern North Sea · Diagenetic processes in the Brent Group (Middle Jurassic) reservoirs of the North Sea · The reservoir properties and diagenesis of the Brent Group: a regional perspective · The fate of feldspars in Brent Group reservoirs, North Sea: a regional synthesis of diagenesis in shallow, intermediate and deep burial environments · Burial diagenesis of Brent sandstones: a study of Statfjord, Hutton and Lyoll fields · K-Ar dating of illites in Brent Group reservoirs: a regional perspective · Open and restricted hydrologies in Brent Group diagenesis: North Sea · Cathodoluminescence of quartz cements in Brent Group sandstones, Alwyn South, UK North Sea · Migration of hydrocarbons into Brent Group reservoirs: some observations from the Gullfaks Field, Tampen Spur area, North Sea · Integration of petroleum engineering studies of producing Brent Group fields to predict reservoir properties in the Pelican Field, North Sea · The petrophysical characteristics of the Brent sandstones

● Coverage of Brent Group geology includes exploration history, structural evolution, sequence stratigraphy, sedimentology, diagenesis, palynology, hydrocarbon generation and migration and petrophysics

● *Geological Society Special Publication No. 61*

● over 500 pages, 300 illustrations, 25 colour pages as well as 3 colour flyouts
ISBN 0-903317-68-0
May 1992

● List price £75/US$125

Please send your order to:
Geological Society Publishing House, Unit 7, Brassmill Enterprise Centre, Brassmill Lane, Bath BA1 3JN, UK.
Tel: 0225 445046.
Fax: 0225 442836.

Please add 10% of order total for overseas delivery

United Kingdom Oil and Gas Fields, 25 Years Commemorative Volume

Edited by Ian L. Abbotts (Clyde Petroleum)

The United Kingdom Oil and Gas Fields has been produced to commemorate the first 25 years of hydrocarbon exploration and production in the United Kingdom North Sea. The result of this exploration has produced many benefits for the UK, its government and industry but above all for geologists and geophysicists.

Articles on the 64 oil and gas fields discovered on the United Kingdom Continental Shelf are given in a standardised layout to provide an easy to use databook for the petroleum geologist and geophysicist.

The producing oil and gas fields have been arranged into: the Viking Graben, the Central Graben and Moray Firth, the Southern Gas Basin and the Morecambe Basin.

Also included are two introductory articles, the first sets the fields in a historical perspective and the second places them in a stratigraphic framework.

- Primary benefits: provides a reference source of data relevant to the discovery and optimum development of the next generation of North Sea fields

- Format provides a succinct description, including relevant figures, of the petroleum geology of each field

- Primary audience: production geologists, geophysicists and managers, exploration geologists

- Secondary audience: oil service companies, oil consultants, academic libraries

- Large format 230mm x 305mm

- *Geological Society Memoir No. 14*

- 574 pages, 400 illustrations, 57 colour plates, hardback
ISBN 0-903317-62-1
1991

- **List price £80/US$134**

Outline of Contents

Part 1: Introduction · 25 Years of UK North Sea Exploration (Enterprise Oil) · Stratigraphy of the oil and gas reservoirs: UK continental shelf (Petroleum Science & Technology Inst., Edinburgh) · **Part 2: The Viking Graben** · The Alwyn North Field (Total) · The Beryl Field, Block 9/13 UK North Sea (Mobil) · The North Brae Field, Block 16/7a, UK North Sea (Marathon) · The Central Brae Field, Block 16/7a, UK North Sea (Marathon) · The South Brae Field, Block 16/7a, UK North Sea (Marathon) · The Brent Field, Block 211/29, UK North Sea (Shell) · The Comorant Field, Blocks 211/21a, 211/26a, UK North Sea (Shell) · The Deveron Field, Block 211/18a, UK North Sea (BP) · The Don Field, Blocks 211/13a, 211/14, 211/18a, 211/19a, UK North Sea (BP) · The Dunlin Field, Blocks 211/23a, 211/24a UK North Sea (Shell) · The Eider Field, Blocks 211/16a, 211/21a, UK North Sea (Shell) · The Emerald Field, Blocks 2/10a, 2/15a, 3/11b UK North Sea (Sovereign) · The Frigg Field, Block 10/1, UK North Sea and 25/1, Norwegian North Sea (Elf) · The Heather Field, Block 2/5, UK North Sea (Union) · The Hutton Field, Blocks 211/28, 211/27, UK North Sea (Conoco) · The Northwest Hutton Field, Block 211/27, UK North Sea (Amoco) · The Magnus Field, Blocks 211/71, 12a, UK North Sea (BP) · The Miller Field, Blocks 16/7b - 16/8b, UK North Sea (BP) · The Murchison Field, Blocks 211/19a, UK North Sea (Conoco) · The Ninian Field, Blocks 3/3 & 3/8, UK North Sea (Chevron) · The Osprey Field, Blocks 211/18a, 211/23a, UK North Sea (Shell) · The Tern Field, Block 210/25a, UK North Sea (Shell) · The Thistle Field, Blocks 211/18a and 211/19, UK North Sea (BP) · **Part 3: The Central Graben and Moray Firth** · The Arbroath and Montrose Field, Block 211/7a, 12a, UK North Sea (Amoco) · The Argyll, Duncan and Innes Fields, Blocks 30/24, 30/25a, UK North Sea (Hamilton Bros.) · The Auk Field Block 30/16, UK North Sea (Shell) · The Balmoral Field, Block 16/21, UK North Sea (Sun) · The Beatrice Field, Block 11/30a, UK North Sea (BP) · The Buchan Field, Blocks 20/5a, 21/1a, UK North Sea (BP) · The Chanter Field, Block 15/17, UK North Sea (Occidental) · The Claymore Field, Block 14/19, UK North Sea (Occidental) · The Clyde Field, Block 30/17b, UK North Sea (BP) · The Crawford Field, Block 9/28a, UK North Sea (Hamilton Bros.) · The Cyrus Field, Block 16/28, UK North Sea (BP) · The Forties Field, Blocks 21/10, 22/6a, UK North Sea (BP) · The Fulmar Field, Blocks 30/16, 30/11b, UK North Sea (Shell) · The Glamis Field, Block 16/21a, UK North Sea (Sun) · The Highlander Field, Block 14/20b, UK North Sea (Texaco) · The Ivanhoe and Rob Roy Fields, Block 15/21a-b, UK North Sea (Amerada Hess) · The Kittiwake Field, Block 21/18, UK North Sea (Shell) · The Maureen Field, Block 16/29a, UK North Sea (Phillips) · The Petronella Field, Block 14/20b, UK North Sea (Texaco) · The Piper Field, Block 15/17, UK North Sea (Occidental) · The Scapa Field, Block 14/19, UK North Sea (Occidental) · The Tartan Field, Block 15/16, UK North Sea (Texaco) · **Part 4: The Southern Gas Basin** · The Amethyst Field, Blocks 47/8a, 47/9a, 47/13a, 47/14a, 47/15a, UK North Sea (BP) · The Barque Field, Blocks 48/13a, 48/14, UK North Sea (Shell) · The Camelot Fields, Blocks 53/1a, 53/2 UK North Sea (Mobil) · The Cleeton Field, Block 42.29, UK North Sea (BP) · The Clipper Field, Blocks 48/19a, 48/19c, UK North Sea (Shell) · The Esmond, Forbes and Gordon Fields, Blocks 43/8a, 48/13a, 48/15a, 48/20a, UK North Sea, (Hamilton Bros.) · The Hewett Field, Blocks 48/28-29-30, 52/4a-5a, UK North Sea (Phillips) · The Indefatigable Field, Blocks 48/18, 48/19, 48/23, 48/24, UK North Sea (Amoco) · The Leman Field, Blocks 49/26, 49/27, 49/28, 53/1, 53/2, UK North Sea (Shell) · The Ravenspurn North Field, Blocks 42/30, 43/26a, UK North Sea (Hamilton Bros.) · The Ravenspurn South Field, Blocks 42/29, 42/30, 43/26, UK North Sea (BP) · The Rough Gas Storage Field, Blocks 47/3d, 47/8b, UK North Sea (British Gas Corporation) · The Sean North and Sean South Fields, Block 49/25a, UK North Sea (Shell) · The Thames, Yare and Bure Fields, Block 49/28, UK North Sea (Arco) · The V-Fields, Blocks 49/16, 49/21, 48/20a, 48/25b, UK North Sea (Conoco) · The Victor Field, Blocks 49/17, 49/22, UK North Sea (Conoco) · The Viking Complex Field, Blocks 49/12a, 49/16, 49/17, UK North Sea (Conoco) · The West Sole Field, Block 48/6, UK North Sea (BP) · **Part 5: The Morecambe Basin** · The South Morecambe Field, Blocks 110/2a, 110/3a, 110/8a, UK East Irish Sea (British Gas Corporation)

Please send your order to:
Geological Society Publishing House, Unit 7 Brassmill Enterprise Centre, Brassmill Lane, Bath BA1 3JN, UK.
Tel: 0225 445046.
Fax: 0225 442836.

Please add 10% of order total for overseas delivery